BIRDS of CANADA

Tyler L. Hoar, Ken De Smet,
R. Wayne Campbell & Gregory Kennedy
with contributions from
Krista Kagume

Lone Pine Publishing

First printed in 2010 10 9 8 7 6 5 4 3 2 1
Printed in China

Lone Pine Publishing
10145 – 81 Avenue
Edmonton, Alberta T6E 1W9

Website: www.lonepinepublishing.com

Library and Archives Canada Cataloguing in Publication

Birds of Canada / Tyler L. Hoar ... [et al.] ; with Krista Kagume.

Includes bibliographical references and index.
ISBN 978-1-55105-589-3 (pbk.) — ISBN 978-1-55105-603-6 (bound)

1. Birds—Canada—Identification. 2. Bird watching—Canada.
I. Hoar, Tyler L., 1969– II. Kagume, Krista III. Title.

QL685.H62 2009 598.0971 C2009-906405-7

Editorial Director: Nancy Foulds
Project Editor: Nicholle Carrière
Editorial: Nicholle Carrière, Tracey Comeau, Carla MacKay, Wendy Pirk, Pat Price, Sheila Quinlan
Production Manager: Gene Longson
Layout & Production: Volker Bodegom, Janina Kuerschner, Lisa Morley
Cover Design: Gerry Dotto
Cover Photo: Snowy Owl, © Jean Chiasson
Photo Management: Gary Whyte
Photo Researcher: Randy Kennedy
Photo Editing: Kamila Kwiatkowska, Lisa Morley, Gary Whyte
Illustrations: Ted Nordhagen, Ewa Pluciennik, Gary Ross
Maps: Volker Bodegom

We acknowledge the financial support of the Government of Canada through the Book Publishing Industry Development Program (BPIDP) for our publishing activities.

Canadian Heritage Patrimoine canadien

PC: 16

CONTENTS

ACKNOWLEDGEMENTS

Over the years, Lone Pine Publishing's field guides have developed into standard references for plants and animals and have become an underlying force for the appreciation and conservation of wildlife on the continent.

Thanks are extended to the growing family of ornithologists and dedicated birders who have offered their inspiration and expertise to help build Lone Pine's expanding library. Thanks go to John Acorn, Chris Fisher, Andy Bezener and Eloise Pulos for their contributions to previous books in this series. In addition, thanks go to Gary Ross, Ted Nordhagen and Ewa Pluciennik, whose skilled illustrations have brought each page to life, and to Michel Gosselin of the Canadian Museum of Nature, who was very helpful in the preparation of the initial species list for this book. Thank you as well to the photographers (listed on page 527) for graciously allowing us to include so many of their images on these pages.

We are very grateful to Nicholle Carrière and the Lone Pine staff for transforming a wide-ranging collection of words, maps, drawings and photographs into another exceptional Lone Pine nature guide that will allow Canadians to learn about and enjoy birds in their backyards and afield.

Greater White-fronted Goose
78 cm • p. 45

Snow Goose
77 cm • p. 46

Ross's Goose
60 cm • p. 47

Brant
65 cm • p. 48

Cackling Goose
70 cm • p. 49

Canada Goose
88 cm • p. 50

Mute Swan
1.5 m • p. 51

Trumpeter Swan
1.6 m • p. 52

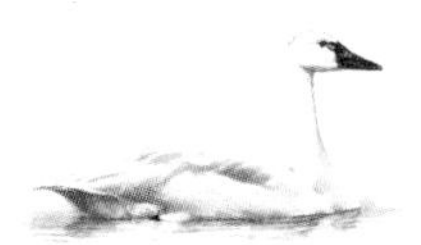
Tundra Swan
1.3 m • p. 53

Wood Duck
46 cm • p. 54

Gadwall
51 cm • p. 55

Eurasian Wigeon
47 cm • p. 56

American Wigeon
52 cm • p. 57

American Black Duck
56 cm • p. 58

Mallard
61 cm • p. 59

Blue-winged Teal
38 cm • p. 60

Cinnamon Teal
40 cm • p. 61

Northern Shoveler
49 cm • p. 62

Northern Pintail
70 cm • p. 63

Green-winged Teal
35 cm • p. 64

Canvasback
52 cm • p. 65

Redhead
51 cm • p. 66

Ring-necked Duck
41 cm • p. 67

Greater Scaup
45 cm • p. 68

Lesser Scaup
42 cm • p. 69

King Eider
56 cm • p. 70

Common Eider
63 cm • p. 71

Harlequin Duck
42 cm • p. 72

WATERFOWL

Surf Scoter
48 cm • p. 73

White-winged Scoter
54 cm • p. 74

Black Scoter
48 cm • p. 75

Long-tailed Duck
47 cm • p. 76

Bufflehead
35 cm • p. 77

Common Goldeneye
46 cm • p. 78

Barrow's Goldeneye
46 cm • p. 79

Hooded Merganser
45 cm • p. 80

Common Merganser
62 cm • p. 81

Red-breasted Merganser
57 cm • p. 82

Ruddy Duck
40 cm • p. 83

GROUSELIKE BIRDS

Mountain Quail
28 cm • p. 84

California Quail
26 cm • p. 85

Northern Bobwhite
25 cm • p. 86

Chukar
33 cm • p. 87

Gray Partridge
32 cm • p. 88

Ring-necked Pheasant
83 cm • p. 89

Ruffed Grouse
43 cm • p. 90

Greater Sage-Grouse
77 cm • p. 91

Spruce Grouse
37 cm • p. 92

Willow Ptarmigan
35 cm • p. 93

Rock Ptarmigan
36 cm • p. 94

White-tailed Ptarmigan
33 cm • p. 95

Dusky Grouse
51 cm • p. 96

Sooty Grouse
51 cm • p. 97

Sharp-tailed Grouse
45 cm • p. 98

Wild Turkey
1.2 m • p. 99

Red-throated Loon
65 cm • p. 100

Pacific Loon
66 cm • p. 101

Common Loon
80 cm • p. 102

Yellow-billed Loon
82 cm • p. 103

Pied-billed Grebe
34 cm • p. 104

Horned Grebe
35 cm • p. 105

Red-necked Grebe
50 cm • p. 106

Eared Grebe
33 cm • p. 107

Western Grebe
56 cm • p. 108

Clark's Grebe
55 cm • p. 109

Black-footed Albatross
72 cm • p. 110

Northern Fulmar
48 cm • p. 111

Pink-footed Shearwater
49 cm • p. 112

Greater Shearwater
48 cm • p. 113

Buller's Shearwater
44 cm • p. 114

Sooty Shearwater
44 cm • p. 115

Short-tailed Shearwater
43 cm • p. 116

Manx Shearwater
34 cm • p. 117

Wilson's Storm-Petrel
17 cm • p. 118

Fork-tailed Storm-Petrel
21 cm • p. 119

Leach's Storm-Petrel
54 cm • p. 120

Northern Gannet
94 cm • p. 121

American White Pelican
1.6 m • p. 122

Brandt's Cormorant
85 cm • p. 123

Double-crested Cormorant
73 cm • p. 124

Great Cormorant
91 cm • p. 125

Pelagic Cormorant
67 cm • p. 126

HERONLIKE BIRDS

American Bittern
64 cm • p. 127

Least Bittern
32 cm • p. 128

Great Blue Heron
1.3 m • p. 129

Great Egret
99 cm • p. 130

Snowy Egret
61 cm • p. 131

Little Blue Heron
61 cm • p. 132

Cattle Egret
61 cm • p. 133

Green Heron
47 cm • p. 134

Black-crowned Night-Heron
62 cm • p. 135

Yellow-crowned Night-Heron
61 cm • p. 136

Glossy Ibis
60 cm • p. 137

White-faced Ibis
58 cm • p. 138

Turkey Vulture
73 cm • p. 139

BIRDS OF PREY

Osprey
60 cm • p. 140

Bald Eagle
92 cm • p. 141

Northern Harrier
51 cm • p. 142

Sharp-shinned Hawk
30 cm • p. 143

Cooper's Hawk
43 cm • p. 144

Northern Goshawk
58 cm • p. 145

Red-shouldered Hawk
48 cm • p. 146

Broad-winged Hawk
41 cm • p. 147

Swainson's Hawk
51 cm • p. 148

Red-tailed Hawk
55 cm • p. 149

Ferruginous Hawk
62 cm • p. 150

Rough-legged Hawk
54 cm • p. 151

Golden Eagle
89 cm • p. 152

American Kestrel
20 cm • p. 153

Merlin
28 cm • p. 154

Gyrfalcon
56 cm • p. 155

Peregrine Falcon
43 cm • p. 156

Prairie Falcon
41 cm • p. 157

Yellow Rail
18 cm • p. 158

King Rail
43 cm • p. 159

Virginia Rail
25 cm • p. 160

Sora
23 cm • p. 161

Common Moorhen
34 cm • p. 162

American Coot
37 cm • p. 163

Sandhill Crane
114 cm • p. 164

Whooping Crane
1.4 m • p. 165

Black-bellied Plover
30 cm • p. 166

American Golden-Plover
25 cm • p. 167

Semipalmated Plover
18 cm • p. 168

Piping Plover
18 cm • p. 169

Killdeer
25 cm • p. 170

Mountain Plover
22 cm • p. 171

American Oystercatcher
47 cm • p. 172

Black Oystercatcher
43 cm • p. 173

Black-necked Stilt
37 cm • p. 174

American Avocet
45 cm • p. 175

Spotted Sandpiper
19 cm • p. 176

Solitary Sandpiper
22 cm • p. 177

Wandering Tattler
25 cm • p. 178

Greater Yellowlegs
35 cm • p. 179

Willet
38 cm • p. 180

Lesser Yellowlegs
27 cm • p. 181

Upland Sandpiper
30 cm • p. 182

Whimbrel
42 cm • p. 183

Long-billed Curlew
58 cm • p. 184

Hudsonian Godwit
38 cm • p. 185

Marbled Godwit
46 cm • p. 186

Ruddy Turnstone
24 cm • p. 187

Black Turnstone
23 cm • p. 188

Surfbird
24 cm • p. 189

Red Knot
27 cm • p. 190

Sanderling
20 cm • p. 191

Semipalmated Sandpiper
16 cm • p. 192

Western Sandpiper
17 cm • p. 193

Least Sandpiper
15 cm • p. 194

White-rumped Sandpiper
19 cm • p. 195

Baird's Sandpiper
18 cm • p. 196

Pectoral Sandpiper
23 cm • p. 197

Purple Sandpiper
23 cm • p. 198

Rock Sandpiper
23 cm • p. 199

Dunlin
21 cm • p. 200

Stilt Sandpiper
22 cm • p. 201

Buff-breasted Sandpiper
20 cm • p. 202

Ruff/Reeve
27 cm • p. 203

Short-billed Dowitcher
29 cm • p. 204

Long-billed Dowitcher
30 cm • p. 205

Wilson's Snipe
28 cm • p. 206

American Woodcock
28 cm • p. 207

Wilson's Phalarope
22 cm • p. 208

Red-necked Phalarope
18 cm • p. 209

Red Phalarope
21 cm • p. 210

Black-legged Kittiwake
43 cm • p. 211

Ivory Gull
41 cm • p. 212

Sabine's Gull
35 cm • p. 213

Bonaparte's Gull
33 cm • p. 214

Black-headed Gull
39 cm • p. 215

Little Gull
27 cm • p. 216

Ross's Gull
32 cm • p. 217

Laughing Gull
41 cm • p. 218

Franklin's Gull
35 cm • p. 219

Heermann's Gull
44 cm • p. 220

Mew Gull
39 cm • p. 221

Ring-billed Gull
48 cm • p. 222

Western Gull
63 cm • p. 223

California Gull
48 cm • p. 224

Herring Gull
62 cm • p. 225

Thayer's Gull
60 cm • p. 226

Iceland Gull
56 cm • p. 227

Lesser Black-backed Gull
53 cm • p. 228

Glaucous-winged Gull
63 cm • p. 229

Glaucous Gull
69 cm • p. 230

Great Black-backed Gull
76 cm • p. 231

Caspian Tern
53 cm • p. 232

Black Tern
24 cm • p. 233

Roseate Tern
32 cm • p. 234

GULLS, TERNS & ALCIDS

Common Tern
37 cm • p. 235

Arctic Tern
40 cm • p. 236

Forster's Tern
38 cm • p. 237

Pomarine Jaeger
54 cm • p. 238

Parasitic Jaeger
44 cm • p. 239

Long-tailed Jaeger
54 cm • p. 240

Dovekie
21 cm • p. 241

Common Murre
42 cm • p. 242

Thick-billed Murre
46 cm • p. 243

Razorbill
43 cm • p. 244

Black Guillemot
33 cm • p. 245

Pigeon Guillemot
33 cm • p. 246

Marbled Murrelet
29 cm • p. 247

Ancient Murrelet
24 cm • p. 248

Cassin's Auklet
22 cm • p. 249

Rhinoceros Auklet
38 cm • p. 250

Atlantic Puffin
32 cm • p. 251

Tufted Puffin
38 cm • p. 252

DOVES

Rock Pigeon
32 cm • p. 253

Band-tailed Pigeon
35 cm • p. 254

Eurasian Collared-Dove
31 cm • p. 255

Mourning Dove
30 cm • p. 256

Black-billed Cuckoo
30 cm • p. 257

Yellow-billed Cuckoo
30 cm • p. 258

Barn Owl
35 cm • p. 259

Flammulated Owl
17 cm • p. 260

Western Screech-Owl
24 cm • p. 261

Eastern Screech-Owl
22 cm • p. 262

Great Horned Owl
55 cm • p. 263

Snowy Owl
62 cm • p. 264

Northern Hawk Owl
40 cm • p. 265

Northern Pygmy-Owl
18 cm • p. 266

Burrowing Owl
25 cm • p. 267

Spotted Owl
44 cm • p. 268

Barred Owl
52 cm • p. 269

Great Gray Owl
72 cm • p. 270

Long-eared Owl
37 cm • p. 271

Short-eared Owl
38 cm • p. 272

Boreal Owl
27 cm • p. 273

Northern Saw-whet Owl
20 cm • p. 274

Common Nighthawk
23 cm • p. 275

Common Poorwill
20 cm • p. 276

Chuck-will's-widow
30 cm • p. 277

Whip-poor-will
24 cm • p. 278

Black Swift
18 cm • p. 279

Chimney Swift
12 cm • p. 280

Vaux's Swift
13 cm • p. 281

White-throated Swift
16 cm • p. 282

Ruby-throated Hummingbird
8.5 cm • p. 283

Black-chinned Hummingbird
9.5 cm • p. 284

Anna's Hummingbird
9 cm • p. 285

Calliope Hummingbird
8 cm • p. 286

Rufous Hummingbird
8 cm • p. 287

Belted Kingfisher
32 cm • p. 288

Lewis's Woodpecker
28 cm • p. 289

Red-headed Woodpecker
22 cm • p. 290

Red-bellied Woodpecker
25 cm • p. 291

Williamson's Sapsucker
23 cm • p. 292

Yellow-bellied Sapsucker
22 cm • p. 293

Red-naped Sapsucker
22 cm • p. 294

Red-breasted Sapsucker
21 cm • p. 295

Downy Woodpecker
17 cm • p. 296

Hairy Woodpecker
21 cm • p. 297

White-headed Woodpecker
22 cm • p. 298

American Three-toed Woodpecker
23 cm • p. 299

Black-backed Woodpecker
24 cm • p. 300

Northern Flicker
32 cm • p. 301

Pileated Woodpecker
45 cm • p. 302

Olive-sided Flycatcher
19 cm • p. 304

Western Wood-Pewee
15 cm • p. 305

Eastern Wood-Pewee
15 cm • p. 306

Yellow-bellied Flycatcher
14 cm • p. 307

Acadian Flycatcher
14 cm • p. 308

Alder Flycatcher
16 cm • p. 309

Willow Flycatcher
15 cm • p. 310

Least Flycatcher
13 cm • p. 311

Hammond's Flycatcher
14 cm • p. 312

Gray Flycatcher
14 cm • p. 313

Dusky Flycatcher
14 cm • p. 314

Pacific-slope Flycatcher
13 cm • p. 315

Cordilleran Flycatcher
14 cm • p. 316

Eastern Phoebe
17 cm • p. 317

Say's Phoebe
19 cm • p. 318

Great Crested Flycatcher
21 cm • p. 319

Western Kingbird
21 cm • p. 320

Eastern Kingbird
22 cm • p. 321

Loggerhead Shrike
23 cm • p. 322

Northern Shrike
25 cm • p. 323

White-eyed Vireo
13 cm • p. 324

Yellow-throated Vireo
14 cm • p. 325

Cassin's Vireo
14 cm • p. 326

Blue-headed Vireo
14 cm • p. 327

Hutton's Vireo
11 cm • p. 328

Warbling Vireo
13 cm • p. 329

Philadelphia Vireo
13 cm • p. 330

Red-eyed Vireo
15 cm • p. 331

Gray Jay
29 cm • p. 332

Steller's Jay
29 cm • p. 333

Blue Jay
28 cm • p. 334

Clark's Nutcracker
31 cm • p. 335

Black-billed Magpie
51 cm • p. 336

American Crow
48 cm • p. 337

Northwestern Crow
41 cm • p. 338

Common Raven
61 cm • p. 339

Sky Lark
18 cm • p. 340

Horned Lark
18 cm • p. 341

Purple Martin
19 cm • p. 342

Tree Swallow
14 cm • p. 343

Violet-green Swallow
13 cm • p. 344

Northern Rough-winged Swallow
14 cm • p. 345

Bank Swallow
13 cm • p. 346

Cliff Swallow
14 cm • p. 347

Barn Swallow
18 cm • p. 348

Black-capped Chickadee
14 cm • p. 349

Mountain Chickadee
13 cm • p. 350

Chestnut-backed Chickadee
12 cm • p. 351

Boreal Chickadee
13 cm • p. 352

Gray-headed Chickadee
13 cm • p. 353

Tufted Titmouse
15 cm • p. 354

Bushtit
11 cm • p. 355

Red-breasted Nuthatch
11 cm • p. 356

White-breasted Nuthatch
15 cm • p. 357

Pygmy Nuthatch
10 cm • p. 358

Brown Creeper
13 cm • p. 359

Rock Wren
15 cm • p. 360

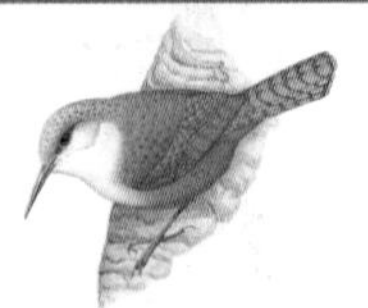
Canyon Wren
14 cm • p. 361

Carolina Wren
14 cm • p. 362

Bewick's Wren
13 cm • p. 363

House Wren
12 cm • p. 364

Winter Wren
10 cm • p. 365

Sedge Wren
10 cm • p. 366

Marsh Wren
13 cm • p. 367

American Dipper
19 cm • p. 368

Golden-crowned Kinglet
10 cm • p. 369

Ruby-crowned Kinglet
10 cm • p. 370

Blue-gray Gnatcatcher
11 cm • p. 371

Northern Wheatear
15 cm • p. 372

Eastern Bluebird
18 cm • p. 373

Western Bluebird
18 cm • p. 374

Mountain Bluebird
18 cm • p. 375

Townsend's Solitaire
22 cm • p. 376

Veery
18 cm • p. 377

Gray-cheeked Thrush
19 cm • p. 378

Bicknell's Thrush
16 cm • p. 379

Swainson's Thrush
18 cm • p. 380

Hermit Thrush
18 cm • p. 381

Wood Thrush
20 cm • p. 382

American Robin
25 cm • p. 383

Varied Thrush
24 cm • p. 384

Gray Catbird
22 cm • p. 385

Northern Mockingbird
25 cm • p. 386

Sage Thrasher
22 cm • p. 387

Brown Thrasher
28 cm • p. 388

European Starling
22 cm • p. 389

American Pipit
16 cm • p. 390

Sprague's Pipit
16 cm • p. 391

Bohemian Waxwing
20 cm • p. 392

Cedar Waxwing
18 cm • p. 393

Blue-winged Warbler
12 cm • p. 394

Golden-winged Warbler
12 cm • p. 395

Tennessee Warbler
11 cm • p. 396

Orange-crowned Warbler
13 cm • p. 397

Nashville Warbler
10 cm • p. 398

Northern Parula
11 cm • p. 399

Yellow Warbler
13 cm • p. 400

Chestnut-sided Warbler
12 cm • p. 401

Magnolia Warbler
11 cm • p. 402

Cape May Warbler
13 cm • p. 403

Black-throated Blue Warbler
13 cm • p. 404

Yellow-rumped Warbler
14 cm • p. 405

Black-throated Gray Warbler
13 cm • p. 406

Black-throated Green Warbler
11 cm • p. 407

Townsend's Warbler
13 cm • p. 408

Blackburnian Warbler
13 cm • p. 409

Pine Warbler
13 cm • p. 410

Kirtland's Warbler
14 cm • p. 411

Prairie Warbler
12 cm • p. 412

Palm Warbler
13 cm • p. 413

Bay-breasted Warbler
14 cm • p. 414

Blackpoll Warbler
14 cm • p. 415

Cerulean Warbler
11 cm • p. 416

Black-and-white Warbler
13 cm • p. 417

American Redstart
13 cm • p. 418

Prothonotary Warbler
14 cm • p. 419

Worm-eating Warbler
13 cm • p. 420

Ovenbird
15 cm • p. 421

Northern Waterthrush
15 cm • p. 422

Louisiana Waterthrush
15 cm • p. 423

Kentucky Warbler
13 cm • p. 424

Connecticut Warbler
14 cm • p. 425

Mourning Warbler
12 cm • p. 426

MacGillivray's Warbler
13 cm • p. 427

Common Yellowthroat
13 cm • p. 428

Hooded Warbler
14 cm • p. 429

Wilson's Warbler
10 cm • p. 430

Canada Warbler
13 cm • p. 431

Yellow-breasted Chat
18 cm • p. 432

Spotted Towhee
19 cm • p. 433

Eastern Towhee
19 cm • p. 434

American Tree Sparrow
15 cm • p. 435

Chipping Sparrow
14 cm • p. 436

Clay-colored Sparrow
13 cm • p. 437

Brewer's Sparrow
14 cm • p. 438

Field Sparrow
14 cm • p. 439

Vesper Sparrow
15 cm • p. 440

Lark Sparrow
15 cm • p. 441

Lark Bunting
16 cm • p. 442

Savannah Sparrow
14 cm • p. 443

Grasshopper Sparrow
12 cm • p. 444

Baird's Sparrow
13 cm • p. 445

Henslow's Sparrow
12 cm • p. 446

Le Conte's Sparrow
12 cm • p. 447

Nelson's Sparrow
13 cm • p. 448

Fox Sparrow
18 cm • p. 449

Song Sparrow
14 cm • p. 450

Lincoln's Sparrow
13 cm • p. 451

Swamp Sparrow
14 cm • p. 452

White-throated Sparrow
16 cm • p. 453

Harris's Sparrow
18 cm • p. 454

White-crowned Sparrow
18 cm • p. 455

Golden-crowned Sparrow
18 cm • p. 456

Dark-eyed Junco
15 cm • p. 457

McCown's Longspur
15 cm • p. 458

Lapland Longspur
15 cm • p. 459

Smith's Longspur
15 cm • p. 460

Chestnut-collared Longspur
15 cm • p. 461

Snow Bunting
16 cm • p. 462

Summer Tanager
18 cm • p. 463

Scarlet Tanager
17 cm • p. 464

Western Tanager
18 cm • p. 465

Northern Cardinal
21 cm • p. 466

Rose-breasted Grosbeak
19 cm • p. 467

Black-headed Grosbeak
19 cm • p. 468

Lazuli Bunting
14 cm • p. 469

Indigo Bunting
14 cm • p. 470

Dickcissel
16 cm • p. 471

Bobolink
17 cm • p. 472

Red-winged Blackbird
20 cm • p. 473

Eastern Meadowlark
23 cm • p. 474

Western Meadowlark
22 cm • p. 475

Yellow-headed Blackbird
24 cm • p. 476

Rusty Blackbird
23 cm • p. 477

Brewer's Blackbird
21 cm • p. 478

Common Grackle
30 cm • p. 479

Brown-headed Cowbird
17 cm • p. 480

Orchard Oriole
16 cm • p. 481

Bullock's Oriole
20 cm • p. 482

Baltimore Oriole
19 cm • p. 483

Gray-crowned Rosy-Finch
15 cm • p. 484

Pine Grosbeak
21 cm • p. 485

Purple Finch
14 cm • p. 486

Cassin's Finch
16 cm • p. 487

House Finch
13 cm • p. 488

Red Crossbill
13 cm • p. 489

White-winged Crossbill
40 cm • p. 490

Common Redpoll
13 cm • p. 491

Hoary Redpoll
13 cm • p. 492

Pine Siskin
11 cm • p. 493

American Goldfinch
14 cm • p. 494

Evening Grosbeak
19 cm • p. 495

House Sparrow
15 cm • p. 496

INTRODUCTION

Common Loon

This book celebrates the rich assortment of birds that live, breed or migrate through Canada, and provides an overview of the birds that regularly occur here. Canada is the second-largest country in the world, with a landmass of 9.9 million square kilometres and three bordering oceans. From British Columbia's coastal rainforests to the rocky shores of Newfoundland, from the western mountains to the southern lowlands and from arctic tundra to prairie grasslands, we have amazingly diverse habitats and climatic conditions that attract a wide variety of bird species.

Over the past 120 years, dedicated professionals and amateurs have contributed a wealth of information about our bird species. The following section focuses on the diversity of Canada's birds, their habitats and some of the conservation stewardship challenges that face bird distribution and abundance.

In 1966, the National Museum of Canada listed 519 species of birds; 658 species are recorded today—a 27 percent increase! Some of the changes are systematic, such as splitting the Blue Grouse into two species (Dusky Grouse and Sooty Grouse) or combining previously separate subspecies, such as the Myrtle Warbler and Audubon's Warbler into the Yellow-rumped Warbler. Other changes result from birds expanding their ranges or, like many of the new additions, from birds straying into new areas from outside their normal ranges. Still other species, such as the House Finch and the European Starling, have been introduced to North America from Eurasia. As the human population increases, roads and railways, urbanization, agricultural practices and forestry have allowed both birds and bird observers to access previously unexplored areas, especially in the boreal forest. Changes to the landscape, whether natural or human-induced, result in species moving in and others dispersing.

Many species are peripheral and reach Canada from well-established ranges. For example, the Gray-headed Chickadee of Eurasia is found in a small corner of the Yukon, and the White-headed Woodpecker, Spotted Owl, California Quail, Sage Thrasher and Kirtland's Warbler are some birds that reach the northern limit of their range along the Canada-U.S. border. Fish Crows from the southeastern United States have expanded northward along river valleys and may soon nest regularly in the Great Lakes area of Ontario. Recovering Great Egret and Snowy Egret populations are expanding their ranges, so nesting in Atlantic Canada may occur in the future. New breeders along the West Coast include the Northern Fulmar, Black-legged Kittiwake, Horned Puffin and Thick-billed Murre from Alaskan populations. Recent additions to the breeding avifauna of the Prairie provinces, or former nesting species that have reestablished in the last decade or two, include the Trumpeter Swan, Cattle Egret, White-faced Ibis, Peregrine Falcon, Black-necked Stilt, Little Gull, Ross's Gull, Eurasian Collared-Dove, Northern Cardinal and House Finch. All these changes have happened over only a few decades, illustrating the dynamic nature of some species in the avian world.

About a quarter of all Canada's nesting species are habitat generalists that breed throughout much of the country. These include the Spotted Sandpiper, Common Raven, American Robin, crows, chickadees, nuthatches, Rock Pigeons and other species that can live in a variety of habitats and adapt to local change. At the other end of the spectrum are habitat specialists, or species that require specialized habitats and are vulnerable to change. Many are colonial in nature, such as cliff-nesting seabirds found along rocky coastal headlands. Another good example is the endangered Kirtland's Warbler, which nests on the ground under the low branches of mature old jack pines.

Canada also supports many bird species of national or international significance that breed, stage, winter or live here year-round.

This book is just a beginning. There are many more things to learn about these feathered wanderers. We hope this book helps you to discover the diverse assortment of birds that occur in Canada.

ECOREGIONS OF CANADA

Canada has one of the most diverse landscapes in the world, ranging from farmland to vast wilderness, from a multitude of rivers, lakes and other wetlands to deserts, from endless prairies to towering mountain ranges, and from temperate forests to arctic barrens. Scientists have categorized these landscapes into communities of plants and animals known as ecosystems or ecoregions. Each ecoregion is distinguished from others by its unique mosaic of plants, wildlife, climate, landforms and human activities. Overall, 15 terrestrial and five marine ecosystems have been identified in Canada.

Canada's terrestrial ecosystems vary greatly in size and shape and cover a broad range of grassland, forested, taiga and arctic ecosystem types. Forested areas range from our largest ecoregion, the vast **Boreal Shield**, which covers nearly one-fifth of Canada's land base, to our smallest ecoregion, the **Mixedwood Plains** of extreme southern Ontario and Québec. Other forested ecoregions are the western mountainous ecoregions (**Montane Cordillera, Boreal Cordillera** and southern **Taiga Cordillera**), the western **Boreal Plains** and **Taiga Plains** ecoregions, and two marine ecoregions (**Pacific Marine** and **Atlantic Marine**). In northern Canada, the **Taiga Shield,** Taiga Plains and **Hudson Plains** ecoregions provide a transition between the boreal forest and the arctic tundra.

The **Pacific Maritimes** is the westernmost ecoregion in Canada and includes Vancouver Island, the Lower Mainland and coastal mountains and ranges north to Mount Logan. The climate in this ecoregion is moderated by its proximity to the Pacific Ocean, and temperatures vary minimally throughout the year with few

Black-necked Stilt

days of freezing or hot temperatures. Being a coastal ecoregion, it supports a rich diversity of shorebirds and seabirds, and its moderate climate in winter results in the largest assemblage of winter birds of any area in Canada.

Three cordilleran ecoregions dominate Canada's western mountains and foothills: the **Montane Cordillera**, which includes most of southern BC and the foothills of southwestern Alberta, the **Boreal Cordillera** of central and northern BC and the southern Yukon, and the **Taiga Cordillera**, which includes mountains and plains along the Yukon-NWT border north to the Beaufort Sea. The Montane Cordillera experiences a great diversity in temperatures and moisture, but is generally mild, with relatively dry summers and wet winters. The Boreal Cordillera is dominated by mountains and plateaus separated by lowlands and valleys. Mountains along its western edge stop much of the precipitation, resulting in this ecoregion being quite dry. The northernmost Taiga Cordillera is also dominated by steep mountains and valleys, but these give way to wetlands, rolling hills and eventually tundra in the north. Its northerly position means that it has long, cold winters, with snow persisting for six to eight months of the year.

The **Taiga Plains** are bordered by mountains to the west, by the Arctic to the northeast, and by boreal plains and forest to the south and east. Like the Taiga Shield immediately east of it, the Taiga Plains have short, cool summers and long, cold winters. Snow and ice last for six to eight months of the year, and permafrost is widespread.

The **Taiga Shield** is Canada's second-largest ecoregion, supporting bedrock from the Canadian Shield from the eastern Arctic to Labrador, and separated by Hudson Bay and the Hudson Plains ecoregion in the middle. This ecoregion has an average temperature just below freezing, and a short summer with mean temperatures of about 10°C. The landscape has been influenced by numerous advancing and retreating glaciers that have scraped the ground down to bedrock in many places, leaving behind numerous glacial depressions that are now lakes.

The **Hudson Plains** ecoregion extends from Manitoba to Québec along the southern edge of Hudson Bay. Flat terrain and poor drainage have created a vast network of continuous wetlands and bogs that dominate this area. Summer temperatures are moderated by Hudson Bay, but in winter, the average temperature for this region drops to almost –20°C.

The **Boreal Plains** ecoregion occupies much of central and northern Alberta, with a thinner band south of the Boreal Shield in Saskatchewan and Manitoba. Its moisture is limited by mountains to the west. Summers are short and warm; winters are cold. Through the centuries, the landscape in this ecoregion has been levelled by numerous glacial deposits, which have also left behind an abundance of glacial meltwater lakes.

Canada's southern Prairie provinces are dominated by the **Prairies** ecoregion. This northernmost extension of the Northern Great Plains represents by far the most altered ecoregion in Canada, with over 95 percent of the prairies having been converted to agriculture. Mountains to the west tend to block out much of the moisture in the western prairies, but precipitation does increase in the east. The prairies also tend to be windier than other areas, with extremes in temperature ranging from an average of –10°C in winter to 15°C in summer.

The **Boreal Shield** stretches in a broad band across parts of six provinces from northern Saskatchewan to Newfoundland. The Boreal Shield provides the classic image of Canada's wilderness with its exposed Precambrian Shield bedrock, vast boreal forests and rushing rivers. Long, cold winters and short summers are typical here, although the Great Lakes moderate the climate in parts of central Ontario and the Atlantic Ocean influences the climate and increases precipitation levels in Newfoundland.

Newfoundland is wholly contained in the Boreal Shield ecoregion, but the remaining Atlantic provinces and the Appalachian region of southern Québec make up the **Atlantic Maritime** ecoregion. Moderate temperatures and a moist climate characterize this ecoregion. Both conifers and deciduous trees grow in the productive, mixedwood forests. Much of the forest is second-growth—over the centuries, the native trees have been harvested or have burned.

Most of extreme southern Ontario and the St. Lawrence Lowlands of Québec are within the **Mixedwood Plains** ecoregion. Although it is the smallest of Canada's terrestrial ecoregions, it supports

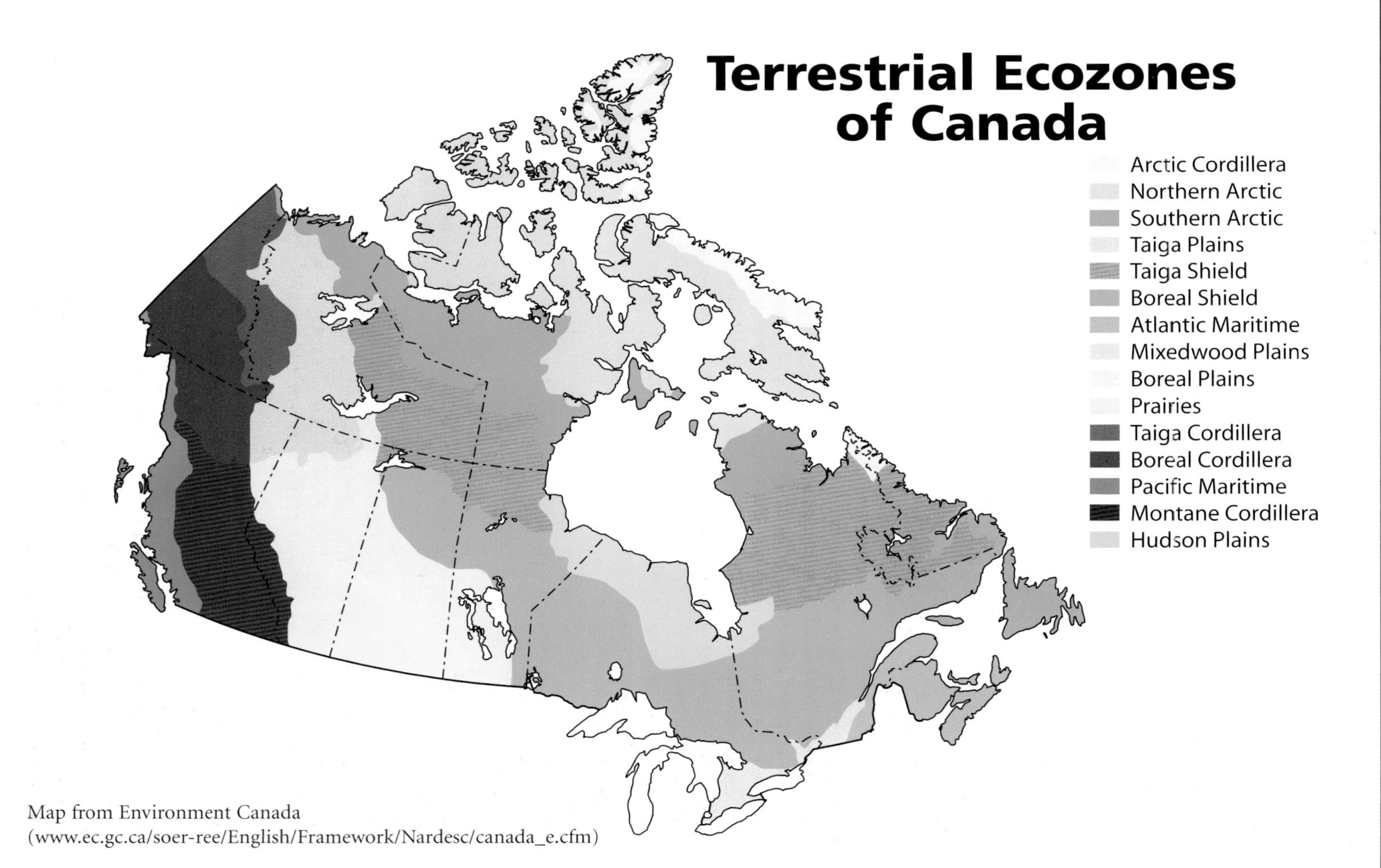

Map from Environment Canada
(www.ec.gc.ca/soer-ree/English/Framework/Nardesc/canada_e.cfm)

American Avocet

half of Canada's human population. Its key geographic location, proximity to waterways, productive soils, warm climate and plentiful precipitation make it a heavily used area. At one time, this ecoregion sustained more tree species than anywhere else in Canada, but today less than 10 percent of the forest cover remains. Forests vary from mixedwood stands of pine, eastern hemlock, oak and maple to the richly diverse Carolinian forest of sugar maple, basswood, walnut and hickory. Rare species such as tulip trees, cucumber-trees, sassafras and sycamores reach the northern limit of their range here.

Canada's treeless Arctic is composed of three large ecoregions; from south to north these are the Southern Arctic, Northern Arctic and Arctic Cordillera. Collectively, these ecoregions encompass one-quarter of Canada's land base. The **Southern Arctic** ecoregion extends through much of the northern continental portions of Nunavut, the Northwest Territories and Québec, bordered by the treeline and the Taiga Plains to the south and the Northern Arctic ecosystem to the north. With an average summer temperature of 5°C and winter averages dropping to as low as nearly –30°C in some areas, permafrost is common throughout this ecoregion. The surface of the land is dominated by Precambrian granite bedrock, with occasional pilings of soil and rocks known as glacial erratics dotting the landscape, giving evidence of the numerous glaciers that shaped this landscape in the past.

Most of the archipelago north of Nunavut and the Northwest Territories is included in the **Northern Arctic** ecoregion. It is the coldest and driest part of the country. Precipitation is so low

Harlequin Ducks

that this area is often referred to as an arctic desert. Average annual temperatures are below 0°C in much of this ecoregion, leading to year-round permafrost that can extend up to one kilometre deep. Snow covers the ground through much of the year, disappearing briefly in July and August.

Great Egret

The **Arctic Cordillera** runs along the north-eastern fringe of Nunavut and Labrador, and includes much of Ellesmere and Baffin Islands. Comprising desolate valleys and mountain peaks that extend up to two kilometres in height, this ecoregion is one of the most inhospitable locations on the planet. Like other areas of our Far North, winters are long and dark, but the short growing season is aided by continuous daylight.

Together, these diverse ecoregions, ranging from arctic tundra in the north to lush Carolinian forest in the south, provide varied habitat for a broad range of wildlife and plant species.

CONSERVATION AND PROTECTED SPACES

While Canada is a land renowned for its spectacular scenery and wealth of wildlife, many of our most significant scenic and nesting sites are protected spaces. The federal government manages much of the lands contained in Canada's protected areas including national parks, national wildlife areas, marine wildlife areas, migratory bird sanctuaries. The Canadian Wildlife Service, for instance, administers a network of protected areas that protects nearly 12 million hectares of wildlife habitat, an area more than twice the size of Nova Scotia. Some of the most significant federally protected nesting areas for birds include the Queen Maud Gulf Migratory Bird Sanctuary in Nunavut, where over 90 percent of the world's population of Ross's Geese nest, and Wood Buffalo National Park in the Northwest Territories, which supports the world's entire wild nesting population of Whooping Cranes. Many significant seabird colonies are protected in marine wildlife areas—the proposed Scott's Island MWA in British Columbia, for instance, would protect more than two million seabirds and over half of the world's Cassin's Auklets.

About half of the protected areas in Canada are administered by provincial, territorial or aboriginal governments (i.e., parks, provincial and territorial wildlife areas and ecological reserves) and by non-governmental organizations (i.e., Nature Conservancy, Canadian Wildlife Federation and others).

Other efforts to recognize and protect birds in Canada include the Important Bird Areas program (IBA), a global effort of BirdLife International to identify key habitats for the world's birds that includes nearly 600 IBAs in Canada. In recent decades, the North American Bird Conservation Initiative (NABCI) has made great strides toward coordinating conservation efforts across North America by identifying key sites and species within Bird Conservation Regions (BCRs) which align closely with terrestrial ecosystems for the continent.

Western Bluebird

WHERE TO BIRD IN CANADA

Suggested Birding Sites

With such an incredible diversity of birds and so many unique and relatively unspoiled birding areas, Canada is truly a haven for birding. Sites such as Point Pelee National Park in southern Ontario and Churchill in Manitoba are often referred to when discussing top birding destinations in North America for bird diversity and unique birds. But every province in Canada has an abundance of premiere birding sites. Significant birding destinations for each province and territory include the following:

BRITISH COLUMBIA

100 Mile House Town Sanctuary
Active Pass/Gulf Islands NPR
Brackendale Bald Eagle Reserve
Creston Valley
Delkatla Wildlife Sanctuary, Masset
Eskers PP
George C. Reifel Bird Sanctuary and South Arm Marshes WMA
Gitnadoix River PRA/Lower Skeena River
Goldstream NP
Gwaii Haanas NPR (Haida Gwaii)
Iona Beach RP
Manning PP
Mount Revelstoke NP
Muncho Lake PP
Nazko Lake PP/Chilcotin Plateau
Okanagan Oxbows
Pacific Rim NPR
Parksville-Qualicum Beach WMA
Pend d'Oreille Valley
Pitt-Addington Marsh/Pitt Lake
Somenos Marsh, Duncan
Tranquille WMA
Vanderhoof Bird Sanctuary
Vaseux Lake Migratory Bird Sanctuary

ALBERTA

Aspen Beach PP
Banff National Park
Beaverhill Lake Bird Observatory
Clifford E. Lee Nature Sanctuary
Cold Lake PP
Cypress Hills PP
Dinosaur PP
Edmonton's Saskatchewan River valley
Elk Island NP
Jasper National Park
Kananaskis Country
Kininvie Marsh
Kootenay Plains
Lesser Slave Lake PP
Porcupine Hills
Saskatoon Island PP
Sheep River Wildlife Sanctuary
Sir Winston Churchill PP
Slack Slough, Red Deer
Waterton Lakes NP
Wildcat Hills
Wood Buffalo NP
Writing-on-Stone PP

American White Pelican

Great Blue Heron

SASKATCHEWAN

- Buffalo Pound PP and Nicole Flats
- Cypress Hills PP
- Danielson PP and Gardiner Dam
- Duck Mountain PP
- Grasslands NP
- Katepwa Point PP
- Lac La Ronge PP
- Last Mountain Lake NWA
- Pike Lake PP
- Prince Albert NP
- Quill Lakes

MANITOBA

- Assiniboine Park and Fort Whyte
- Churchill
- Clearwater Lake PP
- Grindstone PP
- Mixed-grass Prairie Preserve and Broomhill WMA
- Oak Hammock Marsh
- Pisew Flats PP
- Riding Mountain NP
- Spruce Woods PP
- Turtle Mountain PP
- Whiteshell PP and Pinawa area
- Whitewater Lake

Blackpoll Warbler

ONTARIO

- Algonquin PP
- Amherst Island
- Bon Echo PP
- Carden Plain
- Dundas Marsh
- Grundy Lake PP and Killamey PP
- Lake Superior PP
- Long Point PP
- Lynde Shores Conservation Area and Cranberry Marsh
- Niagara Falls
- Ottawa River PP
- Point Pelee NP
- Polar Bear PP
- Presqu'ile PP
- Quetico PP
- Rainy River and Lake of the Woods
- Rondeau PP
- Rouge River PP
- Sleeping Giant PP
- The Bruce Peninsula
- Tommy Thompson Park and Toronto Islands
- Trans-Canada North (Hwy 11)

QUÉBEC

- Anticosti Island
- Baie du Febvre/Lac Saint Pierre
- Basse-Côte-Nord
- Beauharnois Dam
- Bois de l'Île Bizard Nature Park
- Cap Tourmente NWA
- Côté Ste-Catherine Locks
- Côté-Nord (North Shore, St. Lawrence River)
- Digges Islands
- Forillon NP/ Bonaventure Island
- Gaspé Penninsula
- George Montgomery Bird Sanctuary
- Îles de la Madeleine
- La Mauricie NP
- Lac Saint-François NWA
- Longeuil shoreline
- Manicouagan Reservoir
- Mingan Archipelago NPR
- Mont St-Bruno
- Mont St-Hilaire/Gault Natural Reserve
- Montréal Botanical Gardens
- Mount Royal Park and Mount Royal Cemetery
- Nuns' Island/Île des Soeurs
- Oka PP
- Rimouski/Bas-Saint-Laurent region
- St-Lazare Sandpits

NEW BRUNSWICK

Cape Jourimain
Fundy Bay NP
Grand Manan Island
Kouchibouguac NP
Machias Seal Island
Odell Park, Fredericton
Plaster Rock-Nictau

Snowy Owl

PRINCE EDWARD ISLAND

Brackley-Covehead Marshes
Malepeque Bay
Prince Edward Island NP

NOVA SCOTIA

Bay of Fundy
Cape Breton Highlands NP
Kejimkujik National Park
Point Pleasant Park, Halifax
Wallace Bay

NEWFOUNDLAND/LABRADOR

Cape St. Mary's
Codroy Valley
Gros Morne NP
St. John's Metropolitan Area
Terra Nova NP
Witless Bay Islands

YUKON TERRITORY

Babbage and Spring River Deltas
Blow River Delta
Dempster Highway
Kluane National Park
Lake Laberge, southwest end, including Shallow Bay, Big Slough and Swan Lake
Nisultin River Delta
Old Crow Flats
Tagish Narrows

Gyrfalcon

NORTHWEST TERRITORIES

- Anderson River Delta
- Banks Island Migratory Bird Sanctuary
- Cape Parry
- Kugaluk River
- Great Slave Lake, North Arm and South Shore
- Mackenzie River Delta
- Mills Lake
- Nahanni NP
- Thomsen River
- Whooping Crane Nesting Area/Wood Buffalo NP

NUNAVUT

- Queen Maud Lowlands
- Bathurst Inlet
- Boas River and wetlands, Southampton Island
- Great Plain of the Koukdjuak
- Foxe Basin Islands
- Langcaster Sound
- McConnell River
- Thelon Wildlife Sanctuary (straddles NT, NV border)

Wild Turkey

Migration

Many birds pass through the country during migration, heading north in the spring to summer breeding grounds, or heading south in autumn to wintering grounds. These feathered wanderers link our country to the global bio-economy and remind us that how we protect our habitats has ramifications far beyond our borders. Very simply, birds time their migration cycles so that their young are born when there is an abundance of food and the weather is at its warmest season. They return south while the climate is still tolerable, so that they may find enough food along the way to sustain them during their long journey. Some species follow simple migration routes; others travel long distances over complicated pathways. Each species starts at different times, follows its own, unique path, travels at its own speed and migrates between different latitudes, so many migration routes are used over Canadian territory.

Generally speaking, birds typically migrate in a north-south movement, with the heaviest

Red-breasted Sapsucker

concentrations occurring along coastlines, mountain ranges and large river valleys. In North America, these topographical features basically lie in a north-south direction.

The term flyways is used to refer to the wide migration "highways," or broad areas, where several migration routes overlap in specific geographic areas. Four major flyways have been identified in North America: the Pacific, the Central, the Mississippi and the Atlantic flyways. Although birds do follow coastlines, the flyway boundaries are not always distinct, and may overlap in southern wintering or northern breeding grounds. In the narrow strip of land where North and South America meet, all four flyways combine into one.

Northern Cardinal

The Pacific Flyway includes the western Arctic, the Pacific coast and the Rocky Mountains, then follows south to Mexico. The route is used by gulls, ducks, waterbirds and many landbirds. Notably, half of the North American population of "Lesser" Sandhill Cranes follows the Pacific flyway to pass through northeastern British Columbia each spring. The coastal mudflats of the Fraser River delta are used as a staging area each spring and autumn for most of the world's Western Sandpipers. Mountain towns such as Valemont and Brackendale, BC, and Canmore, Alberta, host eagle-watching festivals as raptors fly south, high over the mountain ranges.

In Canada, the Central Flyway begins on the northwest Arctic coast and follows the eastern base of the Rocky Mountains, with roughly parallel pathways through the prairies. Huge flocks of waterfowl, Sandhill Cranes, numerous shorebird species and a variety of raptors pass through, and tens of thousands of Snow Geese and Canada Geese arrive between April and May, reappearing again in late September and October. Staging areas include the Tofield/Camrose area of Alberta, where the Pacific, Central and Mississippi flyways converge.

The longest migrations in the Western Hemisphere follow the Mississippi Flyway, travelling between the Alaska coast and the southern tip of Patagonia. From the mouth of the Mackenzie River, NWT, birds may follows a 4800-kilometre southward route, well forested and uninterrupted by mountain ranges, to the Mississippi River delta in the Gulf of Mexico. Bound by Lake Erie and the Mississippi the east, but less well defined in the west, this route is important to waterfowl, shorebirds and many passerines. Lake Diefenbaker and Last Mountain Lake, Saskatchewan, and Oak Hammock Marsh, near Winnipeg, Manitoba, are important staging areas on this flyway.

The Mississippi and Atlantic flyways converge at Ontario's Point Pelee National Park, a world-famous stopover for neotropical migrants. Approximately 370 species have been recorded in Point Pelee's Carolinian Forests, including some species rare to Canada. The Atlantic Flyway follows the offshore waters of the U.S. Atlantic coast to the Allegheny Mountains, where it curves inland across the Great Lakes, through the central and northern prairie provinces to the Arctic coast. This flyway is important for waterfowl, including Lesser Scaups and Redheads, as well as many other species of birds.

For more information on local migration festivals, staging areas or peak migration times, contact your local nature society.

Winter

Less-showy wintering species may receive less attention, but each region of Canada supports many regular species. Christmas bird counts report a huge range in the number of species, from three species recorded in frigid Nunavut to 180 or so in temperate areas such as BC's Fraser River delta! Year after year, the **Lower Mainland** near Vancouver reports the highest counts in Canada for dozens of species of waterfowl, diving birds (including loons, grebes and cormorants), herons, birds of prey, shorebirds, gulls, pelagic birds, owls and woodpeckers. **Atlantic Canada** follows with approximately 175 species, including many diving birds, waterfowl and gulls such as Black-legged Kittiwakes, Razorbills, Dovekies, gulls and Harlequin Ducks. In winter, open water attracts a wider diversity of birds, and the **Great Lakes–St. Lawrence region** of southern Ontario and Québec hosts large concentrations of gulls and waterfowl such as goldeneyes, mergansers, scoters, scaups, eiders and Long-tailed Ducks. On the prairies, waterfowl feed at open sewage lagoons, attracting predatory raptors.

A few noteworthy species that breed in the territories but spend the winter in southern Canada include the Snowy Owl, Gyrfalcon and Rough-legged Hawk. Backyard feeders in most every province attract Downy Woodpeckers, Hairy Woodpeckers, Black-capped Chickadees, nuthatches and Common Redpolls, but probably our most famous winter visitors are the huge flocks of beautiful Bohemian Waxwings that descend upon fruit trees in urban and suburban areas across the country.

Canadian World Record Setters

MOST FEATHERS: 25,216 (Tundra Swan)

FEWEST FEATHERS: 940 (Ruby-throated Hummingbird)

FASTEST BIRD: Peregrine Falcon in a dive at 320 km/h

FASTEST WINGBEATS: Hummingbirds (60–90/sec)

SLOWEST WINGBEATS: Vultures (1/sec)

MOST AERIAL BIRD: Sooty Tern (flies 3–10 years without landing)

LONGEST ANNUAL MIGRATION: 17,700 km (Arctic Terns nest in the Arctic and winter in Antarctica)

LONGEST NONSTOP DISTANCE FOR A SMALL BIRD: 800 km (Ruby-throated Hummingbirds migrate across the Caribbean Sea)

MOST VOCAL BIRD: Red-eyed Vireo (22,197 song phrases in 10 hours)

COLDEST TEMPERATURE TOLERATED: –62.5°C (Snowy Owl)

MOST NORTHERLY NEST: Ivory Gull, on pack ice 650 km south of the North Pole

BEGINNING TO LEARN THE BIRDS

The Challenge of Birding

Birding (also known as "birdwatching") can be enjoyable, if not downright thrilling. It can also be extremely challenging to master the art of birdwatching, and learning to recognize all the birds in Canada could take a lifetime of careful study. But fear not! This guide will get you get started.

Do not expect to become an expert overnight. After all, only a small number of birders and ornithologists can identify all of our species with confidence.

Classification: The Order of Things

Hooded Mergansers

To an ornithologist (a biologist who studies birds), the species is the fundamental unit of classification because the members of a single species look most alike and they naturally interbreed with one another. Each species has a scientific name, usually derived from Latin or Greek, that designates genus and species (which is always underlined or italicized) and a single accredited common name, so that the different vernacular names of a species do not cause confusion. A bird has been properly identified only when it has been identified "to species," and most ornithologists use the accredited common name. For example, "American Coot" is an accredited common name, even though some people call this bird "Mudhen." *Fulica americana* is the American Coot's scientific name. (*Fulica* is the genus, or generic name, and *americana* is the species, or specific name).

To help make sense of the hundreds of bird species in our region, scientifically oriented birders lump species into recognizable groups. The most commonly used groupings, in order of increasing scope, are genus, family and order. The American Coot and Common Moorhen are different species that do not share a genus (their generic names are different), but they are both members of the family Rallidae (the rail family). The rail, limpkin and crane families are in turn grouped within the order Gruiformes, which comprises the chickenlike birds.

Ornithologists have arranged all of the orders to make a standard sequence. It begins with the waterfowl (order Anseriformes), which are thought to be most like the evolutionary ancestors of modern birds. This sequence ends with those species thought to have been most strongly modified by evolutionary change and most departed from the ancestral form. We have organized this book according to this standard sequence.

Tree Swallows

At first, the evolutionary sequence might not make much sense. Birders, however, know that all books of this sort begin with waterfowl, followed by grouselike birds, diving birds, birds of prey, and other birds that look more and more like songbirds (formally known as "passerines"). Still, many readers will tell us that we should have arranged this book alphabetically. Although alphabetical organization may seem logical, it assumes that you already know all of the up-to-date, accredited names of the birds. In practice, the tried-and-true method of grouping birds according to similarities and differences provides the best format for learning.

TECHNIQUES OF BIRDING

Finding the Birds

Being in the right place at the right time to see birds in action involves both skill and luck. The more you know about a bird—its range, preferred habitat, food preferences and hours and seasons of activity—the better your chances will be of seeing it. It is much easier to find a Boreal Owl in the boreal forest than elsewhere, especially at night in spring, when adults are calling for mates. Snowy Owls, however, are most often seen on fence posts or in fields during the day in winter.

Generally, spring and fall are the busiest birding times. Temperatures are moderate then, and a great number of birds are on the move, often heavily populating small patches of habitat before moving on. Male songbirds are easy to identify on spring mornings as they belt out their courtship songs. Throughout much of the year, diurnal birds are most visible in the early morning hours when they are foraging, but during winter they are often more active in the day when milder temperatures prevail. Timing is crucial because summer foliage often conceals birds and cold weather drives many species south of our region for winter. Birding also involves a great deal of luck.

Binoculars

The small size, fine details and wary behaviour of many birds make binoculars an essential piece of equipment for birding. Binoculars can cost anywhere from $50 to $1500, and at times it may seem that there are as many kinds of binoculars as there are species of birds. Most beginners pay less than $200 for their first pair. Compact binoculars are popular because they are small and lightweight, but they are not necessarily cheap—some of the most expensive binoculars are compact models.

Binoculars come in two basic types: porro-prism (in which there is a distinct, angular bend in the body of the binoculars) and roof-prism (in which the body is straight). Good porro-prism binoculars are less expensive than good roof-prism binoculars: a first-rate pair of "porros" costs $300 to $400; good roof-prism binoculars, which are often waterproof and fog-resistant (nitrogen-filled), can cost $800 or more. Expensive binoculars usually have better optics and generally stand up better to abuse.

The optical power of binoculars is described with a two-number code. For example, a compact pair of binoculars might be "8×21," while a larger pair might have "7×40" stamped on it. In each case, the first number states the magnification, while the second number indicates the diameter, in millimetres, of the front lenses. Seven-power binoculars are the easiest to hold and to use for finding birds; 10-power binoculars give a shakier but more magnified view. Larger lenses gather more light, so a 40 mm or 50 mm lens will perform much better at dusk than

Trumpeter Swans

Western Tanager

a 20 mm or 30 mm lens of the same magnification. For a beginner, eight-power, porro-prism binoculars with front lenses at least 35 mm in diameter (thus 8×35 or 8×40) are suitable. Some binoculars have a wider field of view than others, even if the two-number code is identical. We recommend the wider field of view, because many beginners have trouble finding birds in compact, narrow-view binoculars.

Higher magnification binoculars bring the birds closer but they also magnify vibration. To counteract this effect, image-stabilized binoculars can greatly steady the view, especially as your arms become tired and shaky. Binoculars with this feature usually cost more, but the extra stability may well be worth that extra cost.

Look at many types of binoculars before making a purchase. Talk to other birders about their binoculars and ask to try them out. Go to a store that specializes in birding—the sales people there will know which models perform best in the field.

When birding, lift the binoculars up to your eyes without taking your eyes off the bird. This way you will not lose the bird in the magnified view. You can also note an obvious landmark near the bird (a bright flower or a dead branch, for example) and then use it as a reference point to find the bird with the binoculars.

Spotting Scopes and Cameras

The spotting scope (a small telescope with a sturdy tripod) is designed to help you view birds that are beyond the range of binoculars. Most spotting scopes are capable of magnification by a factor of 20 or more. Some scopes will even allow you to take photographs through them.

If you intend to photograph birds, you can expect to dig deep into your wallet for a decent digital single-lens reflex (DSLR) camera and a high-quality telephoto or zoom lens. Birds are small and difficult to approach, so you will need a telephoto lens of at least 200 mm focal length for the popular smaller-format APS-C sized DSLRs, or at least 300 mm focal length for the more expensive

Ring-necked Duck

35 mm-equivalent, full-format DSLRs (or for a 35 mm film camera). Image stabilization—with stabilization either in the sensor or in the lens—will work wonders by allowing solid, blur-free, handheld images with long lenses. And modern DSLR sensors are excellent at producing clean, low-noise images at higher ISO speeds, thus allowing faster shutter speeds to give more stability in low light and with longer lenses. But to provide real versatility for super-long lenses, a solid tripod can complete your kit. Lightweight carbon-fibre tripods, while expensive, take a huge load off your back in the field.

For casual bird photography, a lightweight, good-quality stuff-it-in-your-large-pocket compact camera with a 10- to 12-times, image-stabilized, optical zoom lens is just the ticket. For the best quality photographs, be sure to set your compact camera settings at full resolution (largest image dimensions) and minimum compression (largest file size), and set your ISO speeds as low as possible (400 ISO or lower; ISO 100 to 200 is preferred for cleaner images). High ISO settings will produce very noisy images with compact cameras because of their small-sized sensors. And one final essential point: never use your camera's digital zoom, which is just a marketing tool designed to substitute for a limited optical zoom. This digital zoom "feature" doesn't zoom closer to the subject—the camera just crops the frame and then enlarges that small centre portion to simulate a zoomed image, giving much-reduced image quality.

To select the perfect camera and lens for your needs, talk to knowledgeable camera sales staff and experienced photographer friends, try out the camera in the store and be prepared to spend some money.

Once you have a good camera, you must develop an equally good technique. There are excellent books on photographing birds and an evening class or two can be very helpful. Then practise, practise, practise—the more you shoot, the better you get. Finally, most successful bird photographs are taken by quiet, patient photographers—few birds stick around to be photographed by noisy, stampeding admirers.

Birding by Ear

Recognizing birds by their songs and calls can greatly enhance your birding experience. When experienced birders conduct breeding bird surveys in the summer, they rely more on their ears than their eyes, because listening is far more efficient. There are numerous tapes and CDs that can help you learn bird songs, and a portable player with headphones can let you quickly compare a live bird with a recording.

The old-fashioned way to remember bird songs is to make up words for them. We have given you some of the classic renderings in the species accounts that follow, such as *who cooks for you? who cooks for you-all?* for the Barred Owl, as well as some nonsense syllables, such as *tsit tsit tsit* for the Blackpoll Warbler. Some of these approximations work better than others; birds often add or delete syllables from their calls, and very few pronounce consonants in a recognizable fashion. Bird songs usually vary from place to place as well, and the words and recordings that have helped you identify birds successfully in one area might not work as well elsewhere.

Black-capped Chickadee

Observing Bird Behaviour

Once you can confidently identify birds and remember their common names, you can begin to appreciate their behaviour. Studying birds involves keeping notes and records. The timing of bird migrations is an easy thing to record, as are details of feeding, courtship and nesting behaviour if you are willing to be patient. Flocking birds can also provide opportunities to observe and note fascinating social interactions, especially when you can recognize individual bird species. Trained ornithologists rely on more standardized, scientific methods of study, but your casual note-taking will contribute greatly toward our knowledge of birds.

Birding, for most people, is a peaceful, non-destructive recreational activity. One of the best ways to watch bird behaviour is to look for a spot rich with avian life, and then sit back and relax. If you become part of the scenery, the birds, at first startled by your approach, will soon resume their activities and allow you into their world.

Osprey

Birding by Habitat

Habitats are communities of plants and animals supported by the infrastructure of water and soil and regulated by the constraints of topography, climate and elevation. Simply put, a bird's habitat is the place in which it normally lives. Some birds prefer the open water, some are found in cattail marshes, others like mature coniferous forest, and still others prefer abandoned agricultural fields overgrown with tall grass and shrubs. Knowledge of a bird's habitat increases the chances of identifying the bird correctly. If you are birding in wetlands, you will not be identifying tanagers or towhees; if you are wandering among the leafy trees of a deciduous forest, do not expect to meet Boreal Owls or Tundra Swans.

Habitats are just like neighbourhoods: if you associate friends with the suburb in which they live, you can just as easily learn to associate specific birds with their preferred habitats. Only in migration, especially during inclement weather, do some birds leave their usual habitat.

Bird Lists

Many birders list the species they have seen at home or during excursions. You can decide what kind of list—systematic or casual—you want to keep, or you may decide not to make a list at all. However, lists may prove rewarding in unexpected ways. For example, after you visit a new area, your list becomes a souvenir of your experiences there. By reviewing the list, you can recall memories and details that you might otherwise have forgotten. Keeping regular, accurate lists of birds in your neighbourhood can also be useful for local researchers. It can be interesting to compare the arrival dates and last sightings of hummingbirds and other seasonal visitors, or to note the first sighting of a new visitor to your area.

Although there are programs available for listing birds on computers, many naturalists simply keep records in field notebooks. Waterproof books and waterproof pens work well on rainy days, though many birders prefer to use a pocket recorder in the field and to transcribe their observations into a dry notebook at home. Find a notebook you like, and personalize it with field sketches, observations or whatever you wish.

BIRDING ACTIVITIES

Getting Involved

We recommend that you join in on such activities as Christmas bird counts, birding festivals and the meetings of your local birding or natural history club. Meeting other people with the same interests can make birding even more pleasurable, and there is always something to be learned when birders of all levels gather. If you are interested in bird protection and conservation and environmental issues, conservation organizations, natural history clubs and conscientious birding stores can keep you informed about the situation in your area and what you can do to help. Bird hotlines provide up-to-date information on the sightings of rarities, which are more common than you might think. It is hoped that more people will learn to appreciate nature through birding, and that those people will do their best to protect the natural areas and their biodiversity. Many bird enthusiasts support groups that directly purchase or enhance land for wildlife to be enjoyed by future generations of nature lovers. Large, active organizations include the following:

Canadian Wildlife Federation
350 Michael Cowpland Drive
Kanata, ON K2M 2W1
Phone: 1-800-563-4953
Website: www.cwf-fcf.org

Nature Canada
300, 75 Albert Street
Ottawa, ON K1P 5E7
Phone: 1-800-267-4088
Website: www.naturecanada.ca

Nature Conservancy of Canada
400, 36 Eglinton Avenue West
Toronto, ON M4R 1A1
Phone: 1-800-465-0029
Website: www.natureconservancy.ca

Bird Studies Canada
(Christmas bird counts, Baillie Bird-a-thon, Breeding Bird Atlas and more)
P.O. Box 160
115 Front Street
Port Rowan, ON N0E 1M0
Phone: 1-888-448-2473
Website: www.bsc-eoc.org

Ducks Unlimited Canada
National Office
Oak Hammock Marsh Conservation Centre
P.O. Box 1160
Stonewall, MB R0C 2Z0
Phone: 1-800-665-DUCK (3825)
Website: www.ducks.ca

Common Murre

Attracting Birds

Landscaping your own property to provide native plant cover and natural foods for birds is an immediate and personal way to ensure the conservation of bird habitat. The cumulative effects of such urban "nature-scaping" can be significant. If your yard is to become a bird sanctuary, you may want to keep cats out—cats kill millions of birds each year. Check with the local Humane Society for methods of protecting both your feline friends and wild birds. Ultimately, for protection of birds, cats are best kept indoors.

Common Moorhen

Backyard Bird Feeders

Many people set up backyard bird feeders or plant native berry- or seed-producing plants in their garden to attract birds to their yard. The kinds of food available will determine which birds visit your yard. Staff at birding stores can suggest which foods will attract specific birds. Hummingbird feeders are popular in summer and are filled with a simple sugar solution made from one part sugar and three to four parts water. Contrary to popular opinion, birds do not become dependent on feeders, nor do they subsequently forget how to forage naturally. Winter is when birds appreciate feeders the most, but it is also difficult to find food in spring before flowers bloom, seeds develop and insects hatch.

Birdbaths will also entice birds to your yard at any time of year, and heated birdbaths are particularly appreciated in the colder months. Avoid birdbaths that have exposed metal parts because wet birds can accidentally freeze to them in winter.

There are many good books written about feeding birds and landscaping your yard to provide natural foods and nest sites.

Nest Boxes

Another popular way to attract birds is to set out nest boxes, especially for House Wrens, Western Bluebirds, Tree Swallows and Purple Martins. Not all birds will use nest boxes: only species that normally use cavities in trees are comfortable in such confined spaces. Larger nest boxes can attract kestrels, owls and cavity-nesting ducks.

Cleaning Feeders and Nest Boxes

Nest boxes and feeding stations must be kept clean to prevent birds from becoming ill or spreading disease. Old nesting material may harbour a number of parasites, as well as their eggs. Once the birds have left for the season, remove the old nesting material and wash and scrub the nest box with detergent or a 10 percent bleach solution (one part bleach to nine parts water). You can also scald the nest box with boiling water. Rinse it well and let it dry thoroughly before you remount it. We advise that you wear rubber gloves and a mask when cleaning nest boxes or feeders.

Feeding stations should be cleaned monthly. Feeders can become mouldy and any seed, fruit or suet that is mouldy or spoiled must be removed. Unclean bird feeders can also be contaminated with salmonellosis and possibly other avian diseases. Clean and disinfect feeding stations with a 10 percent bleach solution, scrubbing thoroughly. Rinse the feeder well and allow it to dry completely before refilling it, and remove any discarded seed and feces on the ground under the feeding.

ABOUT THE SPECIES ACCOUNTS

This book gives detailed accounts of the 451 species of birds that are listed as regular by the official checklist of 658 species issued byAvibase, the World Bird Database; these species can be expected on an annual basis. Fifty occasional species and species of special note are briefly mentioned in an illustrated appendix. Canadian birders can expect to see small numbers of these species every few years, brought here either because of anticipated range expansion, migration or well-documented wandering tendencies. The order of the birds and their common and scientific names follow the American Ornithologists' Union's *Check-list of North American Birds* (7th edition and its supplements).

As well as discussing the identifying features of a bird, each species account attempts to bring a bird to life by describing its various character traits. Personifying a bird helps us to relate to it on a personal level. However, the characterizations presented in this book are based on the human experience and most likely fall short of truly defining the way birds perceive the world. The characterizations should not be mistaken for scientific propositions. Nonetheless, we hope that a lively, engaging text will communicate our scientific knowledge as smoothly and effectively as possible.

One of the challenges of birding is that many species look different in spring and summer than they do in autumn and winter. Many birds have breeding and nonbreeding plumages, and immature birds often look different from their parents. Some birds, such as eagles and gulls, may take at least four years to reach maturity, and during that time they have many transition plumages. This book does not try to describe or illustrate all the different plumages of a species; instead, it focuses on the forms you are most likely to see wherever you are in Canada.

ID

A combination of illustrations and text is used. Where appropriate, the description is subdivided to highlight the differences between male and female, breeding and nonbreeding, and sometimes juvenile, immature and adult birds. The descriptions favour easily understood language instead of heavily technical jargon. Birds may not have "jaw lines" or "eyebrows," but readers understand these terms easily. Some of the most common features of birds are pointed out in the glossary illustration.

Size

The size measurement, the average length of the bird's body from bill to tail, is an approximate measurement of the bird as seen in nature. Ranges are given because there may be variation among individuals, especially in larger birds. Wingspan, from wing tip to wing tip, is also given. Note that birds with long tails often have large measurements that do not necessarily reflect body size.

Habitat

The habitats describe where each species is most commonly found. In most cases it is a generalized description, but if a bird is restricted to a specific habitat, it is described more precisely. Because of the freedom flight gives them, birds can turn up in almost any type of habitat, but they will usually be found in environments that provide the specific food, water, cover and, in some cases, nesting habitat that they need to survive.

Sandhill Cranes

Nesting

The reproductive natural history used by birds varies. In each species account, nest location and structure, clutch-size, incubation period and parental duties are discussed. The nesting behaviour of birds that do not nest in Canada is not described.

Feeding

Birds spend a great deal of time searching for food. If you know what a bird eats and where the food is found, you will have a better chance of finding the bird you are looking for. Birds are frequently encountered while they are foraging; we hope that our description of their feeding methods and diets provides valuable identifying characteristics, as well as interesting dietary facts.

Voice

You will hear many birds, particularly songbirds, that may remain hidden from view. Memorable paraphrases of distinctive sounds will aid you in identifying a species. These paraphrases only loosely resemble the call, song or sound produced by the bird. Should one of our paraphrases not work for you, make up your own—the creative exercise will reinforce your memory of the bird's vocalizations.

Similar Species

Easily confused species that also occur in Canada are briefly discussed. If you concentrate on the most relevant field marks, the subtle differences between species can be reduced to easily identifiable traits. You might find it useful to consult this section when finalizing your identification; knowing the most relevant field marks will speed up the identification process. Even experienced birders can mistake one species for another. Be sure to check range maps of similar species, as some may not occur in your area or may occur seasonally.

Range Maps

The range map for each species represents the overall range of the species in an average year. Most birds will confine their annual movements to this range, though each year some birds wander beyond their traditional boundaries. These maps do not show differences in abundance within the range—areas of a range with good habitat will support a denser population than areas with poorer habitat. These maps also cannot show small pockets within the range where the species may actually be absent or rare or how the range may change from year to year.

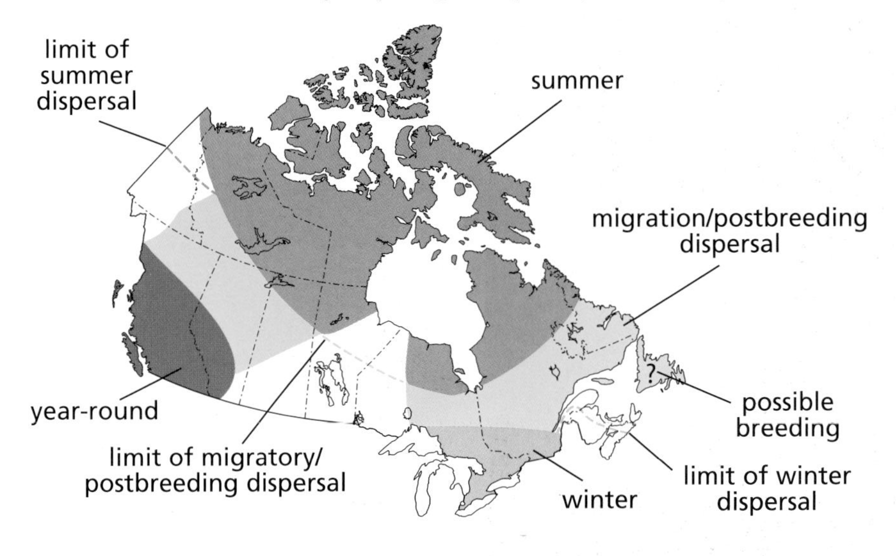

NONPASSERINES

Nonpasserine birds represent 17 of the 18 orders of birds found in Canada, about 60 percent of the total bird species in our country. They are grouped together and called "nonpasserines" because, with few exceptions, they are easily distinguished from the "passerines," or "perching birds," which make up the 18th order. Being from 17 different orders, however, means that nonpasserines vary considerably in their appearance and habits—they include everything from the 1.8-metre-tall Great Blue Heron to the 8-centimetre-long Calliope Hummingbird.

Generally speaking, nonpasserines do not "sing." Instead, their vocalizations are referred to as "calls." There are also other differences in form and structure. For example, whereas the muscles and tendons in the legs of passerines are adapted to grip a perch and the toes of passerines are never webbed, nonpasserines have feet adapted to a variety of uses.

Many nonpasserines are large, so they are among our most notable birds. Waterfowl, raptors, gulls, shorebirds and woodpeckers are easily identified by most people. However, novice birders may mistake small nonpasserines such as doves, swifts and hummingbirds for passerines, thus complicating their identification. With a little practise, it will become easy to recognize the nonpasserines. By learning to separate the nonpasserines from the passerines at a glance, birders effectively reduce by about half the number of possible species for an unidentified bird.

Waterfowl

Grouse & Allies

Diving Birds

Heronlike Birds

Birds of Prey

Rails, Coots & Cranes

Shorebirds

Gulls & Allies

Doves & Cuckoos

Owls

Nightjars, Swifts & Hummingbirds

Woodpeckers

GREATER WHITE-FRONTED GOOSE

Anser albifrons

The Greater White-fronted Goose has a circumpolar Arctic breeding distribution. In North America, there are two distinct populations: the Pacific population, which breeds only in Alaska and travels down the west coast to winter primarily from California to Oregon, and the mid-continental population, which breeds from Alaska across the tundra to the northwest corner of Hudson Bay and winters from the Gulf States to Mexico. During migration, this population travels primarily through the Prairie provinces with small numbers seen east into Ontario. A third population, which breeds in Greenland and winters in Britain, occurs rarely in eastern Canada during migration. • While staging during migration, these geese often mix with flocks of Snow, Canada or Cackling geese. Greater White-fronts can best be distinguished by their bright orange feet, which shine like beacons as the birds stand on frozen spring wetlands and fields. • The Greater White-fronted Goose is named for a white facial patch at the base of its bill. It is also known as "Speckle-belly" because of the black barring on its underside or "Laughing Goose" for its laughing call. • Greater White-fronts are the earliest geese to arrive in autumn migration. They are also among the earliest arrivals on spring staging and nesting areas, making them vulnerable to hunting. The Pacific population is believed to be recovering from steep declines that occurred in the 1970s and '80s. Mid-continent populations also are increasing.

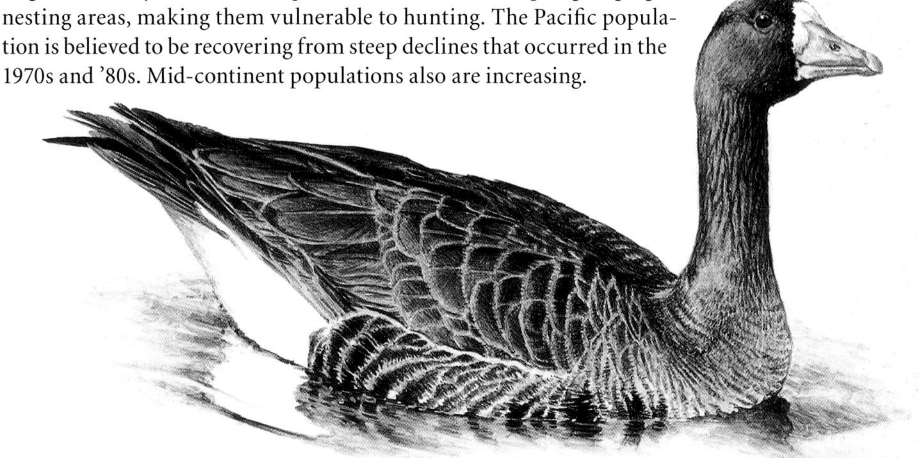

ID: sexes similar; dark brownish grey overall; dense blackish speckling or barring across lower breast and belly; white patch at base of bill extends onto forehead; white hindquarters; orange-pink bill and feet. *Immature:* pale belly lacks dark speckling; no white on face.
Size: *L* 70–85 cm; *W* 1.3–1.5 m.
Habitat: extensive marsh-dotted lakes and reservoirs; also croplands; individuals and small groups may occur nearly anywhere in migration.

Nesting: on the ground; nest scrape is lined with down and plant material; female incubates 5–6 slightly glossy, creamy white eggs for 22–27 days.
Feeding: dabbles in water and gleans the ground for plant shoots, sprouting and waste grain and occasionally acquatic invertebrates.
Voice: flight call, given in flocks, creates a chorus of far-carrying, falsetto *k'LAUW lee-YOK* notes.
Similar Species: *Canada Goose* (p. 50): dark neck; distinctive white "chin strap." *Snow Goose* (p. 46): blue phase has white head and upper neck.

SNOW GOOSE

Chen caerulescens

The arrival of masses of migrant Snow Geese is a birding spectacle not to be missed. Compared to the neat V-formations of some geese, migrating flocks of Snow Geese fly in wavy, disorganized lines. In migration, massive flocks forage in farmers' fields, fuelling up on new sprouts and waste grain. Their strong, serrated bills are well designed for pulling up rootstalks of marsh plants and gripping slippery grasses. • Snow Geese occur in two colour phases, white and blue. Until 1983, the blue morph was considered to be a separate species known as "Blue Goose." • Snow Geese breed across the Arctic from Baffin Island west into northeastern Siberia. There are two subspecies, Lesser and Greater. Greater Snow Geese breed in eastern Nunavut and are the most northern breeding goose in the world. Their migration route to wintering grounds on the mid-Atlantic Coast of the U.S. passes mostly through Québec. In the Lower St. Lawrence region, concentrations of up to half a million birds can be seen. The Lesser Snow Goose breeds from the mid-Arctic west into Siberia. The blue colour phase is much more prevalent in Lesser Snow Goose colonies. • In recent decades, an abundance of food along migratory routes and on wintering grounds has allowed populations to increase several times over. Unfortunately, this burgeoning population is destroying its own tundra nesting grounds through overgrazing.

blue adult

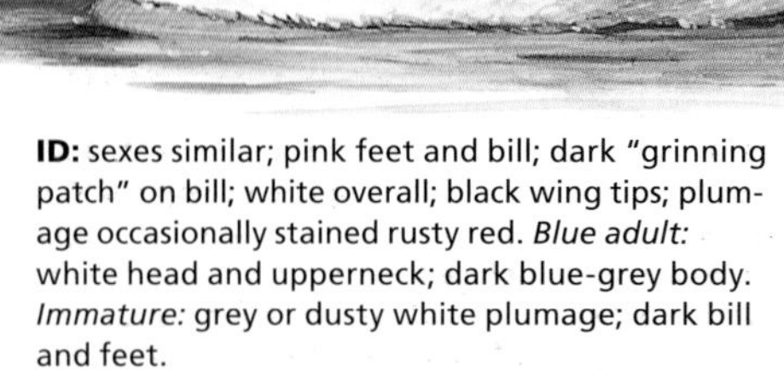

ID: sexes similar; pink feet and bill; dark "grinning patch" on bill; white overall; black wing tips; plumage occasionally stained rusty red. *Blue adult:* white head and upperneck; dark blue-grey body. *Immature:* grey or dusty white plumage; dark bill and feet.

Size: *L* 71–84 cm; *W* 1.4–1.5 m.

Habitat: *Breeding:* tundra near ponds, lake shorelines or on islands. *In migration:* shallow wetlands, lakes and fields.

Nesting: loosely colonial; on a ridge or hummock; female fills a shallow depression with plant material and down; female incubates 3–5 whitish eggs for 22–23 days.

Feeding: grazes on waste grain and new sprouts; also eats aquatic vegetation, grasses, sedges and roots.

Voice: loud, nasal *houk-houk* in flight.

Similar Species: *Ross's Goose* (p. 47): smaller; shorter neck; lacks black "grinning patch." *Tundra* (p. 53), *Trumpeter* (p. 52) and *Mute* (p. 51) *swans:* larger; white wing tips. *American White Pelican* (p. 122): much larger bill and body.

ROSS'S GOOSE

Chen rossii

This small goose looks so similar to the Snow Goose that inexperienced birders can easily confuse the two. The Ross's Goose is often overlooked, particularly when a few individuals are mixed within large flocks of migrating Snow Geese. It is most frequently spotted by birders with spotting scopes, since it might be necessary to scan a large flock of distant Snow Geese to pick out a single Ross's. The Ross's Goose is smaller overall than the Snow Goose, has a shorter neck, a rounder head and a shorter, stubby bill that is somewhat warty at the base and does not have a "grinning patch." • The rare Blue Ross's Goose is a handsome plumage phase that occurs in a tiny minority of the population. • Considered a rare species throughout much of the early 1900s, Ross's Goose numbers increased as a result of closed or restricted hunting seasons from a few thousand birds in the early 1950s to nearly 200,000 nesting in the Queen Maud Gulf Migratory Bird Sanctuary in central Nunavut by the late 20th century. The Queen Maud Gulf nesting area is currently home to about 95 percent of this bird's breeding population. • Ross's Goose is the last goose species to arrive and leave from its breeding grounds.

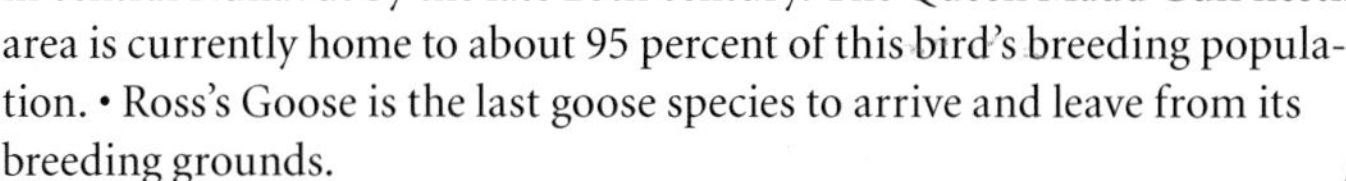

ID: sexes similar; white overall; black wing tips; dark pink feet and bill; stubby, triangular bill; small, bluish or greenish "warts" on base of bill; no "grinning patch"; plumage occasionally stained rusty by iron in the water. *Blue adult:* very rare; white head; blue-grey body. *Immature:* grey plumage; dark bill and feet.
Size: *L* 54–66 cm; *W* 1.2–1.3 m.
Habitat: *Breeding:* tundra along ponds, lake shorelines or on islands. *In migration:* shallow wetlands, lakes and fields.
Nesting: colonial; usually on a sparsely vegetated island or surrounding mainland areas of shallow Arctic lakes; on a ridge or tundra hummock; female fills a shallow depression with plant material and down and incubates 2–6 dull white eggs for 21–23 days.
Feeding: grazes on waste grain and new sprouts; also eats aquatic vegetation, grasses, sedges and roots.
Voice: high-pitched *keek keek keek.*
Similar Species: *Snow Goose* (p. 46): larger; longer neck; dark "grinning patch'" on bill. *Tundra* (p. 53), *Trumpeter* (p. 52) and *Mute* (p. 51) *swans:* much larger; white wing tips. *American White Pelican* (p. 122): much larger bill and body.

BRANT

Branta bernicla

Brant are characterized by their extensive use of coastal habitats outside the breeding season and by their strong flight. These geese are heavily dependent on certain food plants, making the species more vulnerable to occasional breeding failures and heavy losses from starvation than most other geese. • There are two subspecies of Brant in Canada, the "Black" Brant (*B. b. nigricans*), which breeds from western Nunavut to Russia, and the "Light-bellied" Brant (*B. b. hrota*), which breeds in eastern Nunavut. Although the Black Brant generally stays along the coast as it migrates between its breeding grounds and wintering grounds in California or Mexico on the Pacific Coast, the Light-bellied Brant migrates over land between the Atlantic Coast and James Bay, passing over eastern Ontario. • Brant populations in eastern North America declined by almost 90 percent in the 1930s, when a virulent blight killed much of their winter source of saltwater eelgrass along the Atlantic Coast. Fortunately, many birds switched to eating sea lettuce, a strategy that kept the population from disappearing altogether. It is thought that the Atlantic migration route shifted west from the Maritime provinces because of this collapse as the surviving population favoured a more direct route.

ssp. hrota

ssp. nigricans *("Black")*

ID: sexes similar; small, dark sea goose; dark head, neck and upperparts; broken, white collar; extensive white tail coverts; grey and white barring on sides; dark legs. *"Black" race:* dark underparts give overall dark appearance. *"Light-bellied" race:* white underparts; pale brown sides and flanks. *Immature:* lacks distinct barring on sides; no collar.
Size: *L* 59–70 cm; *W* 1.1–1.2 m.
Habitat: marine estuaries, beaches, bays and spits.

Nesting: on coastal tundra; ground hollow is lined with moss, lichens and down; female incubates 3–5 creamy white, elliptical eggs for 23–24 days.
Feeding: tips up and gleans mainly for eelgrass, but also algae growing in brackish waters of intertidal mudflats; terrestrial feeder in the Arctic, grazing on sedges and grasses.
Voice: deep, prolonged croak on the ground; soft clucking in flight.
Similar Species: *Cackling Goose* (p. 49): similar size; paler brown belly; conspicuous, white "chin strap."

CACKLING GOOSE

Branta hutchinsii

In 2004, ornithologists split the Cackling Goose from the Canada Goose. The smaller Canada Goose species that nest in the Arctic have been renamed Cackling Geese, while several of the larger subspecies are still known as Canada Geese. The Cackling Goose can be differentiated from a small Canada Goose by its cackling calls, extremely small size, stubby neck and shorter, stubbier bill, but considerable overlap exists between the two species. Nutrition on the breeding grounds also dramatically influences the overall size of individual geese, further compounding the difficulty of basing identification on size. • Cackling Geese nest on the tundra, preferring pennisulas and islands to avoid predation by arctic foxes. • The Cackling Goose consists of four subspecies, all breeding in North America, though only the *hutchinsii* subspecies breeds in Canada. All four subspecies often travel with flocks of Canada Geese but generally split up into flocks of their own type on the wintering grounds. The three Alaskan subspecies migrate through British Columbia to winter in the western U.S., while *hutchinsii* migrate through Alberta to Ontario to winter in Texas. • The smallest race of Cackling Goose is barely bigger than a Mallard and is only one-quarter of the size of the largest ("Giant") race of Canada Goose.

ID: sexes similar; small goose; generally dark brown; stubby, black bill; rounded head; white "chin strap" encircles throat; relatively long, black neck; white rump and undertail coverts; short, black tail; black legs.
Size: *L* 64–89 cm; *W* 90–120 cm.
Habitat: along water bodies and marshes; also croplands and parks.
Nesting: on tundra near water; prefers peninsulas and islands; female builds a nest of vegetation lined with breast down and incubates 4–7 white eggs for 25–30 days.
Feeding: grazes on new sprouts, aquatic vegetation, grass, roots, berries and seeds (particularly grains); tips up for aquatic roots and tubers.
Voice: high-pitched *ah-honk*.
Similar Species: *Canada Goose* (p. 50): larger; longer neck; larger bill; generally lighter in colour. *Brant* (p. 48): no "chin strap"; more coastal.

CANADA GOOSE

Branta canadensis

Arguably one of our best-known birds, the Canada Goose, with its black head, brown body, white cheeks and loud honk, is especially familiar to city dwellers. For many Canadians, V-shaped migrating flocks usher in spring and symbolize the coming of winter. • This goose can be found breeding in every province and territory of Canada. There are seven subspecies, ranging from the medium-sized "Lesser" Canada, which breeds throughout the boreal and Arctic tundra zones of northwestern Canada, to the "Giant" Canada of the eastern prairies. Subspecies are difficult to distinguish, and most birders use the location to help name them. Each subspecies shows a high degree of individuality in behaviour and use of habitat for breeding, feeding and staging. • Few people realize that at one time, the Canada Goose was hunted almost to extinction. Today, some of the reintroduced populations have become pests on urban waterfronts, picnic sites and golf courses, as well as in city parks and on rural farmland. During winter months, almost anywhere free of ice and snow in refuges, sanctuaries, farmland and city parks may have wintering geese. • Like many geese, adult Canadas mate for life and only take another mate if one member of the pair dies. Young often return and nest in the same area where their parents nested.

ID: sexes similar; large goose; dark brown upperparts; light brown underparts; long, black neck; white "chin strap"; black bill, legs and feet; white undertail coverts; short, black tail.
Size: *L* 55–122 cm; *W* up to 1.8 m.
Habitat: diverse, usually with permanent water and grazing areas; prefers low-lying areas with nearby grassy meadows and abundant wetlands that serve as refuges from land predators.
Nesting: usually close to water; on the ground on an islet, over water in a reed or cattail bed, on a cliff ledge or in an old stick nest; female builds a nest of vegetation lined with breast down and incubates 4–7 white eggs for 25–30 days.
Feeding: grazes on new sprouts, fresh aquatic vegetation, grasses, roots, berries and seeds (particularly grains); also tips up for aquatic roots and tubers.
Voice: familiar, loud *ah-honk*.
Similar Species: *Cackling Goose* (p. 49): smaller; generally darker plumage; smaller bill. *Brant* (p. 48): no "chin strap"; more coastal.

MUTE SWAN

Cygnus olor

Admired for its grace and beauty, this Eurasian native was introduced to eastern North America in the mid-1800s to adorn estates and city parks. Over the years, the Mute Swan has adapted well to the North American environment. By the 1940s and '50s, the species was well established, and this swan is now a common sight at several locations on southeastern Vancouver Island and in Greater Vancouver. In eastern Canada, this swan has established an ever-growing population along the Great Lakes in southern Ontario. Feral populations have continued to expand, and though it is not usually a migratory species, more northerly nesters have established short migratory routes to milder wintering grounds. • The Mute Swan may be quickly distinguished from our two native swans by its orange bill with a black basal knob. Weighing in at 16 kilograms, as much as an average four-year-old child, these large swans are one of Canada's heaviest flying birds. • Beautiful to behold, swans are always a popular sight at local wetlands. Like many non-native species, however, Mute Swans are often fierce competitors for nesting areas and food sources and can be very aggressive toward geese and ducks.

ID: sexes similar; large, all-white swan; orange bill has bulbous, black knob at base; neck often held in S-shape; wings held in arch over back when displaying on water. *Immature:* greyish brown body; frequently has curved neck.
Size: *L* 1.5 m; *W* 2.3–2.5 m.
Habitat: marine bays, harbours and brackish lagoons; freshwater marshes, lakes and urban ponds.
Nesting: on the ground along a shoreline in emergent vegetation; female builds mound of vegetation (male may help gather material) and incubates 5–10 whitish eggs for about 36 days.
Feeding: tips up or dips head below the water's surface for aquatic plants; rarely grazes on land.
Voice: mostly silent; may hiss or issue hoarse barking notes when agitated or during courtship.
Similar Species: *Tundra Swan* (p. 53) and *Trumpeter Swan* (p. 52): lack orange bill with black knob at base; necks usually held straight.

TRUMPETER SWAN

Cygnus buccinator

The majestic Trumpeter Swan, found primarily in western Canada, is the largest of North America's native waterfowl. • This bird was hunted nearly to extinction for its meat and feathers during the early 20th century. Breeding populations in Alaska and Alberta persisted, but eastern populations were less fortunate, and by 1884, the Trumpeter Swan was extirpated in Ontario. The overall population increased substantially in the 1970s because of conservation efforts and reintroduction programs, and today is stable at about 34,000 birds. Reintroduction to Ontario started in 1982 and has resulted in an increasing population of 700-plus birds. During winter, areas of southwestern British Columbia are home to an estimated 10,000 Trumpeter Swans. • Both the common name and the species name *buccinator* refer to the loud, bugling sounds created by the bird's long windpipe.

ID: sexes similar; large swan; all-white plumage; all-black bill with pinkish edges to mandibles; sloping bill appears to continue straight into forehead; black legs and feet. *Immature:* greyish brown plumage; pinkish bill with black tip. *In flight:* straight neck; often flies in family groups.
Size: *L* 1.5–1.8 m; *W* 2.4 m.
Habitat: *Breeding:* shallow, large and small, freshwater lakes with emergent vegetation for nesting and feeding. *Winter:* estuaries, agricultural fields, sloughs, marshes and slow-moving rivers.

Nesting: on a large mound of vegetation on water among cattails, bulrushes or sedges; female incubates 4–6 creamy white eggs for about 33 days.
Feeding: eats mostly plants gleaned from the water's surface or underwater; in upland areas, eats tubers, grasses and leaves; newly hatched cygnets eat insects and other invertebrates.
Voice: loud, resonant, buglelike *koh-hoh*, usually by male in flight; gentle, nasal honking; immature's high-pitched calls deepen over first winter.
Similar Species: *Tundra Swan* (p. 53): slightly smaller; rounder head; slightly concave bill often shows yellow at base; softer, more nasal calls; more commonly in large flocks. *Mute Swan* (p. 51): orange bill has black knob at base; curved neck. *Snow Goose* (p. 46): smaller; stubbier, pink bill; much shorter neck; black wing tips; pink legs.

TUNDRA SWAN

Cygnus columbianus

Before the last winter snow has melted, Tundra Swans can be found foraging in flooded fields and pastures on their migration northward, bringing joy to all who know their distinctive call. • This swan is one of Canada's largest native birds, with adult males weighing on average 7.5 kilograms. • Tundra Swans are primarily seen as spring and autumn migrants throughout most of Canada. They breed across the Arctic and Subarctic from Québec to Alaska. From these regions, Tundra Swans separate into two distinct wintering populations, an eastern and a western. Some birds travel from their breeding grounds in the Yukon across the continent to winter in the U.S. mid-Atlantic states. Hundreds to several thousand Tundra Swans can be heard and seen staging in various locations during migration in March and April and in October and November. • Like the Trumpeter Swan, the Tundra Swan's windpipe loops through the bird's sternum, amplifying its call. • This bird is known as "Whistling Swan" in Europe because of the whistling sound made by the slow, powerful beating of its wings in flight. • Early in the 19th century, the Lewis and Clark exploration team first encountered this bird near the Columbia River, a fact that is reflected in its scientific name.

ID: sexes similar; large swan; whitish plumage; heavy, black bill; yellow "teardrop" in front of eye; large, black feet; white wings; long neck usually held straight. *Immature:* pale greyish wash on plumage; much pink on bill (variable).
Size: *L* 1.2–1.5 m; *W* 1.2 m.
Habitat: permanent, shallow wetlands in open country; prefers areas with clear sight lines in all directions.
Nesting: on the ground on a mound or tundra ridge; nest is a large bowl of down, plant material, lichen and moss; mainly the female incubates 3–5 creamy white eggs for 32 days.

Feeding: tips up, dabbles and gleans the water's surface for aquatic vegetation and invertebrates; grazes for tubers, roots, grasses and waste grain on land.
Voice: muffled, hollow honking or hooting *WHOO!...HOO-wu-WHOO!* call, given frequently in flight and by nervous flocks on the ground or water; wings make a windy, whistling sound at close range.
Similar Species: *Trumpeter Swan* (p. 52): larger; heavier bill; loud, buglelike voice; no yellow "teardrop" on bill. *Mute Swan* (p. 51): orange bill has black knob at base; holds neck in S-shape.

WOOD DUCK

Aix sponsa

The male Wood Duck is one of the most colourful waterbirds in North America. The female is much subtler in plumage but shares the male's crest and sports a large, teardrop-shaped, white patch around each eye. • Despite their flashy attire, Wood Ducks are often heard before they are seen; from high in a tree or in flight, they frequently issue squeaky, high-pitched squeals. • Overhunting and habitat destruction in the early 20th century threatened this species with extinction. Thousands of nest boxes placed across its breeding range have greatly increased populations of this bird. • Because this duck nests in trees, the young face the adventures of life earlier than most. Shortly after hatching, ducklings jump out of their nest cavity, falling 15-plus metres to bounce harmlessly like ping-pong balls on landing. Then they follow their mother to the nearest source of water, which can be up to two kilometres away. Females often return to the same nest site year after year, especially after successfully raising a brood. • Landowners with tree-lined swamps, ponds, sloughs or marshes may attract Wood Ducks by erecting nest boxes over or near the water. These ducks may also nest in urban areas where suitable nest cavities are available. Like many Canadian waterbirds, Wood Ducks are extending their breeding range northward.

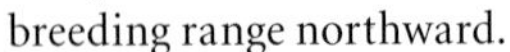

ID: rather large, elongated duck; stocky head; short, thin neck. *Male:* long, slicked-back, greenish crest edged with white; red eyes; white throat, neck ring and facial "spur"; mostly reddish bill; rusty breast with small, white spots; black and white shoulder slashes; golden sides. *Female:* mostly greyish brown upperparts; brown face with white, teardrop-shaped eye patch; mottled brown breast is streaked with white. *In flight:* long, broad, square tail.
Size: *L* 38–53 cm; *W* 76–84 cm.
Habitat: sloughs, marshes, small lakes with marshy areas, freshwater ponds and slow-moving rivers and backwaters, usually bordered by stands of mature deciduous trees.

Nesting: in a natural crevice, tree cavity or nest box, up to 25 m above the ground and usually near water; no materials added except breast down; female incubates 8–13 creamy white to tan eggs for 28–32 days.
Feeding: gleans the water's surface and tips up for aquatic vegetation; feeds in woodlands; also eats fruits, nuts and waste grain.
Voice: *Male:* ascending *whoo-eek* and finchlike *jeee. Female:* slurred, squeaky *crrek, crrek* notes.
Similar Species: *Hooded Merganser* (p. 80): slim, black bill; black-and-white breast; male has black head with white crest patch.

GADWALL

Anas strepera

The male Gadwall, often called "Grey Duck," lacks the striking plumage of most other male puddle ducks, but nevertheless has a dignified appearance and subtle beauty. Once you learn the field marks—a black rump and white wing patches on mostly greyish plumage—this bird is easy to identify. The female resembles female teals and other puddle ducks but may be identified by its orange-sided, brown bill and white wing patches. • Gadwalls favour freshwater ponds, marshes and sloughs with muddy bottoms and abundant aquatic plants, and they may be found in these habitats year-round in southern Canada. • These ducks may start forming pair bonds for the next breeding season as early as late July, with pairs often remaining together in the interim. A preference for breeding on islands or in vegetation immediately adjacent to wetlands has helped prevent Gadwalls from experiencing the declines seen in other species that breed in prairie pothole regions during drought periods. Traditionally birds of the western Canadian prairies, Gadwall numbers have greatly increased in eastern Canada, and this duck has recently expanded its range throughout North America.

ID: generally nondescript, greyish brown duck; white belly. *Male:* dark upperparts; greyish brown head; dark eyes and bill; black-mottled breast; grey flanks; black upper- and undertail coverts; orange legs. *Female:* mottled brown overall; narrow, brown bill with orange sides. *In flight:* white speculum.
Size: *L* 46–56 cm; *W* 74–84 cm.
Habitat: *Breeding:* vegetated shores of marshes, sloughs, lakeshores, sewage lagoons and small islands. *In migration* and *winter:* freshwater ponds, sloughs, shallow lakes, flooded fields and brackish waters.
Nesting: on a wetland shore or small island with dense shrubs and grasses; nest is a grassy, down-lined hollow concealed in tall emergent vegetation; female incubates 7–12 dull creamy white to greyish green eggs for 24–27 days.
Feeding: dabbles and tips up for aquatic plants, seeds and invertebrates; also steals food brought to the surface by American Coots.
Voice: both sexes quack like a Mallard; female's *kaak kaaak kak-kak-kak* oscillates in volume.
Similar Species: greyish brown plumage and black hindquarters of male unique. *American Wigeon* (p. 57): green speculum; male has distinctive head pattern; female has rufous flanks. *Other dabbling ducks* (pp. 56, 58–64): lack white speculum, black hindquarters of male Gadwall and orange-sided bill of female.

EURASIAN WIGEON

Anas penelope

Eurasian Wigeons are probably the most noticeable of our regularly occurring rarities. They consort almost exclusively with American Wigeons, and each year, a few birders will discover a conspicuous, chestnut-headed wigeon while scanning a flock of American Wigeons. • As with most ducks, male wigeons are easily distinguished in their breeding plumage (mid-autumn until late spring), but the females are a bigger challenge to identify. Eurasian Wigeon females are patterned much like their female American Wigeon cousins but have warmer buff flanks, less contrast between the neck and breast and a generally reddish head. • Eurasian Wigeons are regular winter vistors along both coasts, but more of these ducks overwinter on southern Vancouver Island and the Fraser River delta than anywhere else on the continent. Although they are not recorded breeders in North America, the increased number of sightings each spring has convinced some people that there is a small breeding population somewhere in Canada.

ID: black-tipped, pale bluish grey bill; white belly; dark legs and feet. *Male:* rich, orangey brown head; creamy yellow forehead stripe; black-and-white hindquarters; lightly barred, grey sides and back; chestnut breast; green-and-black speculum. *Female:* mottled brown head and breast, usually with rufous tints; rufous tan sides. *In flight:* large, white wing patch; all-grey underwings.
Size: *L* 42–52 cm; *W* 76–84 cm.
Habitat: marshes, sloughs and shallow lakes with abundant floating vegetation; flooded and dry farm and airport fields; marine shores with seaweed; also lawns and ponds in city parks.

Nesting: no North American breeding records; suspected to nest in or near coastal regions and in the Cariboo-Chilcotin region of BC.
Feeding: dabbles or tips up for stems, leaves and seeds of aquatic vegetation; grazes lawns and pastures; frequently steals food from American Coots.
Voice: *Male:* high-pitched, 2-tone *thweeeeeeer* whistle. *Female:* rough quack.
Similar Species: *American Wigeon* (p. 57): white "wing pits"; breeding male has white forehead and crown, iridescent, green facial patch and rusty back and sides; female has greyish head and neck contrasting with browner breast.

AMERICAN WIGEON

Anas americana

American Wigeons are a vocal, charismatic species. These noisy ducks like to graze on young shoots while walking steadily along in dense, formless flocks, each individual no more than a few centimetres from its nearest neighbour. They are good walkers, in contrast to other ducks. • American Wigeons occur in wetlands with abundant surface plants and may spend day and night in favourite feeding sites. City parks, golf courses and airports also provide opportunities for grazing. • This duck is generally vegetarian. Although it frequently dabbles for food, nothing seems to please a wigeon more than the succulent stems and leaves of pond-bottom plants. These plants grow far too deep for a dabbling duck, however, so pirating wigeons often steal from accomplished divers such as American Coots, Canvasbacks, Redheads and scaups. • Amid the flock's orchestra of constant buzzes, quacks and ticks, the male American Wigeon's piping, three-syllable whistle may be the most prominent sound. • The male's bright white crown and forehead has led some people, especially hunters, to call it "Baldpate."

ID: large, rufous brown duck; pale bluish grey bill with black tip. *Male:* rusty breast and sides; grey head with white forehead and crown; wide, iridescent, green patch behind eye; white belly; black undertail coverts. *Female:* greyish head; brown breast and sides; dusky undertail coverts. *In flight:* large, white upperwing patch; green-and-black speculum.
Size: *L* 46–58 cm; *W* 76–84 cm.
Habitat: often grazes on shore. *Breeding:* freshwater marshes, sloughs and ponds. *Winter:* estuaries, agricultural lands, airports and city parks.
Nesting: on dry ground in a brushy site sometimes far from water; well-concealed nest of grasses and leaves is lined with breast down; female incubates 7–11 creamy white eggs for 23–25 days.
Feeding: dabbles and tips up for surface aquatic vegetation as well as insects, small snails and crustaceans; grazes short grasses.
Voice: *Male:* nasal, frequently repeated *whee whee whew* whistle. *Female:* soft, seldom-heard quack.
Similar Species: male's head pattern is distinctive. *Gadwall* (p. 55): white speculum; lacks large, white upperwing patch; male lacks green patch behind eye; female has orange-and-brown bill. *Eurasian Wigeon* (p. 56): grey "wing pits;" male has rufous head, pale forehead and rosy breast, lacks green patch behind eye; female usually has browner head.

AMERICAN BLACK DUCK

Anas rubripes

The American Black Duck breeds in eastern Canada and the northeastern United States and winters throughout much of eastern North America from New Brunswick southward to Florida. • In recent years, the eastern expansion of the Mallard into forested areas has come at the expense of this dark dabbler. Although it was once the most abundant dabbling duck in eastern North America, there are estimated to be half as many American Black Ducks now as there were in the 1950s. A male Mallard will aggressively pursue a female American Black Duck, and if she is unable to find a male of her own species, she will often accept the offer. Hybrid offspring are less fertile, and they are usually unable to reproduce. Attempts to increase wooded swamp habitats, which American Black Ducks favour, have largely failed to thwart this hybridization in eastern Canada because Mallards will nest almost anywhere. Fortunately, the more remote wilderness of rivers and lakes in northern Québec, Newfoundland and Labrador attracts American Black Ducks but few Mallards. In recent decades, American Black Duck populations have stabilized as a result of additional hunting restrictions, as well as protection and enhancement of key breeding and wintering sites. • Like other puddle ducks, American Black Ducks forage by tipping up to reach prey along the bottom of marshes. • Male and female American Black Ducks are remarkably similar in appearance, which is unusual for waterfowl.

ID: blackish body; paler head and neck; streaked throat; orange legs and feet; violet speculum bordered in black. *Male:* dull yellowish bill. *Female:* olive bill. *In flight:* whitish underwings contrast with dark body.

Size: *L* 51–61 cm; *W* 89 cm.

Habitat: marshes, wooded swamps, bogs, flooded agricultural fields and estuaries.

Nesting: usually on the ground near water; among clumps of dense vegetation; female fills a shallow depression with plant material and lines it with down; female incubates 7–11 white to greenish buff eggs for 23–33 days.

Feeding: tips up and dabbles in shallows to reach eelgrass and other plants; also takes molluscs, crustaceans and other small prey.

Voice: *Male:* a croak. *Female:* loud quack.

Similar Species: *Mallard* (p. 59): female is paler overall with unstreaked throat and orange-and-black bill; speculum has narrow, white border; lacks whitish wing linings in flight.

MALLARD

Anas platyrhynchos

The male Mallard, with its shiny green head, chestnut brown breast and stereotypical quack, is one of the best-known ducks. This species is extremely adaptable and has become a semi-tame fixture on suburban ponds. It thrives in a variety of wild habitats, from tiny ponds and irrigation ditches to marshes, rivers and extensive wetlands. • Male Mallards will mate with females of closely related dabbling ducks, including domestic species. The offspring of these matings are always a challenge to identifiy. • Despite being the most heavily hunted waterfowl species, Mallards remain abundant and extremely widespread throughout Canada. These ducks are one of the earliest waterfowl migrants, following the first spring thaws northward. • Like many other male ducks, male Mallards do not participate in incubation duties or the raising of young. After mating, they retreat to larger lakes, where they undergo a moult that leaves them flightless for a short period of time. The resulting, much plainer "eclipse" plumage provides camouflage during the flightless period. Moulting into new breeding plumage usually commences by early August and carries on into the autumn.

ID: *Breeding male:* glossy, green head; bold, white collar; chestnut breast; all-yellow bill; black central tail feathers curl upward. *Female:* mottled brown overall; finely streaked, greyish brown head; orange bill irregularly spattered with black. *Nonbreeding male:* similar to female, but with dull yellow bill and richer brown breast. *In flight:* dark blue speculum bordered with white (purple or green on hybrids).
Size: *L* 51–71 cm; *W* 79–89 cm.
Habitat: almost any large or small wetland including ponds, lakes, marshes and rivers; also seashores.
Nesting: usually near water; on the ground among tall grasses under a bush; nest of grasses and other material is lined with down; female incubates 7–10 white to buffy olive eggs for 26–30 days.
Feeding: tips up and dabbles in shallows for aquatic leaves, stems and seeds; may take aquatic invertebrates; commonly accepts handouts (corn or whole grains are preferred).
Voice: *Male:* deep, quiet quacks. *Female:* loud quacks; very vocal; heard year-round.
Similar Species: *Northern Shoveler* (p. 62): much larger bill; male has white breast. *American Black Duck* (p. 58): darker than female Mallard; olive or greenish yellow bill; purple speculum lacks white borders.

BLUE-WINGED TEAL

Anas discors

Renowned for its aviation skills, the speedy Blue-winged Teal can be identified in flight by its small size, colourful upperwing patches and the sharp twists and turns that it executes with precision. • Not very tolerant of cold, Blue-winged Teals are usually the last dabbling ducks to return in spring, and they leave well before the cold weather sets in. Although a few birds linger late in autumn, they are essentially gone by the end of September. Blue-winged Teals can be found breeding from British Columbia across Canada into the Maritime provinces. Individuals, pairs and small groups frequent almost any wetland but prefer marshes, reed-bordered ponds. • As with other dabbling ducks, the Blue-wing's feet are set near the centre of its body. In contrast, the feet of diving ducks are larger and set farther back, allowing for better propulsion while diving. • Many people are surprised to learn that the Green-winged Teal is not the Blue-winged Teal's closest relative. Blue-wings are more closely related to Cinnamon Teals and Northern Shovelers, species that have broad, flat bills and similar wing patterns—a pale blue forewing and green speculum. • The scientific name *discors* is Latin for "without harmony," which might be in reference to this bird's call as it takes flight.

ID: smallish duck; blackish bill; yellowish legs. *Male:* blue-grey head; bold, white crescent on face; black-spotted, rusty breast and sides. *Female:* mottled brown overall; fairly bold, dark eye line; light throat. *In flight:* bold, pale blue patch on leading edge of upperwing; green speculum; agile flier.
Size: *L* 36–41 cm; *W* 53–58 cm.
Habitat: almost any freshwater or brackish wetland; wetland edges, shallow lakes, flooded fields and ditches; rarely on marine waters.
Nesting: usually near water; on the ground concealed by tall grasses and low shrubs; depression is lined with plant material and down; female incubates 8–11 cream-coloured eggs for 23–24 days.
Feeding: omnivorous; generally feeds in less than 30 cm of water; eats mainly aquatic invertebrates, from mosquito larvae to snails, and seeds and leaves of aquatic plants.
Voice: *Male:* soft *keck-keck-keck*. *Female:* soft quacks.
Similar Species: *Green-winged Teal* (p. 64): smaller; female has smaller bill, black-and-green speculum and no blue forewing patch. *Northern Shoveler* (p. 62): much larger bill; male has yellow eyes, green head and white breast. *Cinnamon Teal* (p. 61): female is virtually identical to female Blue-winged Teal but with larger, more spatulate bill.

CINNAMON TEAL

Anas cyanoptera

If the Stetson is the "hat of the West," then the Cinnamon Teal is the "duck of the West." The principal distribution of both the headwear and the bird define, in a broad sense, that great reach of arid country where the presence of water is dramatic and important. Each spring, the Cinnamon Teal pushes northward from its southwestern and central U.S. wintering grounds to dot reed-fringed ponds and marshes in its northern breeding grounds. This duck breeds primarily in wetlands of the western U.S., with southern BC, southern Alberta and southwestern Saskatchewan marking the northern edge of its range. It is also a rare vagrant into central and eastern Canada. • Birders eagerly await this duck each spring. The intensely reddish brown breeding plumage of the male, accented by his ruby red eyes, is worth an admiring gaze at any time of day. In the low, slanting light of early morning or near sunset, he can be a real showstopper. Despite his spectacular appearance, this duck is relatively shy and is often overlooked as he feeds, moving from one patch of marsh vegetation to another. • Cinnamon Teals do not regularly form large flocks; sightings of more than a few dozen in one area are rare.

ID: broad-ended bill. *Breeding male:* intensely cinnamon red underparts, neck and head; red eyes. *Female:* plain face; brown eyes; mottled, warm brown overall. *Nonbreeding male:* similar to female but with red eyes and rufous tint. *In flight:* conspicuous, large, sky blue forewing patch.

Size: *L* 38–43 cm; *W* 51–56 cm.

Habitat: freshwater ponds, marshes, sloughs, ditches, flooded fields and sewage lagoons with aquatic vegetation.

Nesting: on the ground near water, hidden among sedges, grasses, bulrushes or horsetails; female incubates 7–12 off-white to pale buff eggs for 21–25 days.

Feeding: scoops up floating plants, seeds, snails and insects from the water's surface; often feeds with head partially submerged.

Voice: usually silent. *Male:* whistled *peep*; rough *karr-karr-karr. Female:* soft, rattling *rrrrr*; rather weak *gack-gack-ga-ga*.

Similar Species: *Blue-winged Teal* (p. 60): narrower tip on shorter bill; nonbreeding male has white crescent in front of dark eye and more contrasting greyish brown plumage; female has more distinct facial pattern and lighter blue on forewing. *Ruddy Duck* (p. 83): breeding male has white cheek, large, blue bill and rufous-and-grey upperwing.

NORTHERN SHOVELER

Anas clypeata

With its large, all-green head, the male Northern Shoveler may bring to mind a male Mallard, but the resemblance ends there. The shoveler's large bill allows for immediate identification—no other duck has a bill so proportionately large. • Although it is primarily a western-breeding species, the Northern Shoveler also breeds in various locations scattered across eastern Canada. This species has shown a steady population increase within Canada over the last few decades. • The Northern Shoveler dabbles on the surface for food and often stirs up shallow water with its feet or swings its bill from side to side to strain food from the water, submerging its head to feed. The bird pumps water into and out of its bill with its tongue, using the long, comblike structures, called "lamellae," that line the sides of its bill to filter out food. • Unlike most dabbling ducks, male Northern Shovelers remain on nesting territories throughout the incubation period, and females frequently leave their nests to partake in feeding excursions. Shoveler pairs are monogamous and remain together longer than mated males and females of most other duck species. • The Northern Shoveler is often called "Spoonbill."

ID: medium-sized duck; large, spatulate bill; blue forewing patch; green speculum. *Male:* green head; yellow eyes; black bill; white breast; chestnut flanks. *Female:* mottled brown overall; broad, orange-tinged bill.
Size: *L* 46–51 cm; *W* 69–76 cm.
Habitat: *In migration* and *winter:* sites rich in aquatic plants and invertebrates such as shallow lakeshores, sewage and farm ponds, brackish lagoons and mudflats with algal growth; regularly visits upper reaches of estuaries. *Breeding:* margins of semi-open, shallow wetlands with emergent vegetation, including ponds, marshes and sloughs.

Nesting: usually on dry, grassy uplands close to water; shallow scrape is lined with plant material and down; female incubates 10–12 pale blue eggs for 22–25 days.
Feeding: dabbles and strains out aquatic crustaceans, insect larvae and seeds from wetlands and flooded fields; rarely tips up.
Voice: *Male:* distinctive *took took took* spring courtship call. *Female:* occasionally gives a raspy chuckle or quack.
Similar Species: spatulate bill is distinctive. *Mallard* (p. 59): blue speculum bordered by white; no pale blue forewing patch; male has chestnut breast and white flanks. *Blue-winged Teal* (p. 60): smaller overall; much smaller bill; male has white crescent on face and spotted breast and sides.

NORTHERN PINTAIL

Anas acuta

The trademark of the elegant Northern Pintail is the male's long, tapering tail feathers, often called the "sprig," which make up one-quarter of his body length. The elongated tail feathers are easily seen in flight and point skyward when this duck dabbles. Only the Long-tailed Duck has a similar tail, but the Long-tail is usually found on open salt water or large lakes during migration and winter, whereas the Northern Pintail prefers freshwater marshes, harvested agricultural fields and intertidal habitats. • Northern Pintails are among the earliest nesting ducks, returning to northern nesting areas shortly after ice-out. They have a circumpolar distribution, but within North America, the prairie pothole region of western Canada is their main breeding area. They are now also regular breeders in eastern and northern Canada. • Northern Pintails were once one of the most numerous ducks in North America, but populations have declined by half in recent decades. Drought, wetland drainage and changing agricultural practices are the most serious threats. These ducks, once susceptible to poisoning by inadvertently eating lead shot left behind by hunters, can now feed safely because the lead pellets have been replaced with steel shot.

ID: large duck; long, slender neck; slender, greyish bill. *Male:* greyish above and below; white neck extends onto nape as narrow stripe; brown head; long, tapering, black tail feathers. *Female:* mottled, light brown overall; paler, plain head; "blank" facial expression. *In flight:* slender body; brownish speculum with white trailing edge.
Size: *L* 64–76 cm; *W* 76–86 cm.
Habitat: *Breeding:* dry shores of wetlands with low vegetation and shrubby fields. *In migration:* shallow wetlands, tidal marshes, agricultural fields and estuaries. *Winter:* wide variety of shallow freshwater and intertidal zones and flooded and harvested farm fields.

Nesting: on the ground in brushy cover, usually near water; female lines a small scrape with grasses and down; female incubates 6–8 pale greenish or buff eggs for 22–24 days.
Feeding: tips up and dabbles for seeds; also eats aquatic invertebrates; grazes on waste grain during migration.
Voice: *Male:* soft, 2-note *prrip prrip* whistling call. *Female:* rough quack.
Similar Species: male is distinctive. *Mallard* (p. 59): female is chunkier, with orange-and-black or yellow-orange bill and lacks tapering tail and long, slender neck. *Blue-winged Teal* (p. 60): green speculum; blue forewing patch; female is smaller. *Long-tailed Duck* (p. 76): shorter, black neck; black head; white eye patch.

GREEN-WINGED TEAL

Anas crecca

The name "teal" is applied to 16 of the world's smallest waterfowl species, of which the Green-winged Teal is the best known. This teal is the smallest of North America's dabbling ducks. • The widespread Green-winged Teal exemplifies diminutive build and agile flight and can reach speeds of 70 kilometres per hour. On the wing, its small size, dark plumage and habit of making tight turns in wheeling flocks suggest a shorebird in flight. • This teal differs from other North American dabblers in that it does not breed extensively in the prairie pothole regions. Instead, it is most abundant in river deltas and forest wetlands of Canada and Alaska, where it nests in dense cover, often in shrubs or sedges. • The North American population of Green-winged Teals is estimated at three million birds, with 80 percent breeding within Canada. There are three subspecies with slightly varying plumage. Most Green-wings in Canada are of the *carolinensis* subspecies. During migration, however, keen birders can find a few obvious European *crecca* and even the occasional Alaskan *nimia*. • Because of its very diverse diet, the Green-wing happily forages almost anywhere there is shallow water, including salt water.

ID: small, dark bill; green-and-black speculum. *Male:* chestnut head; green cheek stripe outlined in white; grey flanks; vertical, white shoulder slash; black-spotted, creamy breast; pale grey sides; yellowish undertail coverts; black tail. *Female:* mottled brown overall; dark eye line.
Size: *L* 30–41 cm; *W* 51–58 cm.
Habitat: *Breeding:* beaver ponds, marshes and sloughs near upland sites with dense, low plant cover. *In migration:* tidal mudflats and shallow wetlands with abundant floating and emergent vegetation. *Winter:* wetlands, flooded farmlands, tidal mudflats and estuaries.
Nesting: on the ground; well concealed beneath a tussock of grass or among small bushes; shallow bowl of grasses and leaves is lined with down; female incubates 6–11 creamy white eggs for 20–23 days.
Feeding: dabbles in shallows for seeds of sedges, grasses and other aquatic plants; also takes molluscs, aquatic invertebrates and fish eggs.
Voice: *Male:* crisp whistle; froglike peeping. *Female:* soft quack.
Similar Species: male is unmistakable. *Blue-winged Teal* (p. 60): female has larger bill, blue forewing patch, paler underparts and heavily spotted undertail coverts. *American Wigeon* (p. 57): male lacks white shoulder slash and chestnut head.

CANVASBACK

Aythya valisineria

In profile, the Canvasback casts a noble image—its great, sloping bill is an unmistakable field mark. Despite its proud proboscis, this bird is named for its bright white back, which is usually a better field mark at a distance. The Canvasback occurs in large mixed flocks of diving ducks, and the male stands out from afar because of his bright white body. The much duller female is also readily recognized because of her "ski-slope nose." • Canvasbacks are devoted deep divers and prefer to feed and nest in deep-water marshes, sloughs and sheltered bays of larger lakes. With legs set far back on their bodies, these ducks are clumsy on their rare land excursions but are powerful divers and underwater swimmers. Most members of the genus require a brief running start in order to become airborne, and they fly with rapid, flickering wingbeats. • Canvasbacks breed in the drought-prone areas of the southern prairies, but it is thought that during droughts, these birds may travel farther north to breed. • After decades of decline, Canvasbacks appear to have rebounded in the southern prairies. Even though recent estimates place the overall North American population at about three-quarters of a million breeding adults, the Canvasback remains one of the least abundant duck species in North America.

ID: large duck; unique head shape with forehead sloping into large, black bill. *Male:* very pale body; chestnut head; red eyes; black breast and hindquarters. *Female:* pale body; light brown head and neck; dark eyes.

Size: *L* 48–56 cm; *W* 69–73 cm.

Habitat: *Breeding:* large, shallow marshes, ponds and sloughs and small lakes with emergent vegetation. *In migration* and *winter:* large lakes, lagoons, estuaries, slow-moving rivers and sewage ponds.

Nesting: on water among tall vegetation; platform of emergent vegetation, often cattails, supports a deep inner cup lined with down; female incubates 7–9 olive-grey eggs for 24–29 days.

Feeding: dives several metres to feed on seeds, roots and tubers of aquatic plants; eats molluscs and aquatic invertebrates in summer; occasionally dabbles in shallow areas with surface-feeding ducks.

Voice: generally quiet; wings create a loud whirring noise. *Male:* during courting, male utters moaning, almost dovelike *ik-ik-cooo* call. *Female:* low, quacking *cuk-cuk*.

Similar Species: *Redhead* (p. 66): rounded rather than sloped forehead; male has grey back and black-tipped bluish bill; female is browner overall with light patch near base of bill.

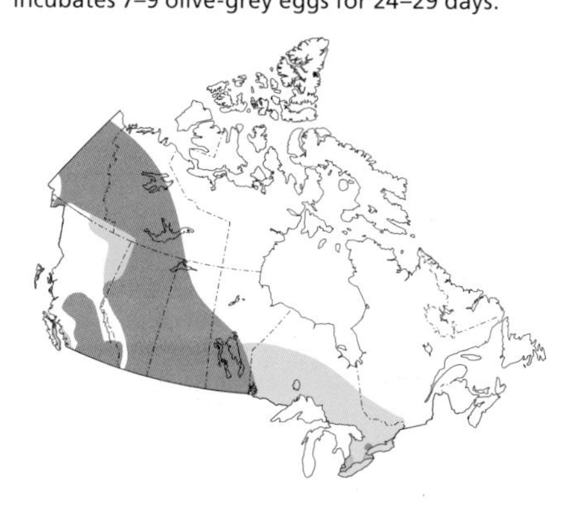

REDHEAD

Aythya americana

During the breeding season, Redhead pairs scatter across shallow wetlands in central and western North America, choosing habitats skirted with plenty of emergent vegetation, as well as lush bottom growth and enough depth to accommodate foraging dives. During winter, Redheads may linger on large lakes in southern Canada. These birds are quite traditional in their choice of wintering locations, often reappearing in about the same numbers year after year, in flocks of hundreds and, rarely, thousands. Despite their local abundance, they are generally uncommon or even rare away from favoured sites and are only usually seen in token numbers. • Female Redheads nest and brood their young as other ducks do, but they also become "brood parasites" at times, laying their eggs in other ducks' nests. Blue-winged Teals, Gadwalls, Ring-necked Ducks and Lesser Scaups are often victims of Redhead egg dumping. • Redheads and Canvasbacks have very similar plumage and habitat preferences. The best way to separate the two species is by head shape. The Redhead has a round head that meets the bill at an angle, whereas the Canvasback has a sloping head that seems to merge with the bill. Another obvious difference is the colour of their backs: the Redhead's is dark, whereas the Canvasback's is pale. • The Redhead is a diving duck, but it will occasionally feed on the surface of a wetland like a dabbler.

ID: large duck; rounded head; black-tipped, blue-grey bill. *Male:* red head; yellow eyes; black breast and hindquarters; grey back and sides. *Female:* dark brown overall; lighter "chin" and cheek patches.
Size: *L* 46–56 cm; *W* 66–74 cm.
Habitat: *Breeding:* shallow freshwater marshes, lakes and sloughs. *Winter:* ponds, sloughs and sewage lagoons on the coast; large, deep lakes in the interior; often found in flocks, frequently with scaups.
Nesting: deep basket of aquatic plants is lined with fine, white down and suspended over water; female incubates 9–13 creamy white eggs for 24–28 days; female sometimes lays eggs in nests of other species.
Feeding: varies with season; mainly aquatic vegetation in summer; more animal food such as invertebrates in winter.
Voice: generally silent. *Male:* nasal courting notes. *Female:* soft, low grunts, harsher quacks and an occasional rolling growl.
Similar Species: *Canvasback* (p. 65): clean white back; bill slopes onto forehead. *Ring-necked Duck* (p. 67): little habitat overlap (exclusively fresh water); female has more prominent, white eye ring, white ring on bill and peaked head. *Lesser Scaup* (p. 69) and *Greater Scaup* (p. 68): males have dark heads and whiter sides; females have white at base of bill.

RING-NECKED DUCK

Aythya collaris

A better name for this species might be "Ring-billed duck," because the ring around the bill stands out more than the faint cinnamon band encircling the male's neck. The male Ring-necked Duck's angular head profile, distinctive white bill markings, uniformly dark upperwings and white shoulder slash all are useful in helping distinguish it from the Greater Scaup and the Lesser Scaup. The female is much less showy and can be an identification challenge; fortunately, it also has a band around its bill, and females nearly always occur with males, an excellent clue as to their identity. • The Ring-necked Duck occurs commonly in forested regions of the Canadian Shield, clearly benefitting from the environmental engineering talents of the industrious beaver. In addition to beaver ponds, they are readily observed on boreal forest bogs and sedge-meadow wetlands. During nonbreeding seasons, small flocks of Ring-necks assemble in coves of lakes, sewage lagoons and vegetated sloughs on southern Pacific and Atlantic coasts. • Ring-necks are diving ducks, like scaups, Redheads and Canvasbacks, but they prefer to feed in shallower shoreline waters. Ring-necks often ride high on the water, and they tend to carry their tails clear of the water's surface. • When alarmed, this duck extends its neck and raises its fluffy crown.

ID: large duck. *Male:* black upperparts; dark purple head; yellow eyes; black breast and hindquarters; grey flanks; vertical, white shoulder slash. *Female:* brown overall; paler head; whitish face; white eye ring extends behind eye as narrow line.

Size: *L* 36–46 cm; *W* 58–63 cm.

Habitat: *Breeding:* shallow, permanent freshwater wetlands, wooded ponds, swamps and flooded beaver ponds with emergent vegetation. *In migration* and *winter:* wide variety of smaller water bodies; rarely estuaries.

Nesting: often on a floating island or in an open marsh; over water on a small platform of cattails, reeds or sedges, often covered with bent-over vegetation; female incubates 8–10 olive-tan eggs for 25–29 days.

Feeding: generalized diet; dives underwater for seeds, roots and tubers; also takes aquatic invertebrates, especially in summer.

Voice: seldom heard. *Male:* low-pitched, hissing whistle. *Female:* growling *churr.*

Similar Species: *Greater Scaup* (p. 68) and *Lesser Scaup* (p. 69): grey back; white flanks; tricoloured bill or shoulder slash; females have broad, clearly defined white face; Greater has rounded head and is typically found in marine waters in winter. *Redhead* (p. 66): rounded head; less white on front of face; female has less prominent eye ring.

GREATER SCAUP

Aythya marila

The Greater Scaup and Lesser Scaup are known to birders chiefly for their remarkable similarity. To make matters more difficult, diagnostic field marks cannot always be seen. When possible, look for the rounder (not peaked), greenish head of the Greater Scaup to help distinguish it from its Lesser relative. • The Greater Scaup is the only *Aythya* duck with a circumpolar breeding range. In Canada, it breeds from tundra regions south into the northern edge of the boreal forest. During migration, this species primarily migrates toward the Atlantic Coast of the northeastern U.S. Large concentrations stage around the Great Lakes alongside various other diving duck species during migration and winter months, when they seem to share the mollusc beds. • Since the introduction of zebra mussels into the Great Lakes, large flocks of Greater Scaups have begun overwintering in southern Ontario, far away from their traditional wintering grounds on the Atlantic Coast. There is concern that consumption of zebra mussels from polluted waters is negatively affecting the species' ability to reproduce and thus leading to population declines. Long-term declines of Greater Scaups are apparent in the species' main wintering range, raising concerns about elevated levels of organic contaminants and heavy metals in the birds' winter habitats, foods and the ducks themselves.

ID: large diving duck; rounded head; yellow eyes. *Male:* iridescent head may have greenish sheen; black-tipped, bluish bill; pale back; white sides; black breast; dark hindquarters. *Female:* brown overall; well-defined white face patch. *In flight:* white wing stripe extends well into primary feathers.

Size: *L* 41–48 cm; *W* 66–74 cm.

Habitat: *Breeding:* tundra, taiga and boreal forest ponds. *In migration* and *winter:* coastal estuaries, bays, harbours, inlets and brackish lagoons; large lakes, slow-moving rivers and large ponds in the interior.

Nesting: only 2 records; in a small, loose colony; on a grass- or sedge-covered shoreline; shallow depression is lined with dry vegetation and down; female incubates 5–11 buffy olive eggs for 24–28 days.

Feeding: dives underwater for aquatic invertebrates and vegetation; takes amphipods, insect larvae and vegetation in summer, freshwater molluscs in winter.

Voice: deep *scaup* alarm call. *Male:* 3-note whistle or soft *wah-hooo*. *Female:* sometimes growls subtly.

Similar Species: *Lesser Scaup* (p. 69): peaked crown; slightly smaller bill; shorter white wing flash in flight; prefers freshwater habitats. *Ring-necked Duck* (p. 67): black back; peaked crown; white shoulder slash; tricoloured bill. *Redhead* (p. 66): male has red head and darker back and sides; female has dark eyes and lacks well-defined, white face patch.

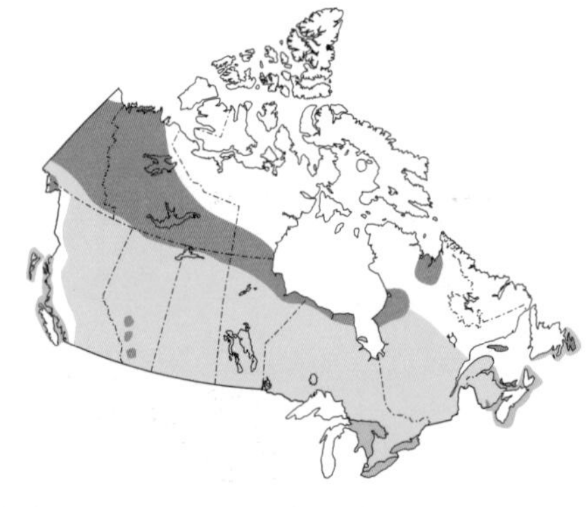

LESSER SCAUP

Aythya affinis

Lesser Scaups are one of the most widespread and abundant diving ducks in North America. The male Lesser Scaup and his close relative, the male Greater Scaup, portray the colour pattern of an Oreo cookie—black at both ends and white in the middle. • The Lesser Scaup breeds primarily from Ontario westward to Alaska but especially in the Prairie provinces, and though it can be found in a variety of habitats and regions, this duck seems to prefer boreal forest and aspen parkland. • Lesser Scaups are one of the latest ducks in Canada to pair and nest, many starting when other duck broods may already be half grown. • During migration and winter, Lesser Scaups favour smaller, less exposed habitats than Greater Scaups. They are frequently seen at city parks, where they are more approachable than in the wild. • Lesser Scaups, also known as "Little Bluebills," leap up neatly before diving underwater, where they propel themselves with powerful strokes of their feet. • The name "scaup" might refer to a preferred winter food of this duck—shellfish beds are called "scalps" in Scotland—or it might be a phonetic imitation of one of this duck's calls.

ID: large diving duck; peaked head; yellow eyes. *Male:* iridescent head may have purplish sheen; black-tipped, bluish bill; greyish back; white sides; black breast and hindquarters. *Female:* dark brown overall; well-defined white patch at base of bill.

Size: *L* 38–46 cm; *W* 58–64 cm.

Habitat: *Breeding:* semi-permanent and shallow, permanent wetlands with emergent vegetation. *In migration* and *winter:* large, open marshes, lakes and coastal bays and estuaries.

Nesting: usually on water in tall vegetation; occasionally on an islet; nest of aquatic grasses is lined with down; female incubates 8–14 buffy olive eggs for 22–25 days.

Feeding: dives for aquatic invertebrates, especially insects, molluscs and crustaceans; seeds and plants are sometimes important.

Voice: deep *scaup* alarm call. *Male:* soft, whistled *whee-oooh* courtship call. *Female:* rough, purring *kwah*.

Similar Species: *Greater Scaup* (p. 68): rounded head; slightly larger bill; longer, white wing edge; male has iridescent, greenish head. *Ring-necked Duck* (p. 67): peaked crown; tricoloured bill; male has white shoulder slash and black back; female has dark eyes and slight eye ring ; little habitat overlap (limited to freshwater marshes and ponds). *Redhead* (p. 66): male has red head and darker back and sides; female has dark eyes and less white at base of bill.

KING EIDER

Somateria spectabilis

If you want to travel to see a truly spectacular duck, the dazzling King Eider is well worth the effort: the male boasts no less than six bold colours on his magnificent head and bill. In the Maritimes, a glimpse of this impressive duck has rescued many cold birding excursions from oblivion. It is not surprising that this bird's favoured wintering sites receive visits from many birders each year. • King Eiders breed along the Arctic coasts of northern Canada and Alaska—few birds nest farther north than this species. Because the Arctic summer is short, adult King Eiders are among the first birds to return north to breeding areas in spring, many arriving before the ice and snow have completely melted. Pushed south in autumn by advancing sea ice, many of these ducks winter at the southern extent of the ice. With the introduction of zebra mussels into the Great Lakes, small numbers are now reported to winter annually on Lake Ontario. • King Eiders are equipped with the finest insulation in the bird world, eiderdown, so they are well adapted for loafing on ice floes and taking deep extended dives into frigid Arctic waters.

♂ ♀ ♂ ♀

ID: *Male:* blue crown; green cheeks; orange nasal disc; red bill; black wings; white neck, breast, back, upperwing patches and flank patches. *Female:* mottled, rich rufous-brown overall; black bill extends into nasal shield; V-shaped markings on sides.

Size: *L* 48–64 cm; *W* 89–102 cm.

Habitat: *Breeding:* Arctic tundra near fresh or salt water. *Winter:* surf lines and beyond off coastal headlands, often farther from shore than Common Eiders and scoters.

Nesting: on an offshore island or dry tundra, usually close to water; nest hollow is sheltered by grass or heather and thickly lined with down; female incubates 4–7 glossy, pale olive eggs for 22–23 days.

Feeding: dive to depths over 45 m; primarily takes aquatic molluscs; also eats small insects, crustaceans, echinoderms and some vegetation, especially in summer.

Voice: *Male:* soft cooing sounds in courtship. *Female:* low, twanging clucks.

Similar Species: *Common Eider* (p. 71): feathering on sides of long, droopy bill extend to nostrils; first-winter male has larger, greyer bill; female has evenly barred sides; immature has white streaking on back. *Scoters* (pp. 73–75): females have all-brown plumage, more bulbous bills; patchier colours on head and lack nasal shield.

COMMON EIDER

Somateria mollissima

The largest and most marine duck in Canada, the Common Eider is very rarely seen far from salt water, and flocks are often observed floating leisurely in rafts of thousands. Like scoters, this eider is well adapted to living in cold, northern seas. • There are four Common Eider subspecies in Canada. The *sedentaria* subspecies is found year-round in Hudson Bay, surviving winter in pockets of open water. Many other populations remain as far north as open water persists in winter, associating with polynyas (openings in the sea ice) and the leeward sides of islands that remain free of moving pack ice. • This bird's high metabolic rate and dense down feathers facilitate its almost entirely marine lifestyle. Long periods of inactivity are punctuated by frantic bouts of feeding or short flights to a nearby area, usually close to a coastal headland. • During the breeding season, like many ducks, female eiders pluck downy feathers from their own bodies to provide insulation and camouflage for their eggs. • Populations of Common Eiders appear to be on the decline throughout the Arctic, while the Maritime population is believed to be stable.

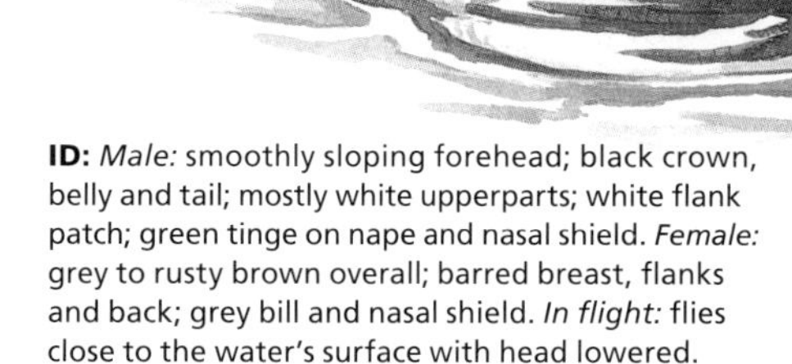

ID: *Male:* smoothly sloping forehead; black crown, belly and tail; mostly white upperparts; white flank patch; green tinge on nape and nasal shield. *Female:* grey to rusty brown overall; barred breast, flanks and back; grey bill and nasal shield. *In flight:* flies close to the water's surface with head lowered.
Size: *L* 58–68 cm; *W* 88–106 cm.
Habitat: shallow coastal waters year-round; occasionally seen on large, freshwater lakes. *Breeding:* rocky shorelines, islands or tundra close to water.
Nesting: colonial; on a rocky shelf or in a rocky depression; shallow depression is lined with plant material and large quantities of down; nest is commonly reused; female incubates 3–6 olive-grey eggs for 24–25 days.

Feeding: may dive deeper than 45 m; pries molluscs, especially blue mussels, from ocean depths and swallows them whole; also takes crustaceans, echinoderms, insects and plant material when available.
Voice: *Male:* raucous, moaning *he-ho-ha-ho* or *a-o-waa-a-o-waa*; courtship calls are *ah-oo* and *k'doo. Female:* Mallard-like *wak-wak-wak-wak-wak*; angry *wh-r-r-r-r*; courtship call is *aw-aw-aw.*
Similar Species: male is distinctive. *American Black Duck* (p. 58): may look similar to female in poor light; much smaller bill; purple speculum; white underwing coverts. *Scoters* (pp. 73–75): lack barring of female and white back and breast of male. *King Eider* (p. 70): male has colourful, blocky head; female has long nasal shield. *Long-tailed Duck* (p. 76), *Bufflehead* (p. 77), *Common Goldeneye* (p. 78) and *Barrow's Goldeneye* (p. 79): lack smoothly sloping facial profile.

HARLEQUIN DUCK

Histrionicus histrionicus

In many parts of the country, birders go to great lengths to see Harlequin Ducks. In British Columbia, these beautiful ducks are found year-round along the rocky Pacific coastline from Haida Gwaii (Queen Charlotte Islands) to the southern tip of Vancouver Island. On the Atlantic Coast, Harlequins are a widespread nesting species from the Gaspé to Baffin Island, but because of overhunting and hydro developments, eastern populations have dwindled to about only 1000 birds and the species is currently listed as endangered. This population is considered to be the source of the small numbers observed every winter on the Great Lakes. • Ocean waters surging around rocky headlands or clusters of intertidal boulders attract Harlequin Ducks, where they seem to enjoy the turbulent roar and tumble of the roughest whitewater. The uniformly brown females are difficult to see among the foam, and when sitting on dark rocks, even the colourful males can be surprisingly difficult to spot. • Tumbling, fast-flowing streams are the favoured breeding habitat of this duck. In the Alberta Rockies, conservationists are concerned that repeated disturbance from river rafting or mining and construction projects could affect the Harlequin's nesting success. With a low reproductive rate, this species is slow to recover from breeding disturbances and population declines. • This duck's distinctive, mouselike squeaks have earned it the local name of "Sea Mouse" in the Maritimes.

ID: small, rounded body; blocky head; short bill; raises and lowers tail while swimming. *Breeding male:* grey-blue body; chestnut sides; white spots and stripes on head, neck and flanks. *Female:* dusky brown overall; light underparts; 2–3 light patches on head. *Nonbreeding male:* similar to female but with darker head, chestnut-tinted flanks and white-marked lower back.
Size: *L* 38–46 cm; *W* 64–69 cm.
Habitat: *Breeding* and *in migration:* shallow, clear, fast-flowing creeks and rivers. *Winter:* on the coast, rocky shorelines and intertidal areas, often with surf lines; in the interior, fast-moving, clear waters with rocks.
Nesting: near a rushing watercourse; on the ground under a bush or on a cliff ledge; nest scrape is lined with plant material and down; female incubates 4–6 creamy to pale buff eggs for 27–29 days.
Feeding: depends on insects and molluscs in summer; eats crustaceans (primarily small crabs), molluscs, chitons, small fish and fish eggs in winter.
Voice: generally silent. *Male:* descending trill and squeaky whistles during courtship. *Female:* harsh *ek-ek-ek* or low croak.
Similar Species: *Bufflehead* (p. 77): never found on fast-flowing water; smaller; female lacks white between eye and bill. *Surf Scoter* (p. 73) and *White-winged Scoter* (p. 74): larger and heavier; females have bulbous bills. *Other diving ducks* (pp. 65–71, 73–83): longer necks; less rounded bodies.

SURF SCOTER

Melanitta perspicillata

Scoters are mostly black-bodied sea ducks with brightly coloured, "swollen" bills. Of the three scoter species, the Surf Scoter is the only one that breeds and winters exclusively in North America. Breeding occurs on or near shallow lakes in pockets scattered across northern Canada. Surf Scoters are most common along the coasts in the nonbreeding seasons, where they feed primarily on shellfish and barnacles plucked from within the zone of steepening swells and breaking surf. Huge rafts of these sturdy, heavily built ducks assemble at frequent intervals off beaches and headlands and around harbor entrances. • Throughout most of spring, Surf Scoters gather in immense numbers anywhere herring are spawning. This is an annual spectacle that birders plan their trips around. Offshore, flock after flock may be watched migrating past a given point over the course of several hours, and a thorough scan of the ocean will reveal a slowly wavering line just above the horizon. • The white patches on the adult male Surf Scoter's head are the reason for its popular nickname "Skunk Head."

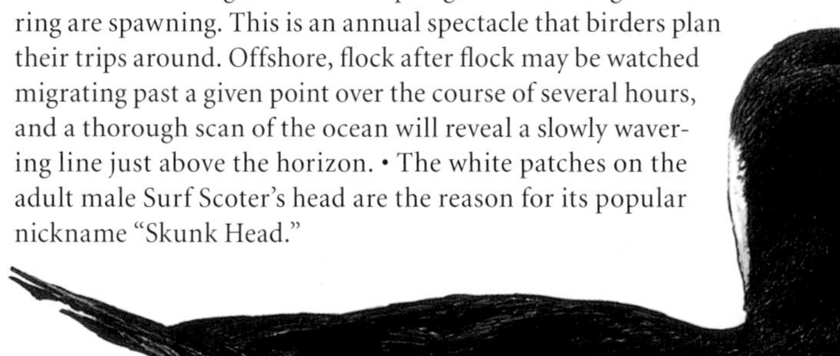

ID: large sea duck; bulky, triangular-shaped bill accentuates sloping forehead. *Male:* black overall; white forehead and nape; gaudy orange bill with large, black spot outlined in white. *Female:* brown overall; 2 whitish patches on each side of face. *In flight:* flies in long, undulating lines, usually near water.
Size: *L* 43–53 cm; *W* 74–79 cm.
Habitat: on the coast, mostly in protected open waters of bays, harbours and lagoons, usually near beaches and spits; lakes and rivers in the interior.
Nesting: on the ground under bushes or fallen branches; nest hollow is lined with plant material and much down; female incubates 5–9 buff-coloured eggs for 28–30 days.
Feeding: obtains food by diving and plucking; on the coast, mainly eats clams and mussels in winter and herring eggs in spring; in the interior, takes freshwater invertebrates, mainly insects and their larvae; rafts of scoters typically gather together and dive for food in a synchronized wave.
Voice: generally quiet; infrequent low, harsh croaks. *Male:* occasional whistles or gargles. *Female:* guttural *krraak krraak*.
Similar Species: *White-winged Scoter* (p. 74): white hindwing patch; male has white around eye only and slimmer, more ridged, less colourful bill with feathered, bulbous base; female has less pronounced forehead and crown angle. *Black Scoter* (p. 75): more smoothly rounded head; shorter bill; male has all-black head and bright yellow to orangey knob on bill; female has uniformly pale cheeks.

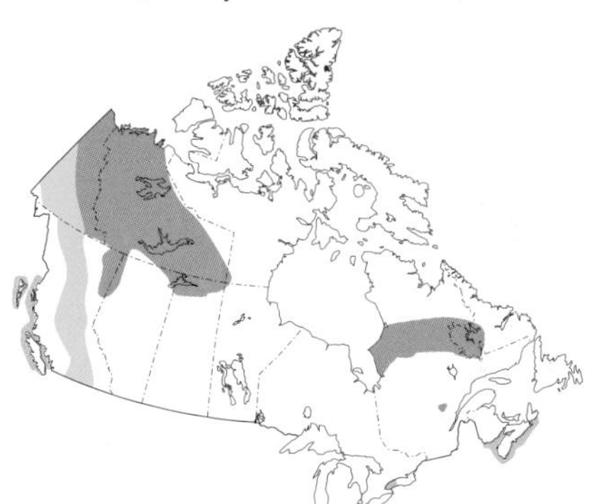

WHITE-WINGED SCOTER

Melanitta fusca

♂ ♀

White-winged Scoters are the largest of the three North American scoters. As mixed scoter flocks take off, White-wings are unmistakeable with their large, white wing patches. • White-wings breed primarily in the taiga-boreal forest from the Yukon to northwestern Québec. On spring evenings on the Great Lakes, lines of scoters can be seen lifting off the water and flying overland toward their breeding grounds as night approaches. • During spring and autumn migration, small lines and bunched strings of White-winged Scoters can be seen from nearly any ocean viewpoint. These scoters winter along both the Pacific and Atlantic coasts, with a small number remaining over summer. With the establishment of zebra mussels in the Great Lakes, White-wings have now become a common wintering species. • Because of their slow population growth and strong tendency to return to previous nesting sites, disturbance during the nesting season and hunting in breeding areas can eliminate local populations. In recent decades, White-wings have been extirpated from large nesting areas in the southern Prairie provinces because of habitat degradation, disturbance, overhunting or a combination of these factors. • The name "scoter" may be derived from the way these birds scoot across the water's surface.

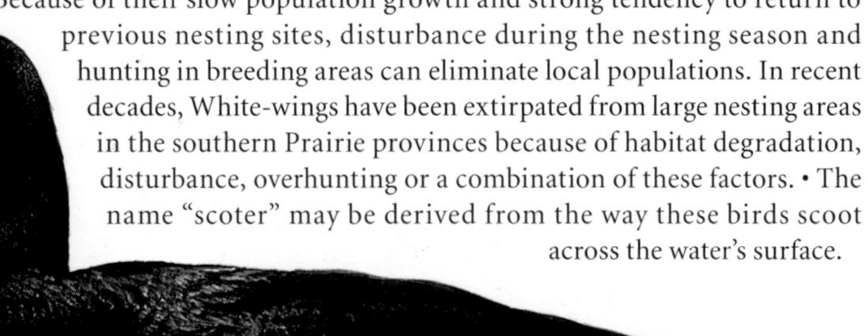

ID: large sea duck; bulbous bill with feathered base; conspicuous, white patch on hindwing. *Male:* mostly black plumage; pale eye within white crescent; orange bill tip. *Female:* dark brownish grey plumage; 2 large, indistinct, pale patches on sides of head.
Size: *L* 48–61 cm; *W* 81–89 cm.
Habitat: *Breeding:* shallow freshwater lakes, ponds and bogs in open country. *In migration* and *winter:* estuaries, bays, inlets, channels and shallow waters over shellfish beds along the coast, often just beyond the surf line; large lakes and slow-moving rivers in the interior.
Nesting: on the ground under dense vegetation, away from water; nest hollow is lined with down and twigs; female incubates 6–14 creamy buff eggs for up to 28 days.
Feeding: may dive up to 40 m for clams and mussels on the coast; eats crustaceans and insects in freshwater habitats.
Voice: courting pair produces guttural and harsh noises, between a *crook* and a quack; otherwise usually silent.
Similar Species: *Surf Scoter* (p. 73): all-dark wings; male has white forehead and nape and more colourful bill; female has more distinct white facial patches. *Black Scoter* (p. 75): male has completely black head and bright yellow to orangey knob on bill; female has light cheek and darkish crown. *American Coot* (p. 163): whitish bill and nasal shield; lacks white patches on wings and around red eyes.

BLACK SCOTER

Melanitta nigra

Except for the conspicuous yellow knob on his bill, the Black Scoter male is entirely black, the only such duck in North America (smaller American Coots are also black, but they are not ducks). • With two separate and remote populations, Black Scoters are one of the least known breeding ducks in North America. Within Canada, they breed in northern Québec, northern Ontario and the Mackenzie Delta in the Northwest Territories. • In winter, look for Black Scoters along both coasts, in outer-coastal surge zones. They are also occasionally seen on the Great Lakes and along the St. Lawrence River, primarily during winter. • Immature or eclipse-plumaged male Surf Scoters are routinely mistaken for adult male Black Scoters. While resting on the water's surface, Black Scoters usually hold their bills parallel to the water, whereas Surf Scoters and White-winged Scoters tend to hold their bills downward. • Scoters use their sturdy bills to wrench shellfish from underwater rocks or the sea floor. They swallow molluscs whole, and the shells are ground up in the birds' muscular gizzards.

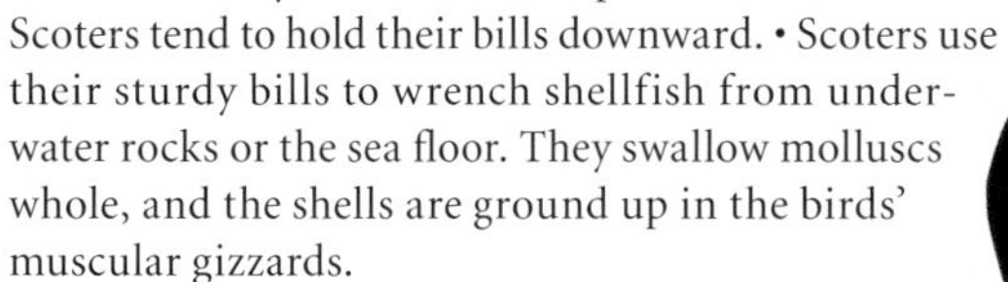

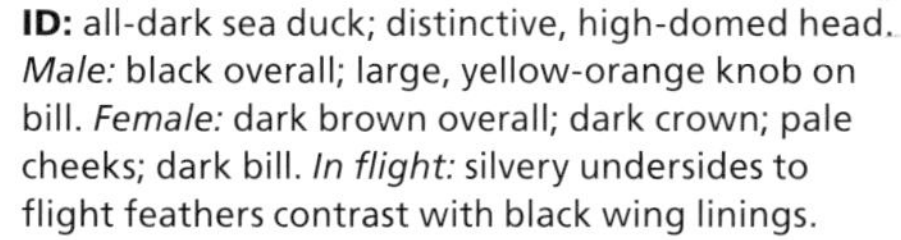

ID: all-dark sea duck; distinctive, high-domed head. *Male:* black overall; large, yellow-orange knob on bill. *Female:* dark brown overall; dark crown; pale cheeks; dark bill. *In flight:* silvery undersides to flight feathers contrast with black wing linings.
Size: *L* 43–53 cm; *W* 68–71 cm.
Habitat: shallow coastal waters over mussel beds (usually less than 11 m deep), including bays, estuaries, harbours and inlets. *Breeding:* large, deep lakes, large rivers and coastal lagoons.
Nesting: on the ground in tall grass within 30 m of water; nest hollow is lined with grass and down; female incubates 6–10 creamy to pinkish buff eggs for 27–31 days.
Feeding: dives for mussels, clams, periwinkles, limpets and barnacles.
Voice: most vocal of the scoters; courting males utter constant mellow, plaintive, whistling sounds; wings whistle in flight.
Similar Species: *White-winged Scoter* (p. 74): white wing patches; male has white slash below eye; female has 2 indistinct, pale patches on face. *Surf Scoter* (p. 73): male has white on head; female has white patches in front of and behind eye. *American Coot* (p. 163): whitish bill and nasal shield; lacks white patches on wings and around red eyes.

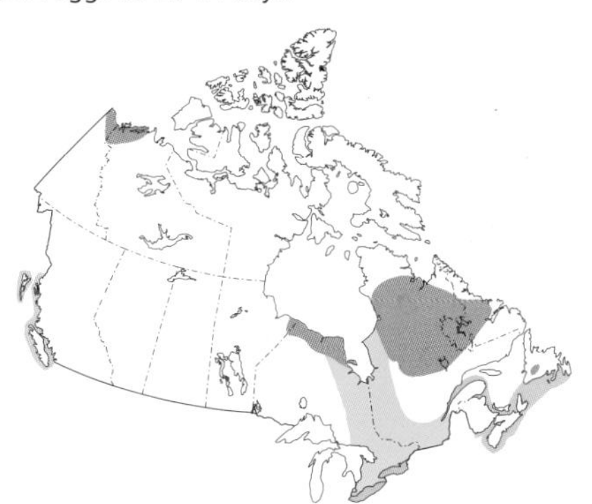

LONG-TAILED DUCK

Clangula hyemalis

This odd-looking but handsome sea duck, formerly known as "Oldsquaw," summers mostly on Arctic tundra but is a common winter resident in patches along both coasts and around the Great Lakes. Small numbers pass through the interior of Canada each year. • Long-tailed Ducks are one of the most abundant ducks above the Arctic Circle, breeding into the northern edges of Ellesmere Island. During winter, they tend to remain well offshore, making them difficult to view. • Long-tails are among the deepest diving ducks, swimming to depths of 60 metres for food and spending the most time underwater relative to surface time. • The breeding and nonbreeding plumages are nearly photo-negatives of each other: the spring breeding plumage is mostly dark with white highlights, and the winter plumage is mostly white with dark patches. Unlike other ducks that have two plumages per year, Long-tailed Ducks moult three times per year and remain distinctive throughout their complex series of plumages. The most distinctive feature is the male's two slim, elongated central tail feathers. • This bird's distinctive call is a nasal-sounding yodel that is audible from quite a distance on the duck's tundra breeding grounds and along the coasts of its wintering grounds. The incessant call has led to the duck's "Cockaree" nickname.

♂ ♀ *breeding*

♀ ♂ *nonbreeding*

ID: medium-sized sea duck; mostly black and white plumage. *Breeding male:* black head, chest and wings; grey patch around eyes; pink bill with dark base; very long central tail feathers. *Breeding female:* dark head and neck; white around eyes; thin, light ear line. *Nonbreeding male:* pale head with dark patch; pale neck and belly; dark breast; long, white patches on back; pink bill with dark base; long, dark central tail feathers. *Nonbreeding female:* short tail feathers; grey bill; dark crown, throat patch, wings and back; white underparts.
Size: *Male: L* 43–53 cm; *W* 66–71 cm.
Female: L 41–43 cm; *W* 66–71 cm.
Habitat: *Breeding:* tundra wetlands and coastal islands. *In migration* and *winter:* coastal waters; can appear some miles out to sea; large, deep lakes and slow-moving rivers in the interior.
Nesting: on islands and tundra, often near water; nest hollow is lined with plant material and down; female incubates 5–9 olive-grey to buffy olive eggs for 24–29 days.
Feeding: dives for molluscs, crustaceans, fish eggs and sometimes small fish; occasionally eats roots and young shoots.
Voice: quiet in early winter but very vocal from February onward. *Male:* musical, throaty *owl-owl-owlet* yodels. *Female:* soft grunts and quacks.
Similar Species: *Northern Pintail* (p. 63): male has brown head, much longer neck and bill, thin, white line extending up sides of neck and mostly pale grey wings; most common on fresh water.

BUFFLEHEAD

Bucephala albeola

Our smallest diving duck, the energetic Bufflehead is confined as a breeder to the boreal forest and aspen parklands of North America. Its small size has probably evolved with its habit of nesting in old Northern Flicker holes, an abundant resource too small to accomodate other, larger cavity-nesting ducks. Buffleheads also use tree cavities abandoned by larger woodpeckers and will use nest boxes if available. • The male Bufflehead's most characteristic feature is a large, white patch on the back of his head and white underparts. The female is sombre but appealing, with a sooty head ornamented by a white cheek patch. • This species is commonly seen wintering in rafts inshore along both ocean coasts and the Great Lakes, with some of the highest densities occurring along southern Vancouver Island. • Buffleheads are a highly territorial species; their nesting density, when not limited by the availability of tree cavities, is limited by their territorial behaviour. • After hatching, the ducklings remain in the nest chamber for up to three days before jumping out and tumbling to the ground. In some areas, it is thought that northern pike predation on ducklings may have a significant impact.

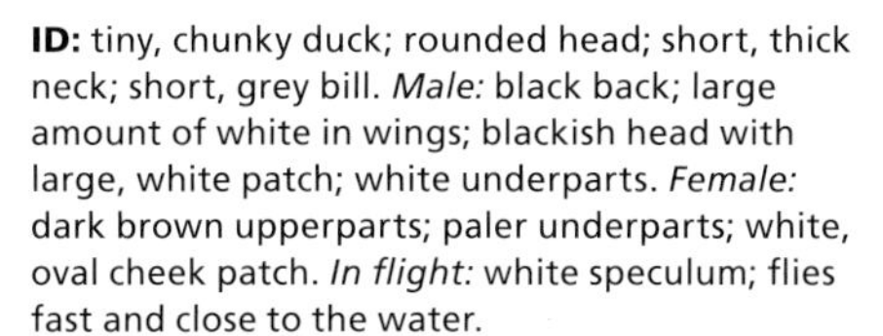

ID: tiny, chunky duck; rounded head; short, thick neck; short, grey bill. *Male:* black back; large amount of white in wings; blackish head with large, white patch; white underparts. *Female:* dark brown upperparts; paler underparts; white, oval cheek patch. *In flight:* white speculum; flies fast and close to the water.

Size: *L* 33–38 cm; *W* 51–53 cm.

Habitat: wide variety of fresh- and saltwater sites. *Breeding:* ponds, marshes, sloughs and small lakes with margins of older deciduous woods. *In migration* and *winter:* protected marine waters, estuaries, lakes, rivers, sewage lagoons and urban ponds.

Nesting: in a cavity in a large trembling aspen or poplar; uses nest boxes; no material added except down; female incubates 6–11 creamy or pale buffy olive eggs for 28–33 days.

Feeding: dives for crustaceans and molluscs in winter; takes insects, larvae and other invertebrates in summer.

Voice: *Male:* growling call. *Female:* harsh quack; *grrk* call when alarmed near the nest or brood.

Similar Species: *Hooded Merganser* (p. 80): larger; longer, slender bill; yellow eyes; brown flanks; white crest outlined in black; female has shaggy crest on wholly dark head. *Common Goldeneye* (p. 78): larger; yellow eyes; male has greenish head with round, white facial patch; female has mostly yellow bill. *Harlequin Duck* (p. 72): prefers turbulent water; female has browner plumage and 3 small, white areas on head.

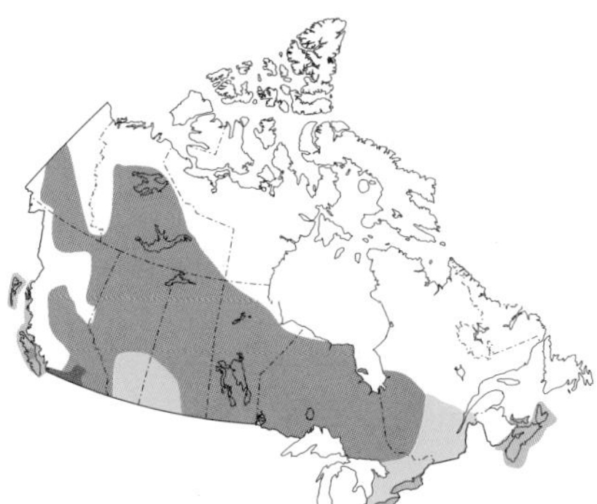

COMMON GOLDENEYE

Bucephala clangula

The Common Goldeneye is sometimes called "Whistler" because the male's wings create a loud, distinctive hum in flight. • In mid-autumn, goldeneyes migrate south from their breeding grounds. These birds are often the last diving ducks to leave the breeding grounds before freeze-up, staying until only narrow channels or pools of open water remain. Common Goldeneyes overwinter on inland fresh water as far north as water is open, being particularly abundant on the St. Lawrence River and the Great Lakes and along the Pacific and Atlantic coasts. • As early as February, testosterone-flooded males begin their crazy courtship dances, often to apparently disinterested females. Emitting low, woodcocklike buzzes, males thrust their heads forward, lunge across the water and kick their brilliant orange feet forward like aquatic breakdancers. • When searching for a nest site, the females hunt high and low for cavities in trees or elevated stumps. Where suitable sites are scarce, they may lay their eggs in other goldeneye nests, a practice known as "brood parasitism." Because female goldeneyes, like many other ducks, often return to the area where they were hatched, these extra eggs often belong to close relatives! • Waterfront development and the logging of cavity trees has reduced Common Goldeneye numbers in some areas of its breeding range.

ID: medium-sized duck; sloping forehead; peaked crown; black wings with large, white patches; yellow eyes. *Male:* black back with extensive white areas; iridescent, green head; round, white facial patch; short, dark bill; white underparts; black hindquarters. *Female:* mostly grey body; rusty head; white collar; dark bill with yellowish tip.
Size: *L* 41–51 cm; *W* 63–66 cm.
Habitat: *Breeding:* lakes, sloughs, marshes, beaver ponds and slow-moving rivers with wooded margins. *In migration* and *winter:* shallow bays, estuaries, harbours and straits along the coast; lakes, rivers and large marshes in the interior.
Nesting: in a tree cavity or nest box, often close to water; cavity is lined with breast down; female incubates 6–15 bluish green eggs for 28–32 days.

Feeding: dives for crustaceans, molluscs and small fish on the coast; takes aquatic insects when breeding.
Voice: usually silent in migration; wings whistle in flight. *Male:* nasal *peent* and hoarse *kraaagh* in late-winter courting. *Female:* harsh croak.
Similar Species: *Barrow's Goldeneye* (p. 79): male has longer, often purple-glossed head (forehead appears steeper), white crescent behind bill and smaller white marks on back; female has yellower bill. *Bufflehead* (p. 77): much smaller; short, grey bill; male has large, white patch on back of head; female has white, oval cheek patch. *Hooded Merganser* (p. 80): male's white crest is outlined in black; slender bill; brown flanks; female has bushy crest and lacks white collar.

BARROW'S GOLDENEYE

Bucephala islandica

Often near the top of an eastern birder's wish list, the Barrow's Goldeneye occurs mostly in northwestern North America, where 90 percent of the world's population breeds. In fact, it has been suggested that 60 percent of the world's population breeds and winters in British Columbia! • Barrow's Goldeneyes are loyal to favoured wintering grounds where they know there is an ample supply of crustaceans and molluscs. Just before the main exodus of birds for spring migration, large groups "fuel up" at Pacific herring spawning sites along the Pacific Coast. • An important breeding location was recently located in the forested regions of the Laurentian Highlands in southeastern Québec. • Although Barrow's Goldeneyes are frequently found in mixed flocks with Common Goldeneyes and both species indulge in acrobatic courtship displays, hybrids are extremely rare. While on their breeding grounds, Barrow's Goldeneyes are aggressive towards other duck species, particularly to similar species such as Common Goldeneyes and Buffleheads. • In addition to nesting in tree cavities, this species has been recorded nesting in crow nests, cliff holes and even marmot burrows. Soon after hatching, the ducklings need little parental care and can fend for themselves. • Females do not breed until they are three years old, and the oldest recorded Barrow's Goldeneye lived for 18 years.

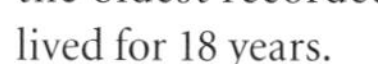

ID: medium-sized duck; long, angled head; bright yellow eyes; stubby bill; yellow legs and feet. *Male:* black head with purplish gloss (may appear greenish); white crescent between eye and black bill; white-spotted, black back; black extends downward from shoulder. *Female:* dark brown head; yellowish orange bill; greyish brown body; white breast, flanks and belly.

Size: *L* 41–51 cm; *W* 69–74 cm.

Habitat: *Breeding:* wooded margins of alkaline lakes, ponds, rivers and creeks to 1800 m elevation. *In migration* and *winter:* along the coast near rocky shores of bays, inlets and harbours where mussels are plentiful; lakes, rivers and ponds in the interior.

Nesting: usually in a tree cavity or nest box; female revisits former nest site; down is added as eggs are laid; female incubates 8–11 whitish or bluish green eggs for 28–30 days; female sometimes dumps eggs in other duck nests.

Feeding: dives to 5 m for invertebrates and some plant material.

Voice: generally silent. *Male:* "mewing" spring call. *Female:* hoarse croaks.

Similar Species: *Common Goldeneye* (p. 78): less-elongated head; male has glossy, often green head with round, white face spot and whiter back markings with more lines than spots; female has dark bill with yellowish tip.

HOODED MERGANSER

Lophodytes cucullatus

Extremely attractive but exceptionally shy, the Hooded Merganser is one of Canada's most sought-after ducks from a birder's perspective. The handsome male, with his flashy headgear, usually gets most of the attention. Much of the time his brilliant crest is held flat, but it quickly unfolds in moments of arousal or agitation. • Mergansers are fish-eating ducks that can be identified by their very narrow, sawtoothed bills, which are used to capture and hold their slippery prey. Of the six merganser species in the world, only the Hooded Merganser is restricted to North America. • Hoodies prefer to breed in small, vegetated, freshwater wetlands surrounded by forests. They are cavity nesters and will readily use a nest box when available. Like many other waterfowl that nest in holes, this species commonly lays its eggs in the nests of other cavity-nesting ducks. • Hooded Mergansers are quite cold hardy, being one of the last ducks to leave in autumn and first to return in spring, often when most areas are still ice covered. • The smallest of the mergansers, Hoodies find prey underwater by sight, and a transparent third eyelid (nictitating membrane) acts like goggles to protect the eyes.

ID: small duck; crested head; narrow, blackish bill. *Male:* black back and wings; bold, white, fan-shaped head crest outlined in black; white breast with 2 vertical, black slashes; rusty sides. *Female:* dusky brown back, wings and head; shaggy, reddish brown crest; white belly. *In flight:* small, white wing patches.

Size: *L* 41–48 cm; *W* 58–61 cm.

Habitat: *Breeding:* shallow wetlands, small lakes, beaver ponds and rivers with wooded shores. *In migration* and *winter:* salmon-spawning streams, lakes, marshes, beaver ponds, sloughs and sewage ponds.

Nesting: above or near water in an old woodpecker hole, natural tree cavity or nest box; eggs are laid in down; female incubates 5–13 almost spherical, white eggs for 35–40 days; female sometimes dumps eggs in other duck nests.

Feeding: dives for small fish, crustaceans and aquatic insects.

Voice: usually silent; low grunts and croaks. *Male:* froglike *crrrrooo* during courtship display. *Female:* occasionally a harsh *gak* or a croaking *croo-croo-crook*.

Similar Species: *Bufflehead* (p. 77): chubbier; male has white sides and lacks black outline to crest and black breast slashes. *Red-breasted Merganser* (p. 82) and *Common Merganser* (p. 81): larger; females have greyer plumage, rufous-coloured heads and much longer, orange bills.

COMMON MERGANSER

Mergus merganser

Like a gleaming white submarine, the male Common Merganser rides low in the water. Noticeably larger than most other ducks, this jumbo can tip the scales at 1.5 kilograms, making it our heaviest commonly occurring duck. • Mergansers have sharply serrated bills, like carving knives, to help them seize and hold their fishy prey. • In Canada, this species nests throughout the boreal forest, aspen parkland and montane forest zones, wherever there are cool, clear, unpolluted lakes and rivers and the trees are large enough to provide cavities. • The Common Merganser is one of the most northerly wintering waterfowl. In winter, any source of open water with a fish-filled shoal may support good numbers of these skilled divers, sometimes in flocks numbering into the thousands. • Merganser ducklings are protected but not fed by their mother. Soon after hatching, the ducklings feed by themselves on aquatic insects, and when they are about 12 days old, they try to catch their first fish. • Also known as "Goosander," a name sometimes restricted to Eurasian populations, the species has a variety of popular North American names that refer to mergansers in general, including "Sawbill," "Fish Duck" and "Sheldrake."

ID: large duck; long, slender, red, hook-tipped bill; dark eyes; red legs. *Breeding male:* mostly white body; glossy, dark green head without crest; black back; black stripe on shoulder and flank. *Female:* grey body; rusty brown head; small, shaggy crest; white "chin" and throat. *In flight:* shallow wingbeats; arrowlike flight.

Size: *L* 56–69 cm; *W* 81–86 cm.

Habitat: *Breeding:* along wooded shores of clear lakes, streams and rivers and on isolated islands. *In migration* and *winter:* on the coast, in fresh and brackish waters close to shores of bays, lagoons, inlets and estuaries; clear rivers, lakes and streams in the interior.

Nesting: adaptable; on an island, lakeshore or riverbank; in a tree cavity or nest box; female incubates 8–11 creamy white, often nest-stained eggs for 30–32 days.

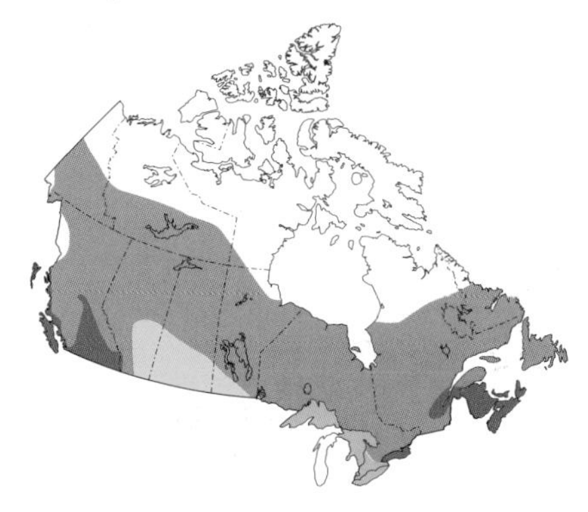

Feeding: eats mainly small fish caught in water up to 10 m deep; opportunistic feeding on insects, crustaceans, worms, frogs and salamanders; even takes small mammals and birds.

Voice: usually silent. *Male:* harsh *uig-a*, like a guitar twang. *Female:* harsh *karr karr*, usually in flight or when tending brood.

Similar Species: *Red-breasted Merganser* (p. 82): smaller; prefers salt water; shaggy, green crest; red eyes; slimmer bill; breeding male has chestnut breast with dark spots, black shoulder patch with white spots and greyish sides; female has brownish "chin." *Common Loon* (p. 102): larger; white-spotted back; dark bill; red eyes.

RED-BREASTED MERGANSER

Mergus serrator

Its glossy, slicked-back crest and wild red eyes give the Red-breasted Merganser the dishevelled, wave-bashed look of an adrenalized windsurfer or windswept pirate. It has a similar distribution and ecology to the Common Merganser, but it breeds farther north, winters farther south, occurs more frequently in salt water and estuaries, typically nests on the ground and has dark-coloured down. • Of the three North American merganser species, the Red-breasted has the largest breeding range, from the shores of the Great Lakes north to Baffin Island and from Newfoundland west into Alaska. It winters along both ocean coasts down into Mexico. Each spring and autumn, the shores of the lower Great Lakes host huge congregations of these ducks. • Red-breasts are usually ground nesters, hiding their nests under stumps, logs and shrubs. During nesting, females frequently lay eggs in the nests of other Red-breasted Mergansers, and broods are often amalgamated. Because females tend to return to former nest sites, they frequently end up babysitting the eggs and young of sisters or close relatives. • Shortly after their mates have begun incubating, the males fly off to join large offshore rafts of seabirds for the summer. During this time, a brief moult makes the males largely indistinguishable from their female counterparts. • This bird was formerly known as "Sawbill" and "Sea-Robin."

ID: large duck; slender body; conspicuous, shaggy crest; red eyes; thin, reddish, serrated bill. *Male:* black-and-white back; green, crested head; white collar; rusty upper breast; black-and-white shoulders; grey flanks; white belly and undertail coverts. *Female:* grey-brown back and wings; reddish, crested head; pale "chin," throat and breast; white belly and undertail coverts. *In flight:* conspicuous, white patch on inner wings.

Size: *L* 48–66 cm; *W* 71–76 cm.

Habitat: nearshore bays, estuaries, lagoons and inlets along the coast; lakes and larger rivers in the interior.

Nesting: on a small marine island, lakeshore or riverbank; on the ground, well concealed under bushes or other dense vegetation; female lines a scrape with down and incubates 7–10 buffy olive eggs for 29–35 days.

Feeding: dives primarily for small fish; also eats insects, crustaceans and tadpoles.

Voice: generally quiet. *Male:* catlike *yeow* when courting or feeding. *Female:* harsh *kho-kha*, mostly when tending her brood.

Similar Species: *Hooded Merganser* (p. 80): female is smaller, with yellow eyes and darker body. *Common Merganser* (p. 81): male lacks crest, has white breast, and red bill is much wider at base; female's rusty foreneck contrasts sharply with white "chin" and especially breast. *Red-necked Grebe* (p. 106): yellow bill; white cheeks; plain back; more white in wings; head lacks crest.

RUDDY DUCK

Oxyura jamaicensis

The male Ruddy Duck in his spring finery, with his bold, rusty red coloration, oversized head, huge, sky-blue bill, chubby body and short, vertically cocked tail, looks like a cartoon character come alive. The Ruddy Duck also displays its courtship rituals with comedic enthusiasm. With great vigour, the brightly coloured male slaps his chest with his bill while making his *plap-plap-plap-plap-plap* courtship call, in a display that increases in speed to a hilarious climax: a spasmodic jerk and sputter. • The nondescript yet appealing female commonly lays up to 15 eggs—a remarkable feat, considering that her eggs are bigger than those of a Mallard, a bird that is twice the size of a Ruddy Duck. The female also occasionally lays eggs in other ducks' nests. • Ruddy Ducks can sink slowly underwater to escape detection; this defence strategy is easier than flying, which requires a laborious running takeoff. • Primarily a breeder in the prairie potholes of western Canada, this duck's preference for deeper wetlands allows it to avoid habitat loss during drought years. • The Ruddy Duck is the only member of the stiff-tailed duck group found in Canada.

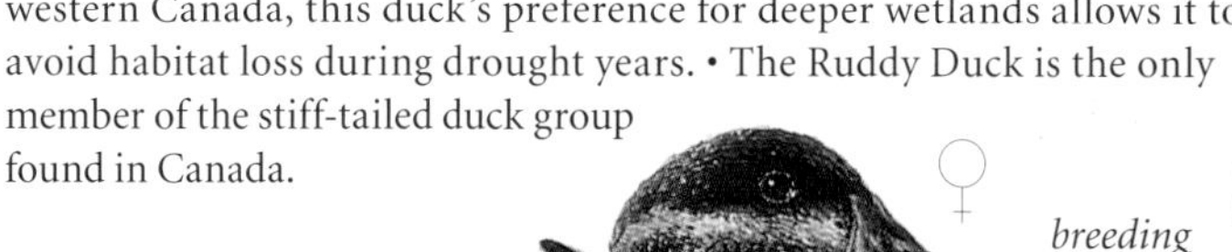

ID: small, pudgy duck; large head; short neck; long tail, often cocked upward. *Breeding male:* bright rusty body; black crown and nape; large, white cheek patch; bright azure bill. *Female:* brownish grey overall; darker crown and back; pale cheek with dark stripe; greyish bill. *Nonbreeding male:* resembles female, but with white cheek.
Size: *L* 38–41 cm; *W* 46–49 cm.
Habitat: *Breeding:* deeper, muddy-bottomed marshes, ponds, sloughs and sewage lagoons skirted with cattails and bulrushes. *In migration* and *winter:* on the coast in protected bays and lagoons, vegetated sloughs, sewage ponds and small, shallow lakes; small, marshy lakes and ponds and sewage lagoons in the interior.

Nesting: on water among cattails, bulrushes or sedges; may use an abandoned duck or coot nest, muskrat lodge or exposed log; platform nest of reeds is built over water and lined with fine materials; female incubates 5–10 rough, whitish eggs for 23–26 days; often lays eggs in other waterbird nests.
Feeding: mainly dives for seeds and tubers of aquatic plants; eats molluscs, crustaceans and infrequently small fish on the coast; feeds on aquatic insects on breeding grounds.
Voice: generally silent. *Male:* courts with *chuck-chuck-chuck-chur-r-r-r* display "song" punctuated with air bubbles on the water.
Similar Species: *Cinnamon Teal* (p. 61): more slender; breeding male has reddish head, red eyes and smaller, dark bill. *Black Scoter* (p. 75): female is larger than Ruddy Duck female and lacks dark line through cheek.

MOUNTAIN QUAIL

Oreortyx pictus

As spring arrives, the foothill and mountain slopes are touched now and again by the resonant, querulous call of the male Mountain Quail. In other seasons, however, Mountain Quail are less readily detected. Family groups seem to vanish in a moment from the roadside, concealing themselves among dense brush and sheltering thickets. • The Mountain Quail is a secretive bird that inhabits dense shrub and forest habitats of the Pacific Coast and western Great Basin. Its life history is poorly documented, but the species appears to differ from our other quail in many ways, including an enhanced ability to use high-elevation habitats by making seasonal long-distance treks and a high degree of herbivory. • Mountain Quail prefer to move on foot, and family groups or small coveys move to lower elevations for winter, staying below the snow line until spring, when pairs migrate back to nesting territories. • Introduced to Canada in the 1860s in southwestern British Columbia, the Mountain Quail is restricted to extreme southern Vancouver Island, but there have been only a handful of reports on the island since 1977, and its long-term status there remains doubtful.

ID: 2 long, straight head plumes; white-edged, chestnut throat; large, white bars on chestnut flanks; variable grey and brown breast and upperparts. *Immature:* shorter head plumes; white "chin"; no chestnut on throat.

Size: *L* 27–29 cm; *W* 41 cm.

Habitat: thickets, dense brush, chaparral, pine-oak woodlands and broken forest in wooded foothills and mountains; remains close to water in hot weather.

Nesting: on the ground under cover of grass, shrubs, thickets or logs; shallow depression is lined with grass, leaves, conifer needles and feathers; pair incubates 6–14 buffy white eggs for about 24 days.

Feeding: digs and scratches among leaf litter and climbs into trees and shrubs to glean food; eats seasonally available seeds, berries, bulbs, leaves, flowers, acorns and some insects and fungi.

Voice: spring call of male is a resounding, upslurred *quee-YARK!*; contact call is a rising series of 4–10 querulous whistles.

Similar Species: *California Quail* (p. 85): forward-curling, teardrop-shaped plume; black throat; "scaly" belly; white streaks on brown sides.

CALIFORNIA QUAIL

Callipepla californica

Although the male's spring courtship call sounds remarkably like the name of America's famous windy city, Chicago, the California Quail rarely strays far from its western homeland. First introduced in the early 1860s to British Columbia near Victoria, where numbers are slowly decreasing, the California Quail has since been released very successfully in various locations in south-central BC, especially the Okanagan Valley. It is now extirpated in the Fraser River delta and remains rare in the Creston area. • Spending most of their time scuttling about in tight, cohesive coveys, California Quail run the constant risk of predation by feral cats, hawks, owls and many medium-sized mammals. • Changes in agricultural practices, urbanization and overhunting have led to declines in California Quail numbers in the 20th century; however, there is some indication that trends have stabilized and that northern populations may be faring better than those in the south. Regularly occurring small fires have been found to increase local populations as a result of increases in food resources in these newly opened spaces. • The distinctive topknot on the head that looks like a single feather is actually comprised of a group of six feathers.

♂

ID: grey breast, back and tail; horizontal, white streaking on brown sides; "scaly" belly and neck. *Male:* white stripes on black face; forward-drooping, black head plume; dark crown; white border to black cheek and throat. *Female:* greyish brown face and throat; small, upright, black plume. *In flight:* grey underwings.

Size: *L* 25–28 cm; *W* 36 cm.

Habitat: edges of mixed woods, residential gardens, orchards, riparian thickets, brushy gullies and farmland.

Nesting: on the ground, well concealed under a shrub, clump of grass, pile of brush or a log; shallow hollow is lined with leaves, grasses and rootlets; female incubates 12–17 brown-blotched, off-white eggs for 22–23 days.

Feeding: gleans and scratches for seeds, leaves, fresh shoots, berries and insects; often eats birdseed.

Voice: soft calls, usually 2 notes; spitting alarm call (mostly by male). *Male:* loud, low-pitched *chi-ca-go* call in courtship and at other times of year.

Similar Species: *Mountain Quail* (p. 84): 2 long, straight head plumes; black and white barring on chestnut sides; chestnut throat.

NORTHERN BOBWHITE

Colinus virginianus

Throughout autumn and winter, Northern Bobwhites typically travel in large family groups called "coveys," collectively seeking out sources of food and huddling together during cold nights. When they huddle, members of the covey all face outward, enabling the group to detect danger from every direction. With the arrival of spring, breeding pairs break away from their coveys to perform elaborate courtship rituals in preparation for another nesting season. The male's characteristic, whistled *bob-white* call is often the only evidence that these birds are present among the dense, tangled vegetation of their rural, woodland home. • The Northern Bobwhite is the only native quail in eastern North America. Unable to survive the cold and heavy snow cover of winter and having lost most of its habitat to intensive farming practices, the few birds that remain in extreme southwestern Ontario, Canada's only resident wild population, were designated endangered in 1994. The Northern Bobwhite's high reproductive rate helps offset its short lifespan and high annual mortality in areas with suitable habitat. Recent data from several sources, however, all point to widespread declines of this bird throughout its range.

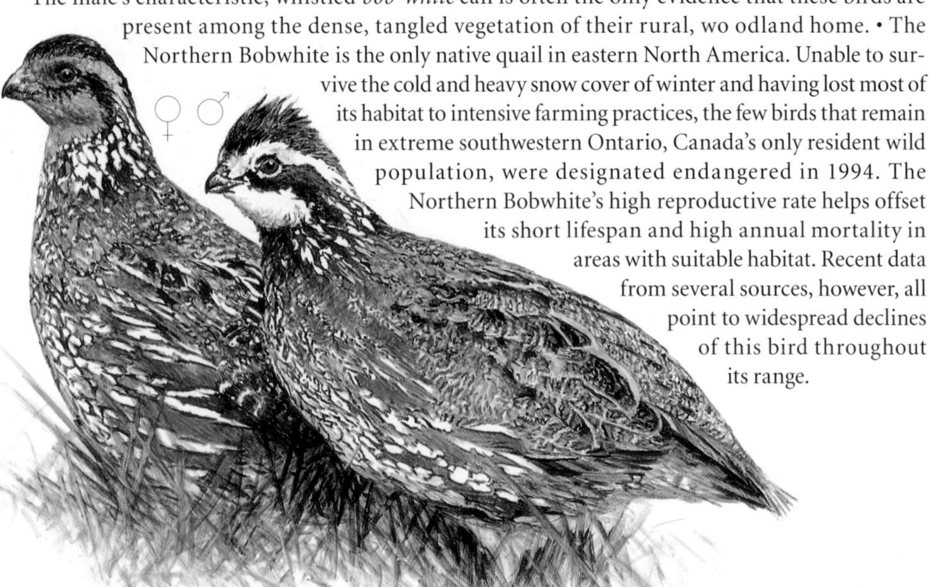

ID: mottled brown, buff and black upperparts; white crescents and spots edged in black on upper breast and chestnut sides; short tail. *Male:* white throat; broad, white eyebrow. *Female:* buff throat and eyebrow. *Immature:* smaller and duller overall; no black on underparts.

Size: *L* 25 cm; *W* 33 cm.

Habitat: farmlands, open woodlands, woodland edges, grassy fencelines, roadside ditches and brushy, open country.

Nesting: on the ground, often concealed by surrounding vegetation or a woven, partial dome; shallow depression is lined with grass and leaves; pair incubates 12–16 white to pale buff eggs for 23 days.

Feeding: eats seasonally available seeds, berries, leaves, roots, nuts, insects and other invertebrates.

Voice: whistled *hoy* is given year round. *Male:* whistled, rising *bob-white* given in spring and summer.

Similar Species: *Ruffed Grouse* (p. 90): long, fan-shaped tail has broad, dark subterminal band; black patches on sides of neck; lacks conspicuous throat patch and broad eyebrow. *Gray Partridge* (p. 88): orange-brown face and throat; grey breast; chestnut outer tail feathers.

CHUKAR

Alectoris chukar

A native of southern Eurasia, Chukars were first introduced to North America in Illinois in 1893, starting a proliferation of introductions. Between 1931 and 1970, nearly 800,000 Chukars were released in the U.S. and 10,000 were released in six Canadian provinces. The Chukar was successfully introduced to BC's Okanagan Valley in 1950–51 when 139 birds were released at Vaseux Creek, and additional birds were released in the Thompson Valley. Failed attempts were made to introduce the species to the Fraser Lowlands and southeastern Vancouver Island. Populations introduced to other provinces failed, and the species is now considered extripated outside of BC. The species has become permanent in the American Great Basin, and its current range extends through Oregon, Idaho, Washington and in a thin line north into south-central BC. • Chukars typically inhabit arid, rocky regions, where they are experts at locating water sources. In winter, these birds stay below elevations of 400 metres and within the sagebush ecosystem. • Females can lay more than 20 eggs in a single clutch. Chukar young can experience significant mortality from late-season snowfalls. In autumn and winter, Chukars gather in coveys consisting of five to 40 adults and their offspring. Coveys break up in late March and early April for the breeding season. • Chukars prefer running to flight. They fly only short distances, usually downhill and only when flushed. • This species was named for its distinctive *chuc-kar* call.

ID: sexes similar; greyish overall; brownish crown and back; reddish eye ring; black line through eye, throat and chest; black and chestnut barring on pale buff sides; rufous undertail coverts; reddish bill and legs. *In flight:* plain, brownish wings; grey rump; red-sided, grey tail; whirring wingbeats.
Size: *L* 33 cm; *W* 51 cm.
Habitat: dry sagebrush habitats; agricultural lands adjacent to rocky canyons; overgrazed, open ranges; sagebrush/bunchgrass benchlands at mid-elevations, often with rocky, steep hillsides.
Nesting: on the ground under bushes; nest scrape is lined with grasses, sagebrush twigs, leaves and feathers; female incubates 10–20 heavily spotted, pale yellow to buff eggs for 23–24 days.
Feeding: in winter gleans the ground for weed and grass seeds, especially cheatgrass; summer diet includes insects and green leaves of forbs.

Voice: short clucking precedes distinctive *chuc-kar chuc-kar chuc-kar*; soft, clucking *whitoo whitoo* when flushed.
Similar Species: *Gray Partridge* (p. 88): buff to rufous face; dark grey bill; chestnut brown barring on greyish flanks; pale legs; no black line through eye, throat and chest. *Mountain Quail* (p. 84): darker overall; 2 long, straight head plumes; black and white barring on chestnut flanks. *California Quail* (p. 85): forward-curling head plume; white streaking on chestnut flanks.

GRAY PARTRIDGE

Perdix perdix

Native to western Eurasia, Gray Partridges were first released in Canada in the early 1900s as game birds. Secretive and shy most of the year, they are rarely seen in the open. Watch for them in the morning or evening, when they venture onto quiet country roads, or in winter, when they sometimes forage beneath rural bird feeders. • When flushed, a covey of Gray Partridges will burst into flight, flapping furiously and dispersing in all directions, a tactic intended to confuse predators. Although Gray Partridges are relatively hardy birds, many perish in harsh weather, and some birds become trapped under layers of hardened snow while taking refuge from the cold. • The Gray Partridge has a high mortality rate and a short lifespan, but it makes up for these with its high reproductive capability. An average clutch size is about 16 eggs, and nests with more than 20 eggs are often found. When incubating her eggs, the female will sit tightly on her nest and risk being stepped on rather than attracting attention. • Like other seed-eating birds, the Gray Partridge regularly swallows small amounts of gravel to help crush hard seeds. The gravel accumulates in the bird's gizzard, a muscular pouch of the digestive system. • This bird was once known as "Hungarian Partridge," and coveys of Gray Partridges are still called "Huns" by many hunters and birding enthusiasts.

ID: small, terrestrial bird; brown eyes; pale greyish blue bill; mottled brown back; grey underparts; chestnut-barred flanks; short, grey, rufous-edged tail; pale greyish legs and feet. *Male:* orangey brown face and throat; dark brown "horseshoe" on belly. *Female:* buff face. *In flight:* low, whirring flight.
Size: *L* 28–36 cm; *W* 20 cm.
Habitat: grassy or weedy fields and croplands with hedgerows or brushy edge cover; prefers roadsides and grassland/grainfield edges.

Nesting: in a hayfield or pasture, or along a weedy fenceline or brushy margin; nest scrape is rimmed and lined with grass; female incubates 10–20 buff, brown or olive eggs for 21–26 days.
Feeding: gleans the ground for waste grain, other seeds and stems; also eats insect adults and larvae.
Voice: flushed covey utters a barrage of cackling *keep* notes. *Male: kshee-rik* call uttered at dawn and dusk sounds like a rusty hinge.
Similar Species: *Chukar* (p. 87): generally grey and brown; creamy cheek and throat; black line through eye, throat and chest; red bill; boldly dark-barred flanks; red legs and feet. *Ruffed Grouse* (p. 90): lacks rusty face and outer tail feathers; generally found in wooded cover.

RING-NECKED PHEASANT

Phasianus colchicus

This spectacular Asian species was first introduced to North America in the 1730s but failed to become established and was not successfully released until the 1880s. Today, aided by continuous reintroductions, small populations exist in the southern reaches of every Canadian province except Newfoundland. Although many Ring-necked Pheasants reproduce successfully in low-snowfall areas, populations are augmented by hatchery-raised birds that are released for hunters each year. This bird can even be fairly common in cities and suburban areas. • The distinctive, loud *krahh-krawk* of the male pheasant is often heard echoing from farmlands, brushy hedgerows, forest edges and marshes, but the birds themselves are less frequently observed. • The male, also known as a cock, establishes a harem of as many as a dozen females in spring, aggressively defending its territory and harem from rival suitors. • Unlike native grouse, Ring-necked Pheasants do not have feathered legs and feet to insulate them during the winter months. They also are unable to survive on native plants alone, requiring grain and corn crops, as well as sheltered hedgerows, woodlots, dense marsh or shrubby vegetation to survive our winters. The most serious limitation in most regions, however, is associated with changes in the agricultural industry from small multicrop farms to large monocultures with clean farming practices.

ID: large, distinctive land bird; very long tail feathers; unfeathered, grey legs. *Male:* glossy, green-and-purple head; red face and wattle; white neck ring; bronze neck and underparts; white-spotted, brownish back; long, neatly barred, bronze tail. *Female:* dull buff overall; brown-mottled back and sides. *In flight:* short, rounded wings; long tail feathers are well spaced at ends.

Size: *Male: L* 76–91 cm; *W* 71–84 cm. *Female: L* 51–66 cm; *W* 49–63 cm.

Habitat: open farm fields, brushy and weedy hedgerows, stubble fields, forest edges, marshes and suburban areas year-round.

Nesting: on the ground at base of grasses, shrubs, cattails and fallen logs; nest is often lined with leaves, grasses and feathers; female incubates 10–12 brownish olive eggs for 23–25 days.

Feeding: gleans the ground for fresh green shoots and a variety of invertebrates; eats mainly seeds, grains, roots and berries in winter.

Voice: hoarse *ka-ka-ka, ka-ka* when flushed or startled. *Male:* loud, raspy, roosterlike *ka-squawk* or *kraah-krawk*, followed by muffled whirring of wings. *Female:* usually clucks.

Similar Species: male is unmistakable. *Sharp-tailed Grouse* (p. 98): female has shorter tail with white outer feathers. *Ruffed Grouse* (p. 90): smaller; feathered legs; fan-shaped tail. *Greater Sage-Grouse* (p. 91): female is darker overall with black belly and feathered legs.

RUFFED GROUSE

Bonasa umbellus

Each spring, the male Ruffed Grouse proclaims his territory, strutting along a fallen log with his tail fanned wide and his neck feathers ruffled, beating the air with accelerating wing strokes. This "drumming" display is primarily restricted to the spring courting season, but Ruffed Grouse also drum for a few weeks in autumn. • The Ruffed Grouse is named for the black "ruffs" on the sides of the male's neck. During his nonvocal display, the male tries to impress females by erecting the ruffs, making them appear larger. • Both rufous and grey colour phases of the Ruffed Grouse occur; grey birds predominate in much of Canada except on the West Coast. • This bird is an important food source for many predators, including the Northern Goshawk. When Ruffed Grouse numbers crash, goshawk numbers follow a similar trend. Grouse populations also are tied to the aging and succession of aspen forests. • As winter approaches, the toes of Ruffed Grouse grow elongated bristles that act like temporary snowshoes. • The Ruffed Grouse is frequently called "Partridge" or "Bush Partridge." This leads to some confusion with the Gray Partridge, though it is never found in bush country.

♂

grey adult

ID: sexes similar; midsized grouse; small crest; greyish or rufous brown overall; black shoulder patches expand into ruff (larger on male) in display or threat; heavy, vertical barring on sides; brown-barred, rufous tail has broad, dark band (incomplete on female) and pale tip. *In flight:* rounded wings; fan-shaped tail; stiff, shallow wingbeats; usually flies low.

Size: *L* 38–48 cm; *W* 54 cm.

Habitat: mixed and hardwood forests, especially alder, aspen and birch, from sea level to over 1500 m elevation.

Nesting: on the ground under low, dense bushes, usually next to a tree, stump or deadfall; bowl-like depression is lined with plant items and feathers; female incubates 9–14 pale buff eggs for 23–24 days.

Feeding: eats mostly plant material such as tree buds, fruits, berries, catkins, leaves and seeds; takes invertebrates and even small snakes and frogs in summer.

Voice: courting male drums to produce a throbbing, accelerating *put put put put purrrrr* sound, somewhat like a far-off motorboat.

Similar Species: *Sharp-tailed Grouse* (p. 98): lighter brown; pointed tail. *Spruce* (p. 92), *Dusky* (p. 96) and *Sooty* (p. 97) *grouse:* females are darker overall with less white and lack crest or ruff.

GREATER SAGE-GROUSE

Centrocercus urophasianus

At dawn in spring, groups of Greater Sage-Grouse assemble at courtship "leks" to perform a traditional mating dance. In March and April, males enter a flat, short-grass arena, inflate their pectoral sacs, spread their pointed tail feathers and strut with vigour to intimidate peers and attract prospective mates. The fittest, most experienced males are found at the centre of the circular lek, whereas immature males are generally poor strutters and are forced to the periphery. The females wield the power, however, acting as the judges of this mating competition. • The Greater Sage-Grouse is our largest grouse and is characteristic of western rangeland dominated by big sagebrush. • By late summer, Greater Sage-Grouse often migrate to higher elevations; some individuals have been sighted at elevations of over 3000 metres. • Road construction, off-road vehicle use and overgrazing have all contributed to the destruction of prime Greater Sage-Grouse habitat throughout its range. This species is now extirpated from BC, and its range has been reduced to healthy sagebrush grasslands in southeastern Alberta and southwestern Saskatchewan. The Canadian populaion was designated as endangered in 1998. It is estimated that fewer than 1000 birds remain.

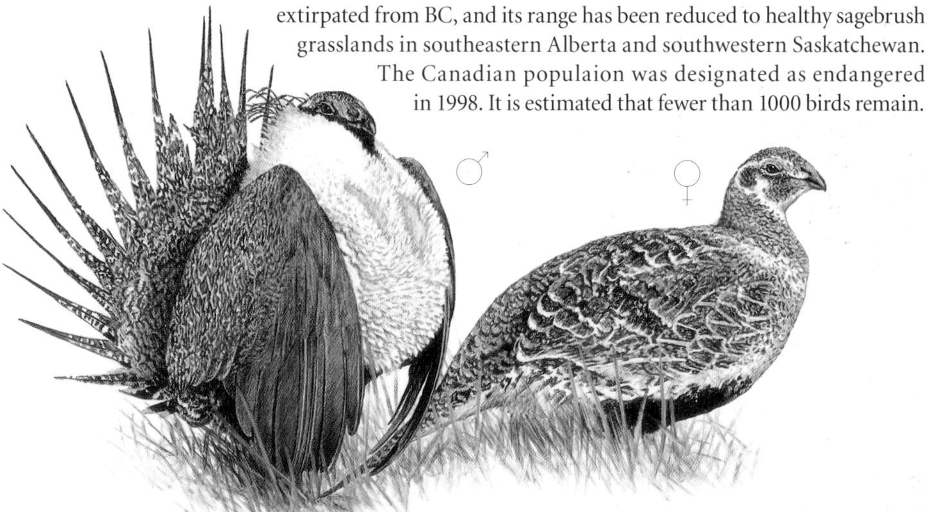

ID: large grouse. *Male:* black belly and "bib"; white breast; feathered legs; long, spiked tail; mottled brown back; yellow comb. *Female:* black belly; mottled brown plumage; very faint yellowish comb.
Size: *Male: L* 68–86 cm; *W* 96 cm. *Female: L* 46–61 cm; *W* 84 cm.
Habitat: treeless plains and rolling hills dominated by sagebrush, grasses and forbs.
Nesting: on the ground, usually under sagebrush; shallow depression is sparsely lined with leaves and grass; female incubates 6–9 variably marked, olive buff eggs for up to 27 days.

Feeding: eats mostly sagebrush leaves; also takes flowers, buds and terrestrial insects during summer.
Voice: generally silent. *Male:* unique, hollow *opp-la-boop* sound on lek as air is released from jiggling air sacs. *Female: quak-quak* call on lek.
Similar Species: *Ring-necked Pheasant* (p. 89): female lacks black belly and has unfeathered legs. *Dusky Grouse* (p. 96) and *Sooty Grouse* (p. 97): fan-shaped tail; lacks black belly.

SPRUCE GROUSE

Falcipennis canadensis

Spruce Grouse spend much of their secretive lives foraging in dark, dense conifer stands. They trust their camouflaged plumage even in open areas, often allowing people to approach within a few metres—which is why they are often called "Fool Hens." Most of the time, however, their strategy seems to work. • The Spruce Grouse is most conspicuous in late May and early June, when the females cackle from trees and the males strut across open areas, or from late July to August, when the females and young visit insect-rich feeding areas. • The male's courtship call is largely beyond human hearing, but he gives an audible double wing-clap after a short flight through the trees, just before he lands and transforms himself from a sombre bird into a red-browed, puff-necked, fan-tailed dandy in display. • In winter, this grouse's digestive organs can increase in size by 75 percent to allow more food to be eaten. • Because of this bird's trusting nature, which makes it easy prey for hunters, and its requirement for extensive tracts of undisturbed forest, it has not always adjusted well to civilization.

ID: medium-sized, heavyset grouse; short neck and tail. *Male:* grey to bluish grey body; red eyebrow; heavily white-marked, dark breast and underparts; black throat and neck; chestnut band at tail tip. *Female:* grey or reddish brown body; heavily barred underparts. *In flight:* short flights to tree branches.
Size: *L* 33–41 cm; *W* 54 cm.
Habitat: dense spruce, fir, cedar and tamarack swamps in the east; prefers higher ground of boreal forests containing black spruce and jack pine in the west.
Nesting: on the ground under dense cover, often at the base of a tree; ground scrape is lined with needles, leaves and some feathers; female incubates 4–8 finely speckled, tawny olive eggs for 22–24 days.
Feeding: on the ground in summer and in trees in winter; summer diet is diverse and omnivorus, including berries, flowers, black spruce buds, moss spore capsules, insects and fungi; conifer needles are winter staple.
Voice: *Male:* extremely low, barely audible hoots accompany whirring flight and 2 wing claps. *Female:* loud, wavering cackle from a tree perch at dawn and dusk.
Similar Species: *Dusky Grouse* (p. 96): female is larger and more bluish grey overall. *Ruffed Grouse* (p. 90): female has spotted rather than barred underparts.

WILLOW PTARMIGAN

Lagopus lagopus

The largest and most common of the three ptarmigan species, the Willow Ptarmigan is found in Arctic, Subarctic and subalpine tundra, breeding just above treeline, then retreating south to shrubby thickets or to subalpine forests in winter. The best time to see this ptarmigan is in mid-June to early July when males are actively displaying and calling. In late autumn, large flocks, sometimes numbering into the thousands, stage at nighttime roosting sites. • Like many Arctic animals, ptarmigans experience significant cyclical fluctuations in their numbers, perhaps explained by predators switching to a diet of ptarmigan when alternative prey is scarce. • Willow Ptarmigans moult almost continuously from April to November, becoming entirely white during winter, mostly mottled brown during summer and a warmer shade of brown in autumn. Other adaptations for living in extreme northern environments include feathered feet that function as snowshoes and the use of snow burrows for shelter. • Winter males lack the black eyeline of the male Rock Ptarmigan. When males moult in spring, their head, neck and upperparts turn a handsome reddish brown, and they reveal their bright red eye combs during courtship and aggression. • The male Willow Ptarmigan is the only grouse in the world that helps raise the family.

♀ breeding ♂

ID: medium-sized grouse; black outer tail feathers; short, rounded, white wings; black bill; fluffy, feathered feet. *Breeding male:* chestnut brown head and neck; mostly white body is splashed with brown on upperparts. *Breeding female:* mottled brown overall; white belly, legs and undertail coverts. *Nonbreeding:* white overall; red eyebrow; black eyes, bill and outer tail feathers.

Size: *L* 33–36 cm; *W* 58 cm.

Habitat: mainly willow and scrub birch thickets in tundra or subalpine; alpine zones at 600–2000 m elevation; heath barrens in Newfoundland.

Nesting: on the ground, usually under a low shrub; shallow scrape is lined with grasses, leaves and feathers; female incubates 6–11 darkly blotched, yellowish eggs for 21–23 days.

Feeding: gleans low vegetation and foliage for buds, flowers and leaves; also eats berries and insects when available.

Voice: loud, crackling *go-back go-back go-back*.

Similar Species: *Rock Ptarmigan* (p. 94) male has dark brown or grey-brown plumage; winter male has black eye line; female's plumage is nearly identical. *White-tailed Ptarmigan* (p. 95): usually found at higher elevations; white tail; lacks chestnut brown head and neck. *Spruce Grouse* (p. 92): female is larger and has no white in wings.

ROCK PTARMIGAN

Lagopus muta

The Rock Ptarmigan is the only bird species that spends its entire life cycle on the tundra, retreating only from extreme northern regions during extended winter darkness. This sedentary, tundra or mountain-loving species is challenging to find, as most of the rocky sites where it lives are inaccessible, even by hiking. Throughout its range in North America, it is know coloquially as "Snow Chicken," and it is the official bird of Nunavut. • Away from the main northern populations, disjunct populations exist in the mountains of northwestern BC and the western heathlands of Newfoundland. Newfoundlanders call older male Willow Ptarmigans "Rockers" because they maintain winter territories among large boulders. • The Rock Ptarmigan has three body moults from spring to autumn. Both sexes are entirely white in winter, but the male takes longer to transform into his summer attire than the female. • Studies have shown that the size and condition of the red comb of the males influences breeding success.

ID: *Breeding male:* red eyebrow (more prominent on male); mottled, brownish grey plumage; white wings; white, feathered legs and feet; black bill. *Breeding female:* similar to male but with very little white in summer and more barring. *Nonbreeding:* white plumage; black tail; male has red eyebrow and black eye stripe.
Size: *L* 36 cm; *W* 56 cm.
Habitat: *Breeding:* rocky Arctic tundra or high mountain slopes in subalpine and alpine zones; heath barrens in Newfoundland. *Winter:* brushy slopes near timberline; may move into shrubby areas at lower altitudes.
Nesting: on the ground in a rocky area; nest scrape is lined with mosses, lichens, grasses and feathers; female incubates 7–9 darkly marked, buff eggs for about 21 days.
Feeding: eats mostly plant buds, catkins, leaves, flowers, berries and seeds; takes insects when available.
Voice: both sexes give clucking notes. *Male:* croaking, throaty rattle followed by a hissing *krrrr-karrrr wsshhh* while displaying.
Similar Species: *Willow Ptarmigan* (p. 93): breeding male's plumage is more reddish; nonbreeding male's plumage lacks black eye stripe; female is very difficult to distinguish in winter, but is browner in summer. *White-tailed Ptarmigan* (p. 95): pure white tail.

WHITE-TAILED PTARMIGAN

Lagopus leucura

The White-tailed Ptarmigan is the smallest of our grouse and the only ptarmigan that occurs exclusively in North America. A permanent resident in the higher elevations of the mountains of western North America, its plumage perfectly matches its surroundings, regardless of the season. • In winter, White-tailed Ptarmigans conserve precious energy by running rather than flying, and to escape the cold, they often tunnel into snow near willow bushes, where they nibble on the buds. Sometimes, if a crust forms on the snow, the birds can perish. Even under the snow, ptarmigans must still defecate undigested plant remains, and up to 49 "pellets" have been found in a single burrow. • An incubating female will remain on her exposed nest even when closely approached, trusting that her cryptic camouflage will protect her. Nevertheless, this is very stressful to her, and sensitive hikers can make a ptarmigan's tough life somewhat easier by keeping a respectful distance. • This bird uses different calls to identify a predator as avian or mammalian, and individuals respond to avian predators by crouching near or under rocks. • The *saxatilis* form found on Vancouver Island, BC, is unique in the world.

♂

breeding

ID: medium-sized grouse; black eyes; dark bill; white outer tail feathers year round; fully feathered feet. *Breeding male:* red eyebrow during courtship (not always visible); streaked, brown and grey upperparts; white underparts; darkly blotched breast. *Female:* finely mottled brown overall; black barring on belly. *Winter:* all-white body; white tail. *In flight:* mostly white wings.

Size: *L* 31–36 cm; *W* 54 cm.

Habitat: remote alpine areas above timberline to 2800 m elevation.

Nesting: on the ground, among small rocks in a snow-free patch; shallow scrape is lined with fine grasses, leaves, lichens and feathers; female incubates 4–8 finely spotted, pale buff eggs for 24–26 days.

Feeding: eats buds, seeds and leaves of plants, especially willows; summer diet is supplemented with fruit, flowers and insects; may swallow grit to help with digestion.

Voice: *Male:* high-pitched *ku-kriii kriii*; low *kuk-kuk-kuk*. *Female:* low clucks around chicks.

Similar Species: *Willow Ptarmigan* (p. 93): black outer tail feathers; breeding male has chestnut head and neck. *Rock Ptarmigan* (p. 94): black outer tail feathers. *Dusky Grouse* (p. 96): much larger; all-dark tail.

DUSKY GROUSE

Dendragapus obscurus

The male Dusky Grouse often begins his low, hooting, owl-like courtship call while patches of snow still remain on the ground. One of the earliest signs of spring, these deep, soft mating calls advertise the bird's presence to rival males and prospective females, but they are essentially inaudible to the human ear beyond 40 metres. • Formerly referred to as the "Interior" subspecies of the Blue Grouse, the Dusky Grouse occurs in the interior coast ranges of the Rockies in British Columbia and along the eastern slopes of the Cascades in Washington. • The Dusky Grouse can be distinguished from the Sooty Grouse in a number of ways: the Dusky Grouse is paler; the bare skin on the side of the male's neck is purplish red; the male's song is usually given from the ground and consists of five soft notes; there are slight differences in the number and shape of the adult's tail feathers; and the chicks are more greyish brown above. • Hormonal changes and food availability cause these birds to make seasonal migrations, but rather than moving north and south, Dusky Grouse move altitudinally, sometimes travelling up to 40 kilometres, usually by walking. • Like other forest-dwelling grouse, the Dusky Grouse spends much of its time on the ground and is often easily approached, earning it the same nickname as the Spruce Grouse—"Fool Hen."

ID: large grouse; long neck and tail; dark eyes and bill; broad, longish tail, often fanned; feathered legs. *Male:* bluish grey crown and nape; brownish upperparts; orange-yellow eye comb; purplish red neck sacs (visible in displays); black tail with light tip. *Female:* greyish brown overall; lightly white-speckled neck and back; banded, brown rump and tail with paler tip.

Size: *Male: L* 43–48 cm; *W* 64–69 cm. *Female: L* 46–56 cm; *W* 58–64 cm.

Habitat: *Breeding:* open, relatively dry, mixed and coniferous forests to the subalpine fir zone. *Winter:* dense coniferous forests.

Nesting: on the ground with some overhead cover; shallow scrape is sparsely lined with dead vegetation and some feathers; female incubates 6–7 finely speckled, pinkish buff eggs for 25–28 days.

Feeding: eats mainly leaves, flowers, buds, berries, conifer needles and insects in summer; depends mainly on conifer needles in winter.

Voice: *Male:* series of 5–8 extremely deep hoots produced in neck sacs; other soft and harsh calls. *Female:* loud whinny and quavering cackles.

Similar Species: *Sooty Grouse* (p. 97): darker; male has yellow neck sacs; call usually given from high in a tree. *Spruce Grouse* (p. 92): male has prominent red eye comb, black upper breast and throat and white-spotted underparts and tail; female has more heavily black-and-white-barred underparts and buffy band at tail tip. *Ruffed Grouse* (p. 90): greyish brown tail with dark band at tip.

SOOTY GROUSE

Dendragapus fuliginosus

The well-known "Blue Grouse" species was split by ornithologists in 2006 into two separate species—the Sooty Grouse of the Pacific Coast and the Dusky Grouse of the Yukon, NWT, British Columbia Interior and western Alberta. • The Sooty Grouse is a solitary species and is rarely found in flocks. During the breeding season, the male often hoots from high in a coniferous tree and less commonly from logging debris. The "song" echoes, making the male very difficult to spot. • Compared to the Dusky Grouse, the Sooty is darker, the male's neck sacs are yellow, its song is louder and higher pitched, and the young have reddish brown upperparts and yellowish underparts. The birds' tails also differ in having 18 rather than 20 feathers that are more rounded at the tip and graduated in length, resulting in a more rounded tail shape. • Because of logging, Sooty Grouse numbers fluctuate greatly, with incredible increases following clearcut logging to much reduced numbers as replacement forests mature.

ID: large grouse; grey terminal band on long, dark tail; dark eyes and bill; feathered legs. *Male:* bluish grey crown and nape; brownish upperparts; red eye comb; yellow neck sacs (visible in displays). *Female:* mottled, dark brown overall; lightly white-speckled neck and back.

Size: *Male: L* 44–49 cm; *W* 65–70 cm. *Female: L* 47–57 cm; *W* 59–65 cm.

Habitat: *Breeding:* wet, mixed coniferous forests from sea level to alpine with ground cover of grasses, forbs and shrubs; also logging burns. *Winter:* dense, mixed and pure coniferous woods, usually higher in elevation than summer habitat.

Nesting: on the ground, usually with overhead cover; shallow scrape is sparsely lined with dead twigs, leaves, needles and a few feathers; female incubates 5–7 finely speckled, pinkish buff eggs for 25–28 days.

Feeding: eats leaves, flowers, buds, berries, conifer needles and insects in summer; depends mainly on conifer needles, especially Douglas-fir, in winter.

Voice: *Male:* series of 5–8 extremely deep hoots produced in neck sacs. *Female:* loud whinny and quavering cackles, especially when with brood.

Similar Species: *Dusky Grouse* (p. 96): paler; male has purplish red neck sacs; male's song usually given from the ground and has 5 softer notes. *Spruce Grouse* (p. 92): male has prominent red comb, black upper breast and throat and white-spotted underparts and tail; female has more heavily barred underparts and buffy band at tail tip. *Ruffed Grouse* (p. 90): greyish brown tail with dark band at tip.

SHARP-TAILED GROUSE

Tympanuchus phasianellus

Each spring, almost as soon as bare patches of ground start to show in the melting snow, male Sharp-tails gather at traditional dancing grounds, called "leks," to perform their mating rituals. With wings drooping at their sides, tails pointing skyward and purple air sacs inflated, males furiously pummel the ground with their feet, vigorously cooing and cackling for a crowd of prospective mates. Each male defends a small stage within the lek with kicks and warning calls, and central positions usually feature the dominant males—the best dancers and those that will attract the interest of the most females. • The courtship display of the male Sharp-tailed Grouse has been emulated in the traditional dance of many First Nations cultures on the prairies. • Because of extensive habitat changes, this grassland species is now much more localized, particularly in southern portions of its historical range, and many northern populations have undergone considerable declines as a result of changing agricultural and forestry practices and livestock grazing. • The word "lek" is derived from the Swedish word for "play."

ID: medium-sized grassland bird; sharply pointed tail; mottled, white-and-brown breast and upperparts; small, dark crescents on white belly; white outer tail feathers. *Male:* yellow eye combs; inflates purplish pink air sacs on neck during courtship displays. *Female:* cryptically patterned, brown-and-white body with dark markings.
Size: *L* 38–51 cm; *W* 64–66 cm.
Habitat: open grasslands with scattered patches of trees and shrubs; retreats to cover of trees and tall shrubs in winter.
Nesting: on the ground, usually under dense shrub or grass cover; shallow depression is lined with grasses, leaves and feathers; female incubates 10–13 finely speckled, light brown eggs for about 24 days.
Feeding: diet changes seasonally; mainly feeds on leaves, green shoots, flowers, grasses, berries, seeds and grasshoppers.
Voice: *Male:* mournful *coo-oo* and cackling *cac-cac-cac-cac* during courtship.
Similar Species: *Ruffed Grouse* (p. 90): slight head crest; black neck patches; fan-shaped tail with broad, dark subterminal band. *Ring-necked Pheasant* (p. 89): larger; unfeathered legs; longer tail; female has paler markings on underparts. *Spruce Grouse* (p. 92): short, fan-shaped tail; female has dark barring on body.

WILD TURKEY

Meleagris gallopavo

The Wild Turkey's reputation as a wary game bird, its fascinating courting behaviour and its photogenic qualities have made it one of the most popular native species among Canadian hunters and wildlife enthusiasts. Once common throughout most of eastern North America, the species suffered habitat loss and overhunting in the late 19th and early 20th centuries. By the early 1900s, the Wild Turkey had vanished throughout much of its range, and numbers had dropped from millions to about 30,000. But the Wild Turkey's popularity prompted widespread restoration efforts throughout North America, which have re-established the species in much of its former range. The North American population is now estimated to be about four million birds. • The Wild Turkey, one of the largest birds in North America, has acute senses and a highly developed social system. In winter, turkeys feed largely in open spaces and form flocks of up to 200 birds. Members of foraging turkey flocks cooperate to elude attacks by stealthy predators. • Although turkeys prefer to feed on the ground and travel by foot, they can fly short distances, and they normally roost high in trees during the night. • This charismatic species is the only widely domesticated native North American animal—the wild ancestors of most other domestic animals came from Europe, Asia or Africa.

♂

ID: large, ground-dwelling bird, usually found in flocks; unfeathered head; dark, glossy, iridescent body plumage; barred, copper-coloured tail. *Male:* long, central "beard" on breast; bluish head; red wattles. *Female:* smaller; blue-grey head; less iridescent body plumage; no "beard."
Size: *Male: L* 1.2–1.3 m; *W* 1.6 cm.
Female: L 89–94 cm; *W* 1.3 cm.
Habitat: open, mixedwood valleys, uplands with shrubby areas and agricultural lands.
Nesting: on the ground under thick cover; nest is lined with a few grasses and leaves; female incubates 8–15 brown-speckled, pale buff eggs for about 28 days.

Feeding: omnivorous; eats mainly seeds, leaves, grasses, corn and berries; also takes insects and other invertebrates, especially in breeding season.
Voice: wide array of sounds; loud *pert* alarm call; gathering call a cluck; loud *keouk-keouk-keouk* contact call. *Male:* gobbles loudly during courtship.
Similar Species: none; all other grouse and grouselike birds are much smaller.

RED-THROATED LOON

Gavia stellata

nonbreeding

Our smallest and lightest loon, the Red-throated Loon is the only member of its family that can take off from small ponds or, in an emergency, directly from land. It is also the only loon that regularly forages away from its nesting pond, flying to larger lakes or the sea to feed or carry fish back to its young. • The Red-throated Loon breeds mainly on remote ponds, primarily in coastal tundra habitat. It winters along both the Atlantic and Pacific coasts in shallow, sheltered, inshore areas and occasionally on the lower Great Lakes. • Although generally awkward on land, Red-throated Loons have been known to travel long distances on the ground, even moving some distance with their chicks when seriously disturbed. • Red-throat numbers have declined in several parts of its range, especially on the western tundra, though the reasons are not clear. The species is very sensitive to human disturbance and will desert breeding lakes if there is excessive human activity. • Unlike other loons, Red-throats do not carry their newly hatched chicks on their backs and both sexes vocalize, often joining to perform a territorial duet. • The Red-throated Loon typically swims low in the water with its bill held high.

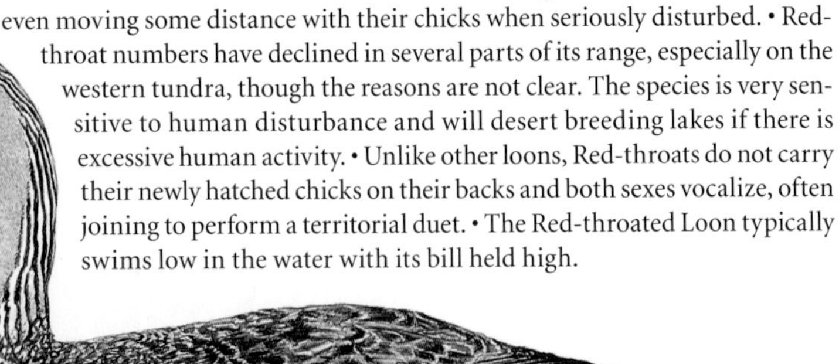

breeding

ID: sexes similar; medium-sized diving bird; smaller and more slender than other loons; slim bill is often held upward; sits low in the water. *Breeding:* grey head; reddish throat; grey face and neck; black and white stripes from nape to back of head; plain brownish back. *Nonbreeding:* white speckles on dark back; white face and throat; dark grey crown and back of head. *In flight:* hunched back; legs trail behind tail; rapid wingbeats.

Size: *L* 60–70 cm; *W* 1.0–1.1 m.

Habitat: protected nearshore shallow waters of ocean bays, estuaries, harbours, inlets and lagoons on the coast; large lakes and slow-moving rivers in the interior. *Breeding:* small lakes near the ocean; large ponds and small tundra lakes in the interior.

Nesting: on the shoreline of a small pond or lake; on an islet or partially submerged logs; nest is a low mass of aquatic vegetation; pair incubates 2 darkly spotted, olive eggs for up to 29 days.

Feeding: mostly feeds on marine fish captured by diving; sometimes eats aquatic insects and amphibians.

Voice: Mallard-like *kwuk-kwuk-kwuk-kwuk* in flight; loud *gayorwork* distraction call; pair issues mournful, gull-like wail during courtship.

Similar Species: *Common Loon* (p. 102): larger; heavier bill; lacks white speckling on back in nonbreeding plumage. *Pacific Loon* (p. 101): larger; sharp separation of black and white on neck; bill held level; purple throat and white speckling on back in breeding plumage; all-dark back in nonbreeding plumage.

PACIFIC LOON

Gavia pacifica

nonbreeding

Sporting twin white racing stripes across the shoulders, a blackish throat and a lustrous grey head and nape, the Pacific Loon is easily recognized in breeding plumage, but more sombrely attired autumn migrants are much less easily detected. Probably the most abundant loon in North America, the Pacific Loon is strictly a marine species, except during its brief three-month breeding season, when it nests on freshwater ponds throughout much of the Arctic and Subarctic tundra and taiga regions of the continent. • Pair formation usually takes place before these birds reach their nesting grounds, with nest construction commencing as soon as thawing ice has provided enough open water for flight. Because it is so awkward on land, this loon generally builds its nest right along the shoreline and will abandon it if receding waters cause the nest to be too far from the edge of the water. • Unlike the Red-throated Loon, the Pacific Loon needs 30 to 50 metres of open water to take flight. It prefers larger, deeper bodies of water for nesting than its Red-throated cousin does. • In winter, Pacific Loons prefer to raft on the open ocean beyond the surf line, often in the immediate shelter of a headland or jetty. They winter exclusively along the Pacific Coast from Alaska to southern California.

breeding

ID: sexes similar; medium-sized diving bird; slender body; high, smoothly rounded crown; thin, straight bill is held level. *Breeding:* gleaming, silver-grey crown and nape; dark back with large, white spots; fine white lines on sides of blackish throat. *Nonbreeding:* dark face; dark grey upperparts; white underparts; white "chin," throat and breast; dark "chin strap." *In flight:* hunched back; legs trail behind tail; rapid wingbeats.

Size: *L* 59–73 cm; *W* 1.0–1.2 m.

Habitat: *Breeding:* small freshwater lakes and ponds. *In migration* and *winter:* larger lakes and slow-moving rivers; open ocean near coast; congregates locally in areas of upwelling, such as marine passages.

Nesting: on an islet, mat of accumulated vegetation or projecting grassy shoreline; nest is a bowl of aquatic plants; pair incubates 1–2 black-spotted, dark olive eggs for 28–30 days.

Feeding: dives deeply for small fish on the ocean; eats mainly aquatic insects and crustaceans in summer.

Voice: largely silent away from breeding sites; calls include doglike yelp, catlike meow and ravenlike croak; issues sharp *kwao* flight call and rising wail at breeding sites.

Similar Species: *Common Loon* (p. 102): larger; heavier bill; breeding bird has black bill, regular rows of white spots on black upperparts and smaller white spots on sides; nonbreeding bird has dark, unmarked grey-brown upperparts, white throat and "indentation" where neck meets body, white underparts, pale eye ring and blue-grey bill. *Red-throated Loon* (p. 100): smaller head; slimmer, upward-tilted bill; faster, deeper wingbeats in flight; breeding bird has plain upperparts and reddish throat; nonbreeding bird has white face and fine, white spotting on back.

COMMON LOON

Gavia immer

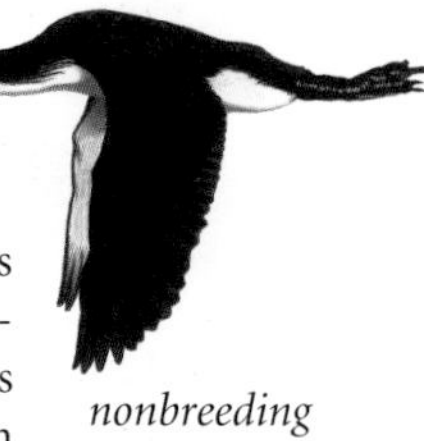

nonbreeding

To many Canadians, the Common Loon is the quintessential symbol of northern wilderness. In summer, its plaintive cries are heard on our larger lakes, most often in sites far removed from human influence. At one time, some First Nations groups believed that the loon's haunting call was the voice of the Great Spirit, Manitou. • These excellent underwater hunters have nearly solid bones that decrease their buoyancy (most birds have hollow bones), and their feet are placed well back on their bodies to aid in underwater propulsion. On land, however, their rear-placed legs make walking awkward, so Common Loons always nest close to the water's edge. • Small gamefish, perch and sunfish are all fair game for this loon, which will chase fish to depths of up to 50 metres, as deep as an Olympic-sized swimming pool is long. • In recent decades, some nesting populations have experienced serious declines as a result of a combination of habitat loss, increased recreational activity near nest sites, mercury pollution and acid rain. The rally of support groups, the initiation of volunteer protectionists known as Loon Rangers and extensive annual censuses and public education efforts turned the tide for many populations and fostered the creation of the North American Loon Fund to unify conservation efforts. • The Common Loon is imprinted on the Canadian dollar coin, which has affectionately become known as the "loonie."

breeding

ID: sexes similar; thick neck; sharp, stout, black bill. *Breeding:* black upperparts with regular rows of bold, white spots; smaller white spots on sides; striped "necklace." *Nonbreeding:* dark, unmarked grey-brown upperparts; white throat and white "indentation" where necks meets body; white underparts. *In flight:* long wings beat constantly; hunchbacked appearance; webbed feet trail beyond tail.
Size: *L* 71–89 cm; *W* 1.2–1.4 m.
Habitat: *Breeding:* larger marshes and sloughs; lakes with clear water. *Winter:* shallow areas close to shore such as inlets, bays, harbours and estuaries.
Nesting: on the shore of a marsh or small island, on a sphagnum hummock, grassy islet or partially submerged log, always close to water; nest is a mound of nearby aquatic plants; pair incubates 1–2 (rarely 3) darkly spotted, olive eggs with for 24–31 days.
Feeding: eats fish usually up to 25 cm long; summer diet may include snails, aquatic insects, leeches, crustaceans and even frogs.
Voice: alarm call is a quavering tremolo; also wails, hoots and yodels; male's territorial call is a complex yodel.
Similar Species: *Pacific Loon* (p. 101): smaller; more rounded crown; slimmer bill; no white "collar"; distinctive silver-grey head and nape in breeding plumage; greyer sides to neck and dark "chin strap" in nonbreeding plumage. *Red-throated Loon* (p. 100): smaller; slender bill; sharply defined white face and white-spotted back in nonbreeding plumage. *Yellow-billed Loon* (p. 103): yellow, upward-angled bill is diagnostic in nonbreeding plumage.

YELLOW-BILLED LOON

Gavia adamsii

nonbreeding

The Yellow-billed Loon is a relatively rare bird that nests in Arctic tundra regions of North America and Eurasia. Closely related to the Common Loon and often confused with it, the Yellow-billed Loon is distinguished by bill shape and colour, and it breeds almost entirely above treeline, north of the range of its smaller and more widespread relative. • In Canada, Yellow-bills nest nearly exclusively in central Nunavut. Nests are located on islands and peninsulas in medium-sized ponds and on lakes of a few hundred hectares where water levels are stable. • The migration of this species is poorly known. Along the coast, most birds apparently travel several hundred metres offshore. In spring, they appear to follow open leads in the sea ice far offshore in the Beaufort Sea toward their nesting grounds. In autumn, Yellow-billed Loons apparently take an overland route over Alaska, as well as a sea route around it, to their wintering grounds. • Yellow-billed Loons winter primarily along the British Columbia coast from southeastern Alaska to Washington. Vagrants occasionally show up on large lakes throughout southern Canada, providing temporary excitement for birders.

breeding

ID: sexes similar; swims with head tilted slightly upward. *Breeding:* yellow bill; black head; partial white collar; large, white spots on back. *Nonbreeding:* pale yellow bill; brownish upperparts; pale brown neck and cheek; more white on head than Common Loon.
Size: *L* 76–89 cm; *W* 1.2–1.4 m.
Habitat: *Breeding:* lakes and slow-moving rivers in low-lying tundra regions. *In migration:* open straits; small and large lakes.
Nesting: on a raised mound close to water; very little or no nesting material is used; pair incubates 2 scantily marked, olive eggs for 27–29 days.
Feeding: pursues prey underwater using feet for propulsion; eats mainly small fish; also takes some invertebrates and vegetation.
Voice: eerie, wailing tremolo, similar to Common Loon, but slightly lower in pitch and executed more slowly.
Similar Species: *Common Loon* (p. 102): lower mandible angles upward less abruptly; upper mandible curves down slightly; swims with bill more parallel to water; breeding bird has black bill; nonbreeding bird has darker edge on upper mandible.

PIED-BILLED GREBE

Podilymbus podiceps

nonbreeding

The smallest and wariest of Canada's grebes, the Pied-billed Grebe is common and widespread but is not always easy to find. When frightened, this grebe will compress its feathers to alter its specific gravity, enabling it to sink slowly until only its eyes and nostrils remain above water. To escape danger in a hurry, these birds also "crash-dive," kicking water several metres into the air. • If other waterfowl enter a Pied-billed's territory, the larger male or both members of a pair will not hesitate to attack. Offensive tactics include hunching itself in a threat posture or launching an underwater diving assault. Perhaps this grebe's aggressiveness is what earned it the nickname "Helldiver." • This secretive grebe has a loud, whooping, junglelike call that is usually heard well before the bird is seen. Like most other grebes, male and female Pied-bills call in duets. • Pied-billed chicks spend most of their first week protected from underwater predators by resting on the backs of their parents, a trait common to other grebe species as well. • This grebe prefers a solitary existence and is rarely found in flocks. • Grebes are unable to walk on land and must run across the surface of the water to become airborne.

breeding

ID: sexes similar but male slightly larger; small, stocky waterbird; short neck; chickenlike bill. *Breeding:* brownish grey body; black ring on pale bill; black throat. *Nonbreeding:* pale bill lacks ring; darker crown.

Size: *L* 30–38 cm; *W* 58 cm.

Habitat: *Breeding:* ponds, marshes, sloughs and lakes, usually with abundant emergent vegetation along shoreline. *Winter:* prefers open, freshwater bays, harbours and brackish lagoons along the coast; may visit the ocean; also uses shallow, permanent water bodies.

Nesting: in thick vegetation near a lake edge; floating platform nest of wet and decaying water plants is anchored in thick vegetation; pair incubates 3–10 bluish or greenish white, often nest-stained eggs for 23–27 days.

Feeding: opportunistic, taking whatever prey is most readily available; dives for small fish and crustaceans; picks insects and other aquatic invertebrates from the water's surface or plants.

Voice: loud, whooping call begins quickly then slows down: *kuk-kuk-kuk cow cow cow cowp cowp cowp*; also whinnies.

Similar Species: *Horned Grebe* (p. 105) and *Eared Grebe* (p. 107): longer, slimmer bills; white on lower part of face, belly and wings in nonbreeding plumage. *American Coot* (p. 163): all-black body; white bill extends onto forehead.

HORNED GREBE

Podiceps auritus

nonbreeding

Cold, mucky wetlands might not seem inviting, but nothing is more appealing to nesting Horned Grebes. Their propensity for these habitats starts early in life—Horned Grebe eggs are incubated on soggy, floating platforms among dense emergent vegetation. • Intensely territorial, these grebes usually nest solitarily, but where desired nest sites are limited, they may form loose aggregations or small colonies. • Horned Grebes are unmistakable in their breeding finery, with a bright rufous neck and golden "ear" tufts. In early autumn, their striking summer plumage fades into a nondescript, black-and-white winter plumage. • Adult grebes often eat their own feathers and feed them to their young. The feathers collect to form a "plug" in the stomach, protecting it from sharp bones, and keeps the bones in the stomach long enough to be digested. • Like loons, grebes appear hunchbacked in flight, with hastily beating wings, head held low and feet trailing behind the stubby tail. • Recent breeding bird surveys have shown that Horned Grebe numbers are declining in the Prairie provinces and throughout much of the species' range. A small, disjunct population on the Magdalene Islands in Québec may be considered for endangered status.

breeding

ID: sexes similar; small diving bird; straight bill; thick neck; rides high in the water. *Breeding:* black head; golden "ear" tufts extend from eyes to back of head; rufous neck and flanks. *Nonbreeding:* black cap and upperparts; white cheeks, foreneck and underparts; no "ear" tufts. *In flight:* wings beat constantly; large, white patch at rear of inner wing.

Size: *L* 30–40 cm; *W* 60 cm.

Habitat: *Breeding:* shallow, well-vegetated wetlands with some open water. *Winter:* inshore bays, harbours, coves and estuaries on the coast; lakes in the interior.

Nesting: singly or in a small, loose colony; floating platform nest of aquatic plants and rotting vegetation is anchored to thick emergent vegetation; pair incubates 4–7 bluish white, often nest-stained eggs for 23–24 days.

Feeding: makes shallow dives and gleans the water's surface for aquatic invertebrates, crustaceans, molluscs, small fish and amphibians; eats primarily small fish in winter.

Voice: during courtship, utters a loud series of croaks and shrieking; shrill *kowee* alarm call; also a sharp *keark keark*.

Similar Species: *Red-necked Grebe* (p. 106): larger; breeding bird has white throat and cheek patches; nonbreeding bird has long, greyish neck and yellowish bill. *Eared Grebe* (p. 107): more crested head; breeding bird has black head and neck; nonbreeding bird has dark crown and "dirty" neck.

RED-NECKED GREBE

Podiceps grisegena

nonbreeding

Nesting Red-necked Grebes are usually noticed because of their frequent, raucous, braying calls, which end in a horselike whinny. • These grebes are highly territorial and aggressive, commonly threatening or making underwater attack dives against other waterbirds that enter their breeding territory. • Like most grebes, Red-necked pairs perform a variety of courtship rituals, many of which involve holding bits of nesting material and simultaneously paddling upright with bills held in the air. These are infrequently observed, however, and are not nearly as showy as the displays of their Western Grebe cousins. • Nests are usually floating platforms of decaying vegetation anchored to emergent vegetation, stumps or a partially submerged log. Incubating Red-necked adults habitually cover their clutch with aquatic plants and quietly slip off the nest, diving to safety at the first sign of approaching danger. • Young Red-necked Grebes ride on their parents' backs for much of the first week or two after hatching. Although the striped young will occasionally remain tucked under the parents' wing feathers when the adults submerge to escape danger, most young eventually pop up to the surface during extended dives.

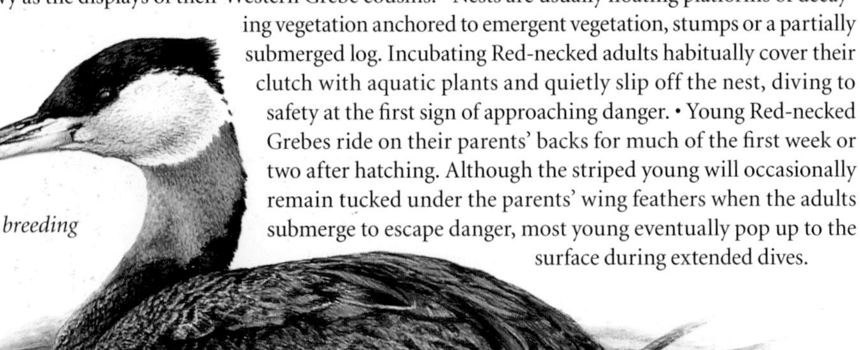

breeding

ID: sexes similar; medium-sized diving bird; brownish grey upperparts; white underparts. *Breeding:* black crown; white cheek; thick, yellow-and-black bill; rufous neck. *Nonbreeding:* dingy brown cheek and neck; dusky yellow bill. *In flight:* 2 white upperwing patches; feet extend beyond tail.

Size: *L* 43–56 cm; *W* 76–84 cm.

Habitat: *Breeding:* small and large lakes, sloughs, marshes, ponds and slow-moving rivers with zones of emergent vegetation. *Winter:* protected bays, inlets, coves, estuaries and harbours along the coast; large lakes in the interior.

Nesting: singly or in a small, loose colony; floating platform nest of aquatic plants and rotting vegetation is anchored to thick emergent vegetation; pair incubates 3–6 light blue to chalky white, often nest-stained eggs for 25–35 days.

Feeding: dives and gleans the water's surface for small fish in winter; eats small fish, aquatic invertebrates, crayfish, leeches and amphibians in summer.

Voice: silent in winter; often repeated, braying *ah-ooo ah-ooo ah-ooo ah-ah-ah-ah-ah* in summer.

Similar Species: *Horned Grebe* (p. 105): smaller; shorter, darker bill with white tip; golden "ear" tufts; rufous flanks in breeding plumage; white lower face and foreneck in nonbreeding plumage. *Western Grebe* (p. 108) and *Clark's Grebe* (p. 109): larger; black-and-white plumage; slender, bright yellow or yellowish orange bill.

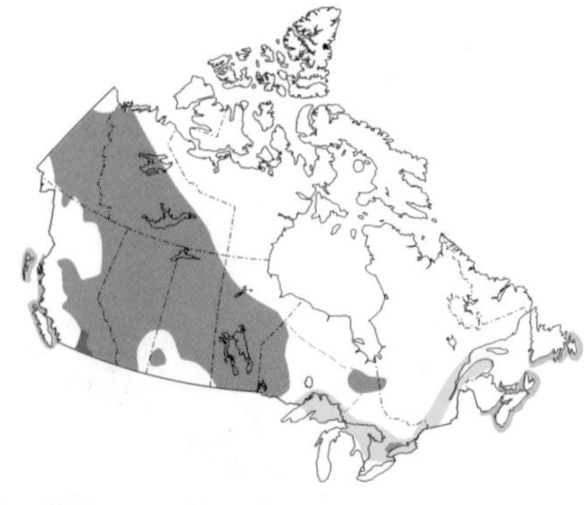

EARED GREBE

Podiceps nigricollis

nonbreeding

The Eared Grebe nests in colonies in western regions of the continent that can number thousands of birds, and it is the most abundant grebe in North America and worldwide. Although Eared Grebe breeding colonies seem to shift unpredictably in location from one year to the next, its behaviour and staging locations during the nonbreeding season are much more predictable. Immediately after the breeding season, most move to highly saline environments such as Mono Lake, California, or Great Salt Lake, Utah, to exploit superabundant crops of brine shrimp and alkali flies in preparation for nonstop migration flights to wintering grounds in the southwestern United States and Mexico. • On staging lakes, adults double in mass and their pectoral muscles atrophy to a flightless condition. These changes in size and proportions, extreme changes in morphology, are unparalleled in the bird world. The changes are reversed in preparation for a nonstop flight to the next staging destination. This cycle may be repeated three to six times annually, making the Eared Grebe essentially flightless for nine to ten months each year.

breeding

ID: sexes similar; small diving bird; peaked head; red eyes; thin bill; black back. *Breeding:* black face and forehead; golden-plumed "ear" tufts; dark bill; rufous flanks; thin, black neck. *Nonbreeding:* white behind ear and on "chin" and foreneck; grey bill; grey-mottled, white flanks; generally dusky neck. *In flight:* white hindwing patch; feet extend beyond tail.

Size: *L* 30–36 cm; *W* 54–58 cm.

Habitat: *Breeding:* freshwater or slightly alkaline shallow lakes and wetlands with floating or emergent vegetation. *Winter:* coastal bays, harbours, estuaries, lagoons and ice-free lakes.

Nesting: strongly colonial; pair builds a shallow, flimsy platform of floating wet and decaying plants on or among thick vegetation in a shallow lake, pond or marsh; pair incubates 4–7 bluish white, often nest-stained eggs for 20–23 days.

Feeding: makes shallow dives and gleans the water's surface for aquatic invertebrates, especially crustaceans, and tiny fish; brine shrimp and alkali flies are important in staging and wintering areas.

Voice: usually silent away from breeding sites; loud, chirping *kowee* threat call; mellow *poo-eee-chk* during courtship.

Similar Species: *Horned Grebe* (p. 105): more rounded head; straighter, chunkier bill with white tip; rufous neck and denser golden "ear" tufts in breeding plumage; solidly white cheek and lighter foreneck in nonbreeding plumage. *Pied-billed Grebe* (p. 104): browner plumage; dark eyes; thicker, shorter bill.

WESTERN GREBE

Aechmophorus occidentalis

Western Grebes are perhaps best known for their elaborate and highly ritualized courtship displays. During the "weed dance," the male and female swim with their torsos and heads held high, caressing each other while holding aquatic vegetation in their bills. The "rushing" display, which Western Grebes are most famous for, involves two or more individuals exploding into a paddling sprint side by side across the water's surface. The grebes stand high, feet paddling furiously, with their wings stretched back and heads and necks held rigid, until the race ends with the birds breaking the water's surface in a graceful, headfirst dive. • Both Western Grebes and Clark's Grebes possess a mechanism in the neck that permits them to thrust the head forward like a spear. This is unique among grebes but well known in herons and anhingas. • Like most grebes, Western Grebe eggs hatch at regular intervals, but unlike other grebes, young Western Grebes are plain, not striped. • Unlike its Red-necked relative, the Western Grebe seems unable to adjust to the disturbances created by cottagers and recreationalists. Many lakes that once supported nesting colonies have been turned into vacation resorts. • Western Grebes winter primarily along the Pacific Coast from Mexico north to southern BC, where flocks of many thousands can be found in some inlets and bays. Small numbers are regularly observed in eastern North America.

ID: sexes similar, but males slightly larger; large diving bird; long neck; long, thin, yellowish bill; dark crown extends below red eye; dark upperparts sharply separated from white underparts. *In flight:* long, pointed body and wings.
Size: *L* 51–61 cm; *W* 76–102 cm.
Habitat: *Breeding:* protected areas in medium-sized freshwater lakes with dense areas of emergent shore vegetation or thick mats of floating aquatic plants. *Winter:* protected inlets, bays, harbours and channels on the coast; mainly lakes and backwaters of rivers in the interior.
Nesting: colonial; pair builds a floating nest of wet vegetation on or among aquatic vegetation; pair incubates 2–6 pale bluish white, often nest-stained eggs for about 24 days.
Feeding: mainly fish year-round, which are captured or speared during dives; also eats insects, molluscs and crustaceans.
Voice: often heard at both breeding and wintering sites; high-pitched, 2-note *crreeet-crreeet* call sounds like a squeaky wheel.
Similar Species: *Clark's Grebe* (p. 109): white surrounds eye; orange-yellow bill; flanks have more white areas; lighter grey back; 1-note call. *Red-necked Grebe* (p. 106): shorter neck; thicker bill; darker sides and neck in nonbreeding plumage.

CLARK'S GREBE

Aechmophorus clarkii

Ornithologists once believed that Western Grebes came in two varieties: those with duller yellow bills and red eyes on a black background, and those with bright yellowish orange bills and red eyes clearly visible on a white face below the black crown. What they eventually realized—something the grebes had apparently known all along—was that there are two separate species. Studies subsequently revealed that the two colour morphs tended to mate with birds of similar coloration, and that the darker-faced birds gave a two-note advertising call, as opposed to the single-note call of the lighter-faced morph. We now know the less-widespread, paler-faced birds as Clark's Grebes, but, particularly in winter, identification remains an intriguing challenge. The two species are often found together, but for the most part they do not interbreed, even when both share a breeding site and very similar courtship rituals. • Clark's Grebe was named in honour of John Henry Clark, a mathematician, surveyor and successful bird collector who procured the first scientific specimen in 1858.

ID: sexes similar; large diving bird; long neck; long, daggerlike, yellowish orange bill; white on face surrounds red eyes; black crown and upperparts; slender black stripe on nape; white underparts. *Male:* slightly larger; longer bill. *In flight:* long, pointed body and wings; conspicuous, white wing stripe.

Size: *L* 52–58 cm; *W* 79–97 cm.

Habitat: *Breeding:* large lakes with dense areas of emergent vegetation or thick mats of floating aquatic plants. *Winter:* sizeable water bodies, including sluggish rivers and lagoons; open coast up to 1.5 km offshore.

Nesting: colonial; floating nest of wet or decaying vegetation is anchored to submerged plants; pair incubates 2–6 bluish white eggs for about 24 days.

Feeding: dives for small fish, invertebrates and other small aquatic or marine prey.

Voice: 1-note *krreek* advertising call.

Similar Species: *Western Grebe* (p. 108): black on face extends below eye; duller yellow to yellowish green bill; broader, black nape stripe; lower-pitched, 2-note call. *Red-necked Grebe* (p. 106): shorter neck; red neck and white cheek in breeding plumage; greyish white cheek and darker, greyer sides and neck in nonbreeding plumage.

BLACK-FOOTED ALBATROSS

Phoebastria nigripes

Albatrosses belong to one of the most pelagic families of birds in the world, the "tubenoses," or Procellariiformes, a group that includes albatrosses, fulmars, petrels, shearwaters and storm-petrels. These birds spend most of their lives at sea, using ocean breezes and updrafts from waves to soar and fly, and they forage by plucking food items from the water's surface. • Although Black-footed Albatrosses visit offshore British Columbia waters year-round, they are mostly within sight of land between May and August. To see one, birders have to travel by boat at least 15 kilometres off the outer West Coast to deep waters. • Most Black-footed Albatrosses breed in the Hawaiian Islands from November to June. Small numbers of immatures remain at sea all year. • Albatrosses and their relatives locate food by smell and are often attracted to fishing vessels with offal. In calm conditions or when loafing, albatrosses will gather together on the water in flocks of hundreds of birds.

ID: sexes similar; very large; ashy brown; plumage becomes paler with age; white on face at base of bill and under eye; heavy, dark bill, darkest at tip; undertail coverts may be white; white crescent at base of tail. *In flight:* long, narrow wings; black feet extend beyond tail.

Size: *L* 69–74 cm; *W* 1.9–2.1 m.

Habitat: open ocean, especially over areas of upwelling and the continental shelf.

Nesting: does not nest in Canada.

Feeding: snatches mainly fish, squids and crustaceans from the water's surface; also scavenges and follows fishing vessels for discarded fish and their remains.

Voice: generally silent; sometimes groans, shrieks or squawks, especially in feeding flocks.

Similar Species: *Shearwaters* (pp. 112–17): much smaller but can be confused at a distance. *Laysan Albatross* (p. 498): very rare vagrant; white head, neck and underparts; pale yellow bill.

NORTHERN FULMAR

Fulmarus glacialis

Northern Fulmars are the only midsized tubenoses that spend all their lives north of the equator. Regularly found well out to sea, Northern Fulmars follow commercial fishing vessels and endlessly search the ocean's surface for food. • Relatively slim-winged, but thick-necked and bull-headed, the Northern Fulmar shares many physical attributes, except for its stubby, pale greenish yellow bill and colour morphs that range from all-dark to all-pale, with the closely related shearwaters. Dark morphs predominate along the BC coast, whereas light morphs are more common in the east. • Not only are fulmars among the longest-lived birds—some have been known to breed for 40 years or more—but they do not breed until they are eight to ten years of age. • "Fulmar" is derived from the Old Norse words meaning "foul gull," a derivation obvious to anyone who has approached one of these birds too closely and received a shot of foul-smelling fish oil.

dark adult

light adult

ID: sexes similar; short, tubed bill; dark eyes; pale greenish yellow to slightly orangey bill; thick neck; stubby, rounded tail. *Dark adult:* blue-tinged, deep brownish grey overall; paler flight feathers. *Light adult:* white overall; patchy, pale bluish grey mantle. *In flight:* long, pointed wings with quick wingbeats and stiff-winged glides.

Size: *L* 44–51 cm; *W* 1.1 m.

Habitat: *Breeding:* steep cliffs on islands with rocky headlands. *Nonbreeding:* open ocean in colder waters, often over upwellings and along the continental shelf.

Nesting: colonial; on a cliff ledge or rocky outcrop; single white egg is laid on bare ground and incubated by both parents for 55–57 days.

Feeding: gleans or makes shallow dives for crustaceans, small fish, squid and jellyfish; cannot feed in flight.

Voice: generally silent; possible low, quacking call when competing for food.

Similar Species: *Shearwaters* (pp. 112–17): slimmer heads and necks; more slender bills. *Black-footed Albatross* (p. 110): much larger; longer, heavier bill; more leisurely, less-flapping flight. *Medium-sized gulls* (pp. 218–22): slimmer bodies and necks; bills not tubed.

PINK-FOOTED SHEARWATER

Puffinus creatopus

Shearwaters are long-winged seabirds that fly with shallow, rapid wingbeats and stiff-winged glides. • Outside the breeding season, Pink-footed Shearwaters are birds of the open ocean, generally seen from land only infrequently. Among the swarms of other shearwaters encountered at sea from April to November, the Pink-footed Shearwater is second only to the Sooty Shearwater as the most common shearwater off Canada's Pacific Coast. Although they tend to stay well offshore, small numbers of Pink-footed Shearwaters may be seen from strategic coastal locations with the aid of a telescope. • Once their nesting duties are complete, Pink-foots migrate north along the continental shelf in search of large schools of fish and squid. They prefer to forage along the edge of the continental shelf where upwellings of cold water occur and marine life is plentiful. • As is true with other members of the shearwater family, Pink-foots assemble with astonishing quickness at any food source. They will closely approach any boat if food of any description is tossed overboard. They regular dive up to three metres below the surface to catch prey, and have even been known to dive to depths of 25 metres in the pursuit of food.

ID: slender, pinkish bill with black, hooked tip; pink legs; dark brown upperparts; white underparts with dark mottling on sides, wing linings and tip of undertail coverts.
Size: *L* 48–50 cm; *W* 1.0–1.1 m.
Habitat: open ocean, usually well out over continental shelf; uncommon within several miles of shore.
Nesting: does not nest in Canada.
Feeding: plunges into water or makes shallow surface dives; swims underwater over short distances; fish, squid and crustaceans probably form bulk of diet.
Voice: generally silent; birds competing for food may make quarrelsome noises.
Similar Species: *Northern Fulmar* (p. 111): stouter body; light adult has stout, greenish bill and unmarked white undersides and head. *Buller's Shearwater* (p. 114): all-black bill; grey-and-black upperwing pattern; bright white underparts, especially narrow-bordered underwings. *Black-vented Shearwater:* smaller; smaller, all-dark bill is; dark undertail coverts; more rapid wingstrokes. *Flesh-footed Shearwater* (p. 498) and *Sooty Shearwater* (p. 115): entirely dark bodies.

GREATER SHEARWATER

Puffinus gravis

Many fishermen are familiar with these gregarious pirates and refer to them as "Hags" or "Bauks." Fishing boats are often mobbed by Greater Shearwaters fighting over the opportunity to grab a free meal. These birds sometimes forage in association with whales and dolphins. • Greater Shearwaters migrate to the North Atlantic for the summer months, but they do not breed here. In fact, considering their abundance here, it is surprising that they breed in such a small area: Gough Island and islands in the Tristan da Cunha group in the South Atlantic. • Greater Shearwaters are most common over the outer part of the continental shelf—they tend to avoid nearshore and mid-ocean areas. • During migration, Greater Shearwaters follow a clockwise migration around the Atlantic Ocean, first passing South America, then North America and then Europe before returning to their breeding islands. However, when capelin and other small fish move to beaches to spawn, Greater Shearwaters will follow them. At such times, and in very calm or foggy weather, huge flocks may appear off headlands, especially in Newfoundland. • In dead calm conditions, these birds can only achieve takeoff with great difficulty, and they are effectively grounded. • Greater Shearwaters eject noxious oil from their nostrils as a form of self-defence.

ID: brown upperparts; white underparts, except for brown belly, brown undertail coverts and incomplete brown collar; usually 1 narrow, white band at base of tail. *In flight:* straight, narrow, pointed wings; dark "wing pits" and wing linings; dark trailing edge of wings; some moulting birds have jaegerlike, white wing patch.
Size: *L* 48 cm; *W* 1.1 m.
Habitat: open ocean; favours cold waters; most common over the outer portion of the continental shelf; drawn inshore by weather conditions and prey abundance.

Nesting: does not nest in Canada.
Feeding: seizes prey from the water's surface or dives to depths of 10 m; eats mainly small fish and squid but may also take crustaceans.
Voice: bleating, lamblike *waaan.*
Similar Species: *Cory's Shearwater* (p. 498): yellow bill; lacks brown collar, belly and undertail coverts. *Manx Shearwater* (p. 117): darker upperparts; white belly and undertail coverts. *Albatrosses:* much larger; few, very deliberate wingbeats. *Jaegers* (pp. 228–40): more prominent wing flashes; longer tails; central tail feathers project in flight.

BULLER'S SHEARWATER

Puffinus bulleri

With a clear "W" pattern on its back, the Buller's Shearwater declares its home turf—the water. This large bird glides effortlessly over the ocean between New Zealand and Canada's Pacific Coast. This shearwater nests in New Zealand and appears in ocean waters offshore of British Columbia from August to October. It can be found along the coast from Alaska southward, often in association with other shearwater species, including Sooty Shearwaters and Pink-footed Shearwaters. • Shearwaters prefer to congregate offshore along the edges of continental shelves. The area near the entrance to the Juan de Fuca Strait is known to attract large numbers of these birds, which come to forage on a variety of prey from small fish to krill and jellyfish. • In flight, Buller's Shearwaters prefer long glides interspersed with only short periods of wing flapping. • This bird's population on the Poor Knights Islands off the northeast coast of New Zealand, one of their main breeding areas, has recovered from near extinction since feral pigs were eradicated. • Shearwaters were given the genus name *Puffinus* in the 15th century when the word was freely used to describe pelagic seabirds such as shearwaters, puffins and razorbills. • Formerly known as "New Zealand Shearwater," Buller's Shearwater was named for Sir Walter Lowry Buller, one of New Zealand's foremost ornithologists.

ID: black bill and cap; gray back with black "W"; bright white underparts; white underwings outlined with black; dark wing and tail tips.
Size: *L* 41–46 cm; *W* 91–100 cm.
Habitat: open ocean; concentrates at upwellings and current edges along the cooler continental shelf.
Nesting: does not nest in Canada.
Feeding: hovers to dip bill in water; submerges head while swimming; small fish, squid and crustaceans.
Voice: silent at sea.
Similar Species: *Pink-footed Shearwater* (p. 112): pink base to bill; pink legs; brown-and-grey upperparts; white underparts with dark smudges; thicker dark outline to underwings; no distinct back pattern.

SOOTY SHEARWATER

Puffinus griseus

Each summer, Sooty Shearwaters come from breeding islands in the Southern Hemisphere in numbers sometimes beyond estimation. During spring and fall, scattered individuals are almost constantly in sight on the open ocean, and concentrations may include tens of thousands of birds. • Various seabird families, including albatrosses, shearwaters and storm-petrels, are categorized widely as "tubenoses"—they all have external tubular nostrils, hooked bills and come ashore only to breed. • The Sooty Shearwater is most abundant along Canadian coastlines from May through October, though a telescope may be necessary to see them because most birds forage well beyond the surf line. Large concentrations of fish can draw these birds closer to land—the popularity of schooling fish often results in multispecies feeding frenzies that include Sooties and other seabirds. Sooties dominate most mixed-seabird foraging flocks over the continental shelf, with scarcer species occurring in their midst. To recognize the less common shearwaters at a distance with assurance, one must first get to know the Sooty. • Currently, Sooty Shearwaters have an estimated global population of 20 million birds. However, research in their breeding colonies has shown this species to be experiencing a persistent decline. • Resident deep-sea fishermen refer to the Sooty Shearwater as "Black Hag."

ID: medium-sized; sooty brown overall; slightly paler underparts; greyer neck and throat; long, slender, black, tubed bill; blackish, rounded tail; black to grey legs and feet. *In flight:* possible silvery flash in underwings; feet extend just beyond tail; strong, direct flight with several deep flaps followed by long glide.

Size: *L* 41–46 cm; *W* 1.0 m.

Habitat: open ocean; concentrates at upwellings and current edges along the cooler continental shelf.

Nesting: does not nest in Canada.

Feeding: gleans the water's surface or snatches prey underwater in shallow dives or plunges; eats mostly fish, squid, crustaceans and jellyfish; gathers in large flocks around commercial fishing vessels.

Voice: usually silent; occasionally utters quarrelsome calls when competing for food.

Similar Species: *Northern Fulmar* (p. 111): dark adult has stocky, thick neck with short, thick, yellowish bill. *Short-tailed Shearwater* (p. 116): slightly smaller; sometimes has pale "chin" and throat; smaller bill; more uniformly coloured underwings. *Flesh-footed Shearwater* (p. 498): black-tipped, pinkish bill; pale pinkish legs and feet; uniformly dark underwings; slower wingbeats.

SHORT-TAILED SHEARWATER

Puffinus tenuirostris

Cruising the pelagic zone with quick wingbeats and short, efficient glides, flocks of Short-tailed Shearwaters are uncommon off the Pacific Coast from fall through early spring. These steel-coloured birds journey to our shores from nesting grounds in Australia. Most are found off the shores of Alaska and the Bering Strait during the summer but move as far south as California in the winter months. • While young birds remain along our coastline part of the year, adults return to Australia to breed, beginning the traverse over the South Pacific's tropical waters in the fall. • This low-flying group of birds has earned the name "shearwater" because, as a bird banks left or right, its lower wing tip appears to slit the surface of the water.

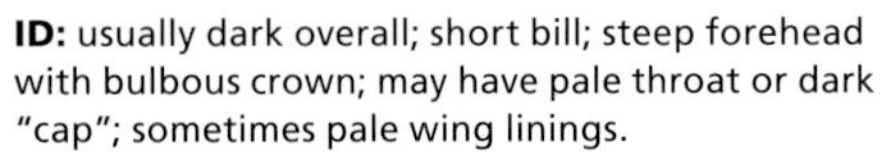

ID: usually dark overall; short bill; steep forehead with bulbous crown; may have pale throat or dark "cap"; sometimes pale wing linings.
Size: *L* 43 cm; *W* 1.0 m.
Habitat: open ocean; continental shelf over cool waters and at upwellings.
Nesting: does not nest in Canada.
Feeding: dives or drops feet first into water; may dive to depths of 18 m; commonly eats fish, crustaceans, squid or octopus; sometimes takes jellyfish, aquatic worms or insects; occasionally forages with whale pods or dolphins.
Voice: generally silent at sea.
Similar Species: *Sooty Shearwater* (p. 115): more prominent, pale wing linings; longer bill; crown not as rounded.

MANX SHEARWATER

Puffinus puffinus

You will likely have to take a pelagic ferry or visit offshore islands to observe this open-ocean bird. Its distribution is mostly European, but there are nesting colonies on islands just off the coast of Newfoundland. Most birds seen in the western Atlantic, including the birds that breed in Atlantic Canada, originate from nesting colonies off the Welsh coast, but there are a few sightings of browner birds that come from the Mediterranean Sea. • Breeding birds dig out burrows, which they may use a few years in a row, into which they lay a single egg per season. Parents may stray as far as 970 kilometres from their nests in search of food, and they feed their nestling by regurgitation. • Large gatherings of shearwaters in Atlantic Canada almost always contain a few Manx Shearwaters. Predominantly black and white in appearance, Manx Shearwaters are sometimes confused with jaegers when they sit on the ocean with their tails up. However, shearwaters dive more frequently and more deeply than jaegers do. • In autumn, these birds leave the Northern Hemisphere to winter off the coast of South America. • Manx Shearwaters have been recorded to have a life span of at least 55 years in the wild.

ID: dark upperparts; white underparts; white wing linings; white on throat extends to white crescent behind eye; long, slender bill with hooked tip.
In flight: white underparts with dark wing border; white undertail coverts; stiff-winged, flap-and-glide flight pattern.
Size: *L* 30–38 cm; *W* 76–89 cm.
Habitat: open ocean, sometimes off headlands and rarely off beaches.
Nesting: loosely colonial; on a cliff or in a rock crevice on a remote island or headland; pair excavates a burrow 0.5–3 m long; nest chamber may be lined with scraps of vegetation, feathers and hair; pair incubates 1 white egg for 47–55 days.

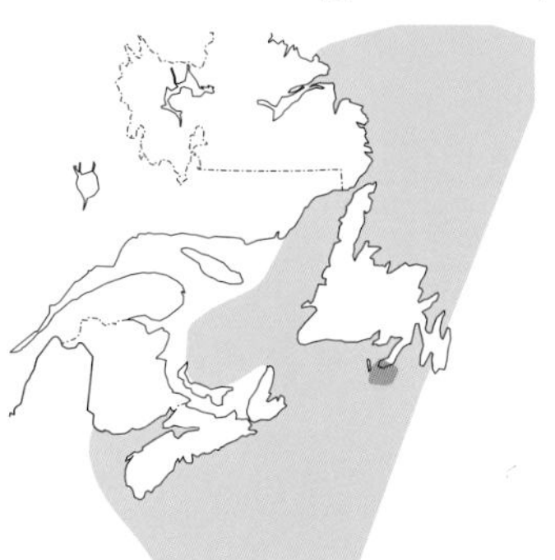

Feeding: feeds on the water's surface while swimming or dives after prey, sometimes from low flight; takes mostly small fish but may also take squid and crustaceans.
Voice: range of crows, coos, croons, howls and screams repeated rapidly; raucous, often piercing or gruff *cack-cack-cack-carr-hoo.*
Similar Species: *Greater Shearwater* (p. 113): larger; browner upperparts; brown belly and undertail coverts. *Cory's Shearwater* (p. 498): larger; browner upperparts; yellow bill. *Jaegers* (pp. 238–40): longer tails; central tail feathers project in flight.

WILSON'S STORM-PETREL

Oceanites oceanicus

This long-legged storm-petrel dances over the surface of the sea, feet daintily pattering as it hovers and forages for small crustaceans and fish. Like many marine birds, it also regularly follows fishing boats, grabbing whatever scraps of food it can glean. You are most likely to see Wilson's Storm-Petrels along the outer portion of the continental shelf, although they follow boats and ferries more readily than other storm-petrels. • Wilson's Storm-Petrel can be distinguished from Leach's Storm-Petrel by its more extensive white rump patch and more swallowlike flight. At close range, its yellow-webbed feet may be seen as they "patter" on the water's surface, presumably to stir up food. • These long-distance migrants travel from Antarctica to the Grand Banks off Newfoundland and back again each year. They nest on islands and cliffs in the Antarctic region and around southern South America. • Despite their small size, storm-petrels may live to be more than 20 years old. • The name "petrel" is derived from St. Peter, and the birds are so named because of their seeming ability to walk on water.

ID: small, dark, tubed bill; blackish brown overall; white rump and undertail coverts; square tail; black legs; yellow toes. *In flight:* flickering, somewhat erratic flight with stiff, shallow wingbeats; toes extend beyond tail.

Size: *L* 17–18 cm; *W* 40–43 cm.

Habitat: open ocean, especially over the continental shelf; occasionally wanders close to land along outer coasts.

Nesting: does not nest in Canada.

Feeding: hovers and skims over the water with its feet "pattering" the surface, grabbing food items from the water's surface; takes shrimp and other crustaceans, small fish, squid and marine worms.

Voice: usually silent; feeding congregations utter soft peeping or chittering calls.

Similar Species: *Leach's Storm-Petrel* (p. 120): divided white rump patch, less extensive than Wilson's Storm-Petrel's; longer wings; slightly forked tail; lacks white undertail coverts; erratic flight. *Shearwaters* (pp. 112–17): much larger; glide more and do not have fluttery flight of the much smaller storm-petrels.

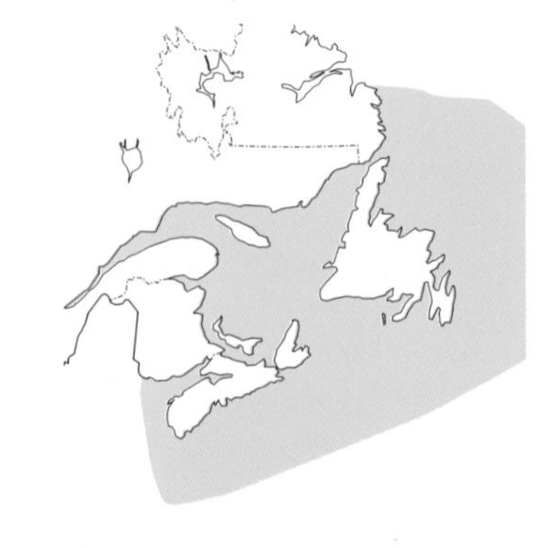

FORK-TAILED STORM-PETREL

Oceanodroma furcata

The silver-tipped wings and deeply forked tail of this small marine bird are good field marks as it dances just above the ocean's surface searching for food. • A pair of Fork-tails will place their single egg at the end of an underground burrow, then the adults maintain a strictly nocturnal schedule involving shift changes and taking turns to feed the nestling out of sight of avian predators. During the day, adults spread out over the open ocean, sometimes travelling hundreds of kilometres in search of food. • Although they sometimes fly alongside ships, Fork-tails are usually attracted only by fishing vessels discarding fish oil and small waste items. • These storm-petrels can appear inshore (but very rarely inland), closely approaching beaches, jetties and piers, especially around fish-processing plants.

ID: sexes similar; small; short, tubed bill; deeply forked tail; dark eye patch; pale bluish grey upperparts, darker at wing tips; pale grey underparts. *In flight:* deeply forked tail; flutters low across the ocean's surface with rapid, shallow wingbeats interspersed with brief glides.

Size: *L* 20–22 cm; *W* 46 cm.

Habitat: *Breeding:* offshore islands with dense shrub and grass cover with soils for burrowing, often wooded; usually intermixed with Leach's Storm-Petrels. *Foraging:* cold, open ocean waters from near shore and beyond; occasionally visits bays and estuaries.

Nesting: colonial; in a self-excavated burrow in firm soil or under drift logs on land; pair incubates 1 white egg for up to 68 days depending on food availability.

Feeding: mostly snatches small fishes, crustaceans and floating animal oils from the surface while hovering.

Voice: usually silent away from breeding sites; low-pitched, trilling calls at the nest.

Similar Species: *Leach's Storm-Petrel* (p. 120): more pelagic; all-dark body with white rump.

LEACH'S STORM-PETREL

Oceanodroma leucorhoa

The Leach's Storm-Petrel is one of the more widespread and abundant storm-petrels in the world, yet it may be the least likely to be seen, and many aspects of its life still remain unknown. • Appearing longer-winged than most storm-petrels, Leach's Storm-Petrel has a buoyant and graceful, weaving flight that incorporates deep wingbeats and very short glides on bowed wings. The subspecies found in the North Atlantic can be distinguished from other storm-petrels by its erratic, fluttering flight. • These birds nest in underground burrows, and parents visit their nestbound young only at night to avoid harassment and predation by marauding diurnal gulls. During the day, these storm-petrels forage far out to sea. They tend to be solitary when feeding, choosing not to follow fishing vessels and rarely pattering or "walking" on the water's surface. • Occasionally, prolonged storms force these birds ashore, where their inability to take off from level ground renders them helpless. • Because of their tiny size and erratic flight, storm-petrels are sometimes called "Sea Swallows." On the Atlantic Coast, all storm-petrels are known locally as "Mother Carey's Chickens," probably a reference to the Virgin Mary (*Mater cara* in Latin).

ID: small; deeply forked tail; dark brown head and underparts; slender, black, tubed bill; whitish rump with grey centre line; black legs and feet. *In flight:* narrow, pointed wings with sharp bend at "wrist"; erratic, zigzagging, bounding flight with alternating flaps and glides; conspicuous lighter diagonal band separates brown and blackish parts of upperwings; long, notched tail.

Size: *L* 49–58 cm; *W* 48 cm.

Habitat: *Breeding:* wooded and shrub-covered islands. *Nonbreeding:* open ocean, well offshore at upwellings and convergence zones between warm and cold waters.

Nesting: strongly colonial; on an island or islet off the outer coast; male uses feet to dig a burrow up to 1 m long in soil under rocks, grass or tree roots; nest chamber is lined with vegetation; pair incubates 1 white egg for 37–50 days depending on food resources.

Feeding: locates food by smell; skims or snatches small fish, squid and crustaceans and floating mammal oils (from carcasses) from the water's surface while hovering; rarely feeds while swimming.

Voice: usually silent; commonly gives purring, chattering and trilling nocturnal notes at the nest site.

Similar Species: *Fork-tailed Storm-Petrel* (p. 119): paler bluish grey and grey; pale flight feathers; dark wing linings; more direct flight. *Wilson's Storm-Petrel* (p. 118): smaller; white, U-shaped rump patch and undertail coverts; yellow feet extend beyond tail; more direct flight pattern. *Sooty* (p. 115), *Short-tailed* (p. 116) and *Flesh-footed shearwaters:* rare; much larger; all-dark rumps; strong, direct flight with short flaps and longer glides.

NORTHERN GANNET

Morus bassanus

Northern Gannets spend most of the year feeding and roosting at sea. Only during the brief summer breeding season do they seek the stability of land to lay eggs and raise young in large sea-cliff colonies. The North American population breeds at only six well-known sites in Newfoundland and the Gulf of St Lawrence. • Unfortunately, the Northern Gannet's spectacular feeding behaviour is often demonstrated far from land. Squadrons of gannets soaring at heights of more than 30 metres above the ocean's surface will suddenly fold their wings back and simultaneously plunge headfirst into the ocean depths in pursuit of schooling fish. The gannet's reinforced skull has evolved to cushion the brain from these diving impacts. • Like most seabirds, Northern Gannets take several years to attain adult plumage, going through various mottled, dark-and-white plumages before becoming pure white with black wing tips at four to five years of age. • At maturity, male gannets begin the long process of acquiring a breeding site in the colony and then attracting a mate. These birds often mate for life, reestablishing pair bonds each year, usually at the same nest site. • The name "gannet" is derived from the Anglo-Saxon word *ganot,* meaning "little goose." This bird was once classified with the geese and is still known as "Solan Goose" in Europe.

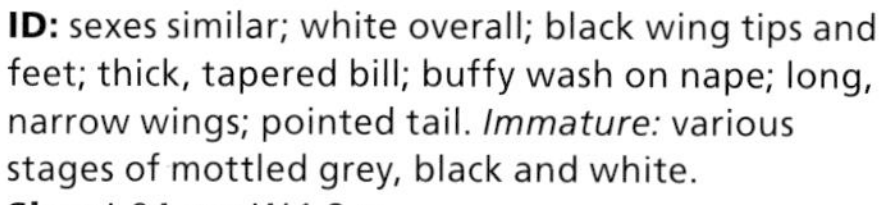

ID: sexes similar; white overall; black wing tips and feet; thick, tapered bill; buffy wash on nape; long, narrow wings; pointed tail. *Immature:* various stages of mottled grey, black and white.
Size: *L* 94 cm; *W* 1.8 m.
Habitat: roosts and feeds in coastal and open ocean waters most of the year; often seen well offshore.
Nesting: on a protected mainland or island sea cliff; male builds a tall mound of seaweed, vegetation, dirt and feathers glued together with droppings; pair incubates 1 pale blue to white egg for about 44 days.
Feeding: dives with closed wings to pursue prey; eats fish, including herring, mackerel and sometimes squid; may steal from other birds or scavenge from fishing boats.
Voice: generally silent at sea or at rest; low, throaty braying at nesting colony.
Similar Species: *Snow Goose* (p. 46): shorter, pinkish bill; broader wings; long, extended neck in flight.

AMERICAN WHITE PELICAN

Pelecanus erythrorhynchos

breeding

Most of us associate pelicans with saltwater habitats, but the American White Pelican is a regular colonial nester on islands and islets in large lakes and a common visitor to smaller lakes and rivers in western Canada. It is often seen flying low over the water, its belly almost touching the surface, or circling high above woodlands or prairies. It is estimated that Saskatchewan and Manitoba host more than one-half of the world's nesting American White Pelicans. North American nesting populations have increased three to four percent annually since the 1980s. These increases have created some conflict with the aquaculture industry, especially during migration and in winter. A nesting pelican will disperse great distances from its colony, occasionally travelling hundreds of kilometres to find productive fishing grounds. • This majestic wetland bird is one of only a few species that feeds cooperatively. Coordinated flocks of swimming pelicans will herd schooling fish into shallow water, then dip their bucketlike bills into the water to capture them. In a single scoop, a pelican can trap more than 12 litres of water and fish in its bill, which is about two to three times as much as its stomach can hold.

nonbreeding

ID: sexes similar; huge, stocky, white bird; long, orange bill and throat pouch; black wing tips; short tail. *Breeding:* horny plate develops on upper mandible. *Immature:* dusky bill; brown-tinged, white plumage. *In flight:* all-white body; black wing tips; S-shaped neck.

Size: *L* 1.4–1.8 m; *W* 2.8 m.

Habitat: *Breeding:* isolated, mostly barren islands and islets on undisturbed lakes. *In migration* and *winter:* lakes, slow-moving rivers and large, open marshes in the interior; brackish lagoons, spits and estuaries on the coast.

Nesting: colonial; nest scrape is bare or lined with twigs and rootlets; pair incubates 2–3 chalky white eggs for 30–33 days; first-hatched nestling is always fed first with regurgitated food; generally only 1 young survives to fledging.

Feeding: forages cooperatively; birds swim side by side to "herd" fish into shallows then dip bills into water to scoop up prey; may also take crayfish and salamanders.

Voice: generally quiet; rarely issues piglike grunts at nesting colony.

Similar Species: no other large, white bird has long bill with pouch. *Brown Pelican* (p. 498): coastal; smaller; much darker; duskier bill. *Snow Goose* (p. 46), *Tundra Swan* (p. 53) and *Trumpeter Swan* (p. 52): fly with necks extended; swans have all-white wings.

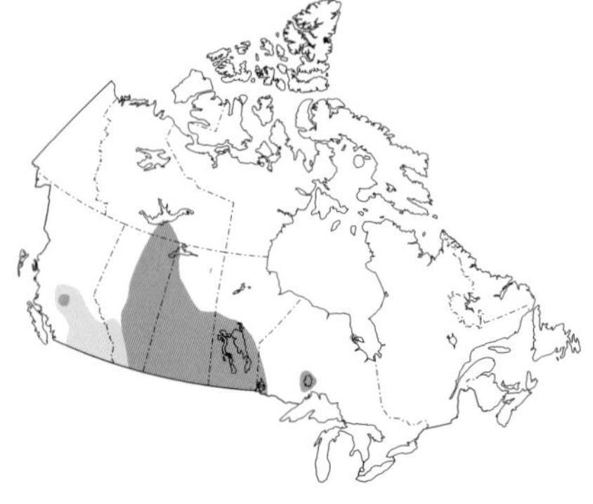

BRANDT'S CORMORANT

Phalacrocorax penicillatus

Cormorants are large, easily identifiable, black seabirds that often perch on buoys, rocky islets, breakwaters and jetties. The Brandt's Cormorant is the largest and stockiest of the four cormorant species found in Canada and is the least vocal. It has a large head and flies with a straight neck low over the water. • These cormorants are gregarious, roosting and feeding together year-round. Food is caught underwater, sometimes to depths of 45 metres. • This bird is a widespread and locally abundant year-round resident in southern BC, especially the Gulf Islands, as well as a common migrant along the entire coast. It breeds on the outer southwest coast of Vancouver Island and in the Juan de Fuca Strait. The first confirmed breeding records in BC date back to 1965; there are rarely more than two colonies active in any given year, with about 100 pairs nesting annually in the province. • Brandt's nesting colonies are often isolated from each other, with breeding adults scattering widely to feed anywhere from within the surf zone to several miles offshore. • Although common, with a worldwide population of about 75,000 pairs, Brandt's Cormorants are threatened by commercial fishing, pollutants and disturbance associated with recreational use of West Coast marine environments.

breeding

ID: sexes similar; dark bluish green eyes; heavy, hooked, blackish brown bill; buff brown throat patch. *Breeding:* green-glossed, blackish plumage; bright blue throat patch; white plumes on neck and upper back. *Nonbreeding:* duller, greyer plumage; less glossy head and rump; light-coloured cheek patches. *In flight:* straight neck.
Size: *L* 81–89 cm; *W* 1.25 m.
Habitat: *Breeding:* bare, rocky islands with gentle slopes. *Foraging:* open ocean and off rocky shorelines. *Roosting:* offshore reefs, rocky headlands, islands and breakwaters.
Nesting: colonial; nest is a sizeable mound of surfgrasses that becomes covered with guano as incubation progresses; pair incubates 3–4 pale bluish white, often nest-stained eggs for 29–31 days.
Feeding: dives underwater for fish and less often for shrimp and small crabs.
Voice: deep, low grunts and croaks at nesting colony.

Similar Species: *Double-crested Cormorant* (p. 124): yellowish to orangey throat region; longer wings and tail; flies with neck bent; more vocal. *Pelagic Cormorant* (p. 126): smaller; slender bill; small, red throat patch and white flank patches in breeding plumage.

DOUBLE-CRESTED CORMORANT

Phalacrocorax auritus

Unlike other cormorant species, the Double-crested Cormorant is equally at home on the coast or inland. This bird is easy to recognize by its thick neck, large head and yellowish throat patch in breeding plumage. • Cormorants can outswim fish, which they capture in underwater dives. Most waterbirds have waterproof feathers, but the structure of a cormorant's feathers allows water in. "Wettable" feathers make this bird less buoyant but a better diver. It has sealed nostrils for diving and therefore must occasionally open its bill while in flight. • Like other cormorants, resting Double-cresteds are often observed holding their wings spread out, which is thought to aid in drying wet flight feathers. They are, however, the only cormorants that carry out this behaviour while perched in trees. • Huge increases in Double-crested Cormorant numbers throughout much of their range since the 1970s have been attributed to reduced pesticide use, altered fish communities and enhanced overwintering survival of adults and young. These increases have resulted in local persecution by commercial fishers and renewed calls to control cormorant numbers. Detailed studies of their diet in several parts of their range, however, reveals that most of the cormorants' prey consists of undesirable, non-commercial fish species.

1st-year juvenile

breeding western race

breeding eastern race

ID: sexes similar; all-black plumage; stocky neck; orange throat patch; dark bill, hooked at tip; blue eyes. *Breeding:* more intense throat patch colour; fine, whitish eyebrow plumes. *Immature:* brown upperparts; pale throat and breast; yellowish throat patch. *In flight:* strong, direct flight; rapid wingbeats; kinked neck.

Size: *L* 66–81 cm; *W* 1.3 m.

Habitat: *Breeding:* bare, rocky offshore islands on coast; rocky islets, islands and stands of mature deciduous trees in interior. *In migration* and *winter:* protected bays, harbours, estuaries and lagoons on the coast; slow-moving rivers, reservoirs and lakes in the interior.

Nesting: colonial, often with or near pelicans, terns, gulls, herons or other cormorants; large stick nest is built on a low-lying island or near the top of a tall, mature deciduous tree; pair incubates 2–7 pale blue, often nest-stained eggs for 25–28 days.

Feeding: dives underwater to depths of 10 m, primarily for "rough" fish; surfaces to swallow prey.

Voice: generally quiet; possible piglike grunts or croaks near breeding colonies.

Similar Species: *Brandt's Cormorant* (p. 123): buff brown throat patch; bluish green eyes; all-black bill; shorter tail; bluish throat patch in breeding plumage; flies with neck straight. *Pelagic Cormorant* (p. 126): smaller; slender, all-black bill; small, red throat patch and white flank patches in breeding plumage.

GREAT CORMORANT

Phalacrocorax carbo

The Great Cormorant is the largest and most cosmopolitan of our cormorants, common throughout much of the Old World from Europe to Australia. In North America, the species is well established in Atlantic Canada. In most of its range, the Great Cormorant is a bird of freshwater lakes and rivers, but in much of the North Atlantic, it is found along sea cliffs and on rocky islands. • Cormorants rely on their superb underwater vision to detect schooling fish. Once a bird locates its prey, it folds its long wings tight against its body and dives down into the murky depths, propelled only by its large, webbed feet. Like owls, the Great Cormorant coughs up any indigestible parts of its prey in the form of a pellet. • At the nest, each chick feeds by sticking its head into the parent's open bill and lapping up the regurgitated food. As with many birds and most colonial waterbirds, the most active, robust chicks are fed first, so that in times of food scarcity, only the strongest nestlings survive to fledging. • *Phalacrocorax* means "bald crow" in Latin and refers to a common European subspecies of Great Cormorant that, like our Bald Eagle, has a white head in breeding plumage. The species name *carbo* refers to the charcoal colour of this bird's plumage.

breeding

ID: sexes similar; all-dark plumage; white "throat strap" borders yellow "chin" patch; thick neck; heavy bill with hooked tip. *Breeding:* white head plumes and flank patches. *Immature:* brown upperparts; white underparts; brown-streaked breast. *In flight:* flocks fly silently in V-formation.

Size: *L* 91 cm; *W* 1.6 m.

Habitat: shallow coastal waters; nests on rocky coastal and island cliffs; winters along sheltered bays, estuaries and jetties.

Nesting: on a sheltered sea cliff or ledge; bulky nest of sticks is lined with seaweed and finer materials; pair incubates 4–5 bluish white eggs for 28–31 days.

Feeding: wide variety of fish caught while diving; dives are usually to depths of 10 m but may be to 30 m.

Voice: generally silent; sometimes utters croaks and low groans at the nest site.

Similar Species: *Double-crested Cormorant* (p. 124): smaller head and body; thinner neck; orange "chin" pouch; lacks white "throat strap" and flank patches; immature is darker below, with unstreaked neck and breast.

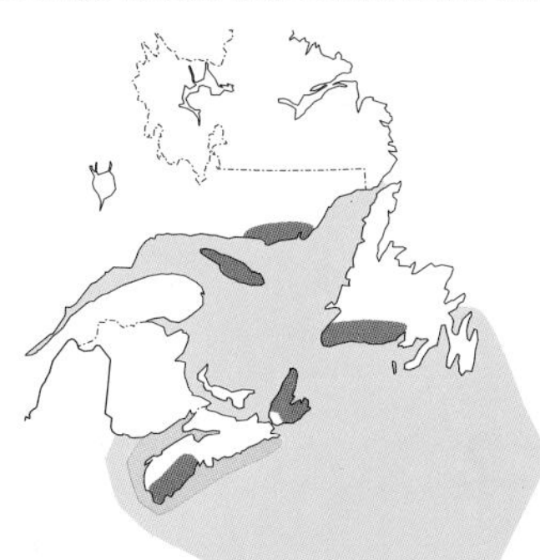

PELAGIC CORMORANT

Phalacrocorax pelagicus

breeding

Pelagic Cormorants are a common sight year-round along the entire inner and outer Pacific Coast and are especially numerous in the Strait of Georgia, off the British Columbia coast. Despite its name, the Pelagic Cormorant prefers the nearshore zone, leaving offshore waters to the Brandt's Cormorant. • The Pelagic Cormorant is the smallest and most widely distributed cormorant species in the North Pacific. It is also among the least gregarious or social, rarely forming feeding flocks or flying in formation, and pairs nest in loose colonies often far from their nearest neighbours. • The smallest of British Columbia's four cormorant species, the Pelagic Cormorant is a rather reptilian-looking bird, sleek and black with a subtle iridescent green and purplish blue gloss at close range. Breeding birds are particularly elegant. • Cormorants have powerful, fully webbed feet that aid not only in swimming and diving, but also in clinging to steep cliffs. They have a naked, extensible throat patch that is used in a panting behaviour known as "gular fluttering," perhaps to cool the body by bringing air across the blood vessels of the throat.

ID: sexes similar; dark, green-glossed plumage; small head; slender, all-black bill; slender neck. *Breeding:* double head crest; small, red throat patch; white plumes on sides of neck; white flank patches. *Nonbreeding:* entirely brownish black body; long tail; no white patches.

Size: *L* 63–71 cm; *W* 90–100 cm.

Habitat: restricted to nearshore marine waters including bays, inlets, channels and harbours. *Breeding* and *roosting:* steep cliffs on offshore islands, concrete bridge abutments and abandoned ship hulls. *Foraging:* nearshore protected waters.

Nesting: only loosely colonial; on a narrow ledge on a vertical cliff; may use a human-made structure; pair builds a nest of grasses, seaweeds, mosses and feathers; pair incubates 3–5 greenish white eggs for 25–33 days.

Feeding: forages by diving up to 40 m deep to catch mainly fish, but also shrimp, crabs, crustaceans and marine worms.

Voice: generally silent; sometimes utters low groans at nest site.

Similar Species: *Brandt's Cormorant* (p. 123): larger; heavier bill; buffy throat patch. *Double-crested Cormorant* (p. 124): larger; larger head; heavier bill; conspicuous pale yellow or yellowish orange throat patch.

AMERICAN BITTERN

Botaurus lentiginosus

The American Bittern is renowned for its cryptic camouflage, its secretive lifestyle and its deep, resonant call. Although the species is a rather common inhabitant of marshes throughout Canada, its tendency to stay hidden within dense vegetation and preference for hunting at dawn or dusk mean that most Canadians have never seen one. • At the approach of an intruder, an American Bittern's first reaction is to freeze with its bill pointed skyward—its vertically streaked, brown plumage blends in well with its marshland surroundings. It will even rock gently from side to side like a reed swaying in the breeze. This defensive reaction can sometimes place the bittern in a humorous situation—it will try to mimic a reed even in an open field or while standing in the middle of a road! • The characteristic deep, "pumping" *pomp-er-lunk* call has resulted in many vernacular names for this species, including "Stake-driver" and "Thunder-pumper." • Breeding bird surveys since the 1960s have revealed long-term annual declines throughout North America, including Canada. Habitat loss and degradation caused by silt, agricultural runoff and acid rain are primary factors in the decline. • At first glance, an adult American Bittern might be mistaken for a young night-heron or Green Heron, both of which have streaked necks.

ID: sexes similar; brown upperparts; brown-streaked neck and breast; straight, stout bill; yellow legs and feet; black "moustache" streak from bill to shoulder. *In flight:* pointed wings and blackish flight feathers are distinctive.
Size: *L* 59–69 cm; *W* 1.1 m.
Habitat: shallow freshwater wetlands and brackish marshes with tall, dense emergent vegetation; migrants may visit ditches, wet fields and urban ponds.
Nesting: above the waterline in a dense cattail or bulrush marsh; nest platform is made of grass, sedges and dead reeds; female incubates 3–6 plain buff eggs for 24–28 days.

Feeding: patient stand-and-wait predator; strikes at small fish, crayfish, frogs and snakes.
Voice: deep, slow, resonant, "pumping" *pomp-er-lunk* or *onk-a-blonk* often heard at dawn or dusk; low *kok-kok-kok* flight call.
Similar Species: *Green Heron* (p. 134): immature is smaller, more heavily marked and has different facial markings. *Black-crowned Night-Heron* (p. 135): juvenile has white-flecked, dark brown upperparts and no "moustache."

LEAST BITTERN

Ixobrychus exilis

The Least Bittern is the smallest of the herons and one of the most reclusive wetland birds in North America. It inhabits marshes where tall, impenetrable stands of cattails conceal its movements. This bird's slender body permits it to move about with ease in its dense marshland habitat. When encountered, the Least Bittern typically scurries into dense vegetation like a rodent, flies away weakly with its legs dangling or "freezes" with its bill pointed upward, feathers compressed and eyes directed forward. • An expert climber, the Least Bittern often perches a metre or more above water, clinging to vertical stems. • Least Bitterns are uncommon and extremely local throughout eastern Canada from southeastern Manitoba to New Brunswick, reaching the northern limit of its North American range. Here it pushes the boundaries of its adaptability, particularly its tolerance for chilly summer nights and rainstorms. Sightings are rare, but knowing the Least Bittern's cuckoolike vocalizations and listening for them in suitable habitat after dark is usually how birders find this elusive species. • The Least Bittern was designated as a threatened species in 2001. Its small, declining population is estimated at less than 1000 pairs for all of Canada. The species depends on high-quality marsh habitat, much of which has unfortunately been greatly altered and is continuously being lost or degraded.

♂

♂

ID: rich buff flanks and sides; streaking on foreneck; white underparts; mostly pale bill; yellowish legs; short tail; dark primary and secondary feathers. *Male:* black crown and back. *Female* and *juvenile:* chestnut brown head and back; juvenile has darker streaking on breast and back. *In flight:* large, buffy shoulder patches.

Size: *L* 28–37 cm; *W* 43 cm.

Habitat: freshwater marshes and swamps where dense, tall emergent vegetation is interspersed with clumps of woody vegetation and open water.

Nesting: mostly the male constructs platform of dry plant stalks on top of bent marsh vegetation; nest site is usually well concealed within dense vegetation; pair incubates 2–6 pale green or blue eggs for 17–20 days.

Feeding: stabs prey with bill; eats mostly small fish but also takes large insects, tadpoles, frogs, small snakes, leeches and crayfish; forages by stalking along emergent vegetation; occasionally constructs hunting platforms.

Voice: *tut-tut* call; *koh* alarm call. *Male:* low, muted *coo-coo-coo*. *Female:* ticking sound.

Similar Species: *American Bittern* (p. 127): larger; bold, brown streaking on underparts; black "moustache" streak extends to shoulder. *Black-crowned Night-Heron* (p. 135) and *Yellow-crowned Night-Heron* (p. 136) juveniles are much larger and have white-flecked, dark brown upperparts. *Green Heron* (p. 134): juvenile has dark brown upperparts.

GREAT BLUE HERON

Ardea herodias

The Great Blue Heron usually hunts near water, where it waits motionlessly for a fish or frog to approach, spears the prey with its bill, and then swallows it whole. This heron also stalks fields and meadows in search of rodents. • Common and widespread, the Great Blue Heron is one of the best-known wading birds in Canada. Because it frequents rivers, lakeshores, beaches, sand flats, piers, parks and a variety of sites that are regularly visited by people, it is seen more often at close range than any other heron. • Great Blue Herons nest in communal treetop nests called "rookeries" that range from a few to hundreds of nests. Nesting herons may be sensitive to human disturbance, so observe the birds' behaviour from a safe distance. • Because herons do not nest every year, and given that nesting birds often forage far from their rookeries, the presence of herons in an area does not necessarily mean that a nesting colony is nearby. • Great Blue Herons are sometimes incorrectly called "cranes." Not closely related but similar in appearance, cranes fly with their necks outstretched, whereas herons fly with their necks folded into their bodies.

ID: sexes similar; large wading bird; long legs; long, curved neck; blue-grey back and wing coverts; large, straight, yellow bill; chestnut thighs. *Breeding:* plumes on crown and throat. *Nonbreeding:* duller colours; no crown or throat plumes. *In flight:* slow, steady wingbeats; folded neck.
Size: *L* 1.3–1.4 m; *W* 1.8 m.
Habitat: almost any freshwater habitat or calm-water intertidal areas where prey can be found, including farmlands and occasionally urban areas.
Nesting: colonial; in a tree, snag, tall bush or marsh vegetation; often on an island or in a wooded swamp; stick-and-twig platform nest up to 1 m in diameter is lined with moss, reeds and dry grass; nest may be refurbished for reuse; pair incubates 3–5 pale blue eggs for 27–29 days.
Feeding: patient stand-and-wait predator; spears fish, snakes, amphibians and even rodents, then swallows prey whole; occasionally scavenges carrion in winter.
Voice: usually quiet away from nest; deep *frahnk-frahnk-frahnk* when startled; utters *roh-roh-roh* when approaching nest.
Similar Species: *Sandhill Crane* (p. 164): unfeathered, red crown; flies with neck outstretched. *Black-crowned Night-Heron* (p. 135) and *Green Heron* (p. 134): much smaller; shorter legs. *Egrets* (pp. 130–33): all are predominately white.

GREAT EGRET

Ardea alba

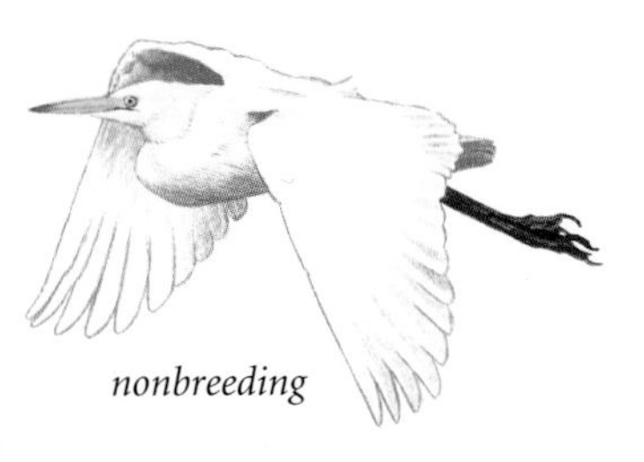

nonbreeding

The plumes of the Great Egret and Snowy Egret were widely used to decorate women's hats in the early 20th century. An ounce of egret feathers cost as much as $32—more than an ounce of gold at the time! As a result, egret populations rapidly plummeted. Some of the first conservation legislation in North America was enacted to outlaw the hunting of these magnificent birds, and conservation organizations such as the National Audubon Society (whose symbol is the Great Egret!) were formed. Great Egret populations have since recovered, and the species now breeds farther north than they did historically. In Canada, Great Egrets nest in only a handful of known sites in southern Saskatchewan, Manitoba, Ontario and Québec, and many of these are used only sporadically. Riding Mountain National Park in Manitoba is considered the most northerly nesting in North America. However, sightings of nonbreeding Great Egrets in southern Canada are becoming more commonplace each year. • Egrets are named for the silky breeding plumes, called "aigrettes," that most species produce during courtship. • The Great Egret can forage in deeper water than other egrets because of its long legs and neck.

breeding

ID: sexes similar; large, all-white wading bird; black legs; yellow bill. *Breeding:* white plumes trail from lower back; green lores. *Nonbreeding:* yellow skin patch between eyes and bill. *In flight:* slow wingbeats; neck folds back over shoulders; legs extend backward.

Size: *L* 94–104 cm; *W* 1.3 m.

Habitat: edges of marshes, lakes, rivers and ponds; flooded agricultural fields.

Nesting: colonial; in a tree or tall shrub; pair builds a platform of sticks and incubates 2–6 pale blue-green eggs for 23–26 days.

Feeding: patient stand-and-wait predator; occasionally stalks slowly; feeds primarily on fish and aquatic invertebrates.

Voice: generally silent away from colonies; calls include a rapid, low-pitched *cuk-cuk-cuk*.

Similar Species: *Snowy Egret* (p. 131): much smaller; black bill; yellow feet. *Little Blue Heron* (p. 132): immature is much smaller with greenish legs and dark, bicoloured bill. *Cattle Egret* (p. 133): much smaller; orange to yellowish legs.

SNOWY EGRET

Egretta thula

nonbreeding

The dainty Snowy Egret is distinguished by its small size, white plumage and, especially, bright yellow feet on black legs. In the late 1800s, it was sought for the plume trade because of its delicate back plumes, called "aigrettes," that were used to adorn women's hats. Like several other wading birds, the Snowy Egret was on the brink of extirpation in many areas by the early 1900s, until outraged citizens forced the passage of laws to outlaw the hunting of these birds. Populations have recovered dramatically, and the birds now occur beyond their historical range limits in North America. • The Snowy has conspicuous white plumage and an active, sometimes frantic, foraging behaviour as it hunts for small fish and crustaceans. By poking its bright yellow feet in the muck of shallow wetlands, the Snowy Egret flushes prey from hiding places. It also pursues fish in shallow waters, and snatches small fish attracted to the shaded spot created by its body as it stands and waits for prey to approach. • A rare but regular visitor to eastern Canada, the Snowy Egret has been only confirmed breeding once, in Hamilton, Ontario. Breeding has been suspected but never confirmed at Bon Portage Island in Nova Scotia.

breeding

ID: medium-sized, all-white wading bird; black bill and legs; bright yellow feet. *Breeding:* long plumes on head, throat and lower back; yellow or orange-red lores. *Nonbreeding:* lacks plumes; less intense colour on lores and feet. *In flight:* long wings; neck folds back over shoulders; legs and feet extend well beyond tail.

Size: *L* 56–66 cm; *W* 1.4 m.

Habitat: flooded agricultural fields and edges of marshes, rivers, lakes and ponds.

Nesting: colonial; in a wetland, in a tree or tall shrub; pair builds a platform of sticks and incubates 3–5 pale blue-green eggs for 23–26 days.

Feeding: stand-and-wait predator; actively chases fish in shallows.

Voice: generally silent away from colonies.

Similar Species: *Great Egret* (p. 130): much larger; yellow bill; black feet. *Cattle Egret* (p. 133): smaller; yellow-orange legs and bill. *Little Blue Heron* (p. 132): immature has bicoloured bill and dark greenish legs.

LITTLE BLUE HERON

Egretta caerulea

breeding

immature

The Little Blue Heron is unique among North America's herons and egrets in having a true white phase: first-year immatures are wholly or nearly wholly white, while adults (birds in their second calendar year and older) have no white at all in their plumage. The blue feathers moult in gradually, giving younger birds a temporary "pied" or "calico" plumage. Immature birds in their all-white plumage can be confused with other egret species. • This bird's dark adult plumage, secretive and solitary feeding habits and small overall numbers mean that it is often overlooked by birders. • Because of its dark plumage and lack of aigrettes, the Little Blue Heron was not slaughtered to the same extent as other egret species by plume hunters in the late 19th and early 20th centuries. Like most other wading birds, Little Blue Heron populations have made a dramatic recovery in the past century. • When foraging, this heron stands in or next to shallow water and waits for prey to pass by, or it slowly walks through water with its neck and bill held outstretched at a 45-degree angle. • Little Blues are vagrant from southern Ontario to Newfoundland, with a few birds reported each year.

ID: medium-sized heron; dark slate blue overall; blue bill, darker at tip. *Breeding:* shaggy, maroon head and neck; dull green legs. *Nonbreeding:* purplish head and neck. *Immature:* all white, usually with dusky-tipped primaries; yellowish green legs; mottled blue and white when moulting to adult plumage. *In flight:* long, narrow wings; neck folds back over shoulders; legs extend well beyond tail.

Size: *L* 61 cm; *W* 1.0 m.

Habitat: edges of marshes, ponds, lakes and ponds; flooded agricultural fields.

Nesting: in a shrub or tree above water; female uses sticks collected by male to build bulky platform nest; pair incubates 3–5 pale greenish blue eggs for 22–24 days.

Feeding: patient stand-and-wait predator; also wades slowly to stalk prey; primarily eats fish, amphibians and aquatic invertebrates.

Voice: usually silent; sometimes makes low clucking or croaking sounds.

Similar Species: *Snowy Egret* (p. 131): black bill; yellow lores; black legs; bright yellow feet. *Cattle Egret* (p. 133): not found near water; stocky yellow bill; yellow legs and feet; immature has black feet.

CATTLE EGRET

Bubulcus ibis

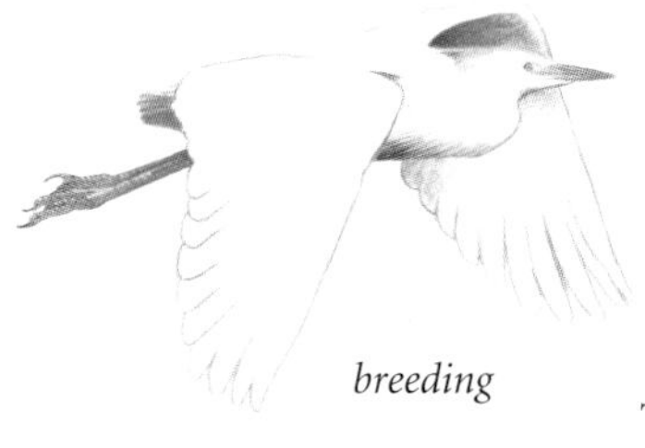

breeding

Cattle Egrets get their name from their habit of following grazing animals to catch the insects and other prey that the ungulates stir up. They sometimes follow field machinery and occasionally scavenge at dumps. Their diet is made up mostly of invertebrates, and most of their food is terrestrial, not aquatic. • Unlike other egrets, Cattle Egrets have a "hunched" posture and an exaggerated, head-pumping strut. They are regularly seen on lawns, fields and pastures. • Originally from Africa, Cattle Egrets crossed the Atlantic Ocean to South America in the first half of the 20th century, then spread throughout North and Central America. The first sight record for Canada occurred in Newfoundland in 1952, and within 10 years, these birds were being seen throughout Southern Ontario and were established as a nesting species there. Expansion of their range in North America has been greatly aided by widespread landscape conversion and the creation of pasturelands, stock ponds, reservoirs, municipal dumps and mechanized farming. Cattle Egrets now outnumber the combined populations of all the other egrets and herons in North America! • The species is still increasing and colonizing new areas; in 2005, large numbers nested in southern Manitoba for the first time.

breeding

ID: medium-sized, stocky wading bird; white plumage; yellow-orange bill and legs; short legs; thick neck; hunched posture. *Breeding:* long plumes on throat and lower back; buff orange throat, rump and crown; orange-red legs and bill. *Nonbreeding:* lacks plumes; yellow-orange bill; dark legs. *Immature:* dark bill turns yellow in late summer; blackish feet; dark bill. *In flight:* rather direct flight with deep wingbeats.
Size: *L* 61 cm; *W* 1.0 m.
Habitat: fields and pastures; also backyards and marshes.

Nesting: colonial with other wading birds; in a tree, tall shrub or emergent vegetation; male supplies sticks and reeds for female to build platform; pair incubates 3–4 pale blue eggs for 21–26 days.
Feeding: picks grasshoppers and other invertebrates from fields; often associated with livestock.
Voice: generally silent away from breeding colony; calls are simple, unmusical and subdued.
Similar Species: *Great Egret* (p. 130): much larger; black legs and feet. *Snowy Egret* (p. 131): slightly larger; dark bill; black legs; yellow feet. *Little Blue Heron* (p. 132): juvenile has blue-grey, bicoloured bill and olive legs.

GREEN HERON

Butorides virescens

Sentinel of the marsh, the ever-vigilant Green Heron sits hunched on a shaded branch at the water's edge. It often perches just above the water's surface along wooded streams, waiting to stab small fish or other prey with its sturdy, daggerlike bill. This heron lurks in dense vegetation along stream, pond and lake margins, and when startled, it takes off with an explosive, somewhat metallic *skyow* call often accompanied by an equally explosive release of feces. • The Green Heron is one of the few birds known to use tools as a hunting strategy. It has been observed dropping feathers, leaves or other small debris into the water as a form of "bait" to attract fish within striking range. • In good light, you may see a glimmer of green on the back and outer wings of adult birds, but most of the time, this greenish iridescence is not apparent. The scientific name *virescens* is Latin for "growing or becoming green," and it refers to this bird's transition from a streaky brown juvenile to a greenish adult. • Unlike most herons, the Green Heron tends to forage solitarily and usually nests singly rather than communally. When disturbed, it characteristically raises its crest and flicks its tail.

ID: small, stocky wading bird; slight crest; blue-grey back and wings with greenish iridescence; greenish black crown; chestnut face and neck; white foreneck and belly; yellow-green legs; short tail. *Breeding:* bright orange legs. *Juvenile:* heavily streaked neck and underparts; dark brown upperparts. *In flight:* bowed wings; otherwise appears crowlike.

Size: *L* 38–56 cm; *W* 66 cm.

Habitat: freshwater and tidal shores, ponds, streamside willows, mudflat edges and similar sheltered or semi-wooded situations.

Nesting: singly or in small, loose groups; male begins and female completes construction of stick platform in a tree, shrub or emergent vegetation, usually close to or over water; pair incubates 3–5 pale blue-green to green eggs for 19–21 days.

Feeding: feeds day or night; stabs prey with bill after slowly stalking or standing and waiting; eats mostly small fish and invertebrates.

Voice: generally silent; loud *skyow* or *kyowk* alarm call.

Similar Species: *Black-crowned Night-Heron* (p. 135): juvenile is much larger and has brownish plumage with white streaks or spots on upperparts. *American Bittern* (p. 127): larger; black "moustache" streak from bill to shoulder; heavy, brown streaking on buff neck and breast; dark flight feathers. *Least Bittern* (p. 128): buffy yellow shoulder patches, sides and flanks.

BLACK-CROWNED NIGHT-HERON

Nycticorax nycticorax

immature

At dusk, hungry Black-crowned Night-Herons start to become active after a day of roosting in dense riparian trees and shrubs. Although widespread and common in North America, their nocturnal feeding habits render them less obvious than diurnal herons. • A popular hunting strategy for Black-crowned Night-Herons is to sit motionless atop a few bent-over cattails. Anything passing below the perch becomes fair game—including ducklings, small shorebirds and young muskrats. These colourful birds use their large, light-sensitive eyes to spot prey lurking in the shallows. During breeding season, Black-crowned Night-Herons sometimes forage during the day. • The adults are boldly patterned, but their counter-shading and relative lack of movement render them inconspicuous in the dappled shadows of their daytime hangouts. Young night-herons are commonly seen around large cattail marshes in autumn. Because of their heavily streaked underparts, juvenile birds are easily confused with other juvenile herons and American Bitterns. • Black-crowned Night-Herons are colonial nesters, and colonies may have as many as 10,000 pairs, though a typical colony has fewer than 50 nests. • A wide-ranging species, Black-crowns breed on every continent except Australia and Antarctica. • This heron's white "ponytail" is present for most of the year but is most noticeable during the nesting season.

breeding

ID: medium-sized, stocky heron; hunched posture; short neck; stout, black bill; large, red eyes; black crown and back; white cheek, foreneck and underparts; greyish neck and wings. *Breeding:* 2 white plumes on crown. *Immature:* lightly streaked underparts; white-flecked, brown upperparts. *In flight:* only tips of feet project beyond tail.

Size: *L* 58–66 cm; *W* 1.1 m.

Habitat: islands and brackish or freshwater sloughs with dense shrubbery or emergent vegetation. *Foraging:* wetlands.

Nesting: colonial; near or over water in a tree, shrub or emergent vegetation; female uses material gathered by male to make loose stick nest; pair incubates 3–5 greenish eggs for 23–26 days.

Feeding: opportunistic stand-and-wait predator; eats mainly fish; also takes amphibians, crustaceans, small snakes, recently fledged birds and small mammals.

Voice: deep, guttural *quark* or *wok*, often on takeoff.

Similar Species: *American Bittern* (p. 127): rarely seen in the open; larger; similar to immature but with black "moustache" streak from bill to shoulder. *Great Blue Heron* (p. 129): much larger; longer legs; juvenile has longer, more heavily marked neck.

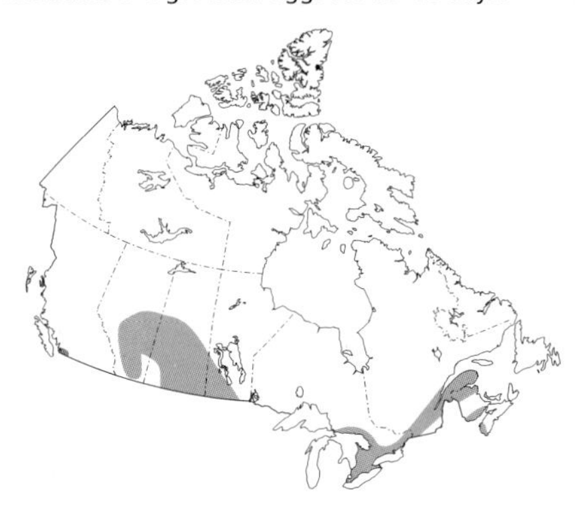

YELLOW-CROWNED NIGHT-HERON

Nyctanassa violacea

immature

Although night-herons are named for their habit of hunting from dusk until dawn, the Yellow-crowned Night-Heron also commonly hunts throughout the daylight hours, particularly when there are hungry mouths to feed. Its secretive habits, choice of dense, concealing habitats and slow, nearly imperceptible hunting methods combine to make this bird a challenge to locate and observe. • This heron specializes in catching crustaceans such as crayfish and crabs. Its thick bill and powerful neck muscles allow it to quickly crack open the hard exoskeleton of its preferred prey. • Primarily a resident of the U.S. Southeast north to Massachusetts, the Yellow-Crowned Night Heron underwent a dramatic northward range expansion from 1925 to 1960. It also established at least one nest on Lake Erie in Ontario during the 1950s and became a regular wanderer into southern Manitoba, southern New Brunswick, Nova Scotia, Newfoundland and parts of southern Ontario.

breeding

ID: chunky, grey heron; black head; slate grey neck and body; white crown and cheeks; yellowish forehead; stout, black bill with pale base to lower mandible; yellow legs. *Breeding:* long, white head plumes extend down back of neck. *Immature:* brown plumage with fine, white spotting; green legs. *In flight:* legs extend well beyond tail.

Size: *L* 61 cm; *W* 1.0 m.

Habitat: forested or shrubby swamps, estuaries, riparian thickets, shallow salt marshes and other aquatic habitats with thick shoreline cover.

Nesting: singly or in a colony with other wading birds; nest platform is built near or over water in a shrub, tree or emergent vegetation; pair incubates 3–6 bluish green eggs for 24–25 days.

Feeding: stands and waits or wades slowly and methodically; eats crustaceans, primarily crayfish and crabs; also takes amphibians, fish, insects and molluscs.

Voice: rarely given call is a high, short *quark*.

Similar Species: *Black-crowned Night-Heron* (p. 135): black crown and back; white underparts and lower half of face; immature has darker back, larger, white flecks, slimmer bill and shorter legs.

GLOSSY IBIS

Plegadis falcinellus

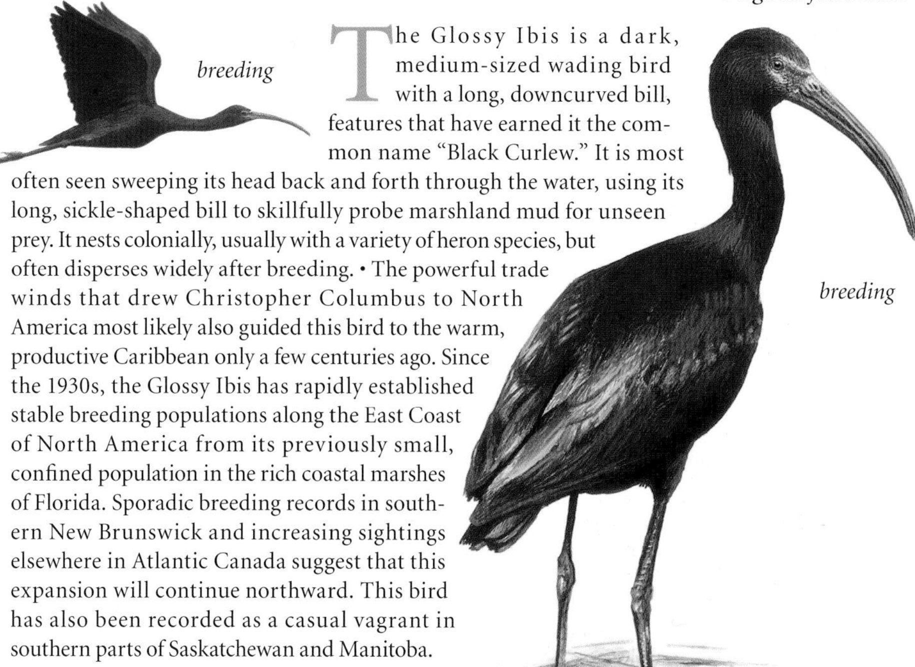
breeding

breeding

The Glossy Ibis is a dark, medium-sized wading bird with a long, downcurved bill, features that have earned it the common name "Black Curlew." It is most often seen sweeping its head back and forth through the water, using its long, sickle-shaped bill to skillfully probe marshland mud for unseen prey. It nests colonially, usually with a variety of heron species, but often disperses widely after breeding. • The powerful trade winds that drew Christopher Columbus to North America most likely also guided this bird to the warm, productive Caribbean only a few centuries ago. Since the 1930s, the Glossy Ibis has rapidly established stable breeding populations along the East Coast of North America from its previously small, confined population in the rich coastal marshes of Florida. Sporadic breeding records in southern New Brunswick and increasing sightings elsewhere in Atlantic Canada suggest that this expansion will continue northward. This bird has also been recorded as a casual vagrant in southern parts of Saskatchewan and Manitoba.

ID: long, downcurved bill; long legs; dark skin in front of brown eye is bordered by 2 pale stripes. *Breeding:* brownish olive bill; grey-green legs with red joints; narrow, cobalt blue border to bare facial skin; chestnut head, neck and sides; green and purple sheen on wings, tail, crown and face. *Nonbreeding:* white streaking on dark greyish brown head and neck. *In flight:* neck fully extended; hunchbacked appearance; legs trail behind tail; flocks fly in lines or V-formation.

Size: *L* 56–64 cm; *W* 91–94 cm.

Habitat: wide variety of inland wetland habitats; also coastal lagoons and estuaries.

Nesting: colonial, often among egrets and herons; bulky platform of marsh vegetation and sticks is built over water, on the ground or in a tall shrub, small tree or emergent vegetation; pair incubates 3–4 pale blue or green eggs for about 21 days.

Feeding: tactile forager in shallow water; probes and gleans for aquatic and terrestrial invertebrates, including insects and crayfish; may also eat amphibians, snakes, leeches, crabs and small fish.

Voice: repeated, gutteral *ka-onk*; low *kruk, kruk*.

Similar Species: *White-faced Ibis* (p. 138): red eyes; reddish bill; all-red legs; reddish skin in front of eyes is bordered by wide, white band of feathers in breeding plumage. *Double-crested Cormorant* (p. 124): much larger; upright gait; straight bill with hooked tip. *Sooty Shearwater* (p. 115): all dark; shorter, straight bill; webbed feet.

WHITE-FACED IBIS

Plegadis chihi

The White-faced Ibis is seen in large, remnant wetlands of western North America, but breeding populations fluctuate markedly from year to year in response to changing habitat conditions. Breeding adults are somewhat nomadic, relocating regularly when drought or wet weather renders traditional sites unusable. • With its long legs, slender toes and long, downcurved bill, this bird is well-adapted to life in the marsh. Wet of flooded meadows and reed-fringed, muddy shorelines are preferred foraging sites, offering soft soil ideal for probing and extracting food items. • This species was probably more common in the past, when permanent or seasonal wetlands were more abundant. Nevertheless, its breeding range and population appear to have expanded considerably in the U.S. since 1970. In Canada, the species was first confirmed to nest in Alberta during the early 1970s, and large numbers have bred regularly in at least two sites in Manitoba since 2005. • Ibises are adept fliers, readily traversing the great distances that may separate secluded nesting colonies from outlying feeding locations. Although White-faced Ibises are migratory, they tend to be restricted to traditional flyways. Because of this, they may be a familiar sight in some regions, but scarcely known a few miles away.

breeding

breeding

ID: dark chestnut overall; long, downcurved bill with thick base; dark red eyes; narrow strip of white feathers borders naked facial patch; iridescent, greenish lower back and wing coverts; long, dark legs. *Breeding:* rich red legs and facial patch. *In flight:* outstretched neck; rapid wingbeats with brief periods of gliding on stiffly bowed wings.

Size: *L* 58 cm; *W* 91 cm.

Habitat: freshwater wetlands, especially cattail and bulrush marshes; feeds in flooded hay meadows, agricultural fields and estuarine wetlands.

Nesting: colonial; in bulrushes or other emergent vegetation; deep cup nest of coarse materials is lined with fine plant matter; pair incubates 3–4 bluish green eggs for about 22 days.

Feeding: probes and gleans soil and shallow water for aquatic invertebrates, amphibians and other small vertebrates.

Voice: generally quiet; occasionally gives a series of low, ducklike quacks.

Similar Species: *Glossy Ibis* (p. 137): breeding bird has brownish olive bill, brown eyes, grey-green legs with red joints and narrow, cobalt blue border to bare facial patch. *Herons* and *egrets* (pp. 129–36): all lack long, thick-based, downcurved bill. *Long-billed Curlew* (p. 184): rich, buffy brown; cinnamon underwing linings; uniformly thin bill.

TURKEY VULTURE

Cathartes aura

The Turkey Vulture appears almost grotesque with its red, featherless head, but this adaptation is essential because it allows the bird to remain relatively clean while digging through carrion and rotting carcasses. • Turkey Vultures rarely flap their wings and rock slightly from side to side as they soar with their wings held in a dihedral, or V-shape. With a wingspan of up to 1.8 metres, these birds regularly use the slightest wind updrafts or thermals to stay airborne when other soaring birds are grounded. • Almost exclusively scavengers, Turkey Vultures can detect the smell of decomposing carrion from great distances, even under a dense forest canopy. Vultures circling above gas pipelines have helped engineers locate leaks—the birds are attracted to ethyl mercaptan, a stinky natural gas additive that smells like rotting flesh. • The ability to regurgitate meals allows parents to transport food over long distances and also enables engorged birds to repulse an attacker or to "lighten up" for an emergency takeoff. • The Turkey Vulture's ability to adapt its diet and choice of nest sites has allowed populations to increase greatly since the 1960s, and it has expanded its range in many parts of south-central Canada. There are two subspecies within Canada that meet at the Manitoba-Ontario border: *C. a. septentrionalis* to the east and *C. a. meridionalis* to the west.

ID: all-black body; pale flight feathers; bare, red head; longish, slender tail; yellow-tipped, red bill. *Immature:* bare, grey head; grey bill. *In flight:* head appears small; silver grey flight feathers contrast with black wing linings; holds wings in shallow "V"; teeters from side to side when soaring.

Size: *L* 66–81 cm; *W* 1.7–1.8 m.

Habitat: farmlands with pasture and abundant carrion close to undisturbed forested areas for perching, roosting and nesting. *Breeding:* open, mixed forests with rocky outcroppings and cliffs. *Foraging:* favours valley edges and foothills for regular thermals; in autumn, groups roost in large, mixed woods.

Nesting: usually on the ground in a dark recess beneath boulders or a cliff ledge; also in a hollow tree, log, stump or abandoned building; no nest is built; pair incubates 2–3 dull or creamy white eggs for 30–40 days.

Feeding: eats fresh or rotting carrion, including sea and land mammals, reptiles, beached fish and large invertebrates.

Voice: generally silent; grunts, hisses and barking sounds used mainly for predator deterrence.

Similar Species: *Golden Eagle* (p. 152) and *immature Bald Eagle* (p. 141): generally brown plumage; larger, feathered head; often show white on underside of wings and tail; wings held flat; do not rock in flight. *Hawks* (pp. 142–51): generally smaller; shorter, broader wings usually held flat when soaring; do not rock in flight; many have distinctive tail colours or patterns.

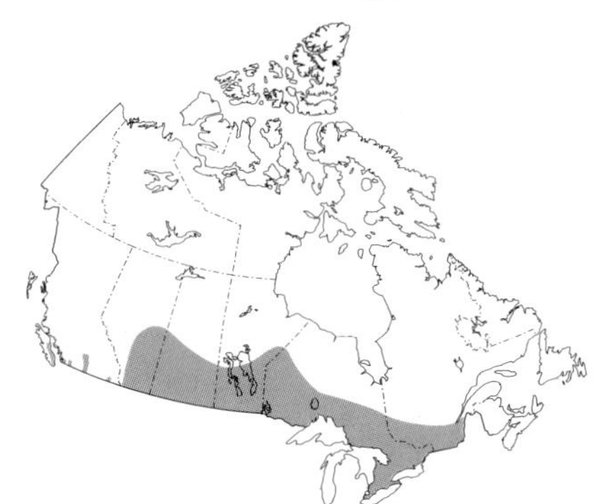

OSPREY

Pandion haliaetus

Often called "Fish Hawk," this bird feeds exclusively on fish, which it captures in dramatic dives. Once prey has been spotted, the Osprey locks itself into a perilous headfirst dive toward its target and thrusts its talons forward an instant before striking the water. In flight, the Osprey positions the still squirming fish so that it faces forward to optimize aerodynamics. Opposable talons and spiny footpads help the bird keep a firm grip on its slippery prey. • As with other raptors, Osprey eggs hatch at differing intervals, sometimes as much as five days apart. This assures that if fish are plentiful, all the young will survive, and if not, at least one will fledge in a healthy state. • During the 1950s to 1970s, Osprey populations crashed as a result of DDT and other contaminants, which caused birds to lay eggs with thin shells, reduced hatching success and lowered the survival rate of young and adults. Nesting success and Osprey numbers have increased greatly in recent years as some of the more deleterious contaminants have gradually disappeared from the environment. • Ospreys are one of the most cosmopolitan of birds, occurring on every continent except Antarctica.

ID: dark brown upperparts; mostly white head; brown nape and cheek patch; yellow eyes. *Female:* indistinct "necklace" across upper breast. *In flight:* white underparts; dark "wrist" marks on underwings; holds wings in a shallow "M."

Size: *L* 56–64 cm; *W* 1.4–1.8 m.

Habitat: primarily open-water habitats of lakes, rivers and sloughs; in coastal areas, primarily in protected bays, inlets and lagoons.

Nesting: in the top of a tree (usually a snag) or on a rocky cliff, the cross arms of a utility pole or a human-made structure; stick nest is refurbished and reused over many years; pair incubates 2–4 yellowish eggs, blotched with reddish brown, for 32–33 days.

Feeding: dramatic, feet-first dives into the water for fish up to 40 cm long.

Voice: melodious, ascending, whistled *chewk-chewk-chewk*; also an often-heard *kip-kip-kip*, uttered when performing aerial courtship displays.

Similar Species: *Bald Eagle* (p. 141): holds wings flatter; immature is larger, with yellow legs, larger bill with yellow base and some black body plumage. *Red-tailed Hawk* (p. 149): smaller; dark head; pale rufous to dark brown underparts; darker wing linings; reddish tail; shorter, broader wings held flat when soaring.

BALD EAGLE

Haliaeetus leucocephalus

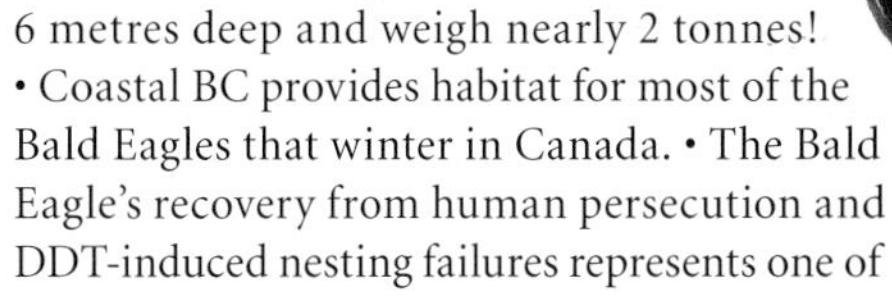

adult

immature

The majestic Bald Eagle is symbolic in both Native and modern cultures, and holds a prominent place in the beliefs and sacred rituals of Canada's First Nations people. • Usually found near water, Bald Eagles catch fish, but they also scavenge dead and dying fish, as well as animal carcasses. To kill and handle prey, the Bald Eagle has a massive beak, large talons and oversized feet equipped with small spikes, called spicules. • Bald Eagles do not mature until their fourth or fifth year, when they develop their white head and tail plumage. They mate for life and reuse their nests year after year, adding to them each season. Nests can grow to mammoth proportions, the largest of any North American bird, and can be up to 2.4 metres wide, 6 metres deep and weigh nearly 2 tonnes! • Coastal BC provides habitat for most of the Bald Eagles that winter in Canada. • The Bald Eagle's recovery from human persecution and DDT-induced nesting failures represents one of the continent's most successful conservation stories. Since the 1970s, populations have increased dramatically as reproductivity returned to normal, human persecution became a thing of the past and extirpated populations were reestablished through reintroductions.

ID: sexes similar; large raptor; white head and tail; dark brown body; yellow bill and feet; soars on broad, flat wings. *Immature* and *juvenile:* variable amounts of white mottling on upperparts, underwings, head and undertail; mostly dark bill.

Size: *L* 76–109 cm; *W* 1.7–2.4 m.

Habitat: *Breeding:* coastal areas near water; along lakeshores and riverbanks and on wooded islands. *Foraging:* almost anywhere including coastal, forested and agricultural areas and garbage dumps.

Nesting: usually in a large tree near water; huge stick nest is reused each year; pair incubates 2 (rarely 3) white eggs for 34–36 days.

Feeding: opportunistic; eats primarily fish; also scavenges various bird and mammal carcasses; takes a variety of waterbirds; frequently pirates fish from Ospreys.

Voice: thin, weak squeal, like a gull's scream broken into a series of notes.

Similar Species: adult is distinctive. *Golden Eagle* (p. 152): dark overall except for golden nape; smaller bill; shorter neck; immature has prominent, white patches on wings and base of tail. *Osprey* (p. 140): dark "wrist" patches; dark bill; M-shaped wings in flight.

NORTHERN HARRIER

Circus cyaneus

The Northern Harrier, a white-rumped raptor of open fields, grasslands and marshlands, may be the easiest bird of prey to identify on the wing—no other hawk flies so close to the ground during hunting forays, routinely grazing the tops of long blades of grass and cattails on wings held in a shallow "V." • With its prominent facial discs to better detect and focus sounds, a perched Northern Harrier looks astonishingly like an owl. • Male harriers are polygamous and can have up to four mates, so they are typically less numerous than females. • In his skydiving courtship flight, the male performs a series of looping dives and barrel rolls in a bid to secure the attention of onlooking females. Britain's Royal Air Force was so impressed by the Northern Harrier's manoeuvrability that it named the Harrier aircraft after this bird. • With the loss of grassland and marshland habitat, Northern Harriers appear to be declining throughout North America. • Once known as "Marsh Hawk" in North America, this bird is still called "Hen Harrier" in Europe.

ID: slim, medium-sized hawk; long wings and tail; conspicuous, white rump. *Male:* grey upperparts; white underparts with faint, rusty streaking; black-banded tail; black wing tips. *Female:* dark brown upperparts; dark brown streaking on buff underparts. *Juvenile:* rich russet brown upperparts; brown-streaked head and upper breast. *In flight:* long, somewhat pointed wings held in shallow "V"; white rump patch; long, narrow tail; flies low over marshes.

Size: *L* 41–61 cm; *W* 1.1–1.2 cm.

Habitat: *Breeding:* open country, grasslands, marshy wetlands, damp shrublands. *Winter:* open country, including fields, marshes, lakeshores, logging burns, forest clear-cuts and alpine meadows.

Nesting: on the ground or over water in dense vegetation; occasionally in a loose colony; nest is made of grasses, sticks and wetland vegetation; female incubates 4–6 pale bluish eggs for 30–32 days while male provides food.

Feeding: opportunistic; eats primarily mice and voles but frequently takes birds, even waterfowl.

Voice: high-pitched *ke-ke-ke-ke-ke-ke* near nest. *Female:* piercing, descending *eeyah eeyah* scream.

Similar Species: *Swainson's Hawk* (p. 148): dark "bib"; dark flight feathers contrast with pale underparts and underwing linings. *Rough-legged Hawk* (p. 151): broader wings; dark "wrist" patches; fan-shaped tail with white base and broad, dark subterminal band; dark belly band. *Red-tailed Hawk* (p. 149): lacks white rump and long, narrow tail.

SHARP-SHINNED HAWK

Accipiter striatus

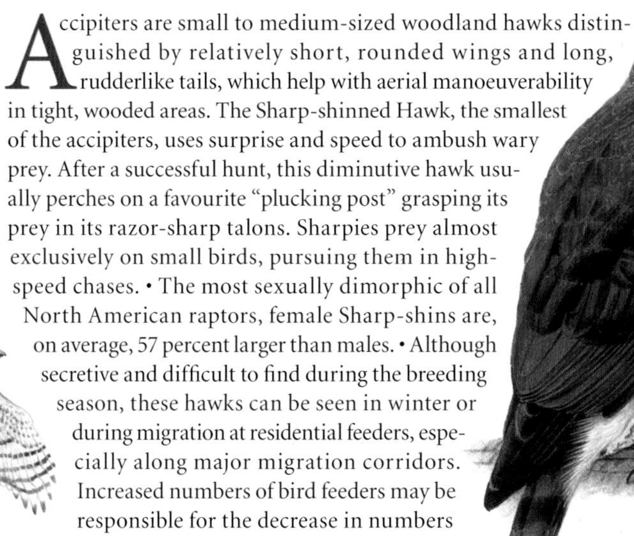

Accipiters are small to medium-sized woodland hawks distinguished by relatively short, rounded wings and long, rudderlike tails, which help with aerial manoeuverability in tight, wooded areas. The Sharp-shinned Hawk, the smallest of the accipiters, uses surprise and speed to ambush wary prey. After a successful hunt, this diminutive hawk usually perches on a favourite "plucking post" grasping its prey in its razor-sharp talons. Sharpies prey almost exclusively on small birds, pursuing them in high-speed chases. • The most sexually dimorphic of all North American raptors, female Sharp-shins are, on average, 57 percent larger than males. • Although secretive and difficult to find during the breeding season, these hawks can be seen in winter or during migration at residential feeders, especially along major migration corridors. Increased numbers of bird feeders may be responsible for the decrease in numbers of Sharp-shinned Hawks counted at traditional migration stations, as many now overwinter farther north. • Common names for the Sharp-shinned Hawk include "Sharpie," "Sharp-shin," "Little Blue Darter" and "Bird Hawk."

ID: sexes similar; small hawk; short, rounded wings; long, square-tipped tail with narrow, white tip; blue-grey upperparts; pale face; dark crown and nape; red eyes; white underparts heavily barred with orange. *Immature:* brown upperparts; yellow eyes; white underparts heavily streaked with brown. *In flight:* flap-and-glide flight.
Size: *Male: L* 25–30 cm; *W* 51–61 cm. *Female: L* 30–36 cm; *W* 61–71 cm.
Habitat: *Breeding:* coniferous and mixed woodlands. *In migration:* woodlands, alpine areas, riparian corridors, wooded lakeshores and shrub thickets. *Winter:* urban and yards with feeders; also estuaries and marine shores.
Nesting: in a tree; may refurbish an abandoned stick nest; female incubates 4–5 (rarely 3–8) heavily brown-blotched, bluish white eggs for 21–35 days.
Feeding: eats small to medium-sized land birds; occasionally takes small mammals, amphibians and insects.
Voice: generally silent; intense, repeated *kik-kik-kik-kik* warns intruders during breeding season; migrating autumn immature utters a clear *tewp.*
Similar Species: *Cooper's Hawk* (p. 144): generally larger; crown is darker than nape and back; more rounded tail tip has broader terminal band. *American Kestrel* (p. 153): pointed wings; 2 black facial stripes; often perches on power lines. *Merlin* (p. 154): rapid wingbeats on pointed wings; 1 dark facial stripe; brown-streaked, buff underparts; dark eyes.

COOPER'S HAWK

Accipiter cooperii

The Cooper's Hawk is a crow-sized accipiter, once known as "Chicken Hawk." This speedy, aggressive raptor is an increasingly common sight in backyards. • A quintessential woodland hawk, this accipiter's short, rounded wings and relatively long tail ensure manoeuverability in dense cover. Bursting from an overhead perch, a Cooper's Hawk will chase a songbird through cover until it can use its long legs and sharp talons to snatch its quarry in midair. • The female Cooper's Hawk is at least one-third bigger than the male. This size difference, common among raptors, allows males and females to hunt different-sized prey. • Distinguishing the Cooper's Hawk from the Sharp-shinned can be challenging. In flight, the Sharpie has a squared-off tail with little white at the tip, whereas the tail of the Cooper's is rounded with a much broader, white tip. Also in flight, the head of a Cooper's Hawk projects beyond the body more than that of a Sharpie. • Since DDT was banned in the 1970s, and with the persecution of the species almost eliminated, this forest raptor has been slowly recolonizing many of its former habitats and has expanded into new ones.

adult

immature

ID: sexes similar; medium-sized raptor; short, rounded wings; long, rounded, white-tipped tail; blue-grey upperparts; pale face; dark crown; pale nape; red eyes; white underparts heavily barred with orange. *Immature:* brown upperparts; yellow eyes; white underparts streaked with brown. *In flight:* flap-and-glide flight.
Size: *Male: L* 38–43 cm; *W* 69–81 cm. *Female: L* 43–48 cm; *W* 81–94 cm.
Habitat: *Breeding:* semi-open, primarily hardwood woodlands, especially in urban and suburban areas; also wooded riparian habitats. *In migration* and *winter:* soars on thermals and along ridgelines; also hunts along edges of mixed forests; often visits feeders.
Nesting: saddled on a branch in a tall tree; stick-and-twig nest is lined with bark flakes; female incubates 3–6 bluish eggs for 30–36 days.
Feeding: mainly eats medium-sized birds and mammals; occasionally takes squirrels and small hares; squeezes with feet to kill prey.
Voice: silent except around nest; fast, flickerlike *cac-cac-cac-cac.*
Similar Species: *Sharp-shinned Hawk* (p. 143): smaller; dark nape; square tail with narrow, white tip. *American Kestrel* (p. 153): long, pointed wings; 2 black facial stripes; typically seen in open country, often perches on power lines. *Merlin* (p. 154): rapid wingbeats on pointed wings; 1 dark facial stripe; brown-streaked, buff underparts; dark eyes.

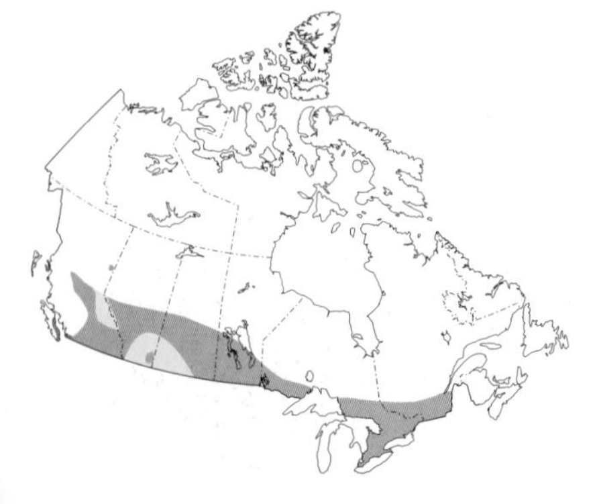

NORTHERN GOSHAWK

Accipiter gentilis

The Northern Goshawk, the largest of the forest hawks, has a legendary disposition as an agile and powerful predator. The female is a devoted parent, often boldly attacking intruders that venture too close to an active nest. • This hawk chases and catches prey in high-speed aerial pursuits and will even crash through brush to chase its quarry on foot. • When prey populations plummet in the northern forests—an event that tends to occur about every 10 years—large numbers of Northern Goshawks move south in search of food, usually during the winter months (known as a "winter irruption"). • These birds require extensive tracts of forest in which to hunt and raise their families. The clearing of forests for agricultural and residential development has caused goshawk populations to decline significantly throughout their range in northern Europe, Asia and parts of North America. • The species name *gentilis* refers to this hawk's use as the preferred "falcon" of the gentry, with the similar-sized Gyrfalcon and smaller Peregrine Falcon reserved for kings and emperors.

adult

immature

ID: sexes similar; large raptor; blue-grey upperparts; dark crown and eye line; white eyebrow; finely barred, grey underparts; white undertail; long, faintly banded tail. *Immature:* pale eyebrow; yellow eyes; brown-streaked, white underparts; dark-banded tail. *In flight:* uniform coloration; short, broad, rounded wings; dark flight feathers on grey upperwings.

Size: *Male: L* 53–58 cm; *W* 1–1.1 m. *Female: L* 58–64 cm; *W* 1.1–1.2 m.

Habitat: *Breeding:* mature forests; prefers combination of mature trees with intermediate canopy coverage and small open areas. *In migration* and *winter:* almost any habitat from sea level to alpine; regularly hunts in estuaries in winter.

Nesting: in deep woods; bulky nest of sticks and twigs is saddled on a branch near the trunk; nest is often refurbished in subsequent years; female incubates 2–4 pale bluish white eggs for 28–32 days.

Feeding: medium to large songbirds, grouse, woodpeckers, squirrels, ground squirrels, chipmunks and snowshoe hares.

Voice: silent, except around nest site; loud, aggressive *kyk-kyk-kyk* and other loud, accipiter-type calls.

Similar Species: *Cooper's Hawk* (p. 144): smaller; fine, reddish breast barring; shorter tail; immature has whiter underparts with narrower streaking. *Buteo hawks* (pp. 146–51): shorter tails; broader wings. *Gyrfalcon* (p. 155): pointed wings; dark eyes.

RED-SHOULDERED HAWK

Buteo lineatus

The Red-shouldered Hawk nests in mature deciduous or mixed deciduous-coniferous woodlands, usually near river bottoms and in lowland tracts of woods along creeks. • As spring approaches and pair bonds are formed, this normally quiet hawk utters loud, shrieking *kee-you, kee-you, kee-you* calls. • Red-shouldered Hawks occasionally build additional alternative nests near a current nest. • During the summer months, the dense cover of this hawk's forested breeding habitat allows few viewing opportunities for hawk watchers, but during spring and autumn migration, this bird can be found hunting from exposed perches. Some individuals may even hunt as far as 800 metres from the nearest stand of trees. Fortunately for birders, the Red-shouldered Hawk's seasonal use of telephone poles and fence posts for hunting gives observers a better glimpse into its otherwise private life. • With the loss of mature hardwood forests over the last century, Red-shouldered Hawks have declined in abundance. Reduced persecution in recent decades, however, has allowed populations to stabilize and even increase in many regions.

ID: sexes similar; chestnut red shoulders on otherwise dark brown upperparts; reddish underwing linings; narrow, white bars on dark tail; barred, reddish breast and belly; reddish undertail coverts. *Immature:* large, brown "teardrop" streaks on white underparts; whitish undertail coverts. *In flight:* light and dark barring on underside of flight feathers and tail; white crescents or "windows" at base of primaries.
Size: *L* 43–61 cm; *W* 94–110 cm.
Habitat: mature deciduous and mixed forests, wooded riparian areas, swampy woodlands and large, mature woodlots.

Nesting: in the crotch of a mature decidious tree; bulky nest of sticks and twigs is often reused; female incubates 2–4 darkly blotched, bluish white eggs for about 33 days.
Feeding: eats small mammals, birds, reptiles and amphibians caught in swooping attacks; occasionally takes crayfish and insects; may catch prey flushed by low flight.
Voice: repeated series of high *kee-you* notes.
Similar Species: *Broad-winged Hawk* (p. 147): lacks reddish shoulders; broader wings paler beneath and outlined with dark brown; wide, white tail bands. *Red-tailed Hawk* (p. 149): no barring in tail; no light "windows" at base of primaries.

BROAD-WINGED HAWK

Buteo platypterus

light adult

dark adult

A common breeder in large, deciduous or mixed-deciduous forests throughout northeastern and north-central North America, the Broad-winged Hawk is as secretive while nesting as it is conspicuous during migration. Nesting pairs spend most of their time beneath the forest canopy, hunting from a perch for insects, amphibians, reptiles, mammals and birds. • During migration, Broad-wings are one of the few North American raptor species that flock, soaring on thermals in "kettles" of hundreds, sometimes thousands, of birds as they migrate more than 7000 kilometres to their wintering grounds in northern South America. Field studies using satellite transmitters have shown that migrating Broad-wings travel at least 111 kilometres each day. • This species' reliance on amphibians for food may have allowed it to escape the drastic declines suffered by other North American raptors because of DDT use in the 1950s and '60s. • A rare dark morph of the Broad-wing is found mainly in northwestern and north-central portions of its range and is common in the western portions of its Canadian breeding range.

light adult

ID: sexes similar; crow-sized raptor; brown upperparts. *Light adult:* white underparts with heavy, rusty barring; broad, black and white tail bands. *Dark adult:* entirely dark upperparts; banded tail; silvery grey remiges on underwings. *Immature:* white underparts with dark brown, teardrop-shaped streaking; buff and dark brown tail bands. *In flight:* short, broad wings; pale underwings outlined with dark brown; white tail bands.
Size: *L* 35–48 cm; *W* 81–100 cm.
Habitat: *Breeding:* semi-open, mixed and deciduous woodlands; usually near water or openings such as roads, trails, wetlands or meadows. *In migration:* semi-open woodlands and forest edges.

Nesting: in a deciduous tree; compact stick nest is lined with bark chips; mostly the female incubates 2–4 brown-spotted, whitish eggs for 28–31 days.
Feeding: swoops down on prey from a perch; feeds on small vertebrates, especially small rodents, amphibians, reptiles and birds.
Voice: high-pitched, whistled *peeeo-wee-ee* on breeding grounds.
Similar Species: *Red-tailed Hawk* (p. 149): red tail without distinct broad, white bands; streaked breast; immature has greyish tail bands. *Other buteos* (pp. 146, 148, 150–51): lack broad banding on tails and broad, dark-edged wings with pointed tips. *Accipiters* (pp. 143–45): long, narrow tails with less distinct banding.

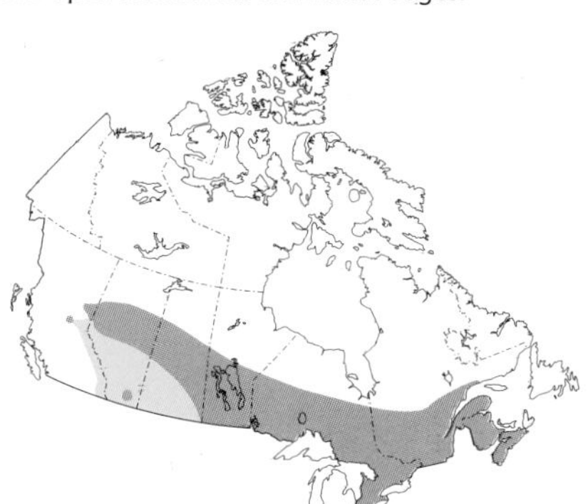

SWAINSON'S HAWK

Buteo swainsoni

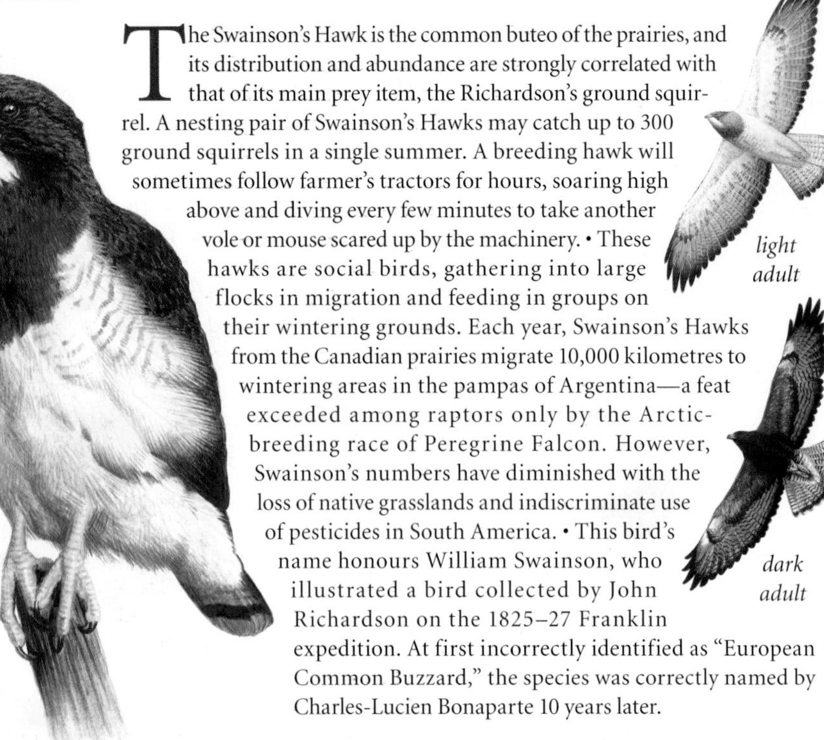

The Swainson's Hawk is the common buteo of the prairies, and its distribution and abundance are strongly correlated with that of its main prey item, the Richardson's ground squirrel. A nesting pair of Swainson's Hawks may catch up to 300 ground squirrels in a single summer. A breeding hawk will sometimes follow farmer's tractors for hours, soaring high above and diving every few minutes to take another vole or mouse scared up by the machinery. • These hawks are social birds, gathering into large flocks in migration and feeding in groups on their wintering grounds. Each year, Swainson's Hawks from the Canadian prairies migrate 10,000 kilometres to wintering areas in the pampas of Argentina—a feat exceeded among raptors only by the Arctic-breeding race of Peregrine Falcon. However, Swainson's numbers have diminished with the loss of native grasslands and indiscriminate use of pesticides in South America. • This bird's name honours William Swainson, who illustrated a bird collected by John Richardson on the 1825–27 Franklin expedition. At first incorrectly identified as "European Common Buzzard," the species was correctly named by Charles-Lucien Bonaparte 10 years later.

ID: sexes similar; medium-sized hawk; dark brown upperparts; long, narrow wings; finely barred, fan-shaped tail. *Light adult:* mostly white face and underparts; dark "bib"; white wing linings contrast with dark flight feathers; white uppertail coverts. *Dark adult:* dark brown underparts; dark wing linings blend with brown flight feathers; brown-flecked, whitish undertail coverts. *Immature:* mottled dark and white; pale edges to feathers on back. *In flight:* holds wings held in shallow "V"; white-tipped tail with dark subterminal band.

Size: *Male: L* 48–56 cm; *W* 1.2–1.3 m. *Female: L* 51–56 cm; *W* 1.3–1.4 m.

Habitat: open fields and grasslands with scattered trees or large shrubs for nesting; open sagebrush grasslands, rangelands and agricultural lands.

Nesting: in a tree or shrub adjacent to open habitat; rarely on a cliff; nest of sticks, twigs and forbs is lined with bark and fresh leaves; pair incubates 2–4 brown-spotted, whitish eggs for 28–35 days.

Feeding: swoops to the ground for voles, mice and ground squirrels; also eats snakes, small birds and large insects.

Voice: not very vocal; typical call on breeding territory a shrill, high-pitched *keeeeeeer;* also a repeated *pi-tik*.

Similar Species: *Red-tailed Hawk* (p. 149): holds wings flat in flight; more rounded wing tips; dark leading edge on underwings; dark belly band. *Other buteos* (pp. 146–47, 150–51): flight feathers are paler than wing linings; lack dark "bib."

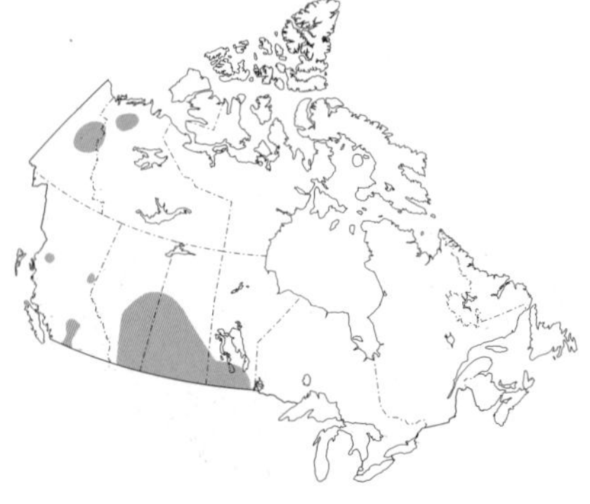

RED-TAILED HAWK

Buteo jamaicensis

"Harlan's"

One of North America's best-known and most widely distributed hawks, the Red-tailed Hawk is famous for its piercing call, which is often used in movies and television commercials. • During the last century, this hawk expanded its range and has largely replaced the Red-shouldered Hawk in much of the east and the Swainson's and Ferruginous hawks on the prairies. This expansion was aided by deforestation in the east and fire suppression in the west; both practices promote a patchwork of wooded areas mixed with large open areas, habitat favoured by the Red-tailed Hawk. This bird has adapted well to urban habitats and is commonly seen around cities. • Red-tailed Hawks are perhaps more variable in coloration than any other hawk, but the rufous to pinkish tail of the adult is a generally reliable field mark. This hawk's tail does not obtain its red coloration until it matures into a breeding adult. The lone exception to tail colour is the "Harlan's" subspecies, which has a mottled white to barred tail, and occurs from northern BC to Alaska and was formerly considered a completely separate species. • Size, coloration and tail markings distinguish up to 16 subspecies of Redtails.

ID: sexes similar; large hawk; dark brown head and upperparts; whitish underparts with belly band of dark brown streaks. *Light adult:* pale rufous breast; brick red tail. *"Harlan's":* dark overall; dark and light breast streaking; pale greyish, possibly banded tail. *Immature:* yellow eyes; whitish outer wings; clearly banded tail. *In flight:* variable dark wing lining with diagnostic darker leading edge; whitish underwing flight feathers; dark wing tips and trailing edges; fan-shaped tail; flies with wings nearly horizontal.

Size: *Male: L* 46–58 cm; *W* 1.1–1.4 m. *Female: L* 51–64 cm; *W* 1.2–1.5 m.

Habitat: almost any open or semi-open habitat with scattered trees and perching sites.

Nesting: usually high in a deciduous tree adjacent to an open area; rarely on a cliff or in a conifer; bulky stick nest is often reused; pair incubates 2–4 brown-blotched, whitish eggs for 28–35 days.

Feeding: opportunistic sit-and-wait predator; eats primarily small rodents but also takes rabbits, birds and snakes.

Voice: powerful, descending *keeearrrr* scream.

Similar Species: *Rough-legged Hawk* (p. 151): white tail base; dark "wrist" patches; broad, dark terminal tail band. *Broad-winged Hawk* (p. 147): smaller; banded tail; broader wings with pointed tips; no belly band. *Swainson's Hawk* (p. 148): white face; dark "bib"; dark flight feathers; pale underwing linings; more pointed wing tips; holds wings in shallow "V" in flight. *Ferruginous Hawk* (p. 150): dark legs; pale underparts.

FERRUGINOUS HAWK

Buteo regalis

Coursing the contours of rolling, grassy hills, circling high above the landscape or sitting alertly in a remote grassland, the Ferruginous Hawk is a bird of open country. Although greatly reduced in numbers and nesting range, this hawk is still the predominant buteo found in the wide-open plains of the Canadian Prairies. It avoids overgrown aspen parkland and areas that have been extensively altered by agricultural cultivation, preferring extensive undisturbed pasture, hayland and prairie complexes. • Ferruginous Hawks were once shot and poisoned because they were considered pests. But, with the realization that much of their diet consists of pesky ground squirrels and pocket gophers, nesting pairs are warmly welcomed by ranchers and farmers. Many landowners have cooperated in the building of artificial nesting platforms to encourage Ferruginous Hawks to establish in areas where nest sites are in short supply. Nevertheless, overall Ferruginous numbers remain low, and the species is considered threatened in Canada. • This hawk has two colour morphs, the common light phase and a rarer dark phase, which occurs in less than 10 percent of the population and is more prevalent in western parts of its nesting range.

ID: sexes similar; dark rufous "leggings." *Light adult:* rusty red shoulder patches; white underparts; pale head; rufous-tipped, pale tail. *Dark adult:* dark brown overall; white tail; dark wing linings; pale flight feathers. *Immature:* very pale underparts and underwing flight feathers; paler leg feathering. *In flight:* holds wings in shallow "V" while soaring; very large wingspan.

Size: *L* 56–69 cm; *W* 1.4 m.

Habitat: open grasslands, badlands and croplands. *In migration:* foothills and mountain ranges.

Nesting: usually in a solitary tree or shrub; more rarely on a cliff, on the ground or on an artificial nesting platform; nests generally reused year after year; massive nest of sticks, weeds and cow dung is lined with finer materials; female and occasionally male incubate 2–5 finely speckled, white eggs for 32–34 days.

Feeding: hunts from the air, a low perch or the ground; eats primarily ground squirrels, rabbits and hares, mice and pocket gophers; rarely takes snakes and birds.

Voice: generally silent; call is a loud, squealing *kaaarr*, usually dropping at the end.

Similar Species: *Red-tailed Hawk* (p. 149): smaller; belly band of dark brown streaks; lacks reddish legs; wings held horizontal in flight. *Swainson's Hawk* (p. 148): dark flight feathers contrast with light wing linings; dark adult is entirely dark underneath. *Rough-legged Hawk* (p. 151): dark "wrist" patches and breast streaking; dark brown band or streaking on belly.

ROUGH-LEGGED HAWK

Buteo lagopus

dark adult

light adult

Rough-legged Hawks breed in tundra or taiga regions of the Canadian Arctic and winter in open country of southern Canada and the northern U.S. Numbers of breeding pairs and their nesting success fluctuate wildly, depending on highs and lows in the cycle of northern voles and lemmings. When lemming and vole numbers are high, Rough-legs can produce up to seven young, but in lean years, a pair is fortunate to raise a single chick. • This hawk can be limited in distribution and numbers by the availability of elevated nest sites, and pairs often nest on a high cliff ledge, in a crevice or on a boulder pile, but occasionally on a tower or abandoned radar station. • Easily identified from afar when foraging, the Rough-legged Hawk often "wind-hovers" to scan the ground below, flapping to maintain a stationary position while facing upwind. • The Rough-leg's plumage is highly variable, with light, dark and intermediate phases. This species is our only buteo with distinctive male and female plumages.

ID: large hawk; pale flight feathers with dark tips; brown eyes; yellow cere; legs feathered down to toes. *Light adult:* white mottling on dark brown upperparts; dark streaking on pale head and breast; wide, dark belly band; in flight. *Dark adult:* dark brown overall; whitish primaries. *In flight:* black "wrist" patches in light adults; whitish tail with wide, black terminal band; frequently hovers.

Size: *L* 48–61 cm; *W* 1.2–1.4 m.

Habitat: open country, including fields, rangeland, grasslands, farmland and marshes; also visits alpine areas in autumn migration.

Nesting: usually on an elevated structure or cliff; occasionally in a tree; large stick nest is often reused; mostly the female incubates 3–4 variably marked, white eggs for 28–31 days.

Feeding: eats mostly lemmings and voles; occasionally takes Arctic ground squirrels and birds, especially ptarmigans.

Voice: catlike *kee-eer* alarm call.

Similar Species: *Other buteos* (pp. 146–50): much smaller; rarely hover; lack dark "wrist" patches and broad, white tail base. *Northern Harrier* (p. 142): smaller; slimmer; facial disc; long, narrow tail; low-level cruising flight. *Golden Eagle* (p. 152): much larger; broader wings; dark, unpatterned underwings; immature's broader white tail has dark band at tip.

GOLDEN EAGLE

Aquila chrysaetos

The Golden Eagle is a huge and very powerful bird of rugged landscapes—the average wingspan of an adult exceeds two metres! • Unfortunately, with the advent of widespread human development and intensive agricultural practices, this noble bird became the victim of a lengthy persecution. Perceived as a threat to livestock, bounties were offered, supporting the shooting and poisoning of this regal bird. • In Canada, Golden Eagles breed primarily in mountainous country in western Canada, with a small number breeding in isolated areas of eastern Canada. • With their remarkable eyesight, these birds can spot prey from 2.5 kilometres away. • While adult Golden Eagles are uniformly coloured, except for the tawny head, immatures have a distinct, sharply defined, white patch in the centre of each wing and at the base of the tail. Immature Bald Eagles have blotchier white patches on the leading edge of their underwings. • In the last few years, naturalists have discovered a mind-boggling migration of thousands of Golden Eagles through a corridor in the Alberta Rockies. They fly high—the reason why they were missed for so long—but good binoculars and spotting scopes have allowed eagle watchers to document the phenomenon.

ID: sexes similar but female is larger; feathered legs; gold-tinted head and neck; yellow cere; greyish bill; lightly banded tail with broad, dark terminal band; yellow feet. *Immature:* prominent white patches on wings and across base of tail. *In flight:* relatively long tail; effortless wingbeats; holds long, broad, flat wings in a slight "V."

Size: *L* 76–102 cm; *W* 2.0–2.3 m.

Habitat: *Breeding:* open, mountainous country with rocky cliffs. *In migration* and *winter:* alpine, mountain slopes, open woodlands, rangelands and agricultural fields.

Nesting: on a cliff ledge or top of a tall tree; usually reuses nest site; huge nest platform, to 3 m across, is built of sticks, branches and rootlets; pair incubates 1–4 brown-marked, creamy buff eggs for 35–45 days.

Feeding: swoops to capture prey with talons; eats mainly ground squirrels, prairie dogs, marmots and rabbits; occasionally takes young ungulates or larger mammals; also eats carrion.

Voice: generally quiet; utters yelps, lower-pitched *chiup* calls and short barks.

Similar Species: *Bald Eagle* (p. 141): immature has larger head, heavier bill, bare legs and more diffuse white patches on wings; lacks white tail base; holds wings flat in flight. *Turkey Vulture* (p. 139): naked head looks small in flight; dark wing linings; silvery flight feathers; rocks and teeters in flight with wings in a sharp "V." *Rough-legged Hawk* (p. 151): smaller; dark adult has pale flight feathers and white tail base.

AMERICAN KESTREL

Falco sparverius

The American Kestrel is the smallest, most numerous and most widespread of our five falcons. It is found from Alaska to southern Argentina, with 18 recognized subspecies. • Kestrels are less inclined toward the all-out aerial attacks that have made other falcons such revered predators. The colourful male has blue-grey wings and a rufous tail with a single, broad, black, subterminal band, while the wings and tail of the female are rufous with black bars. Kestrels are frequently observed along roadsides, either perched on utility wires and fence posts or hovering low over a ditch or meadow searching for ground-dwelling prey. A hovering kestrel appears fixed in space, its wings beating so rapidly that they appear blurred, and its tail adjusting to each eddy in the breeze. • Although formerly known as "Sparrow Hawk," the kestrel's diet consists of insects and small rodents, and rarely includes small birds. • American Kestrels are unique among Canadian raptors, as they nest in tree cavities and nest boxes. Most evidence suggests that kestrel populations are declining significantly in many areas, and lack of nest cavities may be part of the reason.

ID: small falcon; bobs tail when perched; multicoloured head; 2 black facial stripes; grey crown; white face; buffy nape with black patch. *Male*: blue-grey wings; black-spotted underparts; bold, black subterminal band on tail. *Female*: black barring on rufous back; rufous wings; rufous-streaked underparts; rufous and black tail bands. *In flight*: long wings; long, rufous tail; frequently hovers.
Size: *L* 19–20 cm; *W* 51–61 cm.
Habitat: open areas covered by short ground vegetation; attracted to human-modified habitats such as pastures and parkland; also clear-cuts and open, dry forests; often near heavily developed urban areas.

Nesting: in a tree cavity or nest box; mostly the female incubates 3–7 finely speckled, white to pale brown eggs for 29–30 days.
Feeding: swoops from a perch or hovers; eats primarily large insects, especially grasshoppers, crickets and dragonflies; also takes small mammals, birds and reptiles.
Voice: shrill *killy-killy-killy* or *klee-klee-klee*; female's voice is lower pitched.
Similar Species: *Merlin* (p. 154): larger; lacks rufous on back, wings and tail; only 1 facial stripe; does not hover. *Sharp-shinned Hawk* (p. 143): lacks rufous on back, wings and tail; lacks facial stripes; does not hover; flap-and-glide flight.

MERLIN

Falco columbarius

"Prairie"

"Pacific"

Only slightly larger than an American Kestrel, the Merlin considers itself the equal of any falcon, snatching small songbirds and shorebirds in midair at high speed, in classic falcon fashion. The elements of speed and surprise work best in open country close to forest edges, where the Merlin is most at home. • Merlins feed mostly on birds up to the size of Rock Pigeons. Unlike American Kestrels, Merlins seldom perch on power lines and usually hunt from a snag or other conspicuous perch. In western Canada, Merlins are common nesters in urban areas and are on the increase in eastern Canada. • Rural cemeteries and parks with mature spruce trees are good places to look for these birds. • Most Merlins migrate to Central and South America each fall, but many are enticed to stay in southern Canada by mild urban climates and an abundance of overwintering songbirds, which in turn are attracted by numerous fruitbearing trees and backyard feeders. • The former name "Pigeon Hawk" refers to the bird's appearance in flight, not its preferred prey.

♀

♂

"Taiga"

ID: small falcon; 1 vertical facial stripe; dark eye line; heavy brown streaking on pale underparts. *Male:* blue-grey upperparts; lightly streaked, darker crown; long, dark tail with greyish bands. *Female* and *immature:* dark brown upperparts; grey to light brown tail bands. *In flight*: long, narrow, pointed wings; very rapid, shallow wingbeats.

Size: *L* 25–30 cm; *W* 58–66 cm.

Habitat: *Breeding:* semi-open and open coniferous, deciduous or mixed woods. *In migration* and *winter:* farmlands, estuaries, forest edges, urban and suburban yards and alpine areas.

Nesting: in a mature tree, often a conifer; usually uses an abandoned stick or squirrel nest; mostly the female incubates 4–6 chestnut-marked, whitish eggs for 25–32 days;.

Feeding: eats primarily birds, from siskins to medium-sized shorebirds; also takes dragonflies, small rodents and bats.

Voice: loud, noisy, cackling *kek-kek-kek-kek-kek* or *ki-ki-ki-ki* cry, in flight or while perched.

Similar Species: *American Kestrel* (p. 153): smaller; more colourful; 2 facial stripes; often hovers. *Peregrine Falcon* (p. 156): larger; dark "helmet"; pale, unmarked upper breast. *Prairie Falcon* (p. 157): larger; dark "wing pits." *Sharp-shinned Hawk* (p. 143): short, rounded wings; vertical breast streaks. *Rock Pigeon* (p. 253): broader wings in flight; shorter tail; often glides with wings held in a "V."

GYRFALCON

Falco rusticolus

white adult

dark adult

The Gyrfalcon, largest of the falcons, is one of the world's most powerful avian hunters. Unlike the Peregrine Falcon, which often attacks from above in a steep dive, the Gyrfalcon prefers to put on a burst of speed to overtake its prey. It is primarily a ptarmigan specialist, with a breeding distribution that is strikingly similar to that of the Rock Ptarmigan. • Gyrfalcons exhibit pronounced reversed sexual size dimorphism, with males typically weighing one-third less than females. • Gyrfalcons are extremely variable in coloration, occurring in four colour morphs: white, grey, dark and brown (the immature of the grey morph). • Although never common, this bird is not nearly as rare as many believe. The remoteness of its breeding habitat, fluctuations in percentages migrating south in winter, difficulties in identification and the mere rumours of its rarity have all combined to make this species frequently overlooked. • This speedy falcon breeds in the Northern Hemisphere primarily above 60 degrees latitude. A significant percentage of adults, especially males, remain on breeding territories during winter; birds migrating to southern Canada are mostly immatures and subadults.

grey adult

ID: sexes similar but females much larger; heavyset body; broad wings; long, faint, barred "moustache"; long, wide, tapered tail; wings appear shorter than tail on perched bird. *White adult:* white overall; faintly barred back and tail. *Grey adult:* grey upperparts with some mottling; white underparts with brown or grey streaks or blotches; yellowish cere. *Dark adult:* dark upperparts; light underparts with dark streaks. *Immature:* darker grey-brown back; heavily streaked underparts; 2-tone underwings; grey feet and cere. *In flight:* slow, steady wingbeats; pale flight feathers.

Size: *Male: L* 51–56 cm; *W* 1.2–1.3 m.
Female: L 56–64 cm; *W* 1.3–1.4 m.

Habitat: *Breeding:* mountains with cliffs and bluffs usually near rivers. *In migration* and *winter:* open and semi-open areas, including marshes, agricultural fields and open wetlands with prey concentrations; estuaries and brackish cattail marshes on the coast.

Nesting: on ledge of a rocky cliff; in a scrape or old stick nest; may reuse nest without adding material; male helps female incubate 2–4 cinnamon-spotted, white to creamy white eggs for 28–35 days.

Feeding: strikes prey in midair and carries or follows it to the ground; eats mainly ptarmigans and medium-sized waterfowl; also takes other birds, ground squirrels and rabbits.

Voice: loud, harsh *kak-kak-kak* when near nest site; silent in winter.

Similar Species: *Prairie Falcon* (p. 157): dark "wing pits"; shorter tail. *Peregrine Falcon* (p. 156): distinctive, dark "helmet"; shorter tail. *Northern Goshawk* (p. 145): dark cap; light eyebrows; greyer underparts with finer streaking; unstreaked, white undertail coverts; rounded wings in flight. *American Kestrel* (p. 153) and *Merlin* (p. 154): smaller.

PEREGRINE FALCON

Falco peregrinus

The word "peregrine" means "wanderer," and Peregrine Falcons nesting in Canada's north are among long-distance migratory champions, some travelling 25,000 kilometres annually. This bird is widely distributed, occurring on every continent except Antarctica. • When chasing prey in direct pursuit, Peregrine Falcons have been clocked at 112 kilometres per hour, and when stooping in a full dive, they can reach speeds of 320 kilometres per hour, making this species one of the world's fastest flying animals. • Formerly known as "Duck Hawk" for its winter diet, this falcon feeds primarily on medium-sized birds in summer and shorebirds in migration. In urban areas, pigeons, doves and starlings are favoured prey items. • The Peregrine Falcon's legendary speed was no defence against the insidious effects of pesticides such as DDT, and from 1945 to 1970, numbers declined alarmingly as a result of pesticide-related eggshell thinning and reduced reproductivity. Intensive conservation and reintroduction efforts have brought about a heartening recovery. Since the 1970s, these reintroductions combined with increasing numbers in the wild have resulted in numerous urban nestings in city centres throughout Canada.

ssp. pealei

ssp. tundrius

ssp. anatum

ID: sexes similar; crow-sized raptor; black "moustache"; long, pointed wings; narrow tail; blue-grey back and wings; prominent black "helmet"; white face with bold, black "sideburns"; pale underparts with fine, black barring. *Immature:* brown where adult is blue-grey; pale crown; narrower "sideburns"; heavily brown-streaked breast. *In flight*: pointed wings.

Size: *Male: L* 38–43 cm; *W* 94–109 cm. *Female: L* 43–48 cm; *W* 1.1–1.2 m.

Habitat: *Breeding:* open marine, river and lake shorelines with steep rock cliffs; high bridges and skyscrapers in urban areas. *In migration* and *winter:* open country such as lakeshores, seacoasts, estuaries, farmland and large wetlands.

Nesting: on ledge of a rock cliff, steel bridge beam or ledge of a tall building; often reuses nest site; pair incubates 3–4 cinnamon-spotted, yellowish white eggs for 32–36 days.

Feeding: opportunistic; takes a wide variety of avian prey including alcids, ducks, shorebirds, gulls, pigeons and starlings.

Voice: generally silent; loud, harsh, persistent *kak-kak-kak-kak* near the nest.

Similar Species: *Prairie Falcon* (p. 157): smaller; narrower "moustache" is separate from crown patch; sandy brown upperparts; brown-spotted underparts; dark "wing pits" on whiter underwings. *Gyrfalcon* (p. 155): larger; less contrast in plumage; lacks bold facial markings. *Merlin* (p. 154): smaller; lacks bold "moustache" stripe.

PRAIRIE FALCON

Falco mexicanus

Prairie Falcons are mainly summer birds of arid grasslands and sagebrush flats of the southern British Columbia Interior east to southern Saskatchewan, nesting on the sandstone cliffs and bluffs that punctuate open plains and shrub-steppe deserts. A few may overwinter each year within their breeding range. • An efficient and specialized predator of small desert mammals and birds, the Prairie Falcon ranges widely, searching large areas for patchily distributed prey. In spring and summer, it concentrates on hunting small to medium-sized birds as well as ground squirrels. As summer fades to fall, flocks of migrating shorebirds and songbirds capture their attention. • When young falcons first leave the nest, they have all the equipment for spectacular flight but none of the know-how. Like any human pilot learning to fly a plane, recently fledged falcons only get to keep their wings if they survive flight training. Indeed, inexperienced and overeager birds risk serious injury or death when pushing the limits in early hunting forays. • More a "wanderer" than a true "migrant," few individuals fly directly to their wintering areas in the southern Great Plains, most taking a circuitous route well north and east of their nesting range before heading south. • Although it has been estimated that less than 500 pairs breed within Canada, breeding bird surveys have shown significant increases in Prairie Falcon populations throughout North America.

ID: medium-sized, light brown raptor; pale face; narrow, dark brown "moustache"; yellow cere; sandy brown upperparts; brown-spotted, white underparts; lightly banded tail; yellow feet. *In flight:* pointed wings; broad, blackish "wing pits."
Size: *Male: L* 36–38 cm; *W* 94–99 cm. *Female: L* 43–46 cm; *W* 1.0–1.1 m.
Habitat: *Breeding:* arid grasslands, agricultural fields and sagebrush flats with nearby rocky cliffs and promontories. *In migration* and *winter:* alpine, lakeshores, stockyards and pastures.
Nesting: on a rocky cliff ledge or in a crevice on a rocky promontory; nest is usually unlined; mainly the female incubates 2–6 darkly speckled, pale eggs for 29–33 days.
Feeding: high-speed, strike-and-kill by stoop or low flight over open country; eats mammals such as ground squirrels and pocket gophers and mainly ground-dwelling birds such as California Quail, Western Meadowlarks and Horned Larks.
Voice: generally silent; yelping *kik-kik-kik-kik* alarm call.
Similar Species: *Peregrine Falcon* (p. 156): distinctive dark "helmet"; broader, pointier wings; uniform underwing coloration. *Merlin* (p. 154): much smaller; female usually has darker back, upper- and underwings and narrower, dark or heavily banded tail.

YELLOW RAIL

Coturnicops noveboracensis

Under a blanket of darkness, the Yellow Rail slips quietly through tall sedges, grasses and cattails, searching for food. This rather rare bird hides behind a cover of dense, marshy vegetation by day and is most active at night. It is almost always encountered through its distinctive and repetitive call, heard along marsh edges in spring. The male's territorial call, generally heard only in complete darkness, is a unique, repetitive, five-note *tik, tik, tik-tik-tik*, easily imitated by tapping two stones together. But catching a glimpse of these marsh phantoms is the ultimate challenge. When standing motionless, the rail's tawny yellow and dark stripes blend in perfectly with the marsh vegetation. And when required to, this bird can make itself "as thin as a rail," its laterally compressed body allowing it to effortlessly slip through tightly packed stands of marsh vegetation. Its large feet, which help the bird rest atop thin mats of floating plant material, adds to the Yellow Rail's strange appearance. • Agricultural expansion has claimed a large share of this rail's habitat in southern Canada, and concerns regarding low overall numbers and continuing habitat loss throughout its breeding and wintering range has resulted in its recent designation as a species of special concern in Canada.

ID: sexes similar; short, pale bill; white-barred, black and tawny stripes on upperparts and flanks; dark brown crown; broad, dark line through eye; white throat and belly; yellow buff underparts. *Juvenile:* darker overall; pattern on upperparts extends onto breast, sides and flanks. *In flight:* white patch on trailing edge of inner wing.

Size: *L* 17–19 cm; *W* 27–28 cm.

Habitat: damp meadows and marshes with abundant grasses and sedges; prefers a mixture of new growth and dry, dead grasses that form overhanging cover.

Nesting: on the ground or low over water, hidden by overhanging plants; shallow cup nest is made of grasses and sedges; female incubates 8–10 brown-speckled, buff-coloured eggs for 18–23 days.

Feeding: picks food from the ground and aquatic vegetation; takes mostly snails, aquatic insects, spiders and possibly earthworms; occasionally eats seeds.

Voice: *Male:* repeated pattern of short, metallic notes: *tik, tik, tik-tik-tik*.

Similar Species: *Sora* (p. 161): black face and throat; bright yellow bill; lacks white wing patches and stripes on back; different call. *Virginia Rail* (p. 160): rusty breast; grey face; long, reddish bill; lacks white wing patches and stripes on back.

KING RAIL

Rallus elegans

The King Rail is the largest rail in North America, even though it is roughly only the size of a farmyard chicken. Unlike some of the more secretive rails, it is often seen wading through shallow water along the edge of a freshwater marsh, stalking its prey within full view of eager onlookers. Crayfish, crabs, small fish, spiders, beetles, snails, frogs and a whole host of aquatic insects keep this formidable hunter occupied and well fed. • King Rail nests, which are commonly built above shallow water, often include a protective dome of woven vegetation and a well-engineered entrance ramp. Despite these deluxe features, young rails and their attending parents desert the nest mere hours after the eggs hatch. • With the loss and degradation of large, suitable wetlands, King Rail populations have experienced significant declines in the last 100 years. The species is critically imperiled in most of its U.S. range. This bird reaches the northern limit of its range in southern Ontario, where it is believed that fewer than 50 breeding pairs remain. Because this is the only nesting locale in Canada and is believed to be declining, the King Rail is ranked as endangered in Canada.

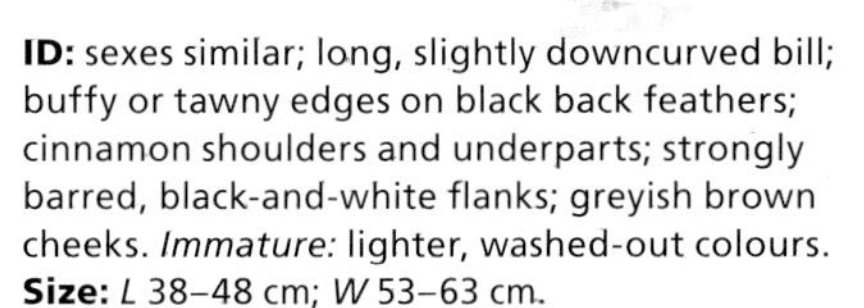

ID: sexes similar; long, slightly downcurved bill; buffy or tawny edges on black back feathers; cinnamon shoulders and underparts; strongly barred, black-and-white flanks; greyish brown cheeks. *Immature:* lighter, washed-out colours.
Size: *L* 38–48 cm; *W* 53–63 cm.
Habitat: freshwater marshes, shrubby swamps, marshy riparian shorelines and flooded fields with shrubby margins.
Nesting: in a clump of grasses or sedges just above the water or ground; mostly the male builds a platform nest with canopy and entrance ramp using marsh vegetation; pair incubates 10–12 brown-spotted, pale buff eggs for about 21–23 days.
Feeding: eats aquatic insects, crustaceans and occasionally seeds; forages in shallow water, often in or near dense plant cover, for small fish and amphibians.
Voice: chattering call is 10 or fewer evenly spaced *kek* notes.
Similar Species: *Virginia Rail* (p. 160): much smaller; brown back feathers; grey face; red bill. *Least Bittern* (p. 128): buff-orange face and wing patches; thicker bill; all-black back feathers lack pale edging.

VIRGINIA RAIL

Rallus limicola

The best way to encounter a Virginia Rail may be to sit alongside a cattail-dominated marsh, clap your hands three or four times and wait patiently. At best, this slim bird might reveal itself for an instant, but on most occasions you will only hear this elusive bird. The Virginia's calls are varied, but excited pairs often issue a curious descending series of grunting *oink* notes. • When pursued by an intruder or predator, a rail will almost always attempt to scurry away through dense, concealing vegetation, rather than risk exposure in a fluttering getaway flight. Rails are very narrow birds that have modified feather tips and flexible vertebrae, all of which allow them to squeeze through the narrow confines of their marshy homes. • For protection against predators and quick transition to a new nest should their initial nest fail, Virginia Rail pairs build numerous "dummy" nests within their territory in addition to their primary nest. • The Virginia Rail and its relative the Sora are often found living in the same marshes. The secret of their successful coexistence is found in their microhabitat preferences and distinct diets. The Virginia Rail typically favours dry shoresides of marshes and feeds on invertebrates, whereas the Sora prefers waterfront property and eats plants and seeds.

ID: sexes similar; long, slightly downcurved, red bill; short, upturned tail; rusty streaking on dark upperparts; dark crown; grey cheeks; rufous throat, breast and upper belly; bold, black and white barring on underparts. *In flight:* rusty shoulders on brownish upperwings; grey underwings with white leading edge; dangling feet.
Size: *L* 23–28 cm; *W* 33 cm.
Habitat: *Breeding:* freshwater and brackish wetlands, especially dense cattail, bulrush and sedge marshes. *In migration* and *winter:* wetland areas with suitable overhead cover, particularly coastal saltwater marshes in winter.

Nesting: over water in emergent marsh vegetation; loose basket nest is made of coarse grasses, cattail stems and leaves or sedges; pair incubates 7–12 brown-spotted, pale buff eggs for 18–20 days.
Feeding: probes soft mud or vegetation for a variety of invertebrates; also eats seeds or small fish.
Voice: telegraph-like *kidick, kidick* or *wak-wak-wak* call; also an accelerating series of *oink* notes that trail off in pitch and energy.
Similar Species: *Sora* (p. 161): black mask; short, yellow bill; greenish yellow legs; straight wings in flight. *King Rail* (p. 159): much larger; dark legs; lacks reddish bill and grey face.

SORA

Porzana carolina

Halfway between a crazed laugh and a horse's whinny, the Sora's call puzzles many visitors to cattail marshes. In addition to the whinnying calls that emanate from a Sora-inhabited marsh, their alternate calls, a loud *kur-ree* or a sharp *keek*, are frequently heard. Although it is the most common and widespread rail in North America, the Sora is seldom seen. Like other rails, its elusive habits and preference for dense marshlands force most observers to settle for a quick glimpse of this odd-looking bird. • Soras are chickenlike birds with large feet that help them walk and rest atop thin mats of floating vegetation. Despite their relatively plump-bodied appearance, Soras share other rails' ability to slip effortlessly through tightly packed stands of bulrushes and cattails, and they swim quite well over short distances. Although it is rarely seen flying, its long migrations, sometimes over thousands of kilometres, prove that it is quite capable of extended flight. • Northern Harriers, Great Horned Owls, mink, weasels, raccoons and even Great Blue Herons are some of the many creatures that occasionally prey on Soras.

breeding

nonbreeding

ID: sexes similar; short, yellow bill; brown nape and crown; grey face, neck, and breast; black mask and throat; white-streaked, black mottled, brown back and tail; black, brown and white barring on underparts; holds tail upright; buffy undertail coverts; yellow legs and feet.
Size: *L* 20–25 cm; *W* 35 cm.
Habitat: *Breeding:* shallow freshwater wetlands with abundant, dense, emergent vegetation. *In migration* and *winter:* freshwater and brackish wetlands and sedge meadows.
Nesting: over or near the water under concealing vegetation; well-built, basketlike nest is made of grasses and dry aquatic vegetation; pair incubates 8–12 brown-spotted, buff eggs for 16–19 days.
Feeding: gleans or probes shallow water or muddy surfaces for aquatic insects, molluscs and seeds.
Voice: several calls, among them a simple *peep*, a *kerwee* and a descending "whinny" that lasts about 3 seconds; also loud *kur-ree* or sharp *keek* alarm calls.
Similar Species: *Virginia Rail* (p. 160): long, downcurved, reddish bill; rusty breast; call is a telegraph-like *kidick, kidick. Yellow Rail* (p. 158): smaller; upperparts streaked black and tawny; white throat; call is a metallic *tik, tik, tik-tik-tik.*

COMMON MOORHEN

Gallinula chloropus

The Common Moorhen, formerly known as "Common Gallinule," is a curious mix of comedy and confusion. This member of the rail family appears to have been assembled by committee: it has the bill of a chicken, the body of a duck and its long legs and huge feet are similar to those of a heron. Moorhens are, in fact, ecologically and behaviourally intermediate between the rails and the American Coot. • Although Moorhens might look gangly and awkward, their strides are executed with a poised determination. As a moorhen strolls around a wetland, its head mirrors the movement of its legs, bobbing back and forth in synchrony with each step and producing a comical, chugging stride. • This bird is a capable swimmer and occasional diver, though its feet are not webbed and it lacks lobed toes. • In addition to their regular nest platforms, Common Moorhens construct a number of other platforms on which adults might brood their young or roost. For moorhens, the responsibilities of parenthood do not end when their eggs have hatched—parents will feed and shelter their young until they are capable of feeding themselves and flying on their own.

breeding

ID: sexes similar; grey-black body; white streak on flank; reddish bill and forehead shield; white undertail coverts; yellow-tipped bill; long, yellowish green legs. *Breeding:* brighter bill and forehead shield. *Juvenile:* browner plumage; duller legs and bill; whitish throat.

Size: *L* 30–38 cm; *W* 51–53 cm.

Habitat: large, freshwater marshes with standing water and broken stands of tall emergent vegetation.

Nesting: in shallow water or along a shoreline; pair builds a shallow nest of bulrushes, cattails and reeds, often with a ramp to the water; pair incubates 6–17 speckled, cinnamon buff eggs for 19–22 days.

Feeding: gleans prey while swimming, walking on land or clambering through marsh vegetation; omnivorous diet of aquatic vegetation, berries, fruits, tadpoles, insects, snails, worms and spiders; may take carrion and eggs.

Voice: various sounds include chickenlike clucks, screams, squeaks and a loud *cup*; courting males give a harsh *ticket-ticket-ticket*.

Similar Species: *American Coot* (p. 163): white bill; lacks white streak on flank.

AMERICAN COOT

Fulica americana

The American Coot is truly an all-terrain bird: in its quest for food, it dives and dabbles like a duck, grazes confidently on land and swims about skillfully with its lobed feet. Sometimes called "Mudhen," the coot is the extrovert of the rail world. While the rest of the clan is furtive, coots swim in open waters and are often mistaken for ducks. • American Coots are raucous and aggressive territorial combatants throughout the nesting season. When a dispute arises, announced by loud cackling, grunting and croaking calls, coots flail their wings and raise their tails, proudly displaying prominent white undertail patches. If the posturing fails to intimidate, rivals run across the water's surface in a charge, strike each other with their bills or attempt to hold their opponent underwater. As the nesting season wanes, coots become less competitive and band together in large groups in preparation for fall migration. • An awkward and rather clumsy flier, the American Coot requires a long running takeoff to become airborne. • A newly hatched coot is easily recognized by its gaudily coloured, reddish orange down and bald, red crown. • American Coot numbers numbers fluctuate widely within its core breeding area—the prairie pothole region of western Canada—in response to wet and dry seasons.

ID: sexes similar; fairly large, chunky waterbird; sooty black body; longish, yellow-green legs and lobed toes; white bill with dark, broken ring near tip; white forehead shield; red eyes. *Immature:* greenish grey legs; lacks forehead shield.

Size: *L* 33–40 cm; *W* 58–70 cm.

Habitat: *Breeding:* shallow marshes, ponds and wetlands with open water and emergent vegetation; also sewage lagoons. *In migration* and *winter:* lakes, protected marine bays, brackish marshes, lagoons and estuaries.

Nesting: on water among emergent vegetation; pair builds a bulky, floating nest of cattails, reeds and grasses; pair incubates 8–12 brown-spotted, pale buff eggs for 21–25 days.

Feeding: opportunistic; mainly eats plant material obtained from grazing on land, gleaning the water's surface or diving to depths of up to 8 m; also eats small fish and aquatic invertebrates.

Voice: very vocal in summer; *kuk-kuk-kuk-kuk-kuk* call; also *kakakakaka* and a variety of croaks, cackles and grunts.

Similar Species: *Common Moorhen* (p. 162): reddish forehead shield; yellow-tipped bill; white streak on flank. *Scoters* (pp. 73–75): larger; males have colourful bills and smaller, webbed, pinkish feet; found on open salt water in winter. *Other ducks* (pp. 54–72, 76–83): lack white, chickenlike bill and uniformly black body. *Grebes* (pp. 104–09): lack white forehead shield and all-dark plumage.

SANDHILL CRANE

Grus canadensis

Deep, resonating, rattling croaks signal the approach of a flock of migrating Sandhill Cranes. The Sandhill Crane's flight style, with a snapping upstroke and slower downstroke, differentiates it from other long-legged, long-necked birds. The distinctive bugling calls are deceiving because the crane's coiled trachea alters the pitch of its voice, making it sound louder but farther away. Migrating cranes sail effortlessly for hours, using thermal updrafts to circle to such great heights that they can scarcely be seen, and then gliding gently back down until they find another rising thermal and repeat the process. The combination of spiralling and gliding allows cranes to fly nonstop for hundreds of kilometres in a single day. • Cranes mate for life and reinforce pair bonds each spring with an elaborate courtship dance in which partners leap high into the air with their wings half spread and then bow like courtiers. • Migrating and wintering cranes often congregate in huge numbers in traditional staging areas, making this species particularly vulnerable to the loss of strategic wetlands. • Sandhill Cranes are sensitive nesters, preferring to raise their young in areas that are isolated from human disturbance.

ID: sexes similar; very large; greyish body (sometimes stained reddish); unfeathered, red crown; white cheeks; long, straight neck and legs; pointed beak; "shaggy" tail; long, blackish bill, legs and feet. *Immature:* greyish brown body; no facial markings. *In flight:* huge wingspan; neck and legs extended; migrant flocks soar or fly in loose lines or V-formation.

Size: *L* 1.0–1.3 m; *W* 1.8–2.1 m.

Habitat: *Breeding:* isolated, open marshes, fens and bogs surrounded by forests or shrubs. *In migration:* agricultural fields and shorelines.

Nesting: on a large mound of aquatic vegetation in shallow water; pair incubates 1–3 brown-splotched, olive buff eggs for 29–32 days.

Feeding: probes and gleans the ground for plant tubers, roots and seeds; also takes invertebrates and a variety of small vertebrates.

Voice: call is a loud, resonant, rattling *gu-rrroo gu-rrroo gurrroo*, audible for great distances.

Similar Species: *Great Blue Heron* (p. 129): lacks red crown patch; S-shaped neck in flight. *Whooping Crane* (p. 165): extremely rare; all-white plumage; black flight feathers.

WHOOPING CRANE

Grus americana

This magnificent bird, the largest in North America, wavered on the brink of extinction in the 1940s, when the total wild population dropped to only 15 birds. Since then, one of the most intensive conservation and public education programs in history has increased that number to 523 birds (as of spring 2008), including 146 in captivity. Today, the only self-sustaining wild population of Whooping Cranes in North America nests in Wood Buffalo National Park on the Alberta–Northwest Territories border and in nearby areas. As well, tremendous efforts have been made to establish a nonmigratory flock and a second migratory flock. The nonmigratory flock was established in central Florida in 1993 and appeared wildly successful, with more than 20 nesting attempts and the first wild-hatched chicks realized during the first 10 years. But, by 2008, facing extremely high predation and mortality, the Florida release program was curtailed. An ambitious effort to establish a Wisconsin–Florida migratory population, using ultralight planes to guide young birds in their initial fall flight, began in 2000. Birds from this flock occasionally become disorientated and wander into Ontario during their spring migration north to Wisconsin. As of spring 2009, this population consisted of 85 birds, and 12 pairs had established nesting territories. Clearly, the future of this species has yet to be decided.

ID: sexes similar; all-white body; unfeathered, red crown; red streak below eye; black primary flight feathers; black bill, legs and feet. *Juvenile:* extensive rusty feathering on upperparts.
Size: *L* 1.3–1.5 m; *W* 2.0–2.3 m.
Habitat: *In migration:* wetlands, croplands and fields.
Nesting: in a fen, bog or isolated marsh; slight nest depression on a large pile of heaped vegetation; pair inclubates 2 variably marked, buff-coloured eggs for 33–35 days; usually only 1 chick survives.

Feeding: picks food from the water or the ground; eats amphibians, small mammals and plant material on breeding grounds; waste grain and some animal material in migration.
Voice: bugling rattle: *ker-loo ker-lee-loo.*
Similar Species: *Sandhill Crane* (p. 164): grey plumage. *American White Pelican* (p. 122): much shorter legs; black secondary feathers. *Snow Goose* (p. 46): much smaller; shorter legs and bill. *Swans* (pp. 51–53): all-white wings.

BLACK-BELLIED PLOVER

Pluvialis squatarola

The plaintive, three-note, ascending whistle of the Black-bellied Plover is a common migration sound on beaches, mudflats, plowed fields and sodfarms. Wary and quick to give its plaintive alarm call, this bird functions as a sentinel for mixed shorebird flocks. • The Black-bellied is the largest of North America's plovers. Its black-and-white breeding plumage is seen briefly in spring as birds race northward to make full use of the short Arctic summer. • Black-bellied Plovers forage with a robinlike, run-and-stop technique, frequently pausing to lift their heads for a reassuring scan of their surroundings. • This plover winters along North and Central American coastlines but can be found inland at favourite foraging sites during migration. Truly one of the widest-ranging shorebirds, the Black-bellied Plover breeds widely in the High Arctic of North America and Eurasia, where it is known as "Grey Plover," and winters over a broad range of latitudes, equally at home in temperate and tropical climates. • Rarer migrants such as Red Knots, Short-billed Dowitchers, Dunlins, and Buff-breasted Sandpipers may be associated with this plover. • Plovers usually have three toes, but the Black-bellied has a fourth toe high on its leg, like most sandpipers.

nonbreeding

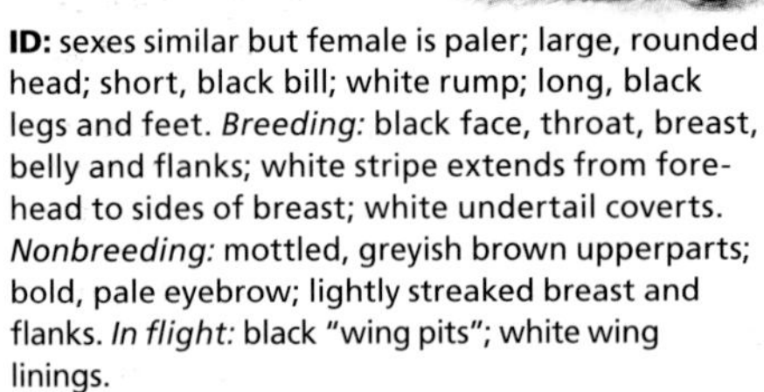

breeding

ID: sexes similar but female is paler; large, rounded head; short, black bill; white rump; long, black legs and feet. *Breeding:* black face, throat, breast, belly and flanks; white stripe extends from forehead to sides of breast; white undertail coverts. *Nonbreeding:* mottled, greyish brown upperparts; bold, pale eyebrow; lightly streaked breast and flanks. *In flight:* black "wing pits"; white wing linings.

Size: *L* 27–33 cm; *W* 74 cm.

Habitat: *Breeding:* relatively dry tundra. *Coastal (winter):* sandy beaches, mudflats, estuaries, agricultural fields, golf courses and airport fields. *Interior (migration):* mudflats, lakeshores, wet meadows, golf courses and fields.

Nesting: on a slightly raised area; shallow hollow is lined with moss or lichens; mainly the male incubates 4 variably marked, pale buff or greyish eggs for about 27 days.

Feeding: picks prey from the water's surface; probes in mud or moist soil; mainly eats small, segmented sea worms, crustaceans and clams.

Voice: rich, plaintive 3-syllable *pee-oo-ee* whistle; flocks may utter mellow, 2-note whistles.

Similar Species: *American Golden-Plover* (p. 167): golden-mottled upperparts; dark crown; all-black underparts in breeding plumage; lacks black "wing pits" in flight. *Red Knot* (p. 190): longer bill; shorter legs; lacks black "wing pits."

AMERICAN GOLDEN-PLOVER

Pluvialis dominica

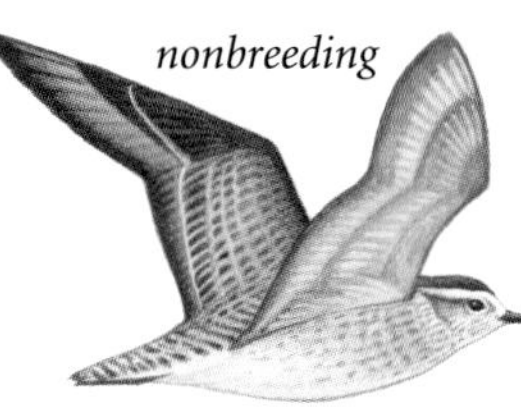

A mere 150 years ago, the American Golden-Plover population was among the largest of any bird in the world, but flocks were decimated by hunting in the late 1800s and early 1900s. Hunted mercilessly in both spring and fall, a single day's shooting often yielded tens of thousands of birds. The Eskimo Curlew (*Numenius borealis*), another Arctic-nesting shorebird that was slaughtered by the millions during the market-hunting era and is now believed to be extinct, once migrated with the American Golden-Plover between the Canadian Arctic and South America. • Few bird migrations in North America rival the route undertaken by American Golden-Plovers. After breeding on the tundra, the birds feed almost continuously for weeks to accumulate extensive reserves of fat. Then they depart eastern Canada and travel southeast over the Atlantic Ocean in a nonstop flight of over 4000 kilometres. Migrating American Golden-Plovers remain aloft for about 50 hours before landing in northern South America to refuel. The birds winter in southern South America. During spring migration, American Golden-Plovers take a more westerly route north, overland through central portions of North America. • This bird's breeding plumage is a fine example of both disruptive and cryptic coloration. Although the bird is boldly marked, the pattern breaks up the image of the bird and confuses the eyes of predators.

breeding

ID: sexes similar but female is duller; medium-sized, chunky shorebird; rounded head; short neck; short, black bill; dark rump; long, black legs. *Breeding:* yellow and black speckling on crown, nape, back and wings; solid black face and underparts; white neck stripe extends along flank. *Nonbreeding:* pale gold mottling on brown-and-grey upperparts; rusty crown; dusky cheek patch; light eyebrow; grey-mottled underparts; light belly. *In flight:* dark underwing coverts with no contrasting "wing pits."

Size: *L* 25 cm; *W* 66 cm.

Habitat: *Breeding:* tundra with low vegetation. *Nonbreeding:* recently plowed fields, overgrazed pastures, golf courses, airport fields, sewage ponds, coastal mudflats, sand beaches and estuaries.

Nesting: on an exposed site; ground scrape is unlined or lined with lichens; mainly the female incubates 4 heavily blotched, creamy white eggs for 26–27 days.

Feeding: gleans short grasses or the ground for insects and other invertebrates; eats some berries; crustaceans and small molluscs form a major part of coastal diet.

Voice: flight call is a soft, melodious, whistled *queedle* or *que-e-a*, dropping at the end.

Similar Species: *Black-bellied Plover* (p. 166): pale crown; white undertail coverts; conspicuous, black "wing pits" in flight; lacks gold speckling on upperparts and white flank stripe. *Pacific Golden-Plover* (p. 500): white stripe extends from face to tail; wings extend to tip of tail.

SEMIPALMATED PLOVER

Charadrius semipalmatus

nonbreeding

On the way to their northern breeding grounds, small flocks of Semipalmated Plovers routinely touch down in southern Canada for a brief stopover. If these birds seem to be in a hurry, they are! There is a tremendous amount of pressure for these long-distance migrants to begin breeding before the end of the short northern summer. If successful, the adults will leave their breeding grounds as early as July to enjoy a prolonged, leisurely migration to the coastlines of the southern U.S., Central America or South America. • Darting here and there in short runs interspersed with deliberate pauses or stabbing probes, the Semipalmated Plover's attention-grabbing feeding methods and small, plump profile distinguish it amid flocks of sandpipers. Plovers also have shorter, sturdier bills than sandpipers, which prevent them from feeding in deeper water but allow them to probe harder ground to retrieve their invertebrate prey. • Shorebird plumage is generally categorized in two forms: cryptic or disruptive coloration. Most sandpipers have cryptic coloration, allowing them to blend into their preferred habitats. Plovers exhibit disruptive coloration, with distinctive banding that breaks their body form into unrecognizable pieces when seen from a distance. Like the stripes of a zebra, the contrast between dark and light makes it difficult for predators to pinpoint the bird's form.

breeding

ID: sexes similar; brown upperparts; white underparts; 1 dark breast band; short bill. *Breeding:* dark forecrown and sides of head; white patch above stubby, black-tipped, yellowish bill; yellow-orange legs and feet. *Nonbreeding:* duller plumage; brown, incomplete breast band; dark legs and bill. *In flight:* head appears large; broad, white upperwing stripe.

Size: *L* 18 cm; *W* 45–49 cm.

Habitat: *Breeding:* tundra, gravel bars of rivers, sandy shores and lakeshores. *Nonbreeding:* mud- and sandflats, beaches, estuaries, brackish lagoons, muddy lakeshores and flooded agricultural fields; sometimes uses sewage lagoons.

Nesting: on sand, gravel or tundra, usually near water; slight depression is sparsely lined with vegetation; pair incubates 4 darkly marked, creamy buff eggs for 23–25 days.

Feeding: run-and-stop foraging technique; gleans or probes the water's surface for small molluscs, crustaceans, worms and insects.

Voice: high-pitched, rising, whistled *tu-wee.*

Similar Species: *Killdeer* (p. 170): much larger and heavier; red eye ring; dark bill; 2 black breast bands; rusty rump; pale, pink-tinged legs. *Piping Plover* (p. 169): much lighter upperparts; lacks dark band through eyes.

PIPING PLOVER

Charadrius melodus

The Piping Plover's plumage blends in perfectly with the open beaches, alkali flats and sandflats where it nests, and the dark bands across its forehead and neckline resemble the shadows cast by scattered pebbles or strips of washed-up vegetation. • Young Piping Plovers are able to run within a day of hatching. These tawny little puffballs match their surroundings perfectly and are even tougher to spot than their parents. • The Piping Plover's camouflage has done little to protect it from habitat loss, increased predation and disturbance by humans. The recreational use of beaches during summer and an increase in human-tolerant predators such as gulls, raccoons and skunks, has impeded this bird's ability to reproduce successfully. Populations have declined in the past few decades, and the species is endangered throughout its relatively small range in North America. The Piping Plover now occurs as three small, disjunct breeding populations in Canada—in the southern regions of the Prairie provinces, the Great Lakes and along the Atlantic Coast. Beachgoers are becoming increasingly familiar with the fencing and warning signs erected around breeding sites that have greatly aided in the recovery of this species. In 2007, Piping Plovers nested on the Canadian side of the Great Lakes for the first time in 30 years.

♂

breeding

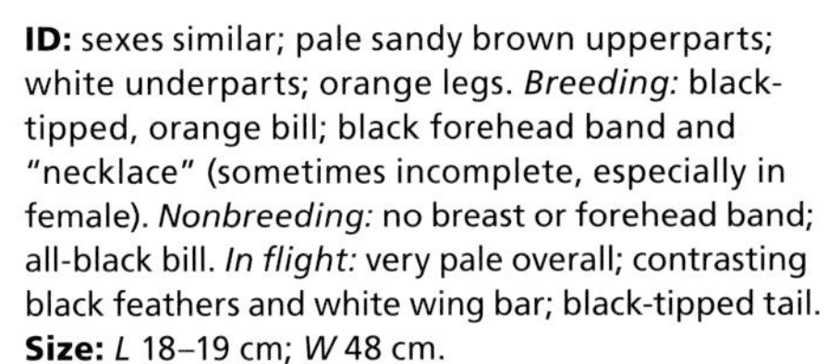

ID: sexes similar; pale sandy brown upperparts; white underparts; orange legs. *Breeding:* black-tipped, orange bill; black forehead band and "necklace" (sometimes incomplete, especially in female). *Nonbreeding:* no breast or forehead band; all-black bill. *In flight:* very pale overall; contrasting black feathers and white wing bar; black-tipped tail.

Size: *L* 18–19 cm; *W* 48 cm.

Habitat: sandy beaches on freshwater lakes; open shorelines on saline lakes.

Nesting: on bare ground along an open shoreline; shallow scrape is lined with pebbles; pair incubates 4 darkly blotched, pale buff eggs for 25–31 days.

Feeding: run-and-stop feeding technique; eats worms, insects and other invertebrates washed up on shore.

Voice: clear, whistled *peep peep peep-lo*.

Similar Species: *Semipalmated Plover* (p. 168): much darker upperparts; dark patch through eyes. *Killdeer* (p. 170): larger; 2 breast bands; much darker upperparts.

KILLDEER

Charadrius vociferus

When an intruder wanders too close to its nest, the crafty Killdeer puts on its "broken wing" distraction display. Feigning injury, one of the Killdeer pair will drag a wing along the ground, issue plaintive calls and expose its rufous rump to lure the intruder away from its eggs or young. • Although the Killdeer is well known for its masterful broken wing act, several ducks and many other shorebird species also practise this defence strategy. • This ubiquitous and widespread plover is often the first shorebird that naturalists learn to identify. Its loud *killdee* or *dee-dee-dee* calls and bright, distinctive plumage make it conspicuous. • The Killdeer has adapted well to urbanization and is often found in habitats many miles from water, including roadside ditches, gravel driveways, golf courses and abandoned industrial areas. Once in serious decline because of hunting, the Killdeer is probably more widespread today than at any time in its history as a result of habitat creation by agriculture, forest clearing and other human-induced changes. Nevertheless, breeding bird surveys have shown a slight decline of this species in Canada, particularly in western and central regions.

ID: sexes similar; brown upperparts; white underparts; 2 breast bands; white neck ring; brown crown, nape and ear patch; white eyebrow and throat; orange rump; dull pinkish legs and feet. *In flight:* longish tail; white underwing with dark trailing edge; white wing stripe; bouncy, somewhat erratic flight.

Size: *L* 22–28 cm; *W* 48–53 cm.

Habitat: *Breeding:* gravel bars, human-altered open spaces, parking lots, fields, rangeland, pastures and lakeshores. *Nonbreeding:* mudflats, lakeshores, open, dry or wet grassy habitats, sewage lagoons, golf courses, airports and muddy estuaries.

Nesting: on open ground; in a shallow depression, usually unlined, but sometimes lined with pebbles, grasses or wood chips; pair incubates 4 darkly blotched, pale buff eggs for 24–28 days; may have 2 broods per season.

Feeding: run-and-stop forager; eats terrestrial insects, spiders, worms, small crustaceans and infrequently seeds.

Voice: distinctive, noisy, repeated *kill-dee*; also a plaintive *dee-dee-dee* and *deer-deer.*

Similar Species: *Semipalmated Plover* (p. 168): smaller; 1 breast band. *Piping Plover* (p. 169): smaller; much lighter upperparts; 1 breast band.

MOUNTAIN PLOVER

Charadrius montanus

nonbreeding

Don't let the name fool you—the Mountain Plover is actually a resident of the dry, open plains and prairies, not the mountains. An uncommon bird with a rather small nesting range, the Mountain Plover breeds in pockets across the Great Plains region from the southern extremes of Alberta and Saskatchewan to Texas. Considered endangered in Canada because of its extremely small numbers and population decline, the species has also disappeared in many areas of the eastern and southern portions of its range. Cultivation of short-grass prairies and habitat loss on both this bird's wintering range and breeding areas is a serious threat. • Like the Upland Sandpiper, the Mountain Plover rarely approaches shorelines, making it more of a "grass-piper" than a sandpiper. • This plover exhibits a most unusual breeding strategy in which the female usually lays eggs in two separate nest sites; the male assumes responsibility for the first clutch, while the female incubates the second. • A rather tame and unwary species, disturbed plovers often crouch or run a short distance, relying upon the disguise of their drab, earthy-coloured plumage for protection.

breeding

ID: sexes similar; sandy brown upperparts; pale underparts; white forehead and eyebrow; black forecrown; thin, black eye line; thin, dark bill.
Size: *L* 20–24 cm; *W* 58 cm.
Habitat: sparse, dry prairies, heavily grazed pastures and mudflats; habitat rejuvenation through burning has produced favourable results.
Nesting: on the ground near a dry hummock or large rock; shallow scrape may be lined with small amounts of grass, cow chips and roots; female often lays eggs in 2 nests; pair incubates (separately) 2 clutches of 3 black-spotted, olive buff eggs for 28–31 days.
Feeding: gleans the ground for insects, especially grasshoppers.
Voice: shrill call note; low whistles during spring and summer.
Similar Species: *Upland Sandpiper* (p. 182): larger; longer neck; long bill. *Killdeer* (p. 170): 2 breast bands. *Buff-breasted Sandpiper* (p. 202): buffy underparts; buffy face.

AMERICAN OYSTERCATCHER

Haematopus palliatus

Wouldn't life be great if you could spend every day wading in the ocean surf, eating fresh, tasty seafood morsels and basking in the Atlantic Coast sunshine? And, of course, there is nothing like rounding out each day with some "wing-surfing" on the salty ocean breeze...if only you were as lucky as those peculiar American Oystercatchers! • These large, stocky shorebirds with their long, razor-sharp, orange bills specialize in prying or hammering open shellfish, including oysters, clams and mussels. When the opportunity arises, they will gladly eat a whole host of other intertidal invertebrates such as limpets, crabs, marine worms, sea urchins, chitons and even jellyfish. • During the summer breeding season, watch for mating pairs performing their loud "piping" courtship display. These strident, calling performances are often given in flight. In winter, they feed along coastal beaches and mudflats in small noisy flocks. • A relatively new breeding species to Canada, the first confirmed breeding record came from Cape Sable Island, Nova Scotia, in 1997.

ID: sexes similar; long, laterally compressed, orange-red bill; black head and neck; brown back; white wing and rump patches; white underparts. *In flight:* black-and-white wing pattern; wings are bowed.
Size: *L* 47 cm; *W* 78–81 cm.
Habitat: coastal marine habitats, including saltwater marshes, sandy beaches and tidal mudflats; will nest on dredge-spoil islands.
Nesting: usually on a gravel beach or ridge in a salt marsh; depression in sand may be lined with shells or pebbles; pair incubates 3 speckled, buffy gray eggs for 27 days; may form a breeding trio of 2 females and 1 male who tend 1–2 nests.
Feeding: finds prey visually or by probing while walking in shallow water, mud or on rocks; takes intertidal invertebrates, including molluscs, crustaceans and marine worms; shellfish form bulk of diet and are hammered open or quickly stabbed through an opening in the shell and cut open.
Voice: loud *wheet* call is often given in series during flight.
Similar Species: none.

BLACK OYSTERCATCHER

Haematopus bachmani

Black Oystercatchers are conspicuous, devoted members of British Columbia's rocky intertidal community. They do not stray far from their rocky shoreline habitat where their meals of limpets, mussels and snails are abundant. Their flamboyant, red, chisel-shaped bills are well adapted for prying open tightly sealed shells, but require practise to use. The young stay with their parents for up to a year while they perfect their foraging technique. Pairs often stray from their territories in winter, forming flocks that can number in the hundreds in areas of high mussel density. Because of their territoriality and the spotty distribution their preferred habitat, Black Oystercatchers are limited in numbers. • Ravens, crows and garter snakes are major predators of this bird's eggs and chicks, but if a Black Oystercatcher survives the perils of its first year, its life at the beach can be as long as 15 years! • The genus name *Haematopus*, Greek for "blood red eye," refers to a related Old World species.

ID: sexes similar; brownish black overall; long, straight, blood-red bill (longer in female); red eye ring; bright yellow eyes; sturdy, pale pink legs. *Immature:* browner body; dark-tipped, orange bill. *In flight:* short tail; broad wings; often calls.

Size: *L* 43 cm; *W* 79 cm.

Habitat: *Breeding:* upper areas of sandy or gravelly marine beaches, bare rock outcrops, gravel roofs and abandoned light beacons. *Nonbreeding:* rocky shorelines, including islands, breakwaters, jetties and reefs.

Nesting: on open ground or bare rocks above the high-water mark; nest scrape may be lined with pebbles and shells; pair incubates 1–3 darkly marked, buff eggs for 26–28 days.

Feeding: forages on falling or rising tides; eats mainly mussels, limpets, barnacles and snails; also takes sea urchins, marine worms, small crabs and other invertebrates; rarely eats oysters.

Voice: loud, piercing, repeated *wheep* or *wik* is usually given in flight; longer series of notes, accelerating into a frantic, uneven trill, in spring or during territorial disputes.

Similar Species: none.

BLACK-NECKED STILT

Himantopus mexicanus

Whether in a small prairie slough, a sewage lagoon or an alkaline pond, the Black-necked Stilt's black-tie plumage always adds a bit of class to the landscape. It is perhaps fortunate that one of North America's most beautiful birds associates with such bleak and dreary environments. With proportionally the longest legs of any North American bird, this shorebird truly deserves the name "stilt." • Black-necked Stilts forage by sight, plucking morsels from the water while using their long legs to explore varying depths. • When nesting, Black-necked Stilt parents are easily provoked and call loudly and incessantly while circling low over their nesting territory. Neighbouring adult stilts participate jointly in antipredator defence; one strategy, called the "popcorn display," consists of adults encircling a ground predator and hopping from side to side while flapping their wings. This loud-voiced bird also serves as a sentinel for other marsh-nesting birds. • In the last 30 years, the Black-necked Stilt's breeding range has expanded northward. Following many summer sightings in western Canada, the first successful nesting of the species was realized in Alberta during the late 1970s, followed by first nesting records for Saskatchewan, British Columbia and finally Manitoba.

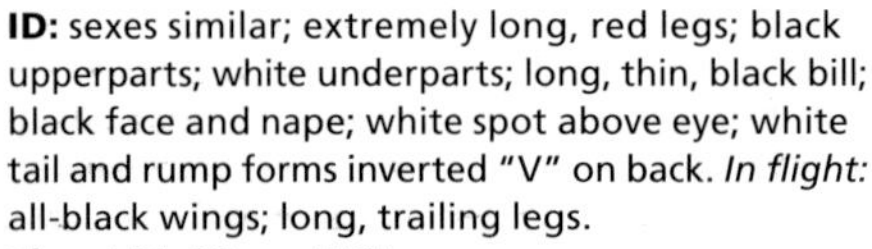

ID: sexes similar; extremely long, red legs; black upperparts; white underparts; long, thin, black bill; black face and nape; white spot above eye; white tail and rump forms inverted "V" on back. *In flight:* all-black wings; long, trailing legs.
Size: *L* 36–38 cm; *W* 74 cm.
Habitat: *Breeding:* flooded fields or marshes with muddy, exposed shorelines. *Nonbreeding:* flooded, shallow wetlands, sewage ponds and lake and marine mudflats.

Nesting: in a shallow depression on slightly raised ground near water; nest scrape is lined with plant materials; pair incubates 4 darkly blotched, buff eggs for 22–26 days.
Feeding: plucks prey, mostly aquatic invertebrates, from the water's surface.
Voice: loud, shrill *yip yap yip yap* at nest site; also *kik-a-rik* or *kek kek kek kek* flight calls.
Similar Species: *American Avocet* (p. 175): bulkier; peach-coloured (breeding) or light grey (nonbreeding) head and neck; upturned black bill; shorter, bluish legs; broad wings with black-and-white upperwing pattern.

AMERICAN AVOCET

Recurvirostra americana

American Avocets in full breeding plumage might just be the most elegant birds in Canada. Indeed, their elegant features, graceful movements and striking colours are unmatched. Avocets are easily identified at a distance by their feeding habit of running in ankle-deep water and whisking their long, upturned bills from side to side to capture invertebrates. They may also wade into deep water, dipping their heads beneath the water's surface or tipping up like a surface-feeding duck. • During courtship, pairs sometimes join in a "circling display," which involves three or more birds forming a circle, heads facing inward and angled toward the water. The birds rotate slowly in one direction and then the other, all the while trumpeting loudly. After mating, pairs may be observed performing a bond reinforcement display, in which males and females cross their bills and walk together in perfect unison. • Nesting avocets are highly territorial and will readily join forces with neighbouring pairs to attack intruders and defend their nests and chicks from birds as large as a Northern Harrier. • Before fall migration, American Avocets stage in large flocks on large, shallow, saline lakes. By this time, their peach-coloured hoods have been replaced by subtler winter greys.

breeding

nonbreeding

ID: sexes similar; black-and-white wing pattern; long, slender, delicately upturned, black bill; white underparts; long, bluish grey legs and feet. *Breeding:* peach-coloured head, neck and upper breast. *Nonbreeding:* pale greyish white head and neck.
Size: *L* 42–48 cm; *W* 68–96 cm.
Habitat: *Breeding:* open, shallow freshwater and alkaline ponds, marshes, lakes and sewage lagoons. *Nonbreeding:* marine mudflats, open, shallow wetlands and flooded fields.
Nesting: semicolonial; on a dried mudflat or exposed shoreline; sometimes on a mound or islet; in a shallow depression to which materials are often added during incubation; pair incubates 3–4 darkly marked, dusky eggs for 22–29 days.
Feeding: scythes bill through shallow water to pick up small fish and crustaceans, aquatic insects and sometimes seeds; sometimes wades or swims into deeper water and pecks, plunges, lunges or tips up.
Voice: noisy aerial pursuits near nest site; loud, shrill, repeated *kleet.*
Similar Species: *Black-necked Stilt* (p. 174): slimmer; straight bill; black pattern on head; mostly white body; mostly black back, wings and nape; reddish legs.

SPOTTED SANDPIPER

Actitis macularius

In a rare case of sexual role reversal, the female Spotted Sandpiper is the dominant partner. She lays the eggs but leaves it to her mate to tend the clutch while she diligently defends her territory. She may mate with several males, competing with other females for mates, and can initiate clutches for up to five males in a single season. This unusual mating system is called "polyandry" and is found in only about one percent of all bird species. • Spotted Sandpipers are solitary birds, preferring to feed singly or in pairs along rocky or sandy shores of rivers and freshwater lakes. And, like Solitary Sandpipers, Spotted Sandpipers are known for their continuous teetering behaviour as they forage, bobbing their hind ends nearly continuously. Their flight style is also distinctive, and they fly low over the water with fluttery, stiff-winged strokes, like a wire under tension that has been "twanged." • Apart from the ubiquitous Killdeer, the charming Spotted Sandpiper is perhaps North America's best-known and most easily identified shorebird. It is one of the few species recorded in every province and territory.

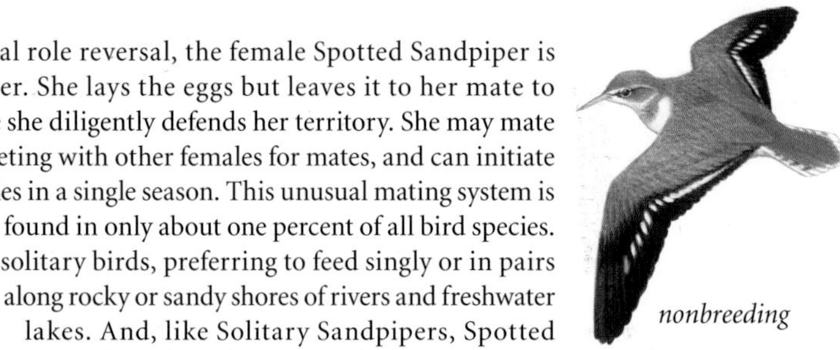

nonbreeding

breeding

ID: sexes similar; small shorebird; pale legs; constantly bobs tail. *Breeding:* plain brown upperparts, tail and rump; white underparts with bold, black spotting; yellow bill. *Nonbreeding:* light brown head; short, white eyebrow; black eye line; whitish throat; brown "spur" on sides of breast; no spots on underparts. *In flight:* white stripe on upperwings.
Size: *L* 18–20 cm; *W* 35–38 cm.
Habitat: *Breeding:* lakeshores, beaches, marshes, streams, roadsides, sewage ponds and rivers, usually within 50 m of fresh water. *Nonbreeding:* mudflats, beaches, sewage ponds, breakwaters, water-filled ditches and, infrequently, intertidal shores.

Nesting: on the ground among vegetation; shallow depression is bare or lined with plant stems; male incubates 4 brown-blotched, buff eggs for 20–24 days.
Feeding: opportunistic; eats mostly invertebrates but occasionally takes small fish.
Voice: sharp *peet-weet* call; series of *weet* notes in flight.
Similar Species: *Solitary Sandpiper* (p. 177): longer bill; dusky face with white eye ring; olive legs and feet; lacks spotted upperparts; plain wings in flight. *"Peep" sandpipers* (pp. 192–96): usually in flocks; do not bob hind ends; lack combination of pale legs and feet and pale bill; all lack spotted underparts in breeding plumage.

SOLITARY SANDPIPER

Tringa solitaria

nonbreeding

For years, this wily shorebird defied all attempts to locate its nest. Finally, in 1903, the Solitary Sandpiper's nesting strategy was revealed when a homesteader in Alberta peered into what he thought was a robin's nest, but found a sandpiper instead. Even though the closely related yellowlegs species regularly call and perch in trees on their nesting territories, the Solitary Sandpiper is North America's only tree-nesting shorebird. • True to its name, the Solitary Sandpiper is somewhat reclusive and does not feed or migrate in large flocks. • Solitary Sandpipers, like other shorebirds, lay very large eggs relative to their body size and incubate them for comparatively long periods of time. Once sandpiper chicks break out of their eggs, they are ready to run, hide and feed on their own. These highly developed hatchlings, known as precocial young, are immediately able to fend for themselves. • The Solitary Sandpiper, like the Wandering Tattler and Spotted Sandpiper, has a distinctive behaviour that distinguishes it from other shorebirds: it slowly bobs its hind end as it forages or rests. • In contrast to the white-rumped yellowlegs, the Solitary can be identified in flight by its dark rump and its white-sided tail.

breeding

ID: sexes similar; bobs hind end; white eye ring. *Breeding:* dark upperparts heavily spotted with white; fine, dark streaking on head, neck and breast; white underparts; olive legs and feet. *Nonbreeding* and *juvenile:* less defined streaking on head, neck and breast. *In flight:* dark wings; dark rump and central tail feathers; white-sided tail.

Size: *L* 19–23 cm; *W* 53–56 cm.

Habitat: *Breeding:* mainly permanent shallow wetlands, often with dead trees, such as swamps, bogs and willow marshes. *In migration:* shallow freshwater ponds, ditches, farm dugouts, sewage treatment lagoons and flooded agricultural fields.

Nesting: in the abandoned nest of another bird species, usually a robin or blackbird; pair incubates 4 variably marked, creamy white eggs for 23–24 days.

Feeding: gleans and probes in water and soft mud for aquatic and terrestrial invertebrates.

Voice: high-pitched, thin *peet-wheet*; clipped *plik*.

Similar Species: *Lesser Yellowlegs* (p. 181): larger; white rump; lacks eye ring; yellow legs and feet; often in flocks. *Spotted Sandpiper* (p. 176): nonbreeding bird has unspotted, greyish brown upperparts, incomplete eye ring, pale bill and buffy wing stripe in flight.

WANDERING TATTLER

Tringa incana

nonbreeding

Named for its migratory prowess, the Wandering Tattler remains little known to most Canadians. Many of these birds head out to sea to reach distant shores during migration, but a small number hug the Pacific coastline and linger at rocky headlands, jetties and tide pools before continuing as far south as Peru, eastern New Guinea and Australia. The Wandering Tattler has been recorded as a vagrant in migration as far east as Ontario. • This bird's breeding sites remained a mystery until 1912, when a geologist exploring river gravel bars discovered the bird's secret. Eventually, nests and chicks were found in northwestern British Columbia, Yukon, Alaska and northeastern Siberia. Although Wandering Tattlers conduct flight displays and much of their daily activity along riverine gravel bars, recent studies found that nests are often situated on tundra hundreds of metres removed from water. • The tattler tends to stand and walk in a horizontal posture while bobbing its tail, and its cryptic plumage makes it difficult to spot when it stands still. • The name "tattler" is derived from this bird's habit of giving a rapid series of clear, hollow alarm whistles at the first sign of any perceived threat.

breeding

ID: sexes similar; short, yellow legs; long, straight bill; dark eye line. *Breeding:* uniformly grey upperparts; grey underparts with dense, blackish barring. *Nonbreeding* and *juvenile:* uniformly grey upperparts; grey underparts; white belly and undertail coverts. *In flight:* 2-tone, dark grey wings; low, flicking flight.

Size: *L* 25 cm; *W* 63–66 cm.

Habitat: *Breeding:* gravel flats along mountain streams; tundra, occasionally far away from rivers and streams. *Nonbreeding:* rocky coastlines, reefs, jetties and breakwaters.

Nesting: in a hollow among rocks or gravel; shallow scrape is lined with plant material; pair incubates 4 brown-spotted, olive eggs for 23–25 days.

Feeding: forages among intertidal rocks and seaweed for crustaceans, molluscs, marine worms and other invertebrates.

Voice: generally silent; short series of rapid-fire whistles: *lididididi*; crisp *klee-ik* alarm call.

Similar Species: *Surfbird* (p. 189): chunkier; browner plumage; stubby, dark-tipped, yellow bill; white wings and tail in flight. *Black Turnstone* (p. 188): black head, breast and upperparts; short, dark bill; complex white pattern on back, wings and tail in flight. *Rock Sandpiper* (p. 199): nonbreeding bird has slightly downcurved bill, dark breast, dark-flecked belly and sides, and white upperwing stripe in flight.

GREATER YELLOWLEGS

Tringa melanoleuca

The Greater Yellowlegs often performs the role of lookout among mixed flocks of shorebirds. Animated and vocal, this large sandpiper begins calling incessantly at the first sign of danger, bobbing its head and moving slowly away. If forced to, the Greater Yellowlegs will usually retreat into deeper water, becoming airborne only as a last resort. • During migration, many shorebirds, including the Greater Yellowlegs, often stand or hop around beachflats on one leg. These stubborn "one-leggers" may be mistaken for crippled individuals, but this stance is an adapation to conserve body heat. • The two species of yellowlegs are medium-sized sandpipers with very similar plumages. The Greater Yellowleg is larger, with a much longer bill and thicker legs with noticeable "knees," and it utters a louder, more strident three-note series of whistles (as opposed to the Lesser Yellowlegs' one- or two-note call). Those who have never seen a yellowlegs except in migration or winter are always shocked to see these birds on their summer territories, perching and calling from the tops of tall trees. • The heavy, black streaking and barring of the breeding plumage, seen on northward-migrating birds in spring, explains the species name of *melanoleuca*, Greek for "black and white."

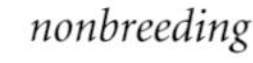

nonbreeding

breeding

ID: sexes similar; medium-sized shorebird; long, yellow legs; white rump; long neck; long bill appears slightly upturned. *Breeding:* brownish upperparts speckled with black and tan; finely streaked head and neck; narrow, white eye ring merges with dark eye line; black-barred breast and flanks; finely barred tail. *Nonbreeding:* grey overall; fine streaks on dusky breast; unmarked, pale underparts. *In flight:* unmarked, grey-brown upperwings; pale underwings; mostly white uppertail with terminal barring.
Size: *L* 33–38 cm; *W* 68–70 cm.
Habitat: *Breeding:* edges of muskeg bogs and marshes in boreal forest. *In migration* and *winter:* wet agricultural lands, standing water in grassy fields, beaches and mudflats; any type of shallow wetland, whether freshwater, brackish, or marine; also estuaries and mudflats.
Nesting: on the ground, usually near water; in a depression on a dry mound or ridge; well-hidden nest is sparsely lined with leaves, moss and grass; female incubates 4 heavily brown-spotted, creamy buff eggs for 23 days.
Feeding: picks prey from mud or the water's surface or sweeps bill back and forth through water; eats small fish, crustaceans, snails and aquatic insects.
Voice: call is a loud, 3-note, whistled *tew-tew-tew.*
Similar Species: *Lesser Yellowlegs* (p. 181): smaller; straight bill is shorter (roughly equal to length of head); quieter, 1–2 note call. *Solitary Sandpiper* (p. 177): greenish legs; barred tail.

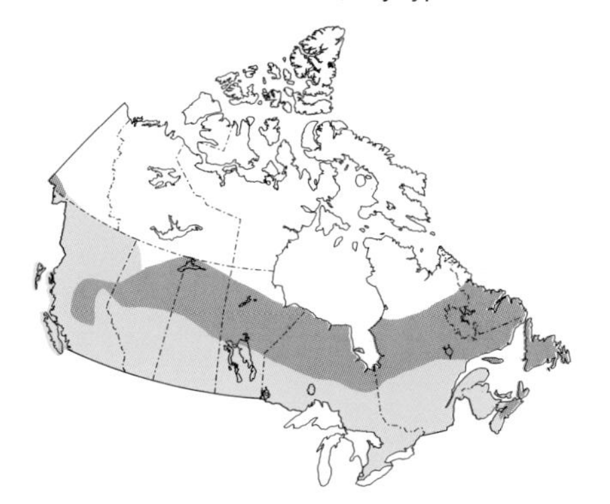

WILLET

Tringa semipalmata

nonbreeding

This large shorebird's greyish plumage makes it seems a rather dull figure as it walks slowly along a shoreline, but the instant it takes flight, its wondrous black-and-white wings flash in harmony with its loud, incessant *pill-will willet* calls. It is widely believed that the bright, bold markings on the Willet's wings serve as a warning to other shorebirds or that they double as a means of intimidating would-be predators when pairs are defending their young. • In spring, most Willets establish territories that are fiercely defended throughout the early summer. Where there are several pairs, they parcel out the area and take turns scolding and dive-bombing potential predators of all sizes. Once young are out of the nest, the aggressive tendencies subside and feeding parties form. • Willets consist of two distinct subspecies, which differ in biology, song and plumage. The "Eastern Willet," *T. s. semipalmata*, is primarily a coastal bird of the Maritime provinces, whereas the "Western Willet," *T. s. inornata*, breeds in the freshwater sloughs of the Prairie provinces and migrates to both ocean coasts to winter. • "Willet" is an onomatopoeic description of this bird's common call.

breeding

ID: sexes similar; large, plump shorebird; heavy, straight, black bill; pale throat and belly; dull bluish grey legs. *Breeding:* grey-brown plumage has intricate dark streaking and barring overall. *Nonbreeding:* lightly mottled, grey-brown plumage; white belly. *In flight:* eye-catching, black-and-white wing pattern.

Size: *L* 38 cm; *W* 66 cm.

Habitat: *Breeding:* wet grassy meadows, usually close to water. *In migration:* wet fields and shores of marshes, ponds and lakes. *Winter:* coastal tidal mudflats.

Nesting: in tall grass; shallow depression is lined with finer grasses; pair incubates 4 brown-blotched, olive-buff to greyish eggs for 22–29 days.

Feeding: probes muddy areas and gleans the ground for insects; may eat shoots and seeds.

Voice: loud, repetitive *pill-will willet*; also a rapidly repeated *kip-kip-kip*; *whee-wee-wee* in flight.

Similar Species: *Marbled Godwit* (p. 186): warm brown plumage; larger, much longer, slightly upturned bill is orange or pinkish with black tip. *Greater Yellowlegs* (p. 179): smaller; long, yellow legs; lacks bold wing pattern in flight.

LESSER YELLOWLEGS

Tringa flavipes

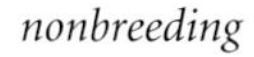

nonbreeding

A daintier version of its larger cousin, the Lesser Yellowlegs nevertheless appears considerably taller than the small sandpipers with which it usually associates in migration. Lesser Yellowlegs are among the first shorebirds to arrive in spring and often the last to leave in fall. • Lesser Yellowlegs and Greater Yellowlegs look very similar, and many birders find it a challenge to distinguish between them. In addition to overall body size, focus on the bill—the Lesser's bill is much finer and roughly equal to the length of its head. • Lesser Yellowlegs breed in wetlands in boreal forests across northern Canada and Alaska. • Although this bird is well known for its vigourous and boisterous defence of its nest and chicks, the female starts her southward migration early, leaving the male to care for the brood when the young are only about 11 days old. • Despite the length of its bill, the Lesser Yellowlegs does not probe for food. • Yellowlegs are sometimes called "telltales" because they alert all the shorebirds on a mudflat to invaders such as birders.

breeding

ID: sexes similar; medium-sized shorebird; long, yellow legs; long neck; long, straight, black bill (equal to length of head); white rump and tail. *Breeding:* brownish upperparts finely speckled with black and white; finely streaked head, neck and upper breast; flanks lightly barred with black. *Nonbreeding:* upperparts less distinctly marked; dusky breast; pale underparts. *In flight:* unmarked greyish wings; pale wing linings; mostly white tail with terminal barring.

Size: *L* 25–28 cm; *W* 58–60 cm.

Habitat: *Breeding:* open spaces near shallow wetlands in boreal forests. *In migration* and *winter:* small ponds, shallow marshes, flooded fields, mudflats, lakeshores and sewage lagoons; also marine estuaries, bays and lagoons.

Nesting: on a dry mound, often in muskeg; shallow depression is sparsely lined with leaves and grasses; pair incubates 3–6 (commonly 4) heavily blotched, greyish eggs for 22–23 days.

Feeding: mostly picks up prey from the water's surface or mud; feeds mainly on insects in summer and small fish and crustaceans at other times.

Voice: call is 1–2 *tew* notes.

Similar Species: *Greater Yellowlegs* (p. 179): larger; longer, slightly upturned bill (greater than length of head); louder, 3-note call. *Solitary Sandpiper* (p. 177): bolder eye ring; dark rump; greenish legs and feet; constantly bobs hind end.

UPLAND SANDPIPER

Bartramia longicauda

Upland Sandpipers are often seen in spring perched atop fence posts and utility poles, gracefully raising and then folding their wings after landing and belting out their unique "wolf-whistle" courtship tunes. Excited males will even launch into the air to perform a courtship flight, combining their wavering song with shallow, fluttering wingbeats. At the height of the breeding season, however, these small-headed, large-eyed, prairie sandpipers are rarely seen, remaining hidden in the tall grass of abandoned fields, hayfields and lightly grazed pastures. • Breeding pairs are faithful to traditional nest sites, and unlike many birds, all the young hatch at the same time. • After a short breeding season, Upland Sandpipers migrate south in late July toward the grasslands of Argentina for their second summer. • During the late 1800s, a high demand for Upland Sandpipers as tablefare led to severe overharvesting and catastrophic declines in populations over much of North America. Their numbers have since rebounded, but healthy populations occur only in areas where grassland habitat persists. • Many old-timers still refer to this species as "Upland Plover"—its name was officially changed in the 1970s.

ID: sexes similar; medium-sized shorebird; small head; mottled, brownish upperparts; lightly streaked breast, sides and flanks; light belly and undertail coverts; straight bill is about as long as head; large, dark eyes; yellowish legs.
Size: *L* 28–32 cm; *W* 47 cm.
Habitat: *Breeding:* clear-cuts, pastures, hayfields, open grasslands and abandoned fields. *Nonbreeding:* short-grass habitats, including prairies, airports and golf courses.

Nesting: in dense grass, usually near a wet site; shallow depression is lined with plant material; pair incubates 4 finely speckled, pale buff to pinkish buff eggs for 22–27 days.
Feeding: gleans the ground for insects.
Voice: airy, whistled *whip-whee-ee you* courtship song; *quip-ip-ip* alarm call.
Similar Species: *Pectoral Sandpiper* (p. 197): streaking ends abruptly at breast; smaller eyes; shorter neck; usually seen in small flocks. *Willet* (p. 180): longer, heavier bill; dark greenish legs; black-and-white wings in flight. *Buff-breasted Sandpiper* (p. 202): buff underparts; shorter neck; daintier bill; lacks streaking on face and foreneck.

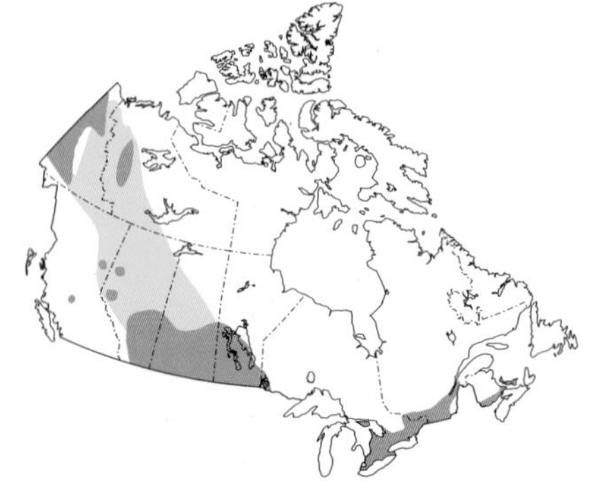

WHIMBREL

Numenius phaeopus

The Whimbrel is a large shorebird with a long, noticeably downcurved bill. It nests in tundra across the northern latitudes of North America, Europe and Asia and spends its winters on the shores of six continents, with some migrants undertaking nonstop ocean flights of up to 4000 kilometres. • Heard long before they come into sight, Whimbrels often fly in V-formation, calling with almost continuous whistles. They are mainly coastal migrants, though some fly overland and make stopovers at inland staging areas. • It is impossible to talk about the Whimbrel without mentioning the Eskimo Curlew (*N. borealis*). Both species suffered devastating losses to their populations during the commercial hunts of the late 1800s. Although the Whimbrel population slowly recovered, the Eskimo Curlew seemed to vanish into thin air—the last confirmed sighting was in 1963. • In Atlantic Canada, Whimbrels are plentiful in fall migration as they gather on the crowberry barrens of Newfoundland and eastern Québec. Here, they gorge on berries before flying to rocky islets or isolated headlands to roost for the night. • The genus name for curlews comes from the Greek words *neos* and *mene*, meaning "new moon"—a clear reference to the crescent-shaped bills of these birds, which conveniently fit the curve of fiddler crab burrows.

ID: sexes similar; long, downcurved bill; dark eye line; long neck and legs; mottled, brown upperparts; striped crown; limited streaking on pale underparts. *In flight:* dark underwings.

Size: *L* 38–45 cm; *W* 78–80 cm.

Habitat: *Breeding:* wet tundra in areas with grassy hummocks. *In migration* and *winter:* estuaries, mudflats, intertidal zones, breakwaters, golf courses and plowed agricultural fields.

Nesting: in loose aggregations; shallow hollow is lined with sparse vegetation; pair incubates 4 brown-speckled, pale green or olive eggs for 27–28 days.

Feeding: probes for insects and berries in summer; eats marine invertebrates, especially fiddler crabs, in winter.

Voice: whistled, 5–7-note *ti-ti-ti-ti-ti-ti* call; incoming flocks utter a distinctive, rippling *pip-pip-pip.*

Similar Species: *Long-billed Curlew* (p. 184): larger; longer bill; cinnamon wing linings; no head stripes. *Eskimo Curlew:* unbarred primaries; pale cinnamon wing linings; shorter, slightly straighter bill; darker upperparts; thought to be extinct.

LONG-BILLED CURLEW

Numenius americanus

Armed with a bill that can reach more than 20 centimetres in length, the Long-billed Curlew, North America's largest sandpiper, is an imposing and remarkable bird. Its long, downcurved bill is a dexterous tool designed for picking grasshoppers from dense prairie grasslands, digging up burrowing earthworms in pastures and capturing shrimp or crabs living in deep burrows on its wintering grounds. • Male curlews engage in spectacular displays over their nesting territory, uttering loud, ringing calls while fluttering high and then gliding down in an undulating flight. Like other curlews, the Long-billed has a ringing *curleeeeeeeeeuuu* call upon which the name is based. • Both parents participate in nesting activities, but like many shorebirds, the female departs less than two weeks after the young hatch, leaving the male to care for the family until the young can forage for themselves. • Habitat loss has led to the decline of the Long-billed Curlew in Canada. Since it breeds primarily on undisturbed interior grasslands, the future of the species is largely tied to the adoption of conservative range and grassland management strategies designed to provide quality nesting habitat. The tidal flats and damp grasslands where curlews winter is also under threat. Recent estimates place the global population of Long-billed Curlews at under 20,000.

ID: sexes similar; large shorebird; very long, downcurved bill (longer in female); long, bluish grey legs; upperparts mottled brown and buffy; buffy brown underparts with limited streaking; crown usually unstreaked. *In flight:* cinnamon underwing linings.
Size: *L* 51–66 cm; *W* 86–89 cm.
Habitat: *Breeding:* mostly dry grasslands; sometimes pastures and ungrazed fields; generally near water. *In migration* and *winter:* tidal mudflats, estuaries, brackish marshes, grasslands, croplands, pastures and golf courses.

Nesting: on the ground in the open; slight depression is sparsely lined with grasses; pair incubates 4 brown-spotted, olive or buff eggs for 27–30 days.
Feeding: probes for invertebrates; eats mostly insects in the interior; takes marine worms, small molluscs and crustaceans on the coast.
Voice: loud, whistling *cur-lee cur-lee cur-lee* call; also a shorter, whistled *kli-li-li-li*; male's display song is a low, whistled *curleeeeeeeuuu*.
Similar Species: *Marbled Godwit* (p. 186): shorter, upturned, bicoloured bill is diagnostic. *Whimbrel* (p. 183): smaller; shorter bill; striped crown; darker upperparts.

HUDSONIAN GODWIT

Limosa haemastica

The Hudsonian Godwit was long regarded as one of the rarest birds on the continent until large migratory flocks were discovered along the shores of James Bay and Hudson Bay in the 1940s. Even today, with a global population estimate of only 50,000 birds, a small, fragmented breeding distribution and with much of the nonbreeding population at risk, it is ranked as a species of particularly high concern. • Although Hudsonian Godwits migrate to and from a few scattered breeding areas in northern Canada and Alaska every spring and fall, it is only in spring that they are reliably observed at inland sites. During the fall passage, most travel south and east, flying nonstop from the Atlantic Coast of Canada to South America. To build up fat reserves for the 4400-kilometre fall migration, Hudsonian Godwits gorge on aquatic plants, an unusual diet for a shorebird. At this time, at least 30 percent of the entire North American population gathers on the north coast of Ontario. • When these birds take to the wing, their broad, white wing bars and two-tone tails give them an extra touch of class. • This species breaks the rules for shorebird sexual dimorphism because male Hudsonian Godwits are more colourful than their mates. This pattern is common in other types of birds but uncommon among shorebirds.

nonbreeding ♂

breeding ♂

ID: sexes similar; large shorebird; slightly upturned, bicoloured bill; black tail; long, bluish legs. *Breeding:* heavily barred, chestnut-red underparts; dark greyish upperparts. *Nonbreeding:* greyish upperparts; whitish underparts may show a few short, black bars. *In flight:* white rump; black "wing pits" and wing linings.
Size: *L* 36–40 cm; *W* 71–73 cm.
Habitat: *Breeding:* sedge meadows and muskeg at tree line. *In migration:* open fresh- and saltwater edges, exposed mudflats and sewage ponds; rarely flooded fields and lakeshores.
Nesting: in a marshy wetland; on a hummock or grassy tussock; well-concealed shallow depression is sparsely lined with willow leaves; pair incubates 4 darkly blotched, olive-brown eggs for 22–25 days.
Feeding: probes deeply into water or mud for invertebrates; also picks earthworms and grubs from plowed fields.
Voice: sharp, rising *god-WIT!*
Similar Species: *Marbled Godwit* (p. 186): larger; mottled brown overall; lacks white rump. *Greater Yellowlegs* (p. 179): shorter, all-dark bill; bright yellow legs; lacks white rump. *Short-billed Dowitcher* (p. 204) and *Long-billed Dowitcher* (p. 205): smaller; straight, all-dark bills; yellow-green legs; mottled, brown-black upperparts in breeding plumage.

MARBLED GODWIT

Limosa fedoa

Its large size, warm brown colour and long, pink-based, slightly upturned bill make the Marbled Godwit easy to recognize. Unlike other North American godwits, the Marbled Godwit's plumage remains much the same throughout the year. • This bird forages in grasslands and wet meadows, but it also feeds on muddy shorelines, sometimes wandering into deep water and dipping its head beneath the water's surface, where it uses its long, sensitive bill to probe deep into soil and mud for worms and molluscs. • Marbled Godwits breed in three highly disjunct regions: on the prairies of the northern U.S. and southern Canada, along the Ontario coast of James Bay and in coastal western Alaska. They draw attention to themselves with their loud, incessant *god-wit!* calls. In fall, flocks of up to 1000 Marbled Godwits stage on large prairie lakes, often in the company of Hudsonian Godwits. Like the Hudsonian, Marbled Godwits are remarkable for foraging almost exclusively on plant tubers during fall migration. • "Marbled" refers to the mottled pattern of dark brown and beige across the godwit's upperparts.

breeding

nonbreeding

ID: sexes similar; large shorebird; long, upturned, bicoloured bill. *Breeding:* mottled dark brown and buffy upperparts; barring across breast; pink base to dark bill. *Nonbreeding:* plain breast and underparts. *In flight:* cinnamon underwing linings.
Size: *L* 41–51 cm; *W* 74 cm.
Habitat: *Breeding:* coastal marshes; prairies or meadows, near water. *In migration* and *winter:* estuaries, mudflats and brackish marshes on coast; lakeshores and exposed mudflats in the interior.
Nesting: on dry ground in grass or sedges, usually near water; shallow hollow is sparsely lined with dry grass; pair incubates 4 variably marked, pale buff or olive eggs for 21–23 days.

Feeding: probes deeply in mud or sand for crustaceans, molluscs and worms.
Voice: ascending *god-wit* call; also an incessant *raddica, raddica.*
Similar Species: *Long-billed Curlew* (p. 184): larger; longer, downcurved bill; bluish grey legs; barred, rufous brown upperwing flight feathers. *Hudsonian Godwit* (p. 185): smaller; nonbreeding bird has black-mottled, greyish upperparts, pale underparts, white tail with broad, black band and white wing stripe. *Short-billed Dowitcher* (p. 204) and *Long-billed Dowitcher* (p. 205): much smaller; stockier; long, straight, dark bills; white back patch; greenish legs; greyish underwings with pale trailing edge.

RUDDY TURNSTONE

Arenaria interpres

nonbreeding

The two turnstones species are stocky, boldly marked shorebirds in both their summer and winter attire. The name "turnstone" is appropriate for these birds, which use their short, stubby, slightly upturned bills to flip over pebbles, shells, seaweed and washed-up vegetation to expose hidden invertebrates. The stocky Ruddy Turnstone walks with a comical, rolling gait and sometimes tilts sideways to counteract the influence of strong winds. • The summer and winter ranges of the Ruddy Turnstone are among the most widely separated of any bird species. It is the most northerly breeding shorebird, nesting on the northern edges of Ellesmere Island and Greenland south to Baffin and Victoria islands. The species overwinters in tropical and temperate habitats well into the Southern Hemisphere. • Ruddy Turnstones are believed to be in decline as a result of habitat disturbance and degradation in wintering and staging areas. Especially critical to turnstones and several other shorebirds is the protection of key staging areas where abundant horseshoe crab or herring eggs are found. The birds rely on the eggs to replenish body stores in preparation for long flights to breeding areas during spring migration.

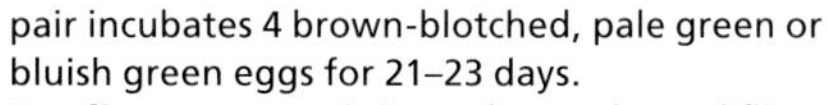

breeding

ID: sexes similar; medium-sized shorebird; white belly; black "bib" curves up to shoulder; stout, black, slightly upturned bill; red-orange legs. *Breeding:* ruddy upperparts (female is slightly paler); white face; black collar; dark, streaky crown. *Nonbreeding:* brownish upperparts and face. *In flight:* boldly patterned wings; white tail with broad, black terminal band; white underwings with black trailing edge; flies close to water in loose flocks.

Size: *L* 23–25 cm; *W* 80 cm.

Habitat: *Breeding:* dry, open tundra near ponds, lakes and streams. *Nonbreeding:* rocky shores and mudflats, seaweed-littered beaches and rocky breakwaters on the coast; rocky islands, newly plowed fields and even garbage dumps in the interior.

Nesting: on open ground near shore, often on an island; shallow hollow is lined with vegetation; pair incubates 4 brown-blotched, pale green or bluish green eggs for 21–23 days.

Feeding: opportunistic; probes under and flips rocks, seaweed and shells for food items; picks, digs and probes for invertebrates, including horseshoe crab eggs, from soil or mud; also eats crabs, barnacles, fish heads, berries, seeds, spiders and carrion.

Voice: clear, rattling, staccato *cut-a-cut* alarm call; lower, repeated contact notes.

Similar Species: *Black Turnstone* (p. 188): black upperparts; whitish edges accent back and wing feathers; more hectic, higher-pitched calls; black-and-white upperwing pattern. *Surfbird* (p. 189): bulkier; stout, dark-tipped, yellowish orange bill; yellow legs; greyish head, black-spotted flanks and paler back in breeding plumage; nonbreeding bird is mostly plain grey. *Other sandpipers* (pp. 176–86, 190–205): all lack the turnstone's bold patterning and flashy wing and tail markings in flight.

BLACK TURNSTONE

Arenaria melanocephala

Black Turnstones are synonymous with rocky, wave-washed Pacific coastlines, which they share with Black Oystercatchers, Surfbirds and Rock Sandpipers. They are visual feeders, methodically searching nooks and crannies in rocks and under seaweed for morsels. Black Turnstones hammer at barnacles and limpets to dislodge them and then extract the soft parts. They also pursue hopping crustaceans over the rocks. • When roosting or resting, thousands of Black Turnstones may gather on log booms in sheltered inlets and bays. Unlike Ruddies, Black Turnstones have a fairly constant plumage year-round. • Feeding flocks are usually small, up to 100 birds, and even when foraging, they show a certain amount of aggression toward each other and to other species. • As with other species that favour rocky coastlines, Black Turnstones are constantly at risk to the threat of oil spills. • The Greek-derived species name *melanocephala* means "black head."

nonbreeding

ID: sexes similar; small to medium-sized shorebird; short, pinkish legs; black head and breast; tapered, black bill; black upperparts with whitish feather edges; white belly. *Breeding:* white eyebrow and "teardrop" in front of eye. *Nonbreeding:* black on body is replaced with dark brown. *In flight:* distinctive, black-and-white pattern on upperwings, back and tail.

Size: *L* 23 cm; *W* 80 cm.

Habitat: *Breeding:* coastal tundra or offshore islands. *In migration* and *winter:* rocky shorelines, breakwaters, jetties and reefs; also beaches with seaweed wracks, mudflats and gravel bars.

Nesting: on wet tundra near water; shallow hollow is lined with grass; pair incubates 4 irregularly marked, pale greenish eggs for 21 days.

Feeding: eats mainly small intertidal barnacles, limpets and chitons.

Voice: shrill, high-pitched *skirrr* call turns into chatter as flock erupts into flight.

Similar Species: *Ruddy Turnstone* (p. 187): reddish to yellowish orange legs; lower-pitched calls; breeding bird has distinctive, black markings on head and breast, and ruddy back and wing bars; nonbreeding bird has lightly mottled, brownish head and upperparts. *Surfbird* (p. 189): larger; dark-tipped, yellow bill; mostly grey upperparts; all-grey back and wing coverts; spotted flanks; yellowish legs.

SURFBIRD

Aphriza virgata

Surfbirds are a characteristic bird of Pacific Coast intertidal zones, spending much of their time feeding on molluscs, limpets, snails, barnacles and other invertebrates just a few metres above the tide line, and constantly navigating the splash and spray of incoming waves. Assembling in favoured feeding sites with Black Turnstones, Ruddy Turnstones and Rock Sandpipers in typical concentrations of 50 to 100 birds, Surfbirds defend foraging and bathing sites from the other "rockpipers" with displays featuring broadly flared wings. • The Surfbird has the longest migration of any breeding North American shorebird. It flies more than 17,000 kilometres from Kodiak Island, Alaska, to Chile's southernmost peninsula, never straying far above the tide line during its journey. Flocks pass through BC from February to mid-May. In Victoria, peak numbers occur from mid-April to early May. In Barkley Sound, peak movement occurs in late April and throughout May. During the breeding season, Surfbirds head inland to mountain ranges in Alaska and the Yukon. In this remote, high-mountain tundra, above treeline, the species' nests and eggs remained undiscovered until 1926.

nonbreeding

ID: sexes similar; small, plump shorebird; dull grey overall; dark eyes; sturdy, dark-tipped, yellow bill; white belly; yellow legs. *Breeding:* fine, white streaking on head; black-marked, golden brown wing coverts; blackish spots and chevrons on flanks and lower breast. *Nonbreeding:* plain breast; grey-spotted flanks. *In flight:* white stripe on grey upperwing; white underwings with grey edges; white tail with black tip.

Size: *L* 23–25 cm; *W* 66 cm.

Habitat: *Breeding:* mountainous tundra. *In migration* and *winter:* outer coastal rocky, wave-washed shorelines, including jetties, breakwaters, islets and rocky beaches.

Nesting: on a dry site on a rocky mountain ridge; shallow hollow is thinly lined with plant material; pair incubates 4 brown-marked, buff eggs; incubation period unknown.

Feeding: pries young mussels, barnacles, snails and limpets from intertidal rocks and swallows them whole, later regurgitating the shells.

Voice: generally silent; sharp *pee-weet* or *key-a weet*; *yif-yif-yif* flight call.

Similar Species: *Ruddy Turnstone* (p. 187) and *Black Turnstone* (p. 188): slightly smaller; boldly marked plumage; pointier, darker bills; orange or dark orange legs; white-striped back in flight; much noisier. *Rock Sandpiper* (p. 199): nonbreeding bird has longer, thinner, darker, slightly downcurved bill, pale legs and white tail with grey centre.

RED KNOT

Calidris canutus

Although the Red Knot's drab winter plumage blends well with the uniform greys and browns of sandy beaches and mudflats, its bright summer wardrobe helps it to avoid detection by predators in a sea of rust-tinged lichens, Arctic grasses, and wildflowers. • In migration, Red Knots concentrate at traditional staging areas to fatten up before embarking on long-distance, non-stop flights, some exceeding 2500 kilometres. One subspecies, *C. c. rufa*, nests in the High Arctic and annually migrates to and from the southern tip of South America, 15,000 kilometres away! During northward migration, this subspecies is reliant on one food source, the eggs of spawning horseshoe crabs, at a single site, Delaware Bay, on the mid-Atlantic Coast. On any given day, 90 percent of the *rufa* population may occur at this site. But faced with a depleted supply of horseshoe crab eggs, *rufa* numbers have declined by 70 percent in the past 15 years, and the subspecies has recently been listed as endangered in Canada. Two other Red Knot subspecies, *C. c. roselaari*, which breeds in the western Arctic, and *C. c. islandica*, which breeds in the High Arctic and migrates through Europe, have been designated as threatened and of special concern, respectively, because of threats to staging areas and recent population declines.

nonbreeding

breeding

ID: chunky body; greenish legs. *Breeding:* rusty face, breast and underparts; brown, black and buff upperparts. *Nonbreeding:* pale grey upperparts; white underparts; faint streaking on upper breast; faint barring grey rump. *Immature:* buffy wash on breast; scaly-looking back. *In flight:* weak, white wing stripe; grey underwings.
Size: *L* 27 cm; *W* 58 cm.
Habitat: *Breeding:* barren tundra. *In migration:* lakeshores, marshes and plowed fields.
Nesting: on bare ground; shallow hollow is lined with lichens; pair incubates 4 variably marked, pale green eggs for 21–23 days.

Feeding: gleans shorelines for molluscs, crustaceans, insects and spiders; probes soft substrates, creating lines of small holes; spring migrants eat horseshoe crab and marine fish eggs, marine worms and snails.
Voice: seldom heard; soft *ker ek* or *wett-wet* in flight.
Similar Species: *Short-billed Dowitcher* (p. 204) and *Long-billed Dowitcher* (p. 205): much longer bills. *Buff-breasted Sandpiper* (p. 202): smaller; shorter bill; dark flecking on sides. *"Peep" sandpipers* (pp. 191–96): smaller; most have black legs; Sanderling and Curlew Sandpiper (rare) have reddish coloration on underparts in breeding plumage.

SANDERLING

Calidris alba

nonbreeding

Lines of lively Sanderlings sprint back and forth, running and cavorting in the surf. Their habit of running ahead of incoming waves and chasing after receding ones has a simple purpose: to snatch washed-up aquatic invertebrates from within the fluid-sand zone before the next wave rolls onto shore. Sanderlings move so fast on their dark legs that they appear to be gliding across the sand. They sometimes feed on wet mudflats, doing so in a manner more typical of other shorebirds. After feeding, the birds slowly reassemble in densely packed roosting flocks, seemingly asleep but ready to escape at any disturbance. • For much of the year, the Sanderling is light grey, paler than any other shorebird. But as the breeding season approaches, its back, head and neck darken, and the male acquires a bright rusty-coloured head and neck. • Sanderlings breed across the Arctic and overwinter on beaches in more hospitable climes on every continent except Antarctica. Like Red Knots and Ruddy Turnstones, Sanderlings forage heavily on horseshoe crab eggs during spring migration, concentrating in a few locations. With reductions in horseshoe crab populations, all three species have experienced significant declines. The number of Sanderlings migrating along the Atlantic Coast has declined by as much as 80 percent since the mid-1970s.

breeding

ID: sexes similar; straight, short, black bill; black legs and feet. *Breeding:* mottled, reddish brown head, breast and upperparts; white underparts. *Nonbreeding:* dark eyes; white face; very pale upperparts; white underparts; black shoulder patch. *In flight:* grey upperwing with dark edge and white stripe; grey-edged, white underwings.
Size: *L* 18–22 cm; *W* 41–43 cm.
Habitat: *Breeding:* coastal tundra. *In migration* and *winter:* sandy beaches, lakeshores, wide mudflats, estuaries, rocky intertidal shores, farm dugouts and flooded fields.

Nesting: on tundra; on the ground, usually near a clump of vegetation; deep cup nest is lined with plant material; pair incubates 4 variably marked, light olive eggs for 23–24 days.
Feeding: probes and picks aquatic invertebrates, including small soft-shelled clams.
Voice: sharp *klit* or *kwit* flight call.
Similar Species: foraging behaviour is unique. *Least Sandpiper* (p. 194): lighter legs. *Semipalmated* (p. 192), *Western* (p. 193) and *Pectoral* (p. 197) *sandpipers:* all lack reddish breeding plumage and very pale nonbreeding plumage. *Dunlin* (p. 200): darker; downcurved bill.

SEMIPALMATED SANDPIPER

Calidris pusilla

The Semipalmated Sandpiper belongs to a group of similar-looking shorebirds collectively known as "peeps" because of their high-pitched calls. Peeps are the relatively nondescript, smaller sandpipers of the *Calidris* genus, a group that includes the Least, Western, White-rumped and Baird's sandpipers, and can be a real challenge to identify in the field. Length of bill and tail, leg colour, overall body colour and feeding behaviour all offer clues to assist in identification. Peeps commonly migrate in large, mixed-species flocks, flying in close formation and wheeling in unison in midflight. Alighting on a shoreline, they quickly scatter, feeding with quick, fidgety movements. • Semipalmated Sandpipers migrate almost the entire length of the Americas, and their migratory pit stops have to provide ample food resources for their journey. In spring, they arrive later than most shorebirds. While currently considered a common species, with the estimated Canadian population at 3.5 million birds, it has recently shown significant annual population declines. • "Semipalmated" refers to the slight webbing between this bird's front toes. However, other sandpipers also have partially webbed toes.

nonbreeding

breeding

ID: sexes similar; small shorebird; black legs and feet; short, black bill with slightly bulbous tip. *Breeding:* mottled, greyish brown upperparts; noticeable streaking on breast. *Nonbreeding:* plain grey-brown upperparts; pale eyebrow and throat; faint streaking on upper breast; pale underparts. *In flight:* narrow, white wing stripe; white underwings; black line through white rump.
Size: *L* 14–18 cm; *W* 36 cm.
Habitat: *Breeding:* grassy, hummocky tundra, often near water. *In migration* and *winter:* mudflats, estuaries and beaches on the coast; lakeshores, mudflats, flooded fields and farm dugouts in the interior.
Nesting: on tundra; shallow scrape is lined with plant materials; pair incubates 4 heavily marked, pale olive to buff eggs for 20–22 days.

Feeding: probes soft substrates and gleans for aquatic insects, worms and crustaceans.
Voice: harsh *cherk* or husky *chrup* flight call.
Similar Species: *Least Sandpiper* (p. 194): thinner bill; yellowish legs and feet; darker upperparts; nonbreeding bird has more conspicuously streaked breast. *Western Sandpiper* (p. 193): longer, slightly downcurved bill lacks bulbous tip; bright rufous wash on crown and "ear" patch. *Sanderling:* (p. 191): nonbreeding bird has pale grey upperparts and blackish trailing edge to flight feathers. *White-rumped Sandpiper* (p. 195): larger; white rump; folded wings extend beyond tail. *Baird's Sandpiper* (p. 196): larger; longer bill; folded wings extend beyond tail.

WESTERN SANDPIPER

Calidris mauri

Although much more common along the Pacific Coast than in the east, the Western Sandpiper is one of the most abundant shorebirds in North America. Significant numbers migrate inland across the continent, but they are infrequently seen in Canada during migration except in BC. • In nonbreeding plumage, the Western Sandpiper is very similar to two other small sandpipers, the Least and the Semipalmated. Although the Western Sandpiper has a somewhat downcurved bill, this characteristic can often be difficult to discern in the field. In breeding plumage, however, Western Sandpipers are much easier to identify because they have noticeable rufous patches on the crown, "ear" patch and scapulars. • An estimated 3.5 to 6.5 million Western Sandpipers pass through the Copper River Delta in Alaska to breed in western Alaska and eastern Siberia. They winter from the southern coastal United States south to central South America and the West Indies. • This sandpiper can be very abundant on tidal flats along the BC coast. Flocks of 100,000 have been reported at favourite stopover points. However, Western Sandpipers are now one of 19 species of shorebirds in North America that are believed to be in decline.

nonbreeding

breeding

ID: sexes similar; small shorebird; short neck; black legs and feet; long, slightly downcurved, black bill; white belly and undertail coverts; black centre to rump and tail. *Breeding:* rufous wash on back, crown and cheek; bold streaking on breast and flanks. *Nonbreeding:* plain grey-brown upperparts; indistinct white eyebrow; white lower face and throat; faint streaking on upper breast; remainder of underparts white. *In flight:* narrow, white wing stripe; white underwings; broad black line through rump and white tail.

Size: *L* 15–18 cm; *W* 36 cm.

Habitat: tidal estuaries and mudflats, shallow brackish marshes, sandy beaches and flooded fields on the coast; mudflats of lakeshores and marshes in the interior.

Nesting: does not nest in Canada.

Feeding: probes or gleans aquatic insects, small molluscs, crustaceans and worms from the water's surface; often submerges head when feeding.

Voice: harsh *cheet* flight call.

Similar Species: *Semipalmated Sandpiper* (p. 192): shorter bill with slightly bulbous tip. *Least Sandpiper* (p. 194): smaller; thinner bill; yellowish legs and feet; more conspicuous breast streaking. *Dunlin* (p. 200): larger; longer, thicker bill. *White-rumped Sandpiper* (p. 195) and *Baird's Sandpiper* (p. 196): larger; folded wings extend beyond tail; lack rufous wing patches in breeding plumage.

LEAST SANDPIPER

Calidris minutilla

The Least Sandpiper is the smallest North American shorebird, but its tiny size doesn't preclude impressive migratory feats. Like most other "peeps," this short-billed, brownish sandpiper migrates most of the length of the globe twice each year, from the Arctic as far as central South America and back again. • Least Sandpipers breed during the brief Arctic summer, and in autumn, the parents start their journey south earlier than their offspring. • Least Sandpipers lay larger eggs than other sandpipers, and the entire clutch might total more than half the weight of the female! Larger eggs allow the young to hatch in an advanced state of development, giving them an early start on preparations for fall migration. • Of all the peeps, the Least Sandpiper is the easiest to identify. Its yellow legs are distinctive; the legs of other small sandpipers are blackish. Although pale legs are a good field mark, dark mud can make the bird's legs look dark, whereas light mud can make other species's legs look light. In all plumages, the Least has a streaked breast that contrasts sharply with its white underparts. • The species name *minutilla* is Latin for "very small."

nonbreeding

breeding

ID: sexes similar; small shorebird; yellow legs and feet; short, slightly downcurved, black bill; clearly defined, white belly; white flanks and undertail coverts. *Breeding:* mottled, buff-brown head, breast and nape; dark brown, mottled back. *Nonbreeding:* darkly marked, dull greyish brown head, breast and upperparts; faint, pale eyebrow. *Juvenile:* similar to breeding adult but brighter with faintly streaked breast. *In flight:* short, dark wings; thin, white upperwing stripe; white rump divided by dark line; stiff, rapid wingbeats.

Size: *L* 13–17 cm; *W* 33 cm.

Habitat: *Breeding:* boggy Subarctic forests or Arctic tundra. *In migration* and *winter:* tidal mudflats, shallow estuaries, edges of brackish marshes and sloughs on the coast; lakeshores, mudflats, beaches, shallow ponds, sewage ponds, dugouts and flooded agricultural fields inland.

Nesting: on tundra; near a hummock or vegetation clump; shallow depression is lined with plant material; pair incubates 4 brown-blotched, pale buff eggs for 19–23 days.

Feeding: probes for aquatic insects, small molluscs, crustaceans and worms.

Voice: high-pitched *prreeep* flight call.

Similar Species: *Other peeps* (pp. 192–93, 195–96): larger; black legs. *Pectoral Sandpiper* (p. 197): much larger; sharp contrast between streaked breast and white underparts.

WHITE-RUMPED SANDPIPER

Calidris fuscicollis

A properly fuelled White-rumped Sandpiper on migration is said to demonstrate incredible stamina, flying for days without landing and covering distances up to 4000 kilometres. In fall, southbound migrants pass through eastern Canada and the Maritimes, with staging areas along the west coast of James Bay and the Bay of Fundy, then fly nonstop over the Atlantic Ocean to northern South America. During spring migration, White-rumps take a route north between the Rockies and Ontario. These long-distance flights are fuelled by fat reserves accumulated at traditional staging areas where food is especially abundant. For many shorebirds, food resources and habitat at these traditional stopovers are crucial if the birds are to survive these phenomenal migrations and arrive on their Arctic breeding grounds in good condition and physiologically ready to nest. • The White-rumped Sandpiper is easiest to identify in flight, when it reveals its white rump—unique among the "peeps." Foraging White-rumps also differ from other peeps in that they stand in deeper water while feeding, walk rather than hop and move about in a more deliberate manner as they pick or probe for food. • Flocks of White-rumped Sandpipers will collectively rush at a predator and then suddenly scatter in its face.

nonbreeding

breeding

ID: sexes similar; medium-sized shorebird; fairly long, straight bill (about as long as head is wide); dark legs; folded wings extend past tail. *Breeding:* brown-mottled upperparts; dark chevron streaks on breast and flanks. *Nonbreeding:* mottled grey upperparts; crisp, dark streaking on breast and flanks. *Juvenile:* similar to breeding; wing feather tips edged in white; brown streaks on neck, upper breast and upper flanks. *In flight:* white rump; dark tail; indistinct wing bar.

Size: *L* 18–20 cm; *W* 40–44 cm.

Habitat: *Breeding:* grassy or mossy tundra; prefers wet areas. *In migration:* sewage lagoons, shores of lakes, ponds and dugouts, coastal mudflats and flooded fields.

Nesting: on tundra; on a raised mound; shallow depression is lined with plant material; female incubates 4 brown-blotched, buff eggs for about 22 days.

Feeding: gleans the ground for invertebrates, especially insects, crustaceans and small molluscs.

Voice: high, mouselike *tzeet* flight call.

Similar Species: *Semipalmated* (p. 192), *Western* (p. 193) and *Least* (p. 194) *sandpipers:* dark line through rump. *Baird's Sandpiper* (p. 196): "scaly" back pattern; no white rump; more noticeable separation of dark and light across chest. *Stilt Sandpiper* (p. 201): much longer legs trail beyond tail in flight.

BAIRD'S SANDPIPER

Calidris bairdii

nonbreeding

The Baird's Sandpiper and the White-rumped Sandpiper are the largest of the small sandpipers commonly referred to as "peeps." At rest, both species have an attenuated look because their folded wings extend well beyond the tips of their tails. • The female Baird's Sandpiper makes one of the largest investments in egg production in the bird world, laying a clutch that is up to 120 percent of her body mass. This achievement is even more amazing considering that she lays her eggs within a few days of arriving in the Arctic, with essentially no stored body fat. • When their chicks can fend for themselves, the parents abandon them and embark on the long, arduous trip south to their wintering grounds. Like White-rumped Sandpipers, Baird's Sandpipers exhibit amazing stamina in migration. After staging briefly in southern Canada or the northern U.S., they travel the full 6000 kilometres or more directly to South America. Some complete the entire 15,000-kilometre journey, from nesting to wintering range, in a month. A few weeks after the parents have left, the young flock together in a second wave of southern migrants. • In both spring and fall, most Baird's Sandpipers migrate through the interior of North America.

breeding

ID: sexes similar; medium-sized shorebird; black bill and legs; grey-brown upperparts; buffy face and neck; faint, brown streaks on buffy breast; white underparts; long wings project well beyond tail when folded. *Breeding:* dark mottling gives "scaly" appearance to back. *Nonbreeding:* indistinctly "scaly." *Juvenile:* brighter, browner upperparts; pale feather edges give "scaly" appearance; light brown upper breast. *In flight:* long wings; indistinct, white wing stripe; grey-brown rump.
Size: *L* 18–19 cm; *W* 43 cm.
Habitat: *Breeding:* dry tundra with low vegetation or shrubs. *In migration:* more terrestrial than other small shorebirds; upper tide lines on ocean beaches, sandy shores, margins of sewage ponds and drier grassy fields; less often on mudflats.
Nesting: on tundra; shallow depression is lined with plant material; female incubates 4 variably marked, creamy buff eggs for 21 days.

Feeding: feeds alone or in small, mixed shorebird groups; gleans aquatic invertebrates, especially larval flies; also eats beetles and grasshoppers; rarely probes; tends to forage out of the water.
Voice: usually silent; soft, reedy *creeeep, creeep* often extended into a trill in flight.
Similar Species: *Pectoral Sandpiper* (p. 197): yellow legs; yellow base to bill; streaky breast band is darker and more clearly defined; undertail coverts are more obvious in flight. *Least Sandpiper* (p. 194): pale, yellowish legs. *Semipalmated Sandpiper* (p. 192): lacks buffy face and neck and "scaly" back. *White-rumped Sandpiper* (p. 195): clean, white rump; breast streaking extends onto flanks. *Western Sandpiper* (p. 193) and *Sanderling* (p. 191): lack streaked, buffy breast.

PECTORAL SANDPIPER

Calidris melanotos

A crisp dividing line between its brown bib and its white underparts make the Pectoral Sandpiper easy to identify. In fact, it is distinctive enough that few birders think of it as a "peep," despite its membership in the genus *Calidris*. • Unlike most sandpipers, the Pectoral exhibits sexual dimorphism—females are only two-thirds the size of the males. These sandpipers are easy to identify thanks to their large size and boldly streaked breast that contrasts with the white belly. • Pectoral Sandpipers get their name from the location of the male's prominent air sacs. On its Arctic breeding grounds, males perform spectacular displays on "leks," inflating their air sacs and emitting a foghornlike, hollow hooting during displays—one of the most unusual sounds heard on the Arctic tundra. • In spring and fall, large flocks of hundreds or even thousands can occur in wet, grassy fields and along shorelines. • This widespread traveller may be found in Siberia as well as the Canadian Arctic, and its epic, annual migrations include destinations such as South America, and sometimes Australia or New Zealand. In spring, most Pectoral Sandpipers take an interior route through central Canada.

breeding

nonbreeding

ID: sexes similar but male is larger; medium-sized shorebird; heavily streaked breast contrasts sharply with white underparts; finely streaked head; dark crown; indistinct, white eyebrow and throat; dark bill droops slightly near tip; reddish brown back with 2 thin lines extending to rump; yellowish legs. *Juvenile:* buffy edges to feathers on upperparts. *In flight:* black line through white tail.
Size: *Male: L* 23 cm; *W* 43–45 cm.
Female: L 15–18 cm; *W* 41–43 cm.
Habitat: *Breeding:* moist, grassy tundra adjacent to wet sedge tundra. *In migration:* wet, grassy meadows, salt marshes, edges of standing water and upland grassy and flooded fields.
Nesting: in a dry site on grassy tundra; shallow ground nest is well lined with plant material; female incubates 4 heavily marked, pale olive eggs for 21–23 days.
Feeding: probes and pecks for small insects and larvae, especially flies, beetles and some grasshoppers; also takes small mollusks and crustaceans.
Voice: sharp *krick* or *drrup* call.
Similar Species: *"Peep" sandpipers* (pp. 192–96): smaller; lack well-defined, streaked breast; all but Least have black legs.

PURPLE SANDPIPER

Calidris maritima

Unlike most shorebirds, which prefer shallow marshy areas, sand beaches or mudflats, Purple Sandpipers forage perilously close to crashing waves along rocky headlands, piers and breakwaters. They also expertly navigate their way across rugged, slippery rocks while hunting for crustaceans, molluscs and insect larvae. They are so well camouflaged that they become visible only when a wave breaks and temporarily dislodges them. • Unlike most adult shorebirds, Purple Sandpiper adults migrate south after moulting, which explains why they do not appear in southern Canada until mid-autumn. Purple Sandpipers breed in the High Arctic from Greenland west into the islands of the Canadian Arctic and winter in eastern North America from Newfoundland to the Carolinas, farther north than other shorebirds. Occasionally, some winter in the lower Great Lakes and even on the Niagara River just above the falls. • The name "purple" was given to this sandpiper for a purplish iridescence that is occasionally observed when winter-plumaged birds are viewed in certain light conditions.

nonbreeding

breeding

ID: sexes similar; medium-sized shorebird; long, slightly drooping, black-tipped bill with yellow-orange base; yellow-orange legs; dull streaking on breast and flanks. *Breeding:* streaked neck; dark streaks on buff crown; dark back feathers with tawny to rusty brown edges. *Nonbreeding:* unstreaked, grey head, neck and upper breast form "hood"; grey spotted, white belly.

Size: *L* 23 cm; *W* 43 cm.

Habitat: *Breeding:* tundra from sea level to 450 m. *In migration* and *winter:* sandy beaches, rocky shorelines, piers and breakwaters.

Nesting: on tundra; deep hollow is lined with dead leaves; mainly the male incubates 4 variably marked, pale olive to bluish green eggs for 21–22 days.

Feeding: moves over rocks and sand to find food visually; takes mostly molluscs, insects, crustaceans and other invertebrates; also eats a variety of plant material.

Voice: soft *prrt-prrt* call; scratchy *keesh* flight call.

Similar Species: *"Peep" sandpipers* (pp. 192–96): lack unstreaked, grey hood in winter, bicoloured bill and yellow-orange legs.

ROCK SANDPIPER

Calidris ptilocnemis

The Rock Sandpiper's mottled, dark grey winter plumage makes it difficult to find among the grey rocks and splashing surf of the coast. This marvellously camouflaged bird is easily overlooked as it feeds on rocky shorelines with its larger and showier rock-feeding associates, Black Turnstones and Surfbirds. Unlike these species, though, Rock Sandpipers do not pry food from the rocks but rather probe crevices for unattached prey. • Rock Sandpipers and Purple Sandpipers are extremely similar in winter plumage; thankfully, they reside on opposite coasts. • Four subspecies of the Rock Sandpiper have evolved, each with a distinct breeding plumage, but most appear similar in winter. In fall migration, Rock Sandpipers arrive later than other shorebirds because adults moult on the breeding grounds before starting their southward migration. • The species name *ptilocnemis* means "feather boot," referring to the protective feathers that cover the bird's legs down to the heel—an adaptation to the cold.

nonbreeding

ID: sexes similar; medium-sized shorebird; dark, slightly drooping bill with dull olive base; short legs; folded wings are shorter than tail; bold, grey spots on sides and undertail coverts. *Breeding:* reddish brown upperparts; streaked neck and breast; rufous crown; dark patch on belly; grey legs. *Nonbreeding:* greyish overall; pale edges to feathers give "scaly" appearance; greenish yellow legs. *Juvenile:* lacks rufous crown and dark belly patch.

Size: *L* 23 cm; *W* 43 cm.

Habitat: *In migration* and *winter:* intertidal rocky shorelines, beaches, jetties and breakwaters.

Nesting: does not nest in Canada.

Feeding: picks prey off rocks; eats small marine invertebrates, especially crustaceans and worms, as well as seaweeds.

Voice: generally silent; scratchy, low *keesh* flight call may change to rougher *cherk;* flocks utter a sharper *kwititit-kwit.*

Similar Species: *Dunlin* (p. 200): slightly longer, more downcurved bill; brownish grey upperparts and plainer underparts in nonbreeding plumage; immature has more, darker breast streaking and black-marked belly. *Surfbird* (p. 189): much heavier; stubbier bill; black-tipped, white tail; yellow legs; paler grey upperparts in nonbreeding plumage.

DUNLIN

Calidris alpina

Previously named "Red-backed Sandpiper," Dunlins in their breeding attire are as distinctive as they are colourful. Spring migrants and breeders have a rusty red back with a black patch across the belly, making them look as though they have been wading belly-deep in a puddle of ink. Spring sightings of Dunlins in breeding plumage are rarer on the coasts, as most adults take a direct route to the Arctic through the interior of the continent. However, in fall, juveniles head to the coasts in large numbers. Outside the breeding season, Dunlins form dazzling, tightly grouped flocks, often consisting of tens of thousands of birds flying wing tip to wing tip, all changing course simultaneously. These tight flocks are generally more exclusive than other shorebird troupes, usually comprised mostly or entirely of their own kind. • The Dunlin is among the swiftest of the shorebird migrants—a flock was once observed passing a small plane at 175 kilometres per hour. • This bird was originally called "Dunling," meaning "small, brown bird," but with the passage of time, the *g* was dropped.

nonbreeding

breeding

ID: sexes similar; medium-sized, chunky shorebird; long, black, downcurved bill; black legs and feet. *Breeding:* distinctive rusty crown, back and wings; streaked, white neck and underparts; large, black belly patch; white undertail coverts. *Nonbreeding:* dull grey-brown head and upperparts; pale eyebrow and throat; lightly streaked, dusky breast. *In flight:* white wing stripe.

Size: *L* 19–23 cm; *W* 43 cm.

Habitat: *Breeding:* tundra, grassy lowlands near water and grassy coastal salt marshes. *In migration* and *winter:* intertidal mudflats, sandy beaches and shallow estuaries on the coast; lake and marsh mudflats, shallow ponds and flooded agricultural fields in the interior.

Nesting: on tundra near water; cuplike hollow on a grassy mound is well lined with plant material; pair incubates 4 brown-marked, pale olive to bluish green eggs for 21–22 days.

Feeding: active forager; probes bill into mud or sand for small molluscs, crustaceans and worms.

Voice: distinct, buzzy *pjeev* or *kreep* flight call; male's display song is a descending, creaky, high-pitched trill.

Similar Species: black belly patch in breeding plumage is distinctive. *Western Sandpiper* (p. 193): smaller; paler back. *Rock Sandpiper* (p. 199) and *Purple Sandpiper* (p. 198): prefer rocky areas; slightly shorter bills; nonbreeding birds have grey upperparts and heavily grey-marked underparts; immature has whitish belly. *Sanderling* (p. 191): paler back; usually seen running in surf.

STILT SANDPIPER

Calidris himantopus

With the silhouette and leg colour of a yellowlegs and the foraging behaviour of a dowitcher, the Stilt Sandpiper is often overlooked by birders. In fact, throughout much of the 19th century, Stilt Sandpipers were called "Bastard Yellowlegs," as they were widely thought to have originated as a hybrid between dowitchers and yellowlegs. • Stilt Sandpipers are one of the most vegetarian of shorebirds; one-third of their diet consists of plant matter. They occasionally take freshwater shrimp, insect larvae or tiny minnows from just below the water's surface, sweeping their bills from side to side like an avocet. Stilt Sandpipers have also been seen holding their bills submerged for prolonged periods in sand or water, waiting for prey to touch the bill and trigger a strike. • The Stilt Sandpiper's main northward and southward migration routes take it through the Prairie provinces and the Great Plains. • Named for its relatively long legs, the Stilt Sandpiper prefers to feed in shallow water, where it probes in an up-and-down, "stitching" fashion similar to a dowitcher, often dunking its head completely underwater. Moving on tall "stilts," this sandpiper will also wade into deep water up to its breast in search of a meal. • Stilt Sandpipers rarely gather in flocks.

nonbreeding

breeding

ID: sexes similar; medium-sized shorebird; long neck; slim head; conspicuous, white eyebrow; long, slender bill droops only at tip; long, yellowish or yellow-green legs. *Breeding:* dark-flecked, brownish upperparts; rusty cap and cheek patch; heavily barred underparts. *Nonbreeding:* pale grey upperparts; faintly streaked breast; remainder of underparts whitish. *Juvenile:* dark brown upperparts; buff wash on throat and breast. *In flight:* plain grey tail; white rump; legs trail behind tail; no wing stripe.

Size: *L* 20–23 cm; *W* 45 cm.

Habitat: *Breeding:* open tundra. *In migration:* freshwater and brackish ponds and marshes; also mudflats and sewage lagoons.

Nesting: on tundra near water or marshy areas; shallow scrape is sparsely lined with plant material; pair incubates 4 variably marked, pale creamy olive or buff eggs for 19–21 days.

Feeding: wades into shallow water and probes mud or sand for crustaceans, worms, molluscs and aquatic insects; also eats seeds, leaves and roots of wetland plants.

Voice: muffled, husky *toof*; sharp *querp* in flight.

Similar Species: *Short-billed Dowitcher* (p. 204) and *Long-billed Dowitcher* (p. 205): white wedge on back; longer, straight bills; shorter legs; chunkier body. *Greater Yellowlegs* (p. 179) and *Lesser Yellowlegs* (p. 181): shorter bills; lack pale eyebrow; pick rather than probe when foraging.

BUFF-BREASTED SANDPIPER

Tryngites subruficollis

Shy in behaviour and humble in appearance, the Buff-breasted Sandpiper is often discovered in the course of scanning a flock of Black-bellied Plovers or American Golden-Plovers. • Courtship begins during migraton, when male Buff-breasts can be seen strutting about and lifting their wings high over their heads. Buff-breasted Sandpipers and Pectoral Sandpipers are unique among North American shorebirds in having a lek mating system. Males display on the leks, trying to interest prospective mates. • From their dry tundra nesting habitat in the central Arctic, Buff-breasts migrate to the pampas grasslands of Argentina for the winter. Most Buff-breasts migrate north in a narrow corridor through the Prairie provinces in mid- to late May and follow much the same route south during late summer and fall, but juveniles disperse over a wider range and are apt to show up in any province. • When feeding, this subtly coloured bird stands motionless, its plumage blending beautifully into its surroundings. Only when it catches sight of moving prey does it become visible, making a short, forward sprint to snatch a fresh meal. • Recent population estimates for Buff-breasts range from 5000 to 15,000 birds. The exact cause for ongoing declines may relate to habitat loss on their wintering grounds in South America.

ID: sexes similar; buffy, unpatterned face and foreneck; large, dark eyes; very thin, straight, black bill; buff underparts; small spots on crown, nape, breast, sides and flanks; "scaly" look to back and upperwings; yellow legs. *In flight:* pure white underwings; no wing stripe.
Size: *L* 19–21 cm; *W* 45 cm.

Habitat: *Breeding:* grassy areas on dry tundra. *In migration:* shores of lakes, reservoirs and marshes; also cultivated and flooded fields.
Nesting: on dry tundra; ground scrape is lined with plant material; pair incubates 4 heavily blotched, creamy white eggs for 23–25 days.
Feeding: gleans the ground and shorelines for insects, spiders and small crustaceans; also eats some seeds.
Voice: usually silent; calls include *chup* or *tick* notes; *preet* flight call.
Similar Species: *Upland Sandpiper* (p. 182): more boldly streaked breast; longer neck; smaller head; larger bill; streaking on cheek and foreneck.

RUFF/REEVE

Philomachus pugnax

nonbreeding

This Eurasian shorebird appears almost yearly, but in tiny numbers, throughout Canada, primarily during spring migration. It might even be more common than we think. This species is known to have nested in Alaska, but nowhere else in North America. The birds seen in western Canada may be migrants that winter in California. • The male of this species is called a "Ruff," while the female is called a "Reeve." It is not known whether the male was named for his neck-feather ruffs, or if ruffs were named after the bird. "Reeve" has even more obscure origins, but some suggest it is linked to the meaning "observer" or "bailiff"—the females oversee the tussling, courting males at the leks on their breeding grounds. • Breeding males occur in different plumage colours (orange, white and black). In this complex species, there is even a type of male that mimics a female in plumage and size. This mimic male sneaks into established territories to breed with the females. This species also has an additional moult in between its winter and summer breeding plumages.

ID: plump body; small head; yellow-green to red legs; yellow or black bill; brownish grey upperparts. *Breeding male:* black, white or orange neck ruff, erected during courtship; dark underparts. *Breeding female:* dark blotches on underparts. *In flight:* thin, white wing stripe; oval, white rump patches; dark line through rump.

Size: *Male: L* 29–32 cm; *W* 54–60 cm. *Female: L* 22–26 cm; *W* 46–49 cm.

Habitat: marshes and flooded fields.

Nesting: does not nest in Canada.

Feeding: probes and picks at the surface of mudflats for aquatic invertebrates.

Voice: rarely vocal; short *tu-whit* call.

Similar Species: *Greater Yellowlegs* (p. 179) and *Lesser Yellowlegs* (p. 181): slimmer bodies; longer, yellower legs; streaked underparts. *Red Knot* (p. 190): shorter legs; "cleaner" breast in nonbreeding plumage.

SHORT-BILLED DOWITCHER

Limnodromus griseus

Dowitchers are large, plump shorebirds with long bills, barred tails, and a unique white rump and wedge on the lower back. • Only in comparison to the Long-billed Dowitcher would anyone ever think of calling this bird "short-billed." In fact, the bill of the Short-billed Dowitcher is only slightly shorter than that of the Long-billed, and it is comparatively longer than that of most other shorebirds. Calls, when the birds oblige, are also a good way to identify them. • Dowitchers tend to be stockier than most shorebirds and generally avoid venturing into deep water. Having short legs, dowitchers feed in mud or shallow pools, where their extra-long, flexible-tipped bills can reach hidden invertebrates unavailable to longer-legged shorebirds with shorter bills. While foraging along shorelines, these birds "stitch" up and down into the mud with a rhythm like a sewing machine, a behaviour that is helpful for long-range identification. • Three subspecies breed in Canada, all in separate areas, and each has slight plumage differences. Only along the west coast of James Bay do two distinct subspecies, *griseus* and *hendersoni*, occur together in the same habitats during breeding season.

nonbreeding

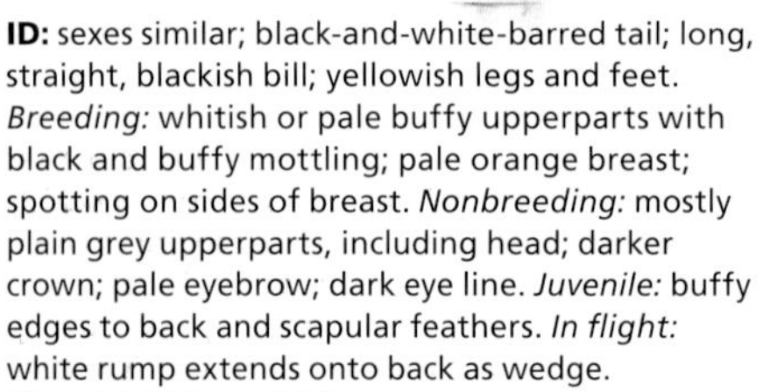

breeding

ID: sexes similar; black-and-white-barred tail; long, straight, blackish bill; yellowish legs and feet. *Breeding:* whitish or pale buffy upperparts with black and buffy mottling; pale orange breast; spotting on sides of breast. *Nonbreeding:* mostly plain grey upperparts, including head; darker crown; pale eyebrow; dark eye line. *Juvenile:* buffy edges to back and scapular feathers. *In flight:* white rump extends onto back as wedge.

Size: *L* 28–30 cm; *W* 48 cm.

Habitat: *Breeding:* coastal tundra, muskeg ponds and fens. *In migration:* intertidal mudflats, estuaries, sewage lagoons, muddy shores of saltwater and brackish marshes and ponds on the coast; muddy shores of wetlands, farm dugouts, shallow ponds and marshes, sewage lagoons and flooded fields in the interior.

Nesting: on coastal tundra or on coniferous muskeg at the base of a small tree; shallow hollow is lined with dry grasses; pair incubates 4 darkly marked, brown to olive-buff eggs for about 21 days.

Feeding: wades in shallow water or probes mud for worms, crustaceans and small molluscs; also eats aquatic insects inland.

Voice: mellow, repeated *tu-tu-tu*, *too-du-lu* or *too-du* flight call.

Similar Species: *Long-billed Dowitcher* (p. 205): barring on sides of breast; high-pitched *keek* call. *Wilson's Snipe* (p. 206): median crown stripe; lacks white wedge on back. *American Woodcock* (p. 207): unmarked, buff underparts; yellow bill; light bars on black crown and nape.

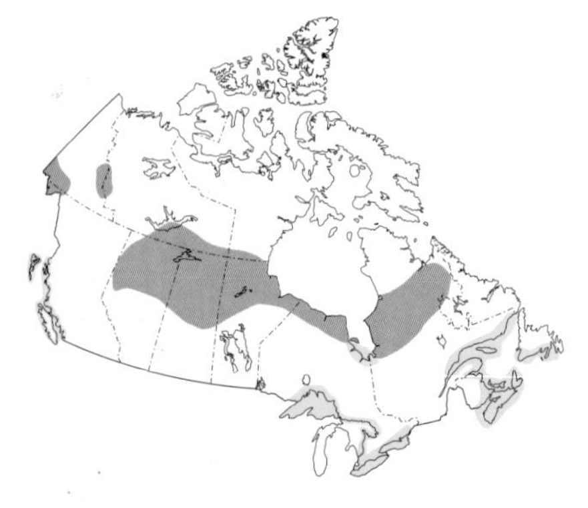

LONG-BILLED DOWITCHER

Limnodromus scolopaceus

In contrast to its slightly shorter-billed cousin, which is a saltwater mudflat specialist, the Long-billed Dowitcher favours freshwater habitats. Even along the coast, the Long-bill prefers exposed mudflats of lakeshores, marshes and flooded pastures, where its slightly longer legs and bill give it a small feeding advantage over the Short-billed Dowitcher. • Long-billed Dowitchers tend to migrate earlier and have a much more compressed migration than many shorebirds, most reaching their Arctic breeding grounds in mid- to late May. • Nesting in northern Alaska and northwestern Canada, Long-bills spread out over the continent in fall. Juveniles may appear as late as November at some coastal sites. • Dowitchers have shorter wings than most long-distance migrant shorebirds, making it more practical for them to take flight from shallow water. • The vocalizations of the two dowitcher species are the surest way to distinguish them in the field. Bill length is *not* a reliable field mark. • In various parts of its range, the Long-billed Dowitcher has earned many different vernacular names, including "Western Red-breasted Snipe," "Greater Long-beaked Snipe," "Jack Snipe," "Red-bellied Snipe," "Western Dowitcher" and "White-tailed Dowitcher."

nonbreeding

breeding

ID: sexes similar; black-and-white-barred tail; long, straight, blackish bill; yellowish legs and feet; barring on sides of breast. *Breeding:* black and buffy speckling on upperparts; wholly orange underparts; black barring on flanks and undertail coverts. *Nonbreeding:* mostly plain grey upperparts, including head; darker crown; white eyebrow; dark eye line. *Juvenile*: buffy edges on dark back and scapular feathers. *In flight:* white rump extends onto back as wedge.

Size: *L* 28–32 cm; *W* 48 cm.

Habitat: *Breeding:* grassy tundra near water; wet meadows. *In migration* and *winter:* coastal mudflats, exposed muddy shores of lakes, marshes and ponds, flooded fields and sewage lagoons.

Nesting: on tundra; on the ground among grasses or sedges; hollow depression is sparsely lined with plant material; pair incubates 4 darkly marked, pale greenish or olive eggs for about 21 days.

Feeding: wades in shallow water, rapidly probing for worms, crustaceans, molluscs and aquatic insects.

Voice: single, loud, piping *keek* call, sometimes given in loose series; unlike Short-billed Dowitcher, calls frequently both in flight and while foraging.

Similar Species: *Short-billed Dowitcher* (p. 204): prefers saltwater or brackish habitats; spotting on sides of breast; call is *too-too-too* or *too-du. Wilson's Snipe* (p. 206): median crown stripe; lacks white wedge on back; different call.

WILSON'S SNIPE

Gallinago delicata

As the courting male Wilson's Snipe performs courtship dives and territorial displays high above its marshland habitat, specialized outer tail feathers vibrate, creating an eerie, hollow winnowing sound. Male snipes display most actively in the early morning, but performances in the evening and after dark are also common. Between flights, the males may be seen scanning their territory from atop a fence post, often issuing *wheat wheat wheat* or *chip-a chip-a chip-a* calls. • The snipe's cryptic plumage camouflages it among vegetation, but when an intruder approaches too closely, the snipe utters a harsh *skape* note as it suddenly flushes and performs a series of evasive zigzags designed to deter the predator as it flies away. Hunters who were skilled enough to shoot a snipe came to be known as "snipers"—a term later adopted by the military. • The name "snipe" comes from *snite*, an old version of "snout," in reference to the bird's long bill and its habit of probing for prey in soft ground. The flexible bill is sensitive to the movement of earthworms and larval insects buried deep within the mud. • This species is found in a variety of freshwater habitats but rarely in saltwater situations. • The Wilson's Snipe was recently recognized as a separate species from the Common Snipe (*G. gallinago*) of the Old World.

ID: sexes similar; medium-sized shorebird; long, straight, bicoloured bill; short legs; buffy head with bold, dark stripes on crown and face; white stripes on black and brown upperparts; black-streaked, buffy breast; dark barring on white sides; unmarked, white belly. *In flight:* zigzag flight pattern; short, orange tail.

Size: *L* 26–29 cm; *W* 44–50 cm.

Habitat: *Breeding:* wet sedge meadows, grassy margins of marshes, sloughs and damp upland fields. *In migration* and *winter:* variety of fresh- and saltwater habitats, including estuaries, grassy roadside ditches, shallow freshwater marshes and flooded agricultural fields.

Nesting: on the ground among short, dry vegetation; shallow depression is lined with grasses, mosses and leaves; female incubates 4 brown-marked, olive buff eggs for 18–20 days.

Feeding: uses flexible bill tip to probe mud for worms, insects, spiders, and other invertebrates; also eats seeds.

Voice: accelerating aerial courtship display sound is a winnowing *woo-woo-woo-woo-woo-woo*; often sings *wheat wheat wheat* or *chip-a chip-a chip-a*; nasal *scaip* alarm call.

Similar Species: *Short-billed Dowitcher* (p. 204) and *Long-billed Dowitcher* (p. 205): usually feed in flocks on open water; nonbreeding birds have paler plumage, white rump and wedge on back and lack streaking on head. *American Woodcock* (p. 207): unmarked, buff underparts; yellowish bill; pale barring on black crown and nape.

AMERICAN WOODCOCK

Scolopax minor

The American Woodcock's behaviour usually mirrors its cryptic and inconspicuous attire. This denizen of moist woodlands and damp thickets usually goes about its business in a quiet and reclusive manner, but during courtship the male woodcock reveals his true character. Before his aerial display, the male will strut provocatively into an open woodland clearing or a brushy, abandoned field, calling out a series of loud *peeent* notes. Suddenly, without warning, the male spirals up into the air, beginning a lengthy circular flight display that includes a variety of high-pitched twittering and whistling sounds. Then he partially folds his wings and plummets to the ground in a zigzag pattern to repeat the performance. • The clearing of forests and draining of woodland swamps has degraded and eliminated large tracts of productive American Woodcock habitat. Although the species persists throughout much of its former range and has even increased in some parts of Canada, in many areas it is not as abundant as it used to be. • The American Woodcock's curious appearance and behaviour have earned it many humorous vernacular names, including "Timberdoodle," "Labrador Twister," "Night Partridge" and "Bog Sucker."

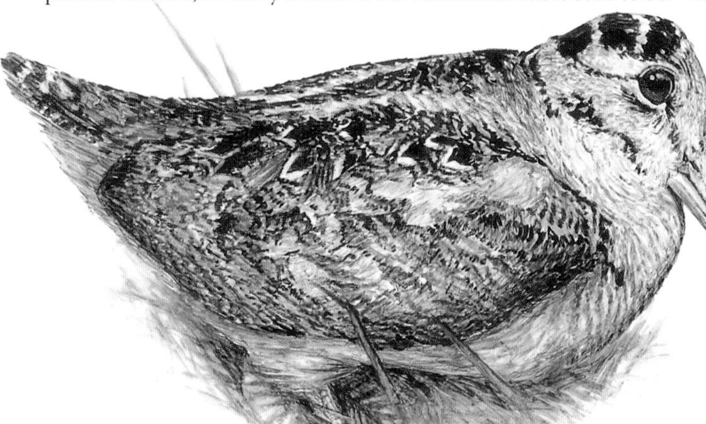

ID: sexes similar; chunky body; short legs; very long, sturdy bill; large head; short neck; large, dark eyes; unmarked, buff underparts; light-coloured bars on black crown and nape. *In flight:* rounded wings; makes twittering sound when flushed from cover.

Size: *L* 27–29 cm; *W* 45–46 cm.

Habitat: moist woodlands and brushy thickets adjacent to grassy clearings or abandoned fields.

Nesting: on the ground in a woodland or overgrown field; nest scrape is lined with dead leaves and other plant debris; female incubates 4 variably marked, pinkish buff eggs for 20–22 days.

Feeding: probes in soft, moist or wet soil for earthworms and insect larvae; also takes spiders, snails, millipedes and some plant material, including seeds, sedges and grasses.

Voice: nasal *peent*; produces high-pitched, twittering, whistling sounds during courtship displays.

Similar Species: *Wilson's Snipe* (p. 206): heavily striped head, back, neck and breast; dark barring on sides and flanks. *Short-billed Dowitcher* (p. 204) and *Long-billed Dowitcher* (p. 205): dark bills; longer legs; lack light-coloured barring on dark crown and nape; usually seen in flocks.

WILSON'S PHALAROPE

Phalaropus tricolor

Phalaropes are among the most colourful and graceful of North American shorebirds. All three of the world's species are found in Canada, but the Wilson's Phalarope is the one most likely to be seen away from salt water. • All phalaropes practise a mating strategy known as "polyandry," in which a female may mate with several males. After laying a clutch, the female usually abandons her partner, leaving him to incubate the eggs and tend the young. She may either continue to "play the field" or act as a lookout for one of her males. Naturally, this reversal of traditional gender roles also includes a reversal of plumage characteristics—females are more brightly coloured than their male counterparts, whose pale plumage allows them to remain inconspicuous while sitting on the nest. Even John J. Audubon was fooled by the phalarope's plumage reversal—he mislabelled the female and male birds in all his phalarope illustrations. • Phalaropes also have an unusual foraging style, using their feet to spin around in tight circles in open water. The spinning motion creates a small vortex that brings prey up to the water's surface.

nonbreeding

breeding

ID: smallish, dainty-looking shorebird; thin, dark bill; white eyebrow and throat; pale underparts; black legs. *Breeding female:* grey cap; chestnut brown sides of neck; black eye line extends down side of neck and onto back. *Breeding male:* lacks black eye line and neck stripe; rufous nape. *Nonbreeding:* all-grey upperparts; grey eye line; white underparts; dark yellowish or greenish legs. *Juvenile:* dark brown upperparts; broad, buffy feather edges give "scaly" look; buffy sides of breast; yellow legs. *In flight:* unmarked gray upperwings; white rump; white underwings.

Size: *L* 21–24 cm; *W* 36–40 cm.

Habitat: *Breeding:* shallow marshes, wet meadows and marshy lakes. *In migration:* alkaline lakes, sewage treatment ponds, shallow marshes and flooded fields.

Nesting: on the ground near water; well-concealed depression is lined with grasses; male incubates 3–4 brown-blotched, buff eggs for 22–25 days.

Feeding: whirls in tight circles in water to stir up prey, then picks out aquatic insects, worms and small crustaceans; on land, makes short jabs to pick up insects, crustaceans and some seeds.

Voice: deep, grunting *work work* or *wu wu wu*, usually in flight on breeding grounds.

Similar Species: *Red-necked Phalarope* (p. 209): shorter bill; dark cap; white upperwing stripe; breeding bird has grey breast and red (female) or buff (male) sides of neck; dark nape and eye line in nonbreeding plumage. *Red Phalarope* (p. 210): stouter bill; all-red neck, breast and underparts in breeding plumage; dark nape and broad, dark line behind eye in nonbreeding plumage.

RED-NECKED PHALAROPE

Phalaropus lobatus

nonbreeding

The Red-necked Phalarope is the smallest of the phalaropes and visits land only in migration and to nest. Seeming all but lost among the ocean troughs, the species rides the waves for nine months of the year. Unlike the Red Phalarope, Rednecks can often be seen from land as they follow tidal rips and convergences along both coasts. • Most of these Arctic breeders migrate along the coast or well out to sea, but for a few weeks in spring and fall, Red-necked Phalaropes can be reasonably abundant at inland locations. They stage at wetlands, storing up reserves for their long haul to and from Arctic nesting grounds and their wintering range at sea off western South America. • In eastern Canada, huge flocks totalling millions used to pass through the Bay of Fundy during fall migration. Recently, these flocks have all but disappeared for reasons that are not clear. Since 1980, large declines have also been noted among nesting populations at La Pérouse Bay, Manitoba, possibly related to degradation of breeding habitat caused by explosive increases in Snow Goose numbers. • "Phalarope" is the Greek word for "coot's foot." Like coots and grebes, phalaropes have individually webbed, or "lobed," toes, a feature that makes them proficient swimmers. • Formerly known as the "Northern Phalarope," this species breeds widely across Holarctic regions.

♀ ♂

breeding

ID: small shorebird; long, thin, straight, dark bill; longish black legs; dark centre in rump; white "chin" and belly. *Breeding female:* chestnut stripe on neck and throat; blue-black head; incomplete, white eye ring; light grey breast; 2 rusty-buff stripes on each upperwing. *Breeding male:* white eyebrow; less intense colours than female. *Nonbreeding:* white underparts; dark nape; black cap; broad, dark band from eye to ear; whitish stripes on blue-grey upperparts. *Juvenile:* similar to nonbreeding adult, but blacker above with bright buff stripes and darker forehead. *In flight:* dark upperwings with bold, white wing stripe; dark markings on white underwings; flies in loose flocks, often settling on water.

Size: *L* 18 cm; *W* 38 cm.

Habitat: *Breeding:* marshy tundra wetlands. *In migration:* open ocean (especially upwellings and tide rips), estuaries, brackish lagoons, shallow marshes, tidal channels, alkaline lakes and sewage ponds.

Nesting: on the ground near water; well-concealed depression on a grassy tussock is lined with grasses; male incubates 3–4 brown-blotched, buff eggs for 18–20 days.

Feeding: whirls in tight circles in shallow or deep water, picking insects, small crustaceans and small molluscs from the water; on land, makes short jabs to pick up insects.

Voice: often noisy; soft *krit krit krit* in migration.

Similar Species: *Red Phalarope* (p. 210): chunkier; heavier bill; breeding bird has rufous or cinnamon underparts; paler, plainer upperparts in nonbreeding plumage. *Wilson's Phalarope* (p. 208): longer, thinner bill; plainer wings; breeding female has black on neck and cinnamon foreneck; male and nonbreeding female have mostly white faces and pale grey upperparts.

RED PHALAROPE

Phalaropus fulicarius

The bulk of Red Phalarope migration in spring and fall takes place well out at sea, and the species spends up to 11 months of the year in marine habitats. • The Red Phalarope passes through interior areas of Canada in small numbers during its annual fall migration toward the coasts. Its dull fall wardrobe does little to make it stand out among the much more abundant Red-necked Phalaropes. Still, the Red Phalarope's heavier bill and larger size provide sufficient clues for a sharp-eyed birder to distinguish between these species in the field. The Red Phalarope's nonbreeding plumage also reveals why it is known as "Grey Phalarope" in other parts of the world. • When incubating the eggs, male phalaropes shed feathers on their abdomen and develop an area of thick skin, which becomes swollen with blood. This "brood patch" provides the required temperature for incubation and is a feature that usually develops in females in other species. • Red Phalaropes have occasionally been observed tending surfaced whales, quickly snatching parasites living on the mammals' backs. They have even been observed feeding on crustaceans in plumes created by foraging grey and bowhead whales, a habit that prompted whalers to call them "Bowhead Birds."

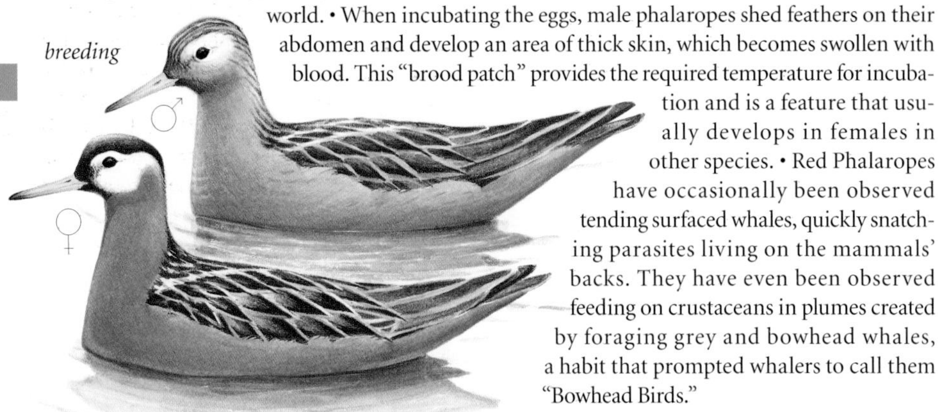

ID: small shorebird. *Breeding female:* chestnut red throat, neck and underparts; white face; black crown and forehead; black-tipped, yellow bill. *Breeding male:* similar to female but much paler; mottled brown crown; duller face and underparts. *Nonbreeding:* blue-grey, unstreaked upperparts; white head, neck and underparts; black nape, eye patch and bill. *Juvenile:* similar to nonbreeding adult; buff-colored overall; dark streaking on upperparts. *In flight:* grey wings; broad, white wing stripe; mostly white underwings; black central tail feathers.
Size: *L* 21.5 cm; *W* 43 cm.
Habitat: *Breeding:* coastal High Arctic tundra close to water. *Nonbreeding:* open ocean waters well away from shore; prefers upwellings and current edges; occasionally seen at inshore bays, lagoons, estuaries, tide rips and salt evaporation ponds; rarely seen well inland.
Nesting: on the ground on a slightly raised site; shallow hollow is lined with plant material; male incubates 4 darkly blotched, light olive or greenish eggs for 19 days.
Feeding: gleans from the water's surface, usually while swimming in tight, spinning circles; eats small crustaceans, molluscs, insects and other invertebrates; rarely takes vegetation or small fish.
Voice: high-pitched *wit* or *creep*; low-pitched *clink clink*.
Similar Species: *Red-necked Phalarope* (p. 209): smaller; thinner bill; breeding bird lacks all-red underparts; nonbreeding bird has thin, white upperwing stripe. *Wilson's Phalarope* (p. 208): breeding bird lacks all-red underparts; nonbreeding bird lacks black eye patch.

BLACK-LEGGED KITTIWAKE

Rissa tridactyla

The ocean-loving Black-legged Kittiwake is one of the only true "pelagic seagulls" found along Canadian coastlines. Most kittiwakes breed on coastal cliffs, generally at higher latitudes, and spend the balance of the year wandering on the open ocean. • Black-legged Kittiwakes are open-ocean fishers and scavengers—unlike most gulls, they do not join in "feeding frenzies" at garbage dumps. Kittiwakes cruise high over the waves, darting downward in an instant to snap up a food item. They are exceptionally rare away from the ocean, scarce even in the lower reaches of estuaries. Some immatures, however, may roost on coastal rocky shores and headlands from spring to fall each year. • Late fall and winter storms that persist for several days may tire these ocean wanderers, at which time they may be found in nearshore waters. The unique combination of short, black legs, wing tips "dipped in ink" and yellow bill readily identify the Black-legged Kittiwake as a "bird of a different feather." • The name "kittiwake" comes from the call, a shrill *kittee-wa-aaake*, this bird makes at nesting colonies. In Newfoundland, this bird is known as "Tickle-Ace."

1st-winter juvenile

breeding

nonbreeding

ID: sexes similar; medium-sized pelagic gull; small, unmarked, yellow bill; short, black legs. *Breeding:* white head and underparts; dark eyes; medium grey upperparts. *Nonbreeding:* smudgy, greyish nape and ear patch. *Juvenile:* dark edges on wings and back of head. *Immature:* heavy, black *M* on upperwings; black-tipped tail. *In flight:* stiff wingbeats; black wing tips.

Size: *L* 41–46 cm; *W* 91 cm.

Habitat: mainly open ocean, often near upwellings; roosts on rocky headlands and islands, sandy beaches and peninsulas.

Nesting: in a large colony, usually on cliff ledges; neat cup nest of seaweed, grass and moss is added to each summer; pair incubates 1–3 variably marked, greenish or bluish buff eggs for 25–30 days.

Feeding: plunges, dips and surface feeds for small fish, krill and other marine organisms; can dive to 8 m below the water's surface.

Voice: generally silent; shrill, repetive *kittee-wa-aaake* at nesting colonies.

Similar Species: *Sabine's Gull* (p. 213): much smaller; yellow-tipped, black bill; more black on upper wing tip and less below; grey wing coverts; wider tail; nonbreeding bird has partial dark "hood." *Bonaparte's Gull* (p. 214): much smaller; small, black bill; black-tipped, white wing tips; rounded tail.

IVORY GULL

Pagophila eburnea

This dainty gull is easy to differentiate from other gulls by its snowy white appearance, black legs and two-tone bill. Immature birds are also distinctive, with a blackish facial patch and black markings on wings and tail. The Ivory Gull's ternlike call and flight also differentiate it from other small gulls. • Confined to the central and eastern Canadian Arctic during the breeding season, Ivory Gulls are found in Atlantic Canada in winter only, when they venture down to the coast of Labrador. It is also a rare vagrant into the Great Lakes region during winter. • Rarely found away from drifting pack ice, Ivory Gulls feed on fish, as well as the droppings of polar bears, whales, walruses and seals! This bird is best known, however, for its preference for seal meat. Birders in northern Newfoundland will even willingly part with cash to obtain a seal carcass to attract a flock of these predominantly offshore gulls. • The Ivory Gull is ranked as endangered in Canada, with the population declining by 80 percent in the last 20 years. Studies have found Ivory Gulls to suffer from higher contaminment levels than most Canadian seabirds. Recent surveys placed the global population of Ivory Gulls at around 14,000 pairs, with 10,000 pairs in Russia and fewer than 1200 pairs in Canada.

nonbreeding

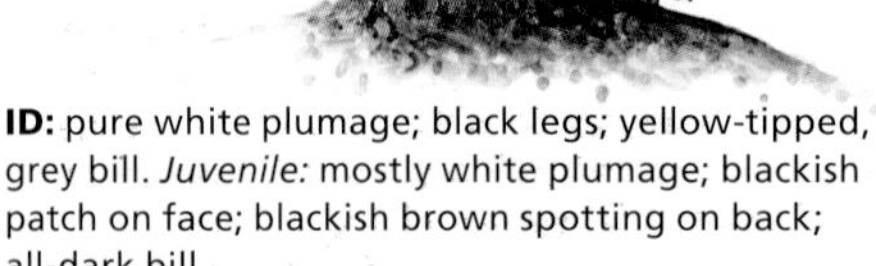

breeding

ID: pure white plumage; black legs; yellow-tipped, grey bill. *Juvenile:* mostly white plumage; blackish patch on face; blackish brown spotting on back; all-dark bill.
Size: *L* 40–43 cm; *W* 1.1–1.2 m.
Habitat: consolidated pack ice, drift ice and the ice edge; occasionally ventures into harbours.

Nesting: colonial; on a cliff ledge or rocky ground; nest scrape is lined with mosses, grasses and some feathers; pair incubates 1–3 variably marked olive to buff eggs for 24–26 days.
Feeding: catches fish on the water's surface or makes shallow dives; also takes marine invertebrates, dung and carrion, especially seal meat.
Voice: harsh, ternlike *keeuur.*
Similar Species: *Iceland Gull* (p. 227) and *Glaucous Gull* (p. 230): larger; pinkish legs.

SABINE'S GULL

Xema sabini

Sabine's Gull is an unusual and distinctive Arctic gull that breeds at high latitudes but winters in coastal upwelling zones of the tropics and subtropics. • This attractive, ternlike gull is a truly stunning bird. It is one of only two gulls with a yellow-tipped, black bill and a notched tail. A rarity in much of Canada, Sabine's Gull can best be observered in autumn, when it trickles southward from its Arctic breeding grounds. Most birds migrate along the Pacific Coast, with smaller numbers along the Atlantic and even fewer across interior regions—the latter primarily in fall. • The Sabine's Gull shares the same size, shape and head colour pattern of the common, black-headed gulls. Like the terns, this gull features a buoyant, dipping flight pattern and a forked tail. • Sabine's Gulls feed while in flight, gently dipping down to the water's surface to snatch up prey without landing. • Sir Edward Sabine was a distinguished military man whose primary interests were astronomy and terrestrial magnetism. He joined an expedition to explore the Arctic islands, and it was near Cape York that he collected the gull that was to be named in his honour.

nonbreeding

breeding

ID: yellow-tipped, black bill; dark grey mantle; black feet. *Breeding:* dark slate grey "hood" trimmed with black. *Nonbreeding:* white head; dark grey nape. *Juvenile*: brown, scalloped upperparts; white head with brown "hood"; white underparts; black-tipped, white tail; boldly marked black, brown and white *M* pattern on upperwing. *In flight:* 3 triangles of black, grey and white on upperwing; shallowly forked tail.

Size: *L* 33–36 cm; *W* 91 cm.

Habitat: *Breeding:* small tundra ponds and lakes. *In migration:* ocean coasts, lakes and large rivers.

Nesting: colonial; on a raised area on swampy tundra; nest scrape is sparsely lined with mosses, grasses and feathers; pair incubates 1–4 variably marked, olive green eggs for 23–26 days.

Feeding: dips or swoops to the water's surface; gleans small fishes and crustaceans while swimming; if forced to shore by storms, picks marine worms, crustaceans and insects from the water's surface or scavenges dead or dying fish.

Voice: ternlike *kee-kee*; not frequently heard in migration.

Similar Species: *Bonaparte's* (p. 214), *Franklin's* (p. 219), *Little* (p. 216), *Laughing* (p. 218) and *Black-headed* (p. 215) *gulls:* lack boldly patterned upperwing, forked tail and yellow-tipped bill.

BONAPARTE'S GULL

Chroicocephalus philadelphia

nonbreeding

The graceful, reserved Bonaparte's Gull is nothing like its larger, aggressive relatives that fight over food scraps. Delicate in plumage and behaviour, this small gull avoids landfills, preferring to dine on insects captured in flight or from the water's surface. Only when a flock of Bonaparte's Gulls spies a school of fish, an abundance of krill or a swarm of aerial insects do they raise their soft, scratchy voices in excitement. • Bonaparte's Gulls are also unlike any other gulls or terns in that they nest in the upper branches of coniferous trees. Nesting solitarily or in very loose colonies, they often choose sites that are close to water. • Spring migrants and breeders are resplendent with black heads, as if the birds had been dunked in dark paint. By late fall, the black is gone, but the birds still display conspicuous, white, primary wedges on their wings and fly with a tern-like grace lacking in larger gulls. • This elegant gull was named after the French emperor's nephew, zoologist Charles Lucien Bonaparte, who collected the first specimen for naming in Philadelphia.

breeding

ID: sexes similar; small gull, often seen in single-species flocks; white wedge on leading edge of wing; slender, black bill; reddish pink legs; dark grey mantle; white underparts. *Breeding:* black "hood"; incomplete, white eye ring. *Nonbreeding:* dusky white head; black ear spot. *Juvenile:* similar to nonbreeding; black tail band; dusky wings show dark *M* pattern in flight. *In flight:* black edges on white primaries.

Size: *L* 30–36 cm; *W* 84 cm.

Habitat: *Breeding:* coniferous forests bordering lakes, large marshes and bogs; often on wooded islands with conifers. *In migration* and *winter:* nearshore coastal waters, including upwellings and channels and estuaries; inland on lakes, marshes, rivers and sewage lagoons.

Nesting: semicolonial; pair builds a twig, moss and lichen nest in a conifer; pair incubates 2–3 variably marked, olive buff eggs for 23–24 days.

Feeding: picks food from the water's surface and flycatches in midair; feeds mainly on small fish, krill and insects.

Voice: utters a scratchy, soft *cheer* or *cherr*; large flocks can be very noisy when feeding.

Similar Species: *Franklin's Gull* (p. 219): larger; darker back; dark legs; prominent, white spots on black wing tips; heavier bill is red in breeding plumage; mottled crown in nonbreeding plumage. *Little Gull* (p. 216): daintier bill; white wing tips; black "hood" lacks white eye ring and extends over nape; nonbreeding bird has white cap. *Black-headed Gull* (p. 215): larger; large, red bill; dark underwing primaries; reddish legs; brownish "hood" in breeding plumage.

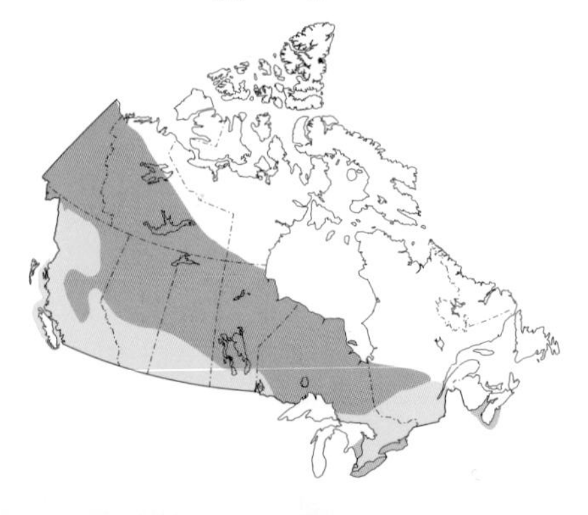

BLACK-HEADED GULL

Chroicocephalus ridibundus

nonbreeding

Black-headed Gulls are a regular, and increasing, attraction for birders in Atlantic Canada. The odd one can be found cavorting with large gatherings of Bonaparte's Gulls in autumn, and numbers increase after their smaller relatives leave. The often-raucous concentrations of gulls in southern Newfoundland's harbours and field roosts usually have a small number of these wanderers. This gull is also a rare visitor to the Great Lakes region during migration. • The Black-headed Gull is a relative newcomer to Canada. The first sighting in North America was recorded in the 1920s, and the first Canadian nesting record occurred in 1977. Since then, small nesting colonies have been established on islands off the coast of Newfoundland, Nova Scotia and Québec, and it may be just a matter of time before the species nests in Ontario. • In North America, there are many gulls with black heads. Ironically, the Black-headed Gull has a chocolate brown head, which can easily be mistaken for black in the misty conditions that sometimes envelop Atlantic Canada's coasts.

breeding

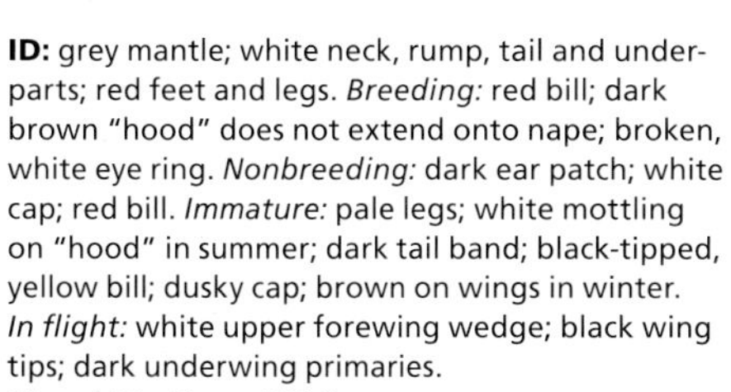

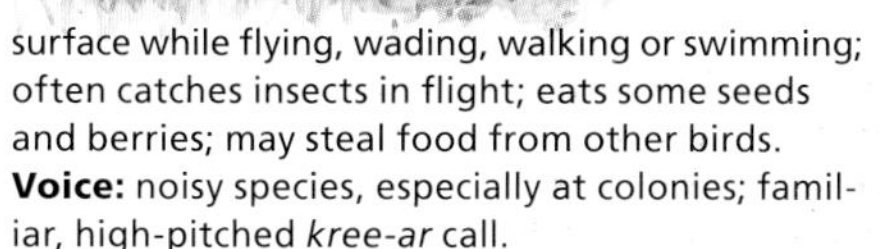

ID: grey mantle; white neck, rump, tail and underparts; red feet and legs. *Breeding:* red bill; dark brown "hood" does not extend onto nape; broken, white eye ring. *Nonbreeding:* dark ear patch; white cap; red bill. *Immature:* pale legs; white mottling on "hood" in summer; dark tail band; black-tipped, yellow bill; dusky cap; brown on wings in winter. *In flight:* white upper forewing wedge; black wing tips; dark underwing primaries.

Size: *L* 36–41 cm; *W* 1.0 m.

Habitat: coastal marshes, lakeshores and offshore islets.

Nesting: colonial; in an offshore islet or coastal marsh; shallow scrape on the ground or in low vegetation is lined with plant materials; pair incubates 2–4 variably marked, light olive to dark brown eggs for 22–24 days.

Feeding: gleans insects, small molluscs, crustaceans, spiders and small fish from the ground or the water's surface while flying, wading, walking or swimming; often catches insects in flight; eats some seeds and berries; may steal food from other birds.

Voice: noisy species, especially at colonies; familiar, high-pitched *kree-ar* call.

Similar Species: *Bonaparte's Gull* (p. 214): smaller overall; daintier, black bill; mostly white underwing primaries; more orange on legs; black "hood" on breeding bird extends onto upper nape. *Little Gull* (p. 216): primary feathers lack black tips; breeding bird has all-black head, smaller bill and no eye ring; nonbreeding bird has dark cap and black bill. *Franklin's Gull* (p. 219) and *Laughing Gull* (p. 218): black crescent on wing tips; lack white triangle on upper leading edge of wings; black "hood" on breeding birds extends over nape; nonbreeding Franklin's has black mask and bill. *Sabine's Gull* (p. 213): yellow-tipped, black bill; black upper forewing wedge; pure white hindwing.

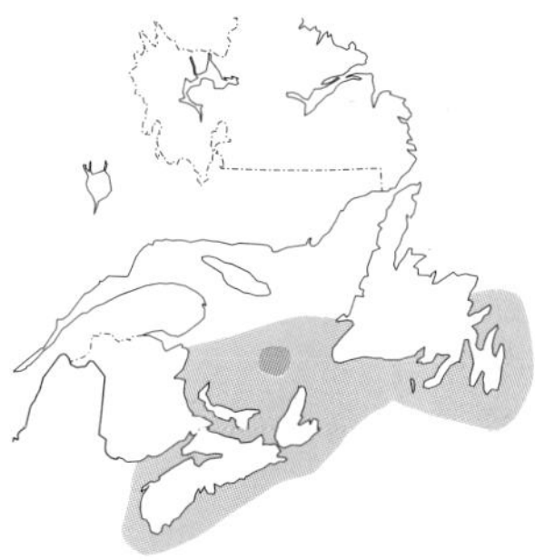

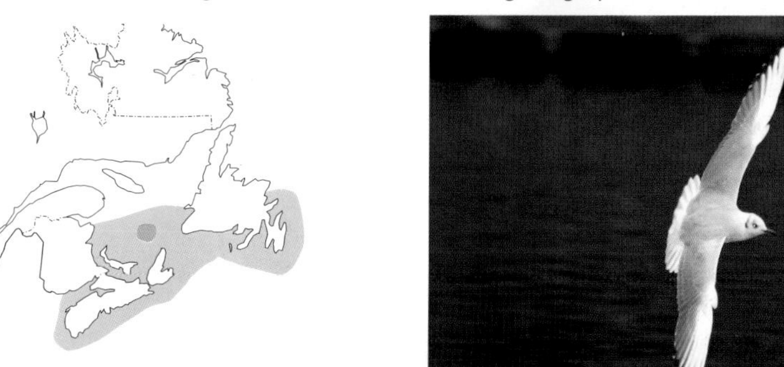

LITTLE GULL

Hydrocoloeus minutus

nonbreeding

The Little Gull is the smallest gull in the world and is unique in having small, somewhat rounded wings with dark grey to black undersides. This common Eurasian bird was first identified in North America during the first Franklin expedition in the early 19th century. It was considered an exceptionally rare vagrant until 1962, when the first documented nest in the New World was discovered in Oshawa, Ontario. The species has since been found nesting at several sites in the Great Lakes–St. Lawrence basin and in a few locations in the Hudson Bay and James Bay lowlands of Manitoba and Ontario. Little Gulls never nest at a given site for more than a few years. • In southern Canada, Little Gulls have often been found nesting in association with Common, Forster's or Black terns, but those in the north are more inclined to be with or near Arctic Terns. During migration, Little Gulls associate with flocks of Bonaparte's Gulls. • The estimated North American population of this gull is only 400 individuals. This species is vulnerable to pollution events because it prefers to congregate in only a few locations during migration.

breeding

ID: grey back and wings; orange-red feet and legs. *Breeding:* black head; dark red bill. *Nonbreeding:* black bill; dark ear spot and cap. *Immature:* pinkish legs; brown and black in wings and tail; dark cap and spot behind eye. *In flight:* short, rather rounded wings; white wing tips and trailing edge of wings; dark grey to black underwings.

Size: *L* 25–29 cm; *W* 61 cm.

Habitat: *Breeding:* marshy areas of freshwater lakes, among fresh or brackish marshes and along coastlines. *In migration:* large lakes, rivers, marshes, ponds and beaches.

Nesting: colonial but occasionally in isolated pairs; on the ground near water; pair lines a shallow depression with vegetation or builds a raised mound; pair incubates 1–5 brown-and-grey-marked, olive to buff eggs for 23–25 days.

Feeding: gleans insects from the ground or from the water's surface while flying, wading, walking or floating; may also take small molluscs and fish, crustaceans, marine worms and spiders.

Voice: repeated *e-key*; low *kek-kek-kek*.

Similar Species: *Bonaparte's Gull* (p. 214): black-tipped primary feathers; breeding bird has broken, white eye ring, larger, black bill and white nape; nonbreeding bird has white cap. *Black-headed Gull* (p. 215): black-tipped primary feathers; breeding bird has broken, white eye ring, larger bill and white nape; nonbreeding bird has white cap and red bill. *Franklin's Gull* (p. 219): black wing tips; darker mantle; breeding bird has broken, white eye ring, white nape and brighter red bill; nonbreeding bird has black mask. *Laughing Gull* (p. 218): black wing tips; darker mantle; black legs; breeding bird has broken, white eye ring and much larger bill. *Sabine's Gull* (p. 213): black forewing wedge; black legs; yellow-tipped bill.

ROSS'S GULL

Rhodostethia rosea

Ross's Gull is one of the species that draws birders from far away when it shows up south of its breeding grounds. Ross's Gull is a small, white-and-grey gull that has a dark collar and pale pink underparts usually only in the breeding season. Those wanting to see it in breeding plumage must travel up to Churchill, Manitoba. • This gull breeds across the Arctic regions of the world, with a population of only 50,000 birds. A small population breeds in a handful of locations in Canada, including Churchill and various Nunavut locations. The small colonies consist of only one to five pairs, and entire colonies can be lost during years of bad weather or high predation. There is a black market for Ross's Gull eggs, and collecting can cause a colony to fail and leave. • Like the Ivory Gull, Ross's Gulls remain in polar regions, feeding along the edges of pack ice and open water. • With its small population, limited known breeding locations and the harshness of its environment, Ross's Gull has been classified as threatened in Canada.

nonbreeding

ID: sexes similar; small gull; black bill; red feet; dark eyes; grey back and wings. *Breeding:* narrow, black collar; pinkish wash on white face and underparts. *Nonbreeding:* black collar absent; face and underparts less rosy and more whitish grey; grey smudges on rear of head. *Juvenile:* similar to nonbreeding; black tips on wings and tail; black ear spot; pink legs. *In flight:* blue-grey underwings; white, wedge-shaped tail.
Size: *L* 31–34 cm; *W* 74 cm.
Habitat: *Breeding:* small tundra ponds and lakes. *In migration* and *winter:* consolidated pack ice, drift ice, ice edge in polar regions, ocean coasts, lakes and large rivers.
Nesting: singly or in a small colony; on tundra near water, often with Arctic Terns; slight scrape is lined with dry grass, sedge, leaves and moss; pair incubates 1–3 finely marked, olive to buff eggs for 21–22 days.
Feeding: plunges, dips and surface feeds; often feeds like a shorebird plucking food off mudflats; eats mostly insects, crustaceans, fish and small molluscs.
Voice: harsh *miaw*; usually silent in winter.
Similar Species: pinkish breast, dark bill and black collar are distinctive. *Bonaparte's Gull* (p. 214): black-tipped primary feathers; breeding bird has broken, white eye ring, larger, black bill and white nape; nonbreeding bird has white cap. *Little Gull* (p. 216): nonbreeding bird has white cap and black underwings.

LAUGHING GULL

Leucophaeus atricilla

nonbreeding

Laughing Gulls are well known to summer beachgoers on the Atlantic Coast. With their striking appearance, bold behaviour, graceful flight and noisy, laughlike call—a boisterously repeated *ha ha ha ha haah haah haah*—these appropriately named gulls are difficult to ignore. • The Laughing Gull has adapted well to the urban landscape, gathering in harbours and parks to "panhandle" for food scraps. • In the late 19th century, high commercial demand for eggs for collections and feathers for women's hats resulted in the extirpation of this gull as a breeding species in many parts of its Atlantic Coast range. Today, East Coast populations are gradually assuming their former abundance, and the Laughing Gull is now found in the Maritimes and Great Lakes on a regular basis. Nevertheless, nesting in eastern Canada is currently limited to one small colony on Machias Seal Island, New Brunswick. • Although this gull may appear in any month of the year, the months of May, June, August and September support the most sightings.

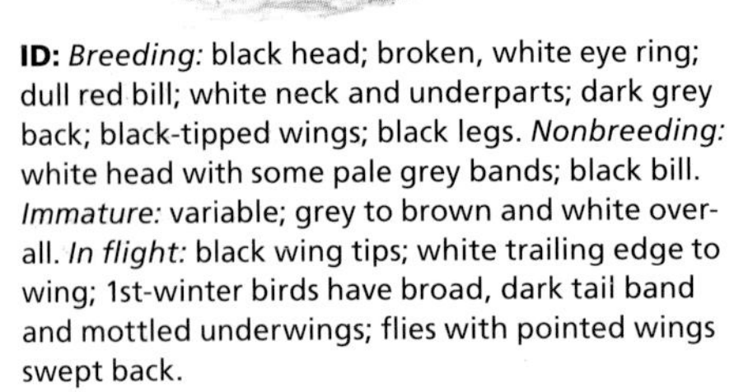

breeding

ID: *Breeding:* black head; broken, white eye ring; dull red bill; white neck and underparts; dark grey back; black-tipped wings; black legs. *Nonbreeding:* white head with some pale grey bands; black bill. *Immature:* variable; grey to brown and white overall. *In flight:* black wing tips; white trailing edge to wing; 1st-winter birds have broad, dark tail band and mottled underwings; flies with pointed wings swept back.

Size: *L* 38–43 cm; *W* 1.0 m.

Habitat: *Breeding:* along coastlines and offshore islands. *In migration* and *winter:* coastal habitats; shorelines of larger lakes and rivers, urban landfills and fast-food restaurant parking lots.

Nesting: rare; in a small colony; ground nest is made of grass and weeds; pair incubates 2–3 brown-blotched, olive to buff eggs for 20–23 days.

Feeding: omnivorous; gleans insects, small molluscs, crustaceans, spiders and small fish from the ground or water while flying, wading, walking or swimming; may steal food from other birds; may eat eggs and nestlings of other birds; often scavenges at landfills.

Voice: loud, nasal, laughlike *ha ha ha ha haah haah haah* call.

Similar Species: *Franklin's Gull* (p. 219): smaller overall; red legs; shorter, slimmer bill; nonbreeding bird has black mask. *Bonaparte's Gull* (p. 214) and *Black-headed Gull* (p. 215): orange or reddish legs; slimmer bill (Bonaparte's has black bill); lighter mantle; white wedge on upper leading edge of wing; black "hood" on breeding bird does not extend over nape; nonbreeding bird has white head and black head spot. *Little Gull* (p. 216): paler mantle; reddish legs; dainty, black bill; lacks eye ring and black wing tips.

FRANKLIN'S GULL

Leucophaeus pipixcan

nonbreeding

Although the Franklin's Gull is often referred to by nonbirders as a "seagull," the species spends much of its life inland and nests in small to very large colonies in prairie marshes. It is sometimes called "Prairie Dove" for its dovelike profile, and it often follows tractors across agricultural fields, snatching up insects from a tractor's path or from freshly tilled soil in much the same way its cousins follow fishing boats. • Franklin's Gull colonies often shift from year to year, sometimes related to drought. Perhaps the largest colony in Canada, with nearly 100,000 pairs, on Whitewater Lake in Manitoba disappeared for a few years during the 1980s when the entire lake dried up. • Franklin's Gull is one of only two gull species that migrate long distances between breeding and wintering grounds and is unique in having two moults per year, one in spring and the other in fall. The majority of Franklin's Gulls overwinter along the Pacific Coast from Guatemala to Chile, but a few occur along either coast as far north as southern BC or Newfoundland. • This gull was named for Sir John Franklin, the British navigator and explorer who led four expeditions to the Canadian Arctic in the 19th century.

breeding

ID: sexes similar; short, straight bill. *Breeding:* all-black head; partial, white crescent above eye; reddish bill; dark grey upperparts; white underparts, often with pink tinge; black-and-white-tipped primaries; dark, red-tinged legs. *Nonbreeding:* white head with dark patch at rear; black bill. *Juvenile:* similar to nonbreeding but with mottled brown upperparts. *In flight:* short wings; grey upperwings with white trailing edge.

Size: *L* 33–38 cm; *W* 94 cm.

Habitat: *Breeding:* marshy shores of inland lakes or large prairie wetlands. *In migration:* on the coast in protected harbours, lagoons, estuaries, bays and sewage ponds; large inland marshes and lakes; forages in plowed fields and landfills.

Nesting: colonial; usually in dense emergent vegetation; floating platform nest is built above water and lined with fine grass and plant down; pair incubates 2–4 variably marked, olive buff eggs for 20–25 days.

Feeding: opportunistic; eats insects, crustaceans, voles, fishes, worms and garbage; also catches flying insects in midair.

Voice: laughlike, strident *ha-ha-ha-ha-ha* call; shrill *kuk-kuk-kuk*.

Similar Species: *Bonaparte's Gull* (p. 214): slimmer; black bill; paler back; pink or red legs; conspicuous white wedge and narrow, black strip on trailing edge above and below wing tips; dark ear spot in nonbreeding plumage. *Little Gull* (p. 216): rare; much smaller; short, rounded wings; lacks black crescent on wing tips and broken, white eye ring. *Sabine's Gull* (p. 213): rare; large black, white and grey triangles on upperwings; dark bill with yellow tip.

HEERMANN'S GULL

Larus heermanni

Coastal birders struggling to identify gulls can be grateful to the Heermann's Gull for providing a sequence of completely distinctive plumages at all ages. The adult Heermann's is unlike any other North American gull in appearance, having a white head that shades abruptly into slate grey on the underparts and a red bill tipped with black. It also has a unique "reverse" or northward migration pattern, drifting from breeding colonies in the Gulf of California in Mexico north to Vancouver Island. Peak numbers arrive in Canada during July and August, with a southward exodus during fall and winter, but a few birds remain year-round. • Although some individuals feed along beaches, in sheltered bays, on rocky promontories or on kelp beds, most join mixed-species assemblages of feeding pelagic birds pursuing schools of herring some distance offshore. • Surprisingly aggressive for their size, Heermann's Gulls, like jaegers, often steal food from other seabirds by bullying them to give up their catch (a behaviour called kleptoparasitism). • Since 1975, Heermann's Gull numbers have nearly tripled, to a current estimate of 150,000 pairs, greatly aided by the Mexican government's establishment of a seabird sanctuary at the species' main colony.

nonbreeding

breeding

ID: sexes similar; medium-sized gull; mostly dark plumage at all ages. *Breeding:* white head; black-tipped, red bill; dark grey back and wings; ashy grey neck and underparts; white-tipped, black tail; black legs; dark eyes. *Nonbreeding:* pale grey head with brownish streaking. *Juvenile:* all dark brown body; pale bill with dark tip. *In flight:* all-dark wings and tail have white trailing edges.
Size: *L* 41–48 cm; *W* 1.4 m.
Habitat: coastal ocean waters, usually close to shore, including bays, beaches, offshore islands, lagoons and coastal creek outfalls.

Nesting: does not nest in Canada.
Feeding: mainly small fish but also a variety of marine invertebrates; frequently steals food from other birds.
Voice: common call is a nasal *kawak*; also a series of whining *ye* notes and low-pitched honks.
Similar Species: *California Gull* (p. 224): larger; juvenile has mottled brown body and dull pink legs and feet. *Pomarine Jaeger* (p. 238) and *Parasitic Jaeger* (p. 239): dark adults resemble juvenile Heermann's but have white primary flashes and narrower tails, often with projecting central feathers.

MEW GULL

Larus canus

nonbreeding

This petite and delicate gull is the smallest of the typical "white-headed" gulls. Mew Gulls in North America breed in western Canada and Alaska. The Mew Gull stages a major invasion of Pacific coastal shorelines from October to December as birds arrive from interior breeding grounds. Flocks often join other gulls at herring and salmon spawning sites. During migration and winter, tens of thousands may be found feeding in flooded fields and pastures of the lower Fraser River valley. • Most of the Mew Gulls seen in Atlantic Canada in winter are of the slightly larger European *canus* subspecies, although smaller North American breeding *brachyrhynchus* individuals may also be seen. Immatures are more problematic to identify, but Europeans, who call this bird "Common Gull," refer to its neater looks, smaller bill and darker terminal tail band. The large eyes and single, subterminal white patch on the black wing tips makes the adult an easy gull to identify. • The Mew Gull breeds in both marine and freshwater habitats and is both a ground and tree nester, being the only "white-headed" gull that regularly uses trees for nesting.

breeding

ID: sexes similar; small to medium-sized gull. *Breeding:* white head, tail and body; dark grey back and wings; dark eyes; short, all-yellow bill; yellow legs. *Nonbreeding:* variable dusky smudging on head; no eye ring; indistinct dark spot on bill. *Juvenile:* mottled brown and grey. *In flight:* buoyant flight; black wing tips with white "mirrors."
Size: *L* 38–40 cm; *W* 1.0–1.1 m.
Habitat: *Breeding:* small, low islands and coniferous-forested islands in lakes. *In migration* and *winter:* tidal waters, estuaries, brackish lagoons, river mouths, flooded agricultural fields and sewage lagoons.
Nesting: semicolonial; on the ground or on a spruce branch; shallow hollow is lined with plant material; pair incubates 3–5 variably marked, olive buff eggs for 23–26 days.
Feeding: opportunistic omnivore; eats invertebrates, fish, small mammals, bird eggs and chicks, carrion and even berries in season; rarely visits garbage dumps or accepts handouts.
Voice: relatively quiet; high, squeaky, nasal notes; loud "mewing" call; high-pitched, coughing *queeoh*.
Similar Species: *Ring-billed Gull* (p. 222): paler eyes; black-ringed, yellow bill (adult) or black-tipped, pink bill (juvenile); paler grey upperparts; nonbreeding bird has less brown on head and more black and less white at wing tips. *California Gull* (p. 224): much larger; red and black marks on lower mandible of yellowish or dull blue-grey bill.

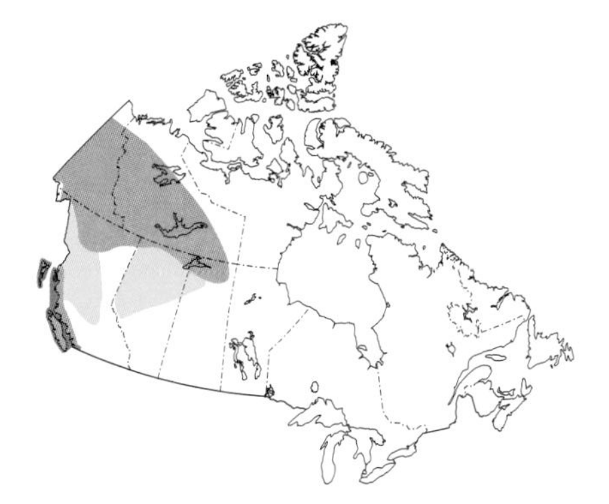

RING-BILLED GULL

Larus delawarensis

The Ring-billed Gull is the most widespread and abundant inland-nesting gull in North America. Industrial farming practices, the proliferation of open landfills and the end of human persecution in the 1920s have allowed this "land gull" to expand exponentially in range and numbers since the 1960s. Having developed a taste for the urban lifestyle, Ring-billed Gulls have become a routine sight at many shopping mall parking lots, sports fields, beaches and city parks. Some colonies have even taken to nesting on the flat-topped, gravelled roofs of urban malls. In rural settings, Ring-bills follow agricultural machinery to feast on crop pests, but they also eat some berry crops and in some places have become a pest themselves. • Ring-bills are a three-year gull, meaning that it takes them three full seasons to acquire the characteristic grey-and-white plumage of adults. Some other gull species require four years to reach adulthood, whereas many of the smaller gulls require only two years. • These gulls typically return to their natal colonies, often nesting in the same general locale as in previous years. Many also return to the same wintering sites year after year.

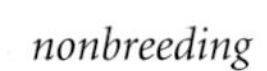

nonbreeding

breeding

ID: sexes similar; medium-sized gull. *Breeding:* medium-grey upperparts; white underparts; white head; large, white spots on black wing tips; yellow eyes; yellow bill with black ring near tip; yellow legs and feet. *Nonbreeding:* dusky streaking on head. *Juvenile:* mottled brown plumage; blackish tail tip; dark wing tips; pale pink legs and feet; pinkish bill with dark tip. *In flight:* mostly pale underwings; white trailing edge to wing.
Size: *L* 46–51 cm; *W* 1.2 m (male is slightly larger).
Habitat: *Breeding:* bare, rocky, shrubby and wooded islands in lakes and rivers; human-made breakwaters, piers and waste grounds; also urban rooftops. *In migration* and *winter:* lakes, beaches, rivers, landfills, golf courses, agricultural fields, coastal bays, estuaries and parks.

Nesting: colonial; on bare rock, among grasses or under low bushes; ground scrape is lined with twigs, grasses and leaves; pair incubates 2–4 brown-blotched, greyish eggs for 25–28 days.
Feeding: opportunistic and varied diet; eats fish, insects, earthworms, small rodents, garbage, sewage and carrion.
Voice: high-pitched *kakakaka-akakaka*; low, laughlike *yook-yook-yook*.
Similar Species: *California Gull* (p. 224): larger; no bill ring; black-and-red spot near tip of lower mandible; dark eyes. *Herring Gull* (p. 225) and *Glaucous Gull* (p. 230): much larger; no bill ring; red spot near tip of lower mandible; pinkish legs. *Mew Gull* (p. 221): smaller; dark eyes; short bill without ring; darker mantle; larger white spots on wing tips.

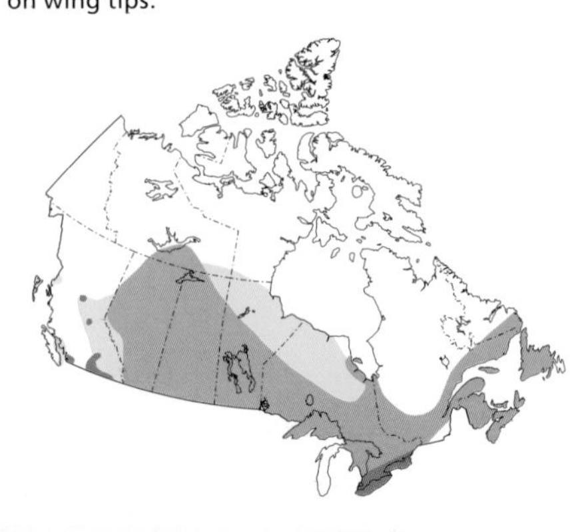

WESTERN GULL

Larus occidentalis

nonbreeding

The Western Gull is a large, white-headed gull that breeds along the Pacific Coast from southern Washington to Baja California at only 200 colonies and winters north to Vancouver Island. Although a familiar and well-known species on the Pacific Coast, the Western Gull is limited in distribution and has a smaller population size than most other North American gulls, with a total population of only about 40,000 pairs nesting at fewer than 200 colony sites. Look for the large, dark-backed Western Gulls among gatherings of Glaucous-winged Gulls along the entire British Columbia coast throughout the year. • The best time to find Western Gulls is from mid-August to mid-September when post-breeding birds visit southern Vancouver Island. Immature birds tend to forage in dumps and harbours, whereas most adults tend to remain offshore in their search for food. • The Western Gull readily hybridizes with the Glaucous-winged Gull, resulting a variety of puzzling plumages.

breeding

ID: sexes similar; large, heavyset gull. *Breeding:* dark grey back and wings; large, white head and underparts; yellow to dark eyes; yellow bill with red spot on lower mandible; pink legs. *Nonbreeding:* dusky streaking on head and neck. *Juvenile:* mottled brown body; dark eyes and bill; grey-pink legs. *In flight:* grey stripe on white underwing; dark wing tips.

Size: *L* 61–66 cm; *W* 1.5 m.

Habitat: *In migration* and *winter:* offshore and nearshore on beaches, rocky outcroppings and intertidal zones; also landfills and less often in agricultural fields.

Nesting: does not nest in Canada.

Feeding: omnivorous; eats various fishes and marine invertebrates; also takes carrion and human refuse.

Voice: most common call is a *keow* note.

Similar Species: *Herring Gull* (p. 225): breeding bird has paler eyes, paler upperparts and whiter underwings; nonbreeding bird has brown-streaked head and neck. *Thayer's Gull* (p. 226): breeding bird has dark eyes, paler upperparts and whiter underwings; juvenile is much paler. *Glaucous-winged Gull* (p. 229): lighter grey back and wings; grey wing tips; slimmer bill.

CALIFORNIA GULL

Larus californicus

nonbreeding

It takes a keen eye to recognize a California Gull as it roosts among Ring-bills, but once you learn how, you'll see them all the time. • California Gulls breed in western Canada from Yellowknife south through the prairies and east to Manitoba, and they winter along the Pacific Coast. • California Gulls tend to nest communally on low-lying islands. Their simple scrape nests are generally placed no closer than the distance two gulls can bridge with aggressive bill jabs from atop their eggs. • A typical four-year gull, the California Gull goes through a series of plumages during its first three years, and most birds do not begin breeding until their fourth year. Subadults generally do not return to breeding sites, preferring to remain on their coastal wintering grounds or else they hang out with Ring-bills and other gulls at colonies until they are old enough to breed. • Considered a beneficial species throughout its range, the California Gull was renowned for having twice saved the early Mormon settlers' crops from plagues of grasshoppers during the mid-1800s. A monument in Salt Lake City honours this prairie gull, and it is the state bird of Utah.

breeding

ID: sexes similar; medium-sized gull. *Breeding:* medium grey upperparts; white head and body; dark eyes; reddish eye ring; yellow bill with diagnostic black-and-red mark on lower mandible; pale greenish yellow legs. *Nonbreeding:* brown-streaked hindcrown and nape. *Juvenile:* mottled brown mantle and wings; black-tipped, pink bill; dark streaking on breast, neck and head; pinkish grey legs; barred tail coverts; dark brown tail. *In flight:* mostly grey upperwings; white spots on black wing tips; white underwings and tail.

Size: *L* 46–51 cm; *W* 1.2–1.4 m.

Habitat: *Breeding:* bare, rocky, shrubby and wooded islands in lakes and rivers. *In migration* and *winter:* open ocean, sandy beaches, rocky seashores, lakes, rivers, landfills, agricultural fields, coastal bays, estuaries and parks.

Nesting: colonial; on bare rock, among grasses or under low bushes; ground depression is lined with grasses, leaves and other plant material; pair incubates 2–3 darkly marked, buff, olive or greenish eggs for 24–27 days.

Feeding: opportunistic; eats fish, crustaceans, insects, earthworms, garbage and carrion.

Voice: loud, ringing *kyow-kyow* barks and "laughlike" notes; also a repeated *kee-yaa.*

Similar Species: *Ring-billed Gull* (p. 222): breeding bird has yellow eyes, shorter bill without red spot, much paler mantle and yellower legs. *Herring Gull* (p. 225): larger; breeding bird has yellowish eyes and pinkish legs. *Mew Gull* (p. 221): smaller; all-yellow bill.

HERRING GULL

Larus argentatus

nonbreeding

Although Herring Gulls are as adept as their smaller Ring-billed relatives at scrounging handouts, they are more likely to be found in wild areas than urban settings. Settling on lakes and large rivers where Ring-bills would never be found, Herring Gulls nest comfortably in large colonies, or they may choose a site kilometres away from any other gulls. • Herring Gulls are skilled hunters, but they are also opportunistic, scavenging for leftovers at landfills, beaches and fast-food restaurant parking lots. Their ability to exploit the offal of human culture, such as at garbage dumps and sewage lagoons, has allowed populations to soar, and they are now found throughout Canada except on islands in the High Arctic. Recent studies on nesting Herring Gulls that forage in landfills found that the consumption of waste junk food containing trans fats has led to a decline in the health of nestlings. • As nesting colonies have increased, so have Herring Gull predations on smaller, more vulnerable species that nest nearby. • Like many gulls, the Herring Gull has a small red spot on the lower mandible. When a chick pecks at the red spot, the parent regurgitates a meal.

breeding

ID: sexes similar; large gull. *Breeding:* pale grey upperparts; white head and body; yellow eyes; large, yellow bill with red spot near tip of lower mandible; pink legs and feet. *Nonbreeding:* brown streaking over most of head and breast. *Juvenile:* mottled brown and white; mostly black bill. *In flight:* white tail and underwings; mostly grey upperwings; white spots on black wing tips.
Size: *L* 58–66 cm; *W* 1.2 m.
Habitat: *Breeding:* rocky and grassy islands, sand dunes and sea cliffs. *In migration* and *winter:* almost any habitat near fresh or salt water; concentrates at landfills, river mouths, flooded agricultural land and beaches; also at sea at upwellings and tidal convergences.

Nesting: colonial or singly; on bare ground or among boulders; shallow scrape is lined with plant material and sticks; pair incubates 3 variably marked, pale olive to buff eggs for 25–32 days.
Feeding: varied diet; eats more "natural" foods than other gulls; also takes waterbird chicks.
Voice: loud, bugling *kleew-kleew* or *hi-yak hi-yak*; loud *kak-kak-kak* alarm call.
Similar Species: *California Gull* (p. 224): smaller; dark eyes; black-and-red mark on bill; darker mantle; greenish yellow legs. *Thayer's Gull* (p. 226): mainly uses coastal habitats; dark eyes; shorter, slimmer bill; pale underwing tips; bright pink legs.

THAYER'S GULL

Larus thayeri

nonbreeding

Just when you thought you had your gulls sorted out, along comes a Thayer's Gull and everything goes out the window. Long considered to be a subspecies of the Herring Gull, the Thayer's Gull was elevated to full species status by the American Ornithologists' Union in 1972. Together with the Herring, Iceland, Glaucous, Glaucous-winged and Western gulls, the Thayer's Gull is part of a group of gulls that all have similar traits, indicating that they are a relatively recently evolved species that could potentially still interbreed. There is still considerable debate among gull researchers as to whether the Thayer's Gull deserves full species status. Some believe that it is actually a subspecies of the Iceland Gull, which shares the same nesting sites in the eastern Canadian Arctic but winters along the Atlantic Coast. • Harvard ornithologist W.S. Brooks named this gull for John Eliot Thayer, the sponsor of the 1913 Alaska expedition on which he collected the first scientific specimen of this bird.

breeding

ID: sexes similar; large gull. *Breeding:* white head and underparts; dark grey mantle; dark eyes; yellow bill with red spot; bright pink legs. *Nonbreeding:* heavy, brown streaking on upper head, neck and breast. *Juvenile:* brown plumage; blackish bill with pink base; blackish brown tail. *In flight:* prominent white spots on black wing tips; grey upperwings with white trailing edges; pale underwings.
Size: *L* 56–64 cm; *W* 1.3–1.4 m.
Habitat: *Breeding:* cliffs on Arctic islands. *In migration* and *winter:* offshore and nearshore coastal waters, including rocky shores, bays and estuaries; also inland lakes and landfills.
Nesting: colonial; on a rocky ledge; deep, bulky cup nest is made of moss and plant material; pair incubates 2–3 variably marked, buff to olive eggs for 31–32 days.
Feeding: opportunistic omnivore; eats fish, crustaceans, molluscs, garbage and carrion.
Voice: generally silent in winter; typically utters short, flat notes.
Similar Species: *Herring Gull* (p. 225): larger; breeding bird has pale yellow eyes, longer, heavier bill and more black on wing tips; immature is more heavily streaked and patterned. *Glaucous-winged Gull* (p. 229): heavier head and bill; breeding bird has dark eyes and grey-and-white wing tips; juvenile is paler. *Iceland Gull* (p. 227): breeding bird has whitish or pure white primaries without dark markings (some grey markings on eastern birds).

ICELAND GULL

Larus glaucoides

breeding

A large, pale, ghostly gull, the Iceland Gull is extremely hardy, nesting in the Arctic and braving the stormy North Atlantic in winter. While most Iceland Gulls appear to travel short distances from their breeding grounds to open waters and ice floes in the North Atlantic in late summer and early fall, immatures tend to move farther south to open coastlines. Like its close relatives the Thayer's Gull and the Glaucous Gull, the Iceland Gull winters from the Great Lakes region to the Maritimes, usually among larger flocks of more common wintering gulls. Birders in Canada are most likely to spot immature birds. • This graceful glider spends much of its time searching for schools of fish over icy, open waters. When fishing proves unrewarding, this opportunistic gull has no qualms about digging through a landfill in search of food. • The Iceland Gull is easily confused with the Thayer's Gull and the Glaucous Gull; the species is very closely related to the Thayer's Gull, which many believe to be a subspecies of the Iceland Gull. Two subspecies of the Iceland Gull are currently recognized: the *kumlieni* subspecies ("Kumlien's Gull"), with grey markings on its wing tips, and the *glaucoides* subspecies, which has pure white wing tips.

nonbreeding

ID: sexes similar; large gull. *Breeding:* white head and underparts; relatively short, yellow bill with red spot on lower mandible; yellow eyes; dark eye ring; pale grey mantle; pink legs. *Nonbreeding:* brown-streaked head and breast. *Immature:* dark eyes; generally pale plumage with beige barring on upperparts; all-black bill or pink bill with black tip. *In flight:* white wing tips with variable amounts of dark grey; whitish to pure white primaries without dark markings.

Size: *L* 56 cm; *W* 1.4 m.

Habitat: *Breeding:* steep, rocky cliffs on eastern Baffin Island. *In migration* and *winter:* harbours and seacoasts; open water on large lakes and rivers.

Nesting: colonial; on a rocky ledge or outcropping; large cup nest is made of turf, mosses and grasses; pair incubates 2–3 darkly marked, very pale olive buff eggs for about 25 days.

Feeding: eats mostly fish; may also take crustaceans, molluscs, carrion and seeds; scavenges at landfills and in harbours; sometimes preys on eggs and chicks of other seabirds.

Voice: high, screechy calls, much less bugling than other large gulls.

Similar Species: *Herring Gull* (p. 225): black wing tips; darker mantle. *Thayer's Gull* (p. 226): more dark grey than white on wing tips; dark eyes. *Glaucous Gull* (p. 230): longer, heavier bill; pure white wing tips. *Ring-billed Gull* (p. 222): smaller; dark ring on yellow bill; yellow legs. *Lesser Black-backed Gull* (p. 228): darker mantle; black wing tips; yellow feet.

LESSER BLACK-BACKED GULL

Larus fuscus

Equipped with long wings for long-distance flights, small numbers of Lesser Black-backed Gulls leave their familiar European and Icelandic surroundings each fall to make their way to North America. Most of these gulls settle along the Atlantic Coast in winter, and it is sometimes a challenge to pick them out among a huge, roosting flock of gulls. However, most Lesser Black-backed Gulls are much daintier and longer-winged than their bulky neighbours. • An increasing number of sightings of this European-nesting species in Atlantic Canada, Québec and Ontario outside of winter suggests that this immigrant may be applying for Canadian status and may soon nest here. Birders are advised to keep a close eye on local cliffs and shoreline rooftops—the most likely locations to support any Lesser Black-backed Gull nesting attempts. Some Lesser Black-backed Gulls have already been found paired with Herring Gulls, which indicates that we may be seeing even more puzzling hybrids.

nonbreeding

breeding

ID: *Breeding:* white head and underparts; dark grey to black mantle; yellow bill with red spot on lower mandible; yellow eyes; yellow legs. *Nonbreeding:* brown-streaked head and neck. *Juvenile:* dark or light eyes; black bill or pale bill with black tip; various plumages with varying amounts of grey on upperparts and brown flecking over entire body. *In flight:* black wing tips with some white spots.

Size: *L* 53 cm; *W* 1.4 m.

Habitat: harbours and seacoasts; also open areas, landfills and open water on large lakes.

Nesting: does not nest in Canada.

Feeding: eats mostly fish, crustaceans, molluscs, insects, small rodents, carrion and seeds; scavenges for garbage.

Voice: laughlike cry.

Similar Species: *Herring Gull* (p. 225): paler mantle; pink legs. *Glaucous Gull* (p. 230) and *Iceland Gull* (p. 227): pale grey mantle; white or grey wing tips; pink legs. *Ring-billed Gull* (p. 222): smaller; dark ring on yellow bill; paler mantle. *Great Black-backed Gull* (p. 231): much larger; black mantle; heavier bill; pale pinkish legs.

GLAUCOUS-WINGED GULL

Larus glaucescens

nonbreeding

Big, bold and mostly married to salt water, the pale-backed Glaucous-winged Gull is the most common gull found along the British Columbia coast. Over the last 50 years, the species has increased by at least three-fold in the province. This large gull is conspicuous, whether sitting atop a piling, visiting a city park, splashing in a river mouth or sailing on a stiff breeze above coastal bluffs. • Glaucous-winged Gulls, like most large gull species, show small, white spots or "mirrors" on their grey wing tips. However, these "mirrors" are inconspicuous, contrasting only slightly because they are set within pale grey, rather than black, outer primaries. • Immatures can been seen at scattered locations along the coast as their drab, brownish, first-winter plumage changes into the handsome adult at four years old. • Over the last few decades, Glaucous-winged Gulls have moved up river systems well beyond the influence of the tides, and some now are regular visitors to the southern Interior of BC. • The Glaucous-winged Gull routinely hybridizes with the Western Gull, resulting in a confusing range of intergrades.

breeding

ID: sexes similar; large gull; pale grey back and wings; darkish eyes; heavy, yellow bill with red spot; pinkish legs. *Breeding:* white head, neck and underparts. *Nonbreeding:* dingy head, neck and upper breast. *Juvenile:* mottled brown plumage; dark eyes and bill. *In flight:* white spots along edge of grey wing tips.

Size: *L* 60–67 cm; *W* 1.3–1.4 m.

Habitat: *Breeding:* mostly low, bare or vegetated islands; also buildings and bridge structures in urban settings. *In migration* and *winter:* marine bays, estuaries and harbours; also beaches, landfills, agricultural fields, salmon-spawning rivers, lakeshores and urban parks.

Nesting: colonial; on the ground on bare rock or in low ground cover on an island; also on a bridge abutment or building roof; loose nest is made of stacked plant materials, string and marine debris; pair incubates 2–4 heavily marked, greenish eggs for 27–28 days.

Feeding: omnivorous; captures prey while walking, wading, swimming or plunging; diet includes mostly fish, molluscs, crustaceans, garbage and carrion.

Voice: squealing *kjau*, high-pitched, repeated *kea* and "mewing" *ma-ah* calls; single, throaty *kwoh* flight call; *eeja-ah* attack call.

Similar Species: *Herring Gull* (p. 225): black-and-white wing tips; breeding bird is darker with pale yellow eyes. *Western Gull* (p. 223): breeding bird has darker upperparts and usually yellow eyes; nonbreeding bird has whiter head and neck. *Thayer's Gull* (p. 226): smaller; slightly darker wing tips; slimmer bill. *Glaucous Gull* (p. 230): heavier build; whiter primary feathers.

GLAUCOUS GULL

Larus hyperboreus

Glaucous Gulls are large, chalky white gulls that breed along Arctic coastlines from Alaska to Labrador. Endowed with great powers of observation and an almost fanatical need to wander, the Glaucous Gull traditionally hunted for small mammals and birds during the northern summer, and in winter fished for its meals, stole food from smaller gulls or followed fishing boats for easy pickings. Some birds even perfected a hunting technique that allowed them to catch alcids trapped in narrow channels in winter ice. More recently, however, wintering Glaucous Gulls can be found teetering atop mounds of freshly bulldozed garbage, securing a chosen food item through an intimidating "mantling" of widespread wings and well-aimed jabs of a snapping bill. • Immature Glaucous Gulls are usually light enough to be distinguished from other immature gulls, and most have a distinctive pale pink, dark-tipped bill. These young gulls traditionally wander farther south than adults and often linger well into summer before drifting north again. • The species name *hyperboreus* means "of the far north," and this great white gull is as much a part of the Arctic landscape as the polar bear, Gyrfalcon and Snowy Owl.

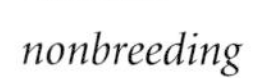

nonbreeding

breeding

ID: sexes similar; large, stout gull; flattened crown profile; very pale grey mantle; heavy, yellow bill with red spot on lower mandible; pink legs. *Breeding:* white head and underparts; yellow eyes. *Nonbreeding:* pale eyes; variable head and neck streaking. *Juvenile:* dark eyes; pale pink, black-tipped bill; varying amounts of brown flecking on body. *In flight:* pure white wing tips.
Size: *L* 69 cm; *W* 1.5 m.
Habitat: *Breeding:* sea cliffs and inshore islands, particularly near human settlements. *In migration* and *winter:* chiefly marine and estuarine habitats including beaches, bays, intertidal zones and industrial waterfronts; also inshore waters with sewage outfalls, landfills, fishing wharves and large inland lakes.

Nesting: solitary or colonial; on a cliff ledge or small island; nest is a large mass of plant material, seaweed and moss; pair incubates 2–3 variably marked, creamy olive or buff eggs for 27–30 days.
Feeding: diverse diet includes marine and freshwater fish and invertebrates, bird eggs and chicks, small mammals, berries, carrion, human refuse and food items pirated from other foraging birds; also feeds on carrion and at landfills.
Voice: high, screechy calls similar to Herring Gull's *kak-kak-kak*.
Similar Species: *Thayer's Gull* (p. 226) and *Iceland Gull* (p. 227): smaller; slightly darker mantle; grey on wing tips. *Herring Gull* (p. 225): slightly smaller; black wing tips; much darker mantle. *Glaucous-winged Gull* (p. 229): grey wing tips; darker eye ring.

GREAT BLACK-BACKED GULL

Larus marinus

nonbreeding

The Great Black-backed Gull's commanding size and bold, aggressive disposition enables it to dominate other gulls, ensuring that it has first dibs at food, whether it is fresh fish or a meal from a landfill. No other gull, with the exception of the Glaucous Gull, is equipped to dispute ownership with this domineering bird. • Like many of the larger gulls, the Great Black-backed Gull is a "four-year gull," meaning that it goes through various plumage stages until its fourth winter, when it develops its refined adult plumage. Most immature gulls have dark streaking, spotting or mottling, which allows them to blend into their surroundings and avoid detection by predators. • Nearly wiped out by feather hunters and egg collectors in the early 1900s, Great Black-backed Gull populations had rebounded in excess of historical numbers by the 1960s. Their numbers have continued to increase in most regions, with recent expansions in colony sizes observed in the Maritimes, along the St. Lawrence River and on Magdalen Island and the Gaspé Peninsula, Québec. However, recent outbreaks of botulism on the Great Lakes have reduced the local breeding population to just a handful of pairs.

breeding

ID: sexes similar; very large gull. *Breeding:* white head and underparts; black mantle; pale pinkish legs; light-coloured eyes; large, yellow bill with red spot on lower mandible. *Nonbreeding:* may have faintly streaked head. *Juvenile:* variable plumage; mottled grayish brown, white and black; black bill or black-tipped, pale bill. *In flight:* grey underwings; large, white terminal spots and narrow, white trailing edge on wings.
Size: *L* 76 cm; *W* 1.6 m.

Habitat: *Breeding:* rocky islands; rarely on beaches. *Winter:* landfills and open water on large lakes and rivers.
Nesting: solitary or in a small colony; on an island, cliff top or beach; pair builds a mound of vegetation and debris on the ground; pair incubates 2–3 brown-blotched, olive to buff eggs for 27–28 days.
Feeding: opportunistic; eats fish, eggs, birds, small mammals, berries, carrion, molluscs, crustaceans, insects and other invertebrates; scavenges at landfills.
Voice: harsh *kyow* or *owk*.
Similar Species: *All other gulls* (pp. 211–30): smaller; lack black mantle.

CASPIAN TERN

Hydroprogne caspia

Terns are graceful waterbirds, more streamlined than gulls, usually with white bodies, black caps, forked tails and colourful, sharply pointed bills. The Caspian Tern is the largest tern in North America and bridges the gulf between the smaller terns and the larger gulls. Because of its gull-like flight and frequent association with gulls on shoreline sandbars and mudflats, many birders confuse the Caspian Tern with a gull, but its heavy, red-orange bill, forked tail and raucous call give away its true identity. • Unlike most gulls, which are omnivores, this tern is strictly a fish-eater. • The Caspian Tern has an amazingly disjunct breeding range in North America, with nesting colonies scattered from the Northwest Territories to the Maritimes and south to coastal California. The species is also widespread, being found on every continent except Antarctica. • Caspian Tern numbers have tripled since the 1960s in the Great Lakes region and in Manitoba, perhaps related to increases in forage fish, resulting from excessive commercial fishing of large, predatory fish. • Juveniles accompany adults in late summer and fall. They follow the adults, begging for food, even several months after fledging. • This species was discovered at the Caspian Sea, hence its name.

breeding

ID: sexes similar; large, gull-like tern; large, red-orange bill; moderately forked tail; black legs and feet; very pale grey mantle; white underparts. *Breeding:* solid black cap includes eyes and forehead. *Nonbreeding:* dusky crown and forehead. *Juvenile:* similar to nonbreeding but upper parts lighter and heavily mottled. *In flight:* short, notched tail; stiff, shallow wingbeats.

Size: *L* 48–58 cm; *W* 1.3–1.4 m.

Habitat: *Breeding:* rocky islands in lakes, sandy marine spits and urban buildings. *In migration* and *winter:* protected estuaries, bays, beaches and mudflats on the coast; lakes, lakeshores and spits in the interior.

Nesting: colonial; in sand or on bare rock; shallow hollow is lined with shells and beach debris; pair incubates 1–3 lightly spotted, buff eggs for 25–28 days.

Feeding: hovers over water and plunges headfirst after small fish near the water's surface.

Voice: low, harsh *ca-arr*; also repeated *kak* notes.

Similar Species: *Forster's* (p. 237), *Arctic* (p. 236) and *Common* (p. 235) *terns:* much smaller; longer, forked tails; orangey, dark-tipped bills (except in the Arctic); orange-red legs on breeding birds; black bill on nonbreeding birds.

BLACK TERN

Chlidonias niger

nonbreeding

Wheeling about in foraging flights, Black Terns pick small minnows from the water's surface and catch flying insects in midair. Even without brilliant colours, these terns are strikingly beautiful birds, unmistakable in their solid black breeding plumage. • These highly social birds breed in loose colonies and forage in scattered flocks. Black Terns are finicky nesters, refusing to return to nesting areas that show even slight changes in the water level or in the density of emergent vegetation. Nests are flimsy, often floating, and are easily destroyed by wind or changing water levels. Black Tern nesting colonies are noisy, and any intruders are met by a barrage of squawks and aerial dive-bomb attacks. Reproductive success varies greatly. • The degradation and destruction of marshes in many parts of the Black Tern's North American range have contributed to a significant decline in populations of this bird over recent decades, but breeding bird surveys reveal that declines are primarily in eastern parts of its range. Significant increases in breeding populations have been observed in central portions of its range, and in BC, numbers are increasing as its range expands northward. • The Black Tern overwinters along tropical coastlines from Mexico to northern South America; it is also found in the Old World.

breeding

ID: sexes similar; small, freshwater-marsh tern; short, slightly forked tail. *Breeding:* black head and body; silvery grey wings; black bill and legs. *Nonbreeding* and *juvenile:* grey back (mottled and "scaly" in juvenile); white head; dusky nape and crown; white breast and belly; dusky "spurs" on each side of breast. *In flight:* buoyant flight; appears black and silver.

Size: *L* 23–25 cm; *W* 58–62 cm.

Habitat: *Breeding:* shallow freshwater marshes, large ponds and lakes with marshy sections. *In migration:* marshes, seacoasts, sewage ponds and brackish lagoons.

Nesting: loosely colonial; on water in a protected opening; small pad of water plants is anchored to vegetation and fallen branches; pair incubates 2–4 darkly blotched, buff to olive eggs for 19–21 days.

Feeding: eats primarily insects, small fish and crustaceans.

Voice: shrill, metallic *kik-kik-kik-kik-kik* greeting call; *kreea* alarm call.

Similar Species: none; breeding plumage is unique.

ROSEATE TERN

Sterna dougallii

The Roseate Tern can be found worldwide. The Canadian breeding population is estimated at fewer than 140 pairs, concentrated on a few islands off the Atlantic Coast of Nova Scotia. Encroachment from expanding populations of gulls and other terns in many areas has resulted in Roseate Terns being forced to move to suboptimal sites and experiencing reduced nesting success. The species has been designated as endangered in Canada. • The Roseate Tern often nests in colonies with other terns, joining forces to threaten and mob invading predators. • Adapted for fast flight and relatively deep diving, the Roseate Tern often submerges completely when pursuing fish and is even said to "fly" short distances under water. • Unmated Roseates sometimes try to help out with the parental duties of a nesting pair. Quite often, these helpers are driven off by the parents, but occasionally they are welcomed with "open wings" to form trios or quartets. • The Roseate Tern was named for the very faint pink tone on its breast in breeding plumage.

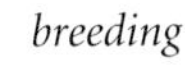

breeding

ID: sexes similar; medium-sized tern. *Breeding:* pale grey upperparts; white underparts; faint pink blush on breast; black crown and nape; black bill (orange with black tip by midsummer); red-orange legs. *Nonbreeding:* white breast; black cap extends forward only to eyes. *In flight:* long, deeply forked tail; relatively short, straight, narrow wings; quick, stiff wingbeats; breeding adult has all-white wings.
Size: *L* 31–34 cm; *W* 70 cm.
Habitat: *Breeding:* sheltered sites on sandy or rocky islands, often with Common Terns. *In migration:* protected bays and estuaries, usually close to shallow water for feeding.

Nesting: colonially with other terns; on the ground under cover of vegetation, rocks or driftwood or in a human-made structure such as a tire or wooden box; pair incubates 1–4 brown-blotched, creamy white eggs for 21–26 days.
Feeding: flies low over water and dives to capture prey; takes mostly small fish and sometimes invertebrates.
Voice: distinctive, 2-syllable *kir-rick* call; sharp *keek* given in flight.
Similar Species: *Caspian Tern* (p. 232) and *Royal Tern:* larger; longer, thicker, all-red bills; black legs. *Forster's* (p. 237), *Arctic* (p. 236) and *Common* (p. 235) *terns:* dark wing tips visible in flight; lack rosy wash on breast. *Gulls* (pp. 211–31): either lack black cap or black cap extends well below eye.

COMMON TERN

Sterna hirundo

nonbreeding

Common Terns are sleek, graceful birds that patrol the shorelines and shallows of lakes, rivers and seacoasts looking for small fish during spring and fall, and settling into large nesting colonies over the summer months. Both males and females perform aerial courtship dances, and the nesting season commences when the female accepts her suitor's gracious fish offerings. • Tern colonies are noisy and chaotic, and they are often associated with even noisier gull colonies. Should an intruder approach a tern nest, the parent will dive repeatedly, often defecating on the offender. • Common Terns characteristically hover over the water, then dive headfirst from heights of 30 metres or more to capture prey. Although over 90 percent of the Common Tern's breeding distribution is in Canada, stretching from Great Slave Lake to the East Coast, it has the most widespread breeding distribution of any tern in North America. • Terns seem to fly effortlessly, and they migrate great distances, spending their winters along the coasts of Central and South America. For decades, the title of "world's greatest long-distance traveller" was credited to the Arctic Tern, but recently, a Common Tern banded in Great Britain was recovered in Australia, setting a new record! • The Common Tern's graceful flight once inspired the name "Sea Swallow."

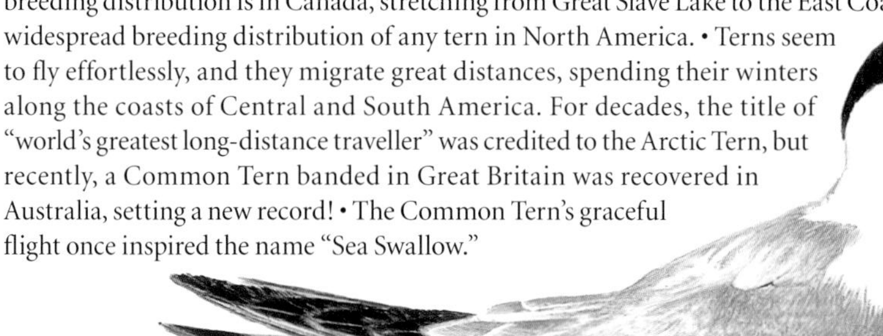

breeding

ID: sexes similar; medium-sized tern; long tail extends beyond folded wingtips. *Breeding:* pale grey upperparts and belly; black cap; black-tipped, red bill; red legs and feet. *Nonbreeding:* dusky brown nape; white forehead; thin, all-black bill; black legs and feet. *Juvenile:* dark shoulder bar; brownish barring on upperparts. *In flight:* narrow, pointed wings; deeply forked, white tail with dark outer tail feathers; dark wedge in spread primaries is diagnostic.

Size: *L* 33–41 cm; *W* 76 cm.

Habitat: bays, harbours and estuaries on the coast; large lakes and marshes in the interior.

Nesting: colonial, often in mixed colonies; nest scrape is lined with pebbles, vegetation, debris or shells; pair incubates 1–3 variably marked, brown eggs for 21–30 days.

Feeding: eats mainly small fish and crustaceans.

Voice: high-pitched, drawn-out *keee-are*; also *kik-kik-kik* or quick *kirri-kirri* calls.

Similar Species: *Arctic Tern* (p. 236): uniformly red bill; longer tail; darker underparts; less black in wings. *Forster's Tern* (p. 237): whiter wings and belly; longer tail.

ARCTIC TERN

Sterna paradisaea

nonbreeding

Arctic Terns that nest in northern Canada make annual, round-trip migrations to foreign lands and oceans that include South America, Antarctica, Europe, Africa and the Indian Ocean. In some years, an Arctic Tern might fly a distance of 32,000 kilometres! Viewed in the context of a potential life span of 34 years or more, it is safe to say that at least some Arctic Terns log over one million kilometres in migrations between their summer and winter homes. • While migrating, these elegant terns rarely land on the sea; instead, they search for something to rest on, such as driftwood or mats of seaweed. • The Arctic Tern is more common than the Common Tern in northern parts of Atlantic Canada. It is also more likely to nest in isolated pairs than its less demonstrative relative. • Arctic Terns are most easily identified on the ground, when their short legs give them a crouched appearance in comparison to the Common Tern. • Because they experience 24 hours of daylight on their northern nesting grounds and long days of sunlight on southern wintering grounds, Arctic Terns perpetually avoid winter and probably experience more daylight in an average year than most living creatures.

breeding

ID: sexes similar; medium-sized tern; short neck. *Breeding:* black cap and nape; all-red bill; very short, red legs; white cheeks; light grey underparts. *Nonbreeding:* white forehead; partial black cap; black bill; white underparts; short, black legs. *Juvenile:* barred upperparts; indistinct, dark shoulder bar. *In flight:* long, deeply forked, white tail; white rump and undertail coverts; long, narrow, grey wings with thin, dark trailing edge to underwings.

Size: *L* 36–43 cm; *W* 74–84 cm.

Habitat: *Breeding:* tundra, muskeg, exposed gravelly shores of islands and bare, rocky islets. *In migration:* mainly open ocean; occasionally seen near bays and channels with feeding gulls.

Nesting: colonial; on sand, gravel or a mossy hummock; shallow nest scrape is usually unlined; pair incubates 2–3 darkly blotched, pale olive to buff eggs for 20–24 days.

Feeding: plunge dives and surface dips for a wide variety of small fish, crustaceans and other invertebrates; also hawks for flying insects.

Voice: harsh, high-pitched, down-slurred *kee kahr*; dry, nasal *raaaaz* attack call; also utters single, short *kip* notes.

Similar Species: *Common Tern* (p. 235): longer legs; shorter tail compared to folded wings; breeding bird has black-tipped bill. *Forster's Tern* (p. 237): longer legs; breeding bird has thicker, black-tipped, orange bill.

FORSTER'S TERN

Sterna forsteri

nonbreeding

Forster's Tern so closely resembles the Common Tern that the two are often indistinguishable to the eyes of many observers, and they were once considered to be the same species. The Forster's is primarily a "marsh tern," breeding in freshwater, brackish and saltwater marshes, including the marshy borders of lakes, islands or streams. Colonies of Forster's Terns are often associated with Black Terns and Yellow-headed Blackbirds. The Common Tern, on the other hand, rarely nests in marshes, preferring to nest in mixed-species associations on islands, barrier beaches or promontories attached to mainland in large lakes. • Most terns are known for their extraordinary ability to catch fish in dramatic headfirst dives, but the Forster's Tern excels at gracefully snatching flying insects in midair. • Forster's is the only exclusively North American tern, but the species bears the name of a man who never visited this continent. Naturalist Johann Reinhold Forster, who lived and worked in England and accompanied Captain Cook on his 1772 voyage around the world, examined tern specimens sent from Hudson Bay. He was the first to recognize this bird as a distinct species, and in 1832, the species was officially named the Forster's Tern. • Forster's Tern primarily breeds across the prairies with a small disjunct population in extreme southwestern Ontario.

breeding

ID: sexes similar; medium-sized tern; long, orange legs. *Breeding:* black crown and nape; thin, black-tipped, orange bill; pale grey mantle; pure white underparts; white rump. *Nonbreeding:* white head, foreneck, breast and belly; black mask through eye and ear; black bill; no black cap. *Juvenile:* brownish back and nape; black mask. *In flight:* long, pointed, mostly white wings; frosty-tipped primaries; long, deeply forked, grey tail with white outer edges.

Size: *L* 36–41 cm; *W* 79 cm.

Habitat: *Breeding:* shallow marsh lakes with dense surface vegetation. *Migration:* rocky seashores, tidal mudflats, marshes and sloughs.

Nesting: colonial; on water, on aquatic plants or windblown cattails; nest is a small mound of water plants; pair incubates 2–5 brown-marked, buff eggs for 20–28 days.

Feeding: eats mainly small fish and crustaceans; opportunistically forages for insects.

Voice: short, nasal *keer keer* flight call; grating *tzaap* repels intruders.

Similar Species: *Common Tern* (p. 235): longer legs; shorter tail compared to folded wings. *Arctic Tern* (p. 236): very short, red legs; grey underparts; white tail with grey outer edges; lacks black-tipped bill. *Caspian Tern* (p. 232): much larger; much heavier, red-orange bill.

POMARINE JAEGER

Stercorarius pomarinus

Jaegers are powerful, swift predators and notorious pirates of vast, open oceans. Pomarine Jaegers forage primarily by scavenging, often stealing food from other birds (a behaviour called "kleptoparasitism") and by predation on small seabirds. They rest only occasionally on the water's surface and only seek the solid footing of land during the nesting season. • Pomarine Jaegers are unique in their dependence on a single species of prey, the brown lemming, for successful reproduction. Lemming cycles typically peak only once every three or four years, and Pomarine Jaegers only reproduce successfully during these peaks. • Jaegers usually occur singly or in loose groups far from land, where ocean currents and upwellings bring food to the surface. However, any concentratioan of feeding terns or Black-legged Kittiwakes at sea will attract pirating Pomarines, often in large numbers. • The Pomarine Jaeger is the largest of the three jaegar species found in Canada. Its wings are broader at the base than the other jaegers, and its flight is more laboured, with slower wingbeats. • An adult, with its twisted central tail feathers, is a magnificent sight, but these retrices are often lost soon after the birds leave the Arctic.

breeding light adult

breeding dark adult

ID: sexes similar; large, heavyset, gull-like seabird; white shafts on outer 4–6 primaries; long, blunt, twisted central tail feathers. *Light adult:* dark brown upperparts; black crown and face; yellowish nape, cheek and throat; white underparts; brown-mottled breast and sides; dark undertail coverts. *Dark adult:* less common than light adult; all dark except for white in primary feathers. *Juvenile:* dark upperparts with light edges to feathers; central tail feathers barely longer than tail; barred wing linings. *In flight:* powerful, steady wingbeats; extensive white patch at base of underwing primaries.

Size: *L* 51–58 cm; *W* 1.2 m.

Habitat: *Breeding:* swampy areas of tundra. *In migration:* highly pelagic; often seen in small numbers over upwellings; regularly seen around fishing vessels.

Nesting: on the ground; shallow depression in moss is unlined or sparsely lined; both parents incubate 2 darkly flecked, brownish eggs for 26–28 days.

Feeding: adaptable; pirates food from seabirds, captures small migrating songbirds and small fish, scavenges mammal carcasses and readily accepts refuse thrown overboard by humans; takes brown lemmings on breeding grounds.

Voice: generally silent; may give sharp, chattering notes or a squeaky, whistled note during migration.

Similar Species: *Parasitic Jaeger* (p. 239): smaller; white shafts restricted to outer 3–5 primaries; pointed central tail feathers. *Long-tailed Jaeger* (p. 240): smaller; buoyant flight; white primary shafts usually restricted to outer 2 primaries; long central tail feathers.

PARASITIC JAEGER

Stercorarius parasiticus

light adult breeding

Although "jaeger" is a German word meaning "hunter," "parasitic" more aptly describes this bird's foraging tactics. Swift and relentless, Parasitic Jaegers hound and intimidate gulls, which predictably regurgitate their meal to placate these aerial pirates. During the breeding season, Parasitic Jaegers have been known to follow hunting mammals and people in the hopes of snatching the eggs and young left unattended by fleeing adult birds of other species. • Like the other two jaeger species, the Parasitic Jaeger breeds in tundra regions of northern Alaska and Canada, as well as in the Old World, where it is known as "Arctic Skua." • Jaegers have two colour phases, but intermediate plumages do occur and are a challenge to identify. • The Parasitic Jaeger is the least numerous of the three jaeger species in the Canadian Arctic, but it is the most widely observed species during migration. On their way southward in fall, Parasitic Jaegers tend to remain near shore, occasionally wandering through the interior of the country. Consequently, of the three jaegers, it is the one most often seen in southern Canada.

breeding light adult

ID: sexes similar; medium-sized, gull-like seabird; brown upperparts; thin bill; white shafts on outer 3–5 primaries. *Light adult:* brown crown and face; yellowish nape, cheek and throat; white underparts; brown breast band. *Dark adult:* less common than light adult; all dark except for white at base of primary feathers. *Juvenile:* central tail feathers barely longer than tail; generally barred; pale edging to body feathers. *In flight:* long, dark, pointed wings; 2 long, pointed central tail feathers.
Size: *L* 38–53 cm; *W* 91 cm.
Habitat: *Breeding:* tundra, cliffs or offshore islands. *In migration:* mostly open ocean; nearshore in bays, harbours, estuaries, channels and heads of inlets; visits lakes in the interior.
Nesting: on the ground; shallow depression in grass or moss is unlined or sparsely lined; pair incubates 2 sparsely marked, brownish eggs for 24–28 days.
Feeding: pirates food from other seabirds; also plucks prey off the water's surface, including fish and crustaceans; readily attracted to discarded human food items.
Voice: generally silent; migrating groups may make shrill calls at feeding sites.
Similar Species: *Pomarine Jaeger* (p. 238): larger; more heavyset; white shafts on outer 4–6 primaries; blunt, twisted central tail feathers. *Long-tailed Jaeger* (p. 240): smaller; more buoyant flight; white primary shafts usually restricted to outer 2 primaries; very long central tail feathers.

LONG-TAILED JAEGER

Stercorarius longicaudus

The Long-tailed Jaeger is the smallest and most abundant jaeger, with the longest central tail feathers. In migration, few come close enough to shore to identify them with certainty, but seasoned observers can use a combination of body build and plumage to distinguish them from the more common jaegers. • Highly pelagic, this jaeger spends more than three-quarters of its life over open sea. • The Long-tailed Jaeger breeds the farthest north of any jaeger, probably as far north as any bird. While breeding, it is almost exclusively terrestrial, and its diet consists primarily of lemmings and voles, which are caught in swooping aerial attacks. In years with low lemming densities, most Long-tailed Jaegers and Pomarine Jaegers do not breed, returning promptly to sea. • The Long-tailed is less inclined than other jaeger species to pirate food from other seabirds, but it never misses an opportunity to fill its belly. • Jaegers are members of the skua family and are known as skuas in other parts of their world range.

juvenile
intermediate adult

breeding
light adult

ID: sexes similar; medium-sized, gull-like seabird; slender wings; small, black bill; no white patch in primary feathers. *Light adult:* grey-brown upperparts; dark cap; yellow collar; white nape; white underparts; dark undertail coverts. *Dark adult:* rare; all-dark plumage. *Juvenile:* generally barred appearance; no dark cap; rounded central tail feathers; whitish feather edging on upperparts. *In flight:* very long central tail feathers during breeding season; long, slender wings; little or no white in upperwings; more buoyant flight than other jaegers.
Size: *L* 51–58 cm; *W* 1.0 m.

Habitat: *Breeding:* tundra or bogs near forests. *In migration:* open ocean far from land; rarely visits lakes and larger marshes along the coast.
Nesting: on the ground; shallow hollow is unlined or sparsely lined with moss; pair incubates 2 darkly marked, dull buff or olive brown eggs for 23–25 days.
Feeding: eats mainly fish and crustaceans at sea; occasionally pirates food from other seabirds; takes lemming and voles on breeding grounds.
Voice: generally silent except on breedng grounds; nasal *kee-ur* call.
Similar Species: *Pomarine Jaeger* (p. 238) and *Parasitic Jaeger* (p. 239): stockier; may have dark breast band; no long tail streamers. *Gulls* (pp. 211–31): no elongated central tail feathers.

DOVEKIE

Alle alle

A member of the alcid (auk) family, the Dovekie is the northern equivalent of a penguin, but it can fly. Its small wings, webbed feet and sleek, waterproof plumage are designed to help it pursue fish underwater. • In general, the best time to look for alcids along coastlines is during late fall and winter storms, when the birds must seek refuge close to shore. • Despite heavy predation and the fact that each pair lays only one egg per year, Dovekies remain common in the northwestern Atlantic. Fewer than 1000 pairs are estimated to breed in North America, but the world population of Dovekies numbers in the tens of millions and could range up to 100 million individuals. • The Dovekie can be found in large concentrations on the Grand Banks and the Scotian Shelf and along the shores of Newfoundland in winter. With only one known Canadian colony near northern Baffin Island, most of the birds wintering along the East Coast likely come from breeding grounds in Greenland. • Dovekie means "little diver" in Anglo Saxon, but it is also called "Little Auk" in Great Britain and "Bullbird" in Newfoundland. Bay Bulls in Newfoundland was named for this bird, which sometimes overwinters in the harbour in large numbers.

nonbreeding

ID: sexes similar; small alcid; stocky body; stubby bill; black upperparts with some white on back and trailing edge of wings; white underparts. *Breeding:* black neck and breast. *Nonbreeding:* white throat and breast; black collar; black rump. *In flight:* dark underwings; "buzzy" flight with rapid wingbeats.

Size: *L* 21 cm; *W* 38 cm.

Habitat: *Breeding:* on or near the coast in Arctic regions. *In migration* and *winter:* often very far from land; sometimes around pack ice; prefers colder waters.

Nesting: colonial; on a cliff; in a hole or crevice among rocks or boulders; pair incubates 1 pale blue egg for about 29 days.

Feeding: dives and swims underwater; most dives are relatively shallow; feeds mostly on small crustaceans; may also take fish, molluscs, marine worms and algae.

Voice: generally silent.

Similar Species: *Other alcids* (pp. 242–52): larger; longer necks and bills.

COMMON MURRE

Uria aalge

nonbreeding

The Common (or "Thin-billed") Murre nests in huge, tightly packed colonies on cliffs and offshore islands on both of Canada's coastlines. Its range overlaps extensively with the Thick-billed Murre in the Pacific, less so in the Atlantic. • At a distance, the Common Murre is often impossible to distinguish from the Thick-billed Murre and the Razorbill. Although both have much thicker bills than the Common Murre, some hybridization occurs with both species. • Common Murres have a unique breeding strategy, with chicks leaving the colony at just three to four weeks of age and swimming to energy-rich feeding sites in the company of the male parent. • In winter, Common Murres may travel several kilometres out into the open ocean, disappearing from nearshore areas altogether. However, the sudden midwinter appearance of large numbers of birds on breeding rocks suggests that many Common Murres also overwinter fairly close to shore. • Like the eggs of many cliff-nesting birds, murre eggs are pointed at one end, which allows them to roll in tight circles and stay on the narrow nesting ledges.

breeding

ID: sexes similar; large alcid; short neck; long, slender, black bill; white wing bar. *Breeding:* dark brown head, neck, back and wings; grey-streaked flanks. *Nonbreeding:* white lower face, "chin" and throat; dark line curves down from eye. *In flight:* flies low over water; stocky body is tilted upward; rapid wingbeats.
Size: *L* 41–44 cm; *W* 66 cm.
Habitat: *Breeding:* offshore rocky islands and headlands. *Nonbreeding:* open ocean; nearshore in bays, straits, channels and passages.

Nesting: colonial; on bare rock on a steep cliff ledge; pair incubates 1 heavily marked, pale egg for 30–35 days.
Feeding: dives mainly for fish; also eats squids and crustaceans.
Voice: utters a low, harsh *murrr* in colonies; dependent immature gives a high-pitched, quavering, repeated *feed-me-now* whistle.
Similar Species: *Marbled Murrelet* (p. 247) and *Ancient Murrelet* (p. 248): much smaller; stocky bodies; stubby bills. *Rhinoceros Auklet* (p. 250): slightly smaller; stockier; generally greyish brown; stubbier, yellowish bill; white belly and undertail. *Thick-billed Murre* (p. 243) and *Razorbill* (p. 244): slightly larger; black on wings and back; white stripe on bill. *Dovekie* (p. 241): much smaller; small, stubby bill.

THICK-BILLED MURRE

Uria lomvia

The Thick-billed Murre is one of the most abundant marine birds in the Northern Hemisphere as well as being one of the deepest diving birds, regularly diving to depths of 100 metres. It occasionally reaches 200 metres and can remain underwater for over three minutes! This bird steers and propels itself with its wings while pursuing marine fish and invertebrates. • Pairs pack themselves into suitable breeding areas, often nesting shoulder to shoulder on the rocky cliffs of their nest sites in colonies that can include up to one million birds. The largest colonies in North America are along the Labrador coast and in the eastern Arctic, but there are also small colonies in Newfoundland and British Columbia. • As with Common Murres and Razorbills, Thick-billed Murres have an unusual rearing strategy in which young leave the colony when only partially grown, using their covert feathers as miniature wings to assist them in gliding from their breeding ledge to the water. The young complete their growth at sea, accompanied by the male parent alone. • Many waterbirds share the countershading pattern of the murre. From above, the bird's dark back blends with the steely sea; from below, predators and prey may be slow to notice this bird's light-coloured underbelly against the shimmering surface of the water.

ID: sexes similar; large alcid. *Breeding:* black upperparts; white underparts; short, thick, black bill with whitish line along gape. *Nonbreeding:* white "chin" and throat; black crown extends below eye. *In flight:* stubby wings; heavy neck; rapid wingbeats; feet trail behind tail.
Size: *L* 46 cm; *W* 71 cm.
Habitat: along coastal and continental shelf waters, usually farther offshore than Common Murres; harbours and coastal bays in winter.
Nesting: in colonies; on a cliff ledge; pair incubates 1 variably marked, pale egg for 30–35 days.
Feeding: dives deeply for fish, crustaceans and occasionally marine worms and squid.
Voice: guttural calls range from a faint *urr* to a loud *aargh*.
Similar Species: *Common Murre* (p. 242): dark brown upperparts; grey-streaked flanks; more white on head and neck in nonbreeding plumage. *Razorbill* (p. 244): thicker bill with vertical, white line; more white on head and neck in nonbreeding plumage. *Dovekie* (p. 241): much smaller; thick neck; very short bill.

RAZORBILL

Alca torda

The Razorbill is a stocky, crow-sized seabird that is widely distributed in the North Atlantic. World populations of the Razorbill are believed to total at least half a million pairs, with an estimated Canadian population of 38,000 pairs breeding at over 100 colonies. • The Razorbill's sharp bill is well adapted for capturing slippery fish. As the bird "flies" underwater with its powerful wings, it can catch several fish in its large mouth in a single dive. One of the larger alcid species, the Razorbill also routinely supplements its diet with fish pirated from other birds. • Like other alcids, the Razorbill moults all of its feathers at once, rendering it flightless at sea for a few weeks following the breeding season. • The Razorbill is believed to be the closest living relative of the extinct Great Auk (*Pinguinus impennis*), a very large alcid that once frequented the waters of northern Europe and North America. Like the Great Auk, Razorbills were heavily persecuted for eggs, meat and feathers in the early 20th century, and many populations were wiped out or greatly reduced. Many colonies in Québec and New Brunswick, however, have been protected as sanctuaries for over 80 years, and their Razorbill populations have dramatically increased in recent decades.

1st-winter juvenile

breeding

ID: sexes similar; large alcid; black upperparts; white underparts; thick, laterally compressed bill with 1 vertical, white line; horizontal white line along top of bill; pointed tail is cocked when swimming. *Breeding:* black extends down neck. *Nonbreeding:* white "cheek" and throat. *In flight:* short, stubby wings; hunchbacked appearance; heavy head and neck; white trailing edge on wings.

Size: *L* 43 cm; *W* 66 cm.

Habitat: concentrates over offshore shoals and ledges; nests on cliffs or rocky shorelines and on islands.

Nesting: colonial; on a cliff ledge or crevice or among boulders along a shoreline; usually no nest material; pair incubates 1 brown-marked, pale egg for 32–39 days.

Feeding: dives to 100 m for small fish, small crustaceans and marine worms; may steal fish from other alcids.

Voice: generally silent; sometimes produces a grunting *urrr* call; also a growling *hey all!*

Similar Species: *Common Murre* (p. 242), *Thick-billed Murre* (p. 243) and *Black Guillemot* (p. 245): thinner bills lack vertical, white line. *Dovekie* (p. 241): much smaller; short bill. *Shearwaters* (pp. 112–17): thinner bills; plumage generally brown or grey instead of black.

BLACK GUILLEMOT

Cepphus grylle

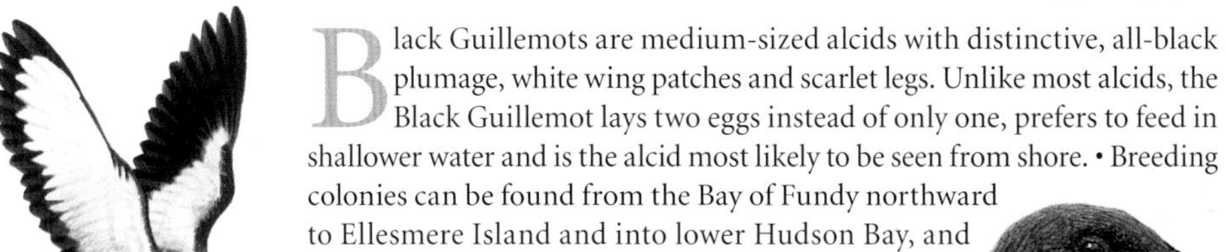

nonbreeding

Black Guillemots are medium-sized alcids with distinctive, all-black plumage, white wing patches and scarlet legs. Unlike most alcids, the Black Guillemot lays two eggs instead of only one, prefers to feed in shallower water and is the alcid most likely to be seen from shore. • Breeding colonies can be found from the Bay of Fundy northward to Ellesmere Island and into lower Hudson Bay, and colony size increases with latitude. Breeding guillemots at lower latitudes are sedentary or make only short-distance migrations to nearby ice-free seas. • Although they remain fairly close to shore year round, in the winter months, Black Guillemots are more pelagic, frequently feeding along pack ice, with some overwintering in polynyas in the Arctic. • Alcid eggs are conical so they roll in a circle and remain on their rocky nesting ledge, potentially avoiding disaster. • In its shallow dives, the Black Guillemot sometimes turns over rocks along the sea floor in search of prey. This species is believed to be more susceptible to pollutants than other alcids because of its inshore and bottomfeeding habits. • This bird's common name is appropriate only during the breeding season—its plumage is primarily white in winter. It is also commonly known as "Sea Pigeon" because of its winter plumage and pigeonlike flight.

breeding

ID: sexes similar; medium-sized alcid; thin, black bill; reddish orange legs and feet. *Breeding:* black overall; large, white wing patch. *Nonbreeding:* whitish overall; some black on back, wings and tail. *In flight:* very broad, rounded wings; white underwings; feet trail behind tail.
Size: *L* 33 cm; *W* 53 cm.
Habitat: usually close to shore in relatively shallow water; sometimes far offshore; may feed on freshwater lakes near the coast.
Nesting: generally in small colonies; along a rocky shore, low cliff or beach; in a hole under or among boulders; nest hollow is unlined; pair incubates 1–2 variably marked, white eggs for 23–39 days.
Feeding: may dive to 30 m but usually only to 9 m; feeds primarily on fish and crustaceans; also takes molluscs, insects, marine worms and some vegetation.
Voice: drawn-out, high-pitched *see-oo* or *swweeeeeer.*
Similar Species: *Pigeon Guillemot* (p. 246): black bar divides white wing patch. *Other alcids* (pp. 241–44, 247–52): lack distinctive, white wing patch; darker upperparts in nonbreeding plumage.

PIGEON GUILLEMOT

Cepphus columba

Pigeon Guillemots are among the most widespread and commonly seen alcids in nearshore waters along the Pacific Coast. In breeding season, they can be spotted flashing their white wing patches among small groups of black-coloured birds or roosting on rocks below sea cliffs. Their distinctively patterned breeding plumage is highlighted with bright red feet and a scarlet mouth lining. • In May and June, courting birds flirt outrageously, waving their flamboyant feet, peering down each other's throats and whistling wheezily. • Pigeon Guillemots will nest in a variety of locations, including the old burrows of other seabird species. It is suspected that breeding Pigeon Guillemots from Alaska head south in winter to the edge of the pack ice, and California birds move north as far as British Columbia, whereas guillemots in the central range area of British Columbia and Oregon may be permanent residents. • The Pigeon Guillemot's much different grey-and-white winter look, which is reminiscent of the Rock Pigeon, inspired its common name "Sea Pigeon."

nonbreeding

breeding

ID: sexes similar; medium-sized alcid; thin, straight bill; longish neck; bright red feet; orange-red inside of throat. *Breeding:* black overall; dark eyes and bill. *Nonbreeding:* mottled white and dark crown and back; white belly; dark eye patch. *In flight:* stubby wings; grey underwings; large, white wing patch divided by black line; conspicuous, trailing red feet.
Size: *L* 31–36 cm; *W* 59 cm.
Habitat: *Breeding:* offshore islands with rocky headlands, bluffs and cliffs, usually with soil or crevices; also human-made structures such as docks, wharves and piers. *Nonbreeding:* rocky and inshore marine waters usually less than 30 m deep.

Nesting: loosely colonial or solitary; on bare ground or rock, inside a rock crevice or boulder rubble, on an artificial structure or in a burrow; pair incubates 1–2 variably blotched, cream white eggs for 29–32 days.
Feeding: dives with help of wings for small fish and crustaceans; also eats Pacific herring eggs.
Voice: distinctive stuttering series of wheezy *peeeeeee* notes at nest site; otherwise silent.
Similar Species: *Black Guillemot* (p. 245): white wing patch without black bar. *Other alcids* (pp. 241–44, 247–52): lack distinctive, white wing patch; darker upperparts in non-breeding plumages.

MARBLED MURRELET

Brachyramphus marmoratus

Marbled Murrelets are one of the most mysterious and least known seabirds on the Pacific Coast. The species was first described in 1789, but this bird's nesting habits remained one of the last great ornithological mysteries until the first verified nest was reported in Asia in 1961. Then, in 1974, a tree-trimmer discovered the first North American nest, high in a Douglas-fir tree in California. • Nesting Marbled Murrelets travel back and forth between the sea and nest sites, which are usually on tree limbs high within mature coniferous forests sometimes up to 70 kilometres inland! The birds visit their nests at night to avoid predators. Farther north, nesting habitat is more variable, and some ground nests have been found in Alaska, occasionally in areas with few trees. • The Marbled Murrelet's dependence on old-growth forests and nearshore coastal habitats often conflicts with human interests. Although numbers in Canada are fairly large, threats to nesting and wintering habitats and to feeding areas have led to the species being designated as threatened in Canada. • Fishers call this bird "Australian Bumble Bee," and loggers have given it the nicknames "Fogbird" and "Fog Lark."

nonbreeding

breeding

ID: sexes similar; small alcid; flat head; short, pointed bill and tail. *Breeding:* mottled, dark brown body; heavily barred underparts; paler throat and undertail coverts. *Nonbreeding:* dark upperparts; white underparts; black "helmet"; white stripe between back and wings. *In flight:* all-dark wings with white strip at base; rocks from side to side.

Size: *L* 23–35 cm; *W* 41 cm.

Habitat: *Breeding:* mature to old-growth, wet coniferous forests, particularly stands of coastal Douglas-fir and Sitka spruce. *Nonbreeding:* protected bays, inlets and harbours.

Nesting: solitary or semi-colonial; in a shallow depression on a large tree limb up to 75 m above the ground; pair incubates 1–2 darkly spotted, pale green eggs for 28–30 days.

Feeding: dives to depths of 60 m for small schooling fish.

Voice: common location call on water or in flight is a shrieking *keer*; various whistles and groans at the nest.

Similar Species: *Ancient Murrelet* (p. 248): black crown; grey back; very stubby, pale bill; lacks white stripe between back and wings. *Cassin's Auklet* (p. 249): sooty grey overall; rounder head; heavier bill; whitish crescent above yellowish eye; light stripe on underwing. *Rhinoceros Auklet* (p. 250): larger; generally greyish brown overall; heavier, yellow bill; white belly and undertail coverts; feeds farther offshore.

ANCIENT MURRELET

Synthliboramphus antiquus

During the breeding season, this proficient mariner's feathery white "eyebrows" and shawl-like grey mantle give it a distinguished, aged or "ancient" look. • Although the breeding range of the Ancient Murrelet extends throughout the north Pacific Rim from China to British Columbia, half of the world's population nests in BC, mostly on Haida Gwaii (Queen Charlotte Islands). Nevertheless, the species has been designated as special concern in Canada based on substantial declines in the Haida Gwaii population as the result of predation from introduced rats and raccoons. Despite predator removal on some islands, murrelet numbers have not recovered. • The Ancient Murrelet is unusual among alcids in that it typically raises two young (most alcids raise one) and its chick departure strategy is unique—the young leave the burrow and follow their parents to be reared at sea after only one to three days. It is also the only known alcid that has extended incubation shifts of two to six days, and it regularly uses a distinctive voice recognized by its mate and young to keep the family together during their period at sea. • The Ancient Murrelet's feet and wings are not suited to diving as deep for food as other small alcids.

breeding

ID: sexes similar; small alcid; black crown; dark grey upperparts; white sides to neck and throat; short, pale yellow bill; white-mottled, grey flanks. *Breeding:* black throat patch; wispy, white eyebrow. *Nonbreeding* and *juvenile:* all or mostly black upper head; white to dusky throat. *In flight:* bright white underwings; no white stripe on scapulars; commonly plunges directly from the air to feed.
Size: *L* 23–26 cm; *W* 44 cm.
Habitat: *Breeding:* coniferous forests on offshore islands and grassy bluffs. *Nonbreeding:* areas of upwellings in passages, straits and channels to the edge of the continental shelf.

Nesting: colonial; in a burrow dug into soil; nest chamber is lined with dry grasses, leaves and small twigs; pair incubates 2 finely speckled, creamy buff eggs for 33–47 days.
Feeding: swims and dives to 15–20 m for krill, other crustaceans and very small fish.
Voice: varied; song consists of 1 or more *chirrups*, connected by well-spaced *chip* notes.
Similar Species: *Marbled Murrelet* (p. 247): grey back; white stripe between back and wings; dark bill; lacks white eyebrow and grey mantle. *Cassin's Auklet* (p. 249): generally sooty grey overall; white crescent over eye; dusky bill with light area at base of lower mandible; white belly and undertail coverts; grey underwings with light stripe.

CASSIN'S AUKLET

Ptychoramphus aleuticus

One of the most numerous breeding seabirds in British Columbia, the small, stout Cassin's Auklet is easily distinguished from its alcid relatives by its somewhat rounded wings and its grey neck and underwings. • Highly secretive, Cassin's Auklets nest in underground burrows and only make feeding visits under the cover of darkness. It is the only alcid that produces two broods in a single breeding season, at least in the southern part of its range, where birds may be seen at nesting colonies throughout the year. This auklet's ground-nesting habits make it extremely vulnerable to predators, including foxes, weasels, cats and rats that are introduced to their remote island nesting colonies. • Although Cassin's Auklets breed all along the Pacific Coast from Alaska's Aleutian Islands to Baja California, more than three-quarters of the world's population nests on offshore islands in British Columbia. Up to half the world's population nests at one site, Triangle Island, off the northwest tip of Vancouver Island! • Like the flightless penguins of the Southern Hemisphere, auklets, murrelets and puffins use their stubby wings to "fly" underwater. • The Cassin's Auklet does not migrate but birds disperse at sea in their constant search for food. • Oil pollution from marine shipping traffic poses a major threat to all seabirds.

breeding

ID: sexes similar; small, plump alcid; sooty grey upperparts; pale grey neck and breast; paler grey underparts; white belly; white crescent above eye; stubby, dark bill with light area at base of lower mandible. *Breeding:* pale eyes. *Nonbreeding* and *juvenile:* dark eyes. *In flight:* rounded wings; grey underwings; flies close to the water's surface.

Size: *L* 21–23 cm; *W* 38 cm.

Habitat: *Breeding:* offshore forested and densely shrubby and grassy islands. *Foraging:* open ocean, especially over upwellings, out to the continental shelf.

Nesting: colonial; in a burrow in soil or in a cavity or crevice among rocks; pair incubates 1 creamy white egg for 37–42 days.

Feeding: swims underwater to depths of 40 m for small crustaceans (krill), squid and small fish.

Voice: usually silent away from nest site; foraging pair may utter short, sharp *krik* or *kreek* location notes.

Similar Species: *Ancient Murrelet* (p. 248): much smaller body and bill; greater contrast between dark upperparts and white underparts; more pointed wing tips. *Marbled Murrelet* (p. 247): slimmer, usually uptilted bill; completely dark underwings; brown, heavily mottled breeding plumage.

RHINOCEROS AUKLET

Cerorhinca monocerata

Rhinoceros Auklets are stocky, medium-sized alcids that are closely related to the puffins in appearance and biology. They are commonly seen in April through August, flying to and from nesting colonies, which can be found along the Pacific Coast from Alaska's Aleutian Islands to California. Three quarters of the North American population breeds on islands on the British Columbia coast, with almost one-third occurring on two adjacent islands, Pine and Storm. • The striking "horn" that gives the Rhinoceros Auklet its name is present only during the breeding season, from March to June. • Year round, auklets stay between shore and the edge of the continental shelf, not venturing far out into the Pacific Ocean. During winter storms, Rhinoceros Auklets drift on mountainous waves, seemingly unaffected by the chilling wind and ocean spray. • While pursuing prey underwater, Rhinos can remain submerged for up to two minutes. Unlike the Cassin's Auklet, which transports food in throat pouches, the Rhinoceros Auklet carries fish in its bill. Rhino adults deliver food to nesting colonies at night mainly to avoid being pirated by gulls.

breeding

ID: sexes similar; medium-sized alcid; drab greyish brown upperparts; white underparts; angular head; stout bill. *Breeding:* orangey yellow bill with vertical, ivory-coloured "horn"; 2 whitish plumes above and below eye. *Nonbreeding:* dull yellow bill; paler eyes; white plumes are shorter. *In flight:* all-dark wings.
Size: *L* 36–41 cm; *W* 56 cm.
Habitat: *Breeding:* forested, shrubby and grassy islands with enough soil for burrowing. *Foraging:* inshore and offshore wherever food is concentrated.
Nesting: colonial; in a burrow up to 6 m long, usually on a grassy slope; nest chamber is lined with grasses and leaves; pair incubates 1 lightly spotted, off-white egg for 42–49 days.
Feeding: dives up to 40 m deep for small schooling fish and crustaceans.
Voice: usually silent; rasping squeak at sea; growling and braying at nest site.
Similar Species: *Common Murre* (p. 242): bulkier; long, dark, pointed bill; white sides and flanks; whitish wing linings. *Cassin's Auklet* (p. 249): much smaller; stubby bill; 2 thick, white crescents around eyes. *Marbled Murrelet* (p. 247) and *Ancient Murrelet* (p. 248): smaller; more extensive white underparts; more rapid wingbeats. *Tufted Puffin* (p. 252): immature has wider bill and dark undertail coverts.

ATLANTIC PUFFIN

Fratercula arctica

With its black-and-white "tuxedo," comical upright posture, gaudily coloured yellow, orange and grey bill and bright orange legs and feet, this smallest of puffins universally delights birders and tourists alike. These "clowns of the sea" strut about their breeding colonies on rocky islands in the North Atlantic, circling nesting islands with rapid wingbeats and diving or bobbing in nearby waters. • When it brings food back to its young, the Atlantic Puffin can line up more than a dozen small fish crosswise in its serrated bill. Captured fish are held against the roof of its mouth by its round tongue, and the slight serrations on its upper mandible help to keep its catch in place. • When the single nestling is about 40 days old, the parents curtail feeding and leave the young to its own resources. Following a week of fasting, the nestling ventures out to the ocean, fully capable of feeding itself. • Atlantic Puffins are now being reintroduced to former nesting locations that were destroyed by overhunting and egg collecting. • The Atlantic Puffin is the provincial bird of Newfoundland and Labrador, with half of North America's puffin population breeding in Witless Bay, Newfoundland. • This bird was previously known as "Common Puffin." Other names include "Sea Parrot," "Labrador Auk" and "Hatchet-Bill."

breeding

ID: sexes similar; medium-sized alcid. *Breeding:* white face; orange bill with triangular grey patch, bordered by yellow; black crown, nape and upperparts; white underparts; orange legs. *Nonbreeding:* grey face; base of grey bill lacks yellow border. *Immature:* similar to nonbreeding, but with smaller, darker bill. *In flight:* short, rounded wings; dark collar and underwings; orange feet trail behind tail; rapid, whirring wingbeats.

Size: *L* 32 cm; *W* 53 cm.

Habitat: *Breeding:* coastal cliffs and offshore islands with turf; coastal and offshore waters. *In migration* and *winter:* open ocean; occasionally inshore.

Nesting: colonial; in a burrow in the ground or in a crevice among rocks; nest chamber is lined with grass and feathers; pair incubates 1 darkly marked, white egg for 39–42 days.

Feeding: dives to 15 m, sometimes as deep as 60 m; takes small fish, crustaceans, molluscs and marine worms.

Voice: usually silent; may give a low, growling *arrr* at breeding sites.

Similar Species: *Dovekie* (p. 241): smaller; tiny bill; black head on breeding bird; black-and-white head pattern on nonbreeding bird.

TUFTED PUFFIN

Fratercula cirrhata

Famous for their flamboyant bills and cavalier head tufts, Tufted Puffins are among the most photographed and painted of birds. Unlike other North American puffins, which display comic charm and innocence, the Tufted Puffin exudes confidence. With its all-black body, stern facial appearance, massive orange bill and curved golden ear tufts, this is one serious-looking bird! • Stubby wings propel these birds with surprising speed and agility underwater, and their large webbed feet help steer, but these features make for awkward running takeoffs and laborious flight. • Tufted Puffins typically eat their food underwater, but when food is gathered for young, a parent can hold and transport over a dozen small fish crosswise in its bill. • During the breeding season, Tufted Puffins can be seen at the entrances to their nesting burrows and on the water near colonies, but they forage well out to sea. In August, the adults abandon their single chick and move offshore. The chick remains in the nest burrow for a week, and then, still flightless, leaps into the water and paddles out to sea, where it will remain until attaining breeding maturity at four or five years of age.

breeding

ID: sexes similar; medium-sized alcid; stocky body; large, rounded head; large, laterally flattened bill; yellow eyes; pinkish orange legs and feet. *Breeding:* white face; long, downcurved, yellow "ear" tufts; red eye ring; reddish orange and yellow bill. *Nonbreeding:* sooty face; short, golden grey "ear" tufts; dull orange bill with dark base. *In flight:* all-dark body; orange feet; large bill.

Size: *L* 37–39 cm; *W* 64 cm.

Habitat: *Breeding:* offshore vegetated islands with steep slopes and enough soil to excavate burrows. *Foraging:* almost any nearshore waters, especially with upwellings.

Nesting: colonial; in a burrow in the ground or in a crevice among rocks; nest chamber is lined with grass and feathers; pair incubates 1 darkly marked, white egg for 39–42 days.

Feeding: dives to great depths, often submerged for up to 2 minutes, to capture small schooling fish; also eats crustaceans and cephalopods.

Voice: generally silent; softly growls and grunts at the nesting colony.

Similar Species: *Rhinoceros Auklet* (p. 250): smaller bill; white undertail coverts. *Surf Scoter* (p. 73): larger; male has white patches on forehead and nape and bulbous, orange, white and black bill.

ROCK PIGEON

Columba livia

The Rock Pigeon is the common city pigeon known to all urbanites, railyard workers and pigeon-racing enthusiasts. Introduced to North America in the 17th century, Rock Pigeons have settled wherever there are cities, towns, farms or grain elevators. In the wild, these birds breed on cliffs, but many have adapted to urban, suburban and farmland habitats and nest on building roofs or ledges, under bridges or highway overpasses and in barns and silos. • First domesticated from Eurasian birds in about 4500 BCE as a source of meat, Rock Pigeons have been widely used by humans as message couriers—both Caesar and Napoleon used them—and as scientific subjects. Much of our understanding of bird migration, endocrinology, sensory perception, flight and other avian traits and biological functions derives from experiments involving these birds. • Rock Pigeons reaching reproductive maturity within six months of hatching, and they nest in any season. • All pigeons and doves feed their young "pigeon milk." It is not true milk, of course, but a nutritious liquid produced in the bird's crop. • No other "wild" bird varies as much in coloration, which is a result of semi-domestication and extensive inbreeding. The most commonly encountered colour variation is the wild phenotype "blue-bar" pigeon, which is mostly grey with two black bars.

ID: sexes similar; typically iridescent, blue-grey plumage, but highly variable (may be red, white or tan); 2 black wing bars; white rump; orange feet; dark-tipped tail. *In flight:* claps wings on takeoff; holds wings in deep "V" while gliding.

Size: *L* 31–33 cm; *W* 71 cm.

Habitat: urban and suburban areas, railway yards, grain terminals, farms and ranches; high cliffs are used by some.

Nesting: on a ledge in a building or barn, or on a cliff, bridge or tower; flimsy nest is made of sticks, grass and assorted vegetation; pair incubates 2 white eggs for 16–19 days; may raise up to 3 broods per year.

Feeding: gleans the ground for waste grain, seeds and fruits; gathers in flocks at sites where food is plentiful, especially grain yards, railway terminals and farm fields; occasionally eats insects.

Voice: soft, gurgling *cooing coorrr-coorrr-coorrr* song; male also vocalizes a soft series of low *coos*; female occasionally clucks.

Similar Species: *Band-tailed Pigeon* (p. 254): yellow bill; white "half-collar"; yellow feet; typically dark underwings; grey rump; dark band at base of paler tail; no wing bars. *Mourning Dove* (p. 256): much slimmer; greyish fawn overall; less grey on head, breast and belly; shorter wings; dark wing spots; long, diamond-shaped tail. *Merlin* (p. 154): slimmer body; longer tail; does not hold wings in "V" or clap on takeoff.

BAND-TAILED PIGEON

Patagioenas fasciata

Clinging to twigs that may barely support their weight, Band-tailed Pigeons reach into adjacent foliage to pick berries, seeds and acorns. Their presence overhead is often revealed by the occasional noisy slap of their broad wings as they shift position among the branches. In spring, the soft, low-pitched cooing of the Band-tailed Pigeon is so low and resonant that it is widely mistaken for the call of an owl. • These large and social pigeons forage and nest together anywhere open, mixed woodlands are found in extreme southwestern British Columbia. Over the past two decades, the species has expanded its range from the coastal nucleus eastward across southern parts of the province. It occurs only periodically in the foothills and mountains of Alberta. Generally fond of acorns, the Band-tailed Pigeon is most numerous in forests with a strong oak component. Flocks regularly visit mineral springs throughout their range, attracted by calcium in the water. In the Lower Mainland, nonbreeding flocks can quickly empty a backyard bird feeder. • This forest-dwelling bird has suffered long-term declines throughout its range in the western mountains of North America; in Canada, it is currently ranked as special concern.

ID: sexes similar; large, all-grey dove; purple-tinged head and breast; dark eyes; iridescent, green nape; white "half-collar"; black-tipped, yellow bill; yellow feet; long, grey tail with broad, dark band at base. *Juvenile:* uniformly grey body; lacks white "half-collar." *In flight:* bicoloured grey underwings; direct flight, often in small flocks.

Size: *L* 33–38 cm; *W* 66 cm.

Habitat: *Breeding:* mainly coniferous and mixed forests; tidal mudflats for minerals. *In migration* and *winter:* mixed forest edges, tall shrubby areas, suburban and agricultural lands.

Nesting: in a tree, usually a conifer; sparse twig nest is saddled on branch; female incubates at night, male by day, 1 all-white egg (rarely 2) for 16–22 days; may raise 2–3 broods per year.

Feeding: eats berries, seeds, nuts and less often tree and shrub buds; also feeds on mineral-rich tidal flats in summer.

Voice: wings make loud slapping noise on takeoff; male's song is 2 repeated, owl-like *hwoo* notes.

Similar Species: *Rock Pigeon* (p. 253): variable build and colour; usually has white rump and dark wing bars; typically pale underwings; no "half-collar." *Mourning Dove* (p. 256): much smaller; greyish fawn overall; less grey on head, breast and belly; dark wing spots; long, diamond-shaped tail.

EURASIAN COLLARED-DOVE

Streptopelia decaocto

The genus name *Streptopelia* is Greek for "twisted dove," and this species seems to be effortlessly winding its way through the continent. Native to India and Southeast Asia, this dove greatly expanded its range during the 20th century to include Europe, Africa and other parts of Asia. Perhaps 50 doves were released in the Bahamas in 1974 and probably reached the southeastern Florida peninsula that same decade, although they were misidentified as a similar species for years and not formally "discovered" until 1986. Eurasian Collared-Doves have now colonized much of the continent. The species is now a permanent resident in various locations from southern British Columbia to Ontario, and breeding has been confirmed in Alberta, Saskatchewan and Manitoba. • In warmer climates, the Eurasian Collared-Dove can breed six times per year, but in most of Canada, it is unlikely to produce more than two broods in a season. The young disperse long distances, which has helped the population spread. A love of suburban areas combined with a willingness to live around humans means there is little doubt that this bird's range will soon include most of urban and suburban North America. • These doves feed on seeds and other grain and are frequent visitors to bird feeders.

ID: sexes similar; large, chunky dove; pale grey overall; black "half-collar" outlined in white on nape; dark wing tips; square tail with white outer tail feathers.
Size: *L* 30–32 cm; *W* 45–50 cm.
Habitat: primarily associated with humans; generally in urban and suburban areas; also parks with both open ground and tree cover.
Nesting: in a tree; female builds a platform of twigs and sticks; pair incubates 2 white eggs for about 14 days; may raise 3 or more broods per season.

Feeding: eats mainly seeds and grain; feeds on the ground or at feeders.
Voice: soft, low *coo-COO-coo*, repeated incessantly throughout the day.
Similar Species: *Mourning Dove* (p. 256): fanned tail; lacks black "half-collar" and dark wing tips. *Rock Pigeon* (p. 253): stockier; white rump; black tail band.

MOURNING DOVE

Zenaida macroura

The soothing, rhythmic cooing of the Mourning Dove is an oft-heard sound emanating from the edges of woodlands, farmlands, suburban parks and gardens across southern Canada. The establishment of urban woodlots and rural shelterbelts has helped increase this bird's numbers, and land clearing for agriculture and forestry has greatly aided the expansion of the species. Bird feeders and feedlots in extreme southern Canada have provided a food source for an increased number of overwintering Mourning Doves. • One of North America's most widespread birds, the Mourning Dove is ranked second only to the Red-winged Blackbird as the most seen and heard bird during national breeding bird surveys. • Despite its fragile appearance, the Mourning Dove is a swift, direct flyer, capable of eluding many winged predators. In courtship or territorial displays, its wings often produce a distinctive whistle as the bird cuts through the air at high speed. • This dove is named for its plaintive, mournful calls, which novice birders sometimes mistake for the sound of a hooting owl. • The genus name *Zenaida* honours Zénaïde, Princess of Naples and wife of zoologist Charles-Lucien Bonaparte, who lived in the early 1800s.

ID: sexes similar; slender body; small head; long, pointed, white-trimmed tail; generally grey-brown plumage; buffy underparts; iridescent nape; dull reddish orange legs; dark bill; black spots on upperwings; nods head while walking. *Juvenile:* patterned face; lightly spotted breast. *In flight:* fast, direct flight; grey underwings.

Size: *L* 28–33 cm; *W* 43–48 cm.

Habitat: *Breeding:* open, mixed woodlands; orchards and forest edges next to grasslands and farmlands; suburban areas and open parks. *In migration* and *winter:* agricultural fields and feedlots; suburban feeders; gravel roadsides for grit.

Nesting: usually in a tree but occasionally on the ground; nest is a flimsy twig platform; pair incubates 2 white eggs for 14 days.

Feeding: gleans the ground and vegetation for seeds, waste grain and berries; frequents feeders and feedlots in winter.

Voice: mournful, soft *co-ooooooh coo-coo-coo.*

Similar Species: *Rock Pigeon* (p. 253): stockier; typically with iridescent neck, white rump and dark wing bars; broad, fan-shaped tail. *Band-tailed Pigeon* (p. 254): much larger; yellow bill; iridescent, green nape; white "half-collar"; yellow feet; broad, fan-shaped tail. *Black-billed Cuckoo* (p. 258): curved bill; larger head; reddish eyes; white underparts; long tail with broad, rounded tip.

YELLOW-BILLED CUCKOO

Coccyzus americanus

A strange, junglelike call emanating from deep within dense tangles and thickets might be the only evidence of your meeting with one of our most elusive birds, the Yellow-billed Cuckoo. Calls are most often heard during the nesting period, often on dark, cloudy days—earning this bird the name "Rain Crow" in some parts of its range. Most of the time, the Yellow-billed Cuckoo skillfully negotiates its tangled home in silence, relying on obscurity for survival. • Cuckoos are one of few birds that routinely gorge on spiny tent caterpillars. When the bird's stomach lining becomes packed with spines, it regurgitates it and grows a replacement. • Cuckoos lay larger clutches when outbreaks of cicadas or tent caterpillars provide abundant food. Unlike distantly related Eurasian cuckoos, which lay their eggs only in other birds' nests, North American cuckoos only occasionally lay eggs in the unattended nests of neighbouring birds. • Some Yellow-billed Cuckoos migrate as far south as Argentina for the winter. Populations of this bird in western North America are in sharp decline, with the British Columbia population already extirpated.

ID: sexes similar; olive brown upperparts; white underparts; downcurved bill has black upper mandible and yellow lower mandible; yellow eye ring; long tail with large, white spots on underside; rufous tinge to primaries. *In flight:* mostly rufous flight feathers; mostly white underwings.
Size: *L* 28–33 cm; *W* 46 cm.
Habitat: semi-open deciduous habitats; dense tangles and thickets at the edges of orchards, urban parks, agricultural fields and roadways.
Nesting: on a horizontal branch in a deciduous shrub or small tree, usually within 2 m of the ground; flimsy platform of twigs is lined with roots and grasses; pair incubates 3–4 pale bluish green eggs for 9–11 days.

Feeding: gleans insect larvae, especially hairy caterpillars, from deciduous vegetation; also eats berries, small fruits, small amphibians and occasionally the eggs of small birds.
Voice: long series of deep, hollow *kuk* notes, slowing near the end: *kuk-kuk-kuk-kuk kuk kop kow kowlp kowlp.*
Similar Species: *Mourning Dove* (p. 256): short, straight bill; pointed, triangular tail; buffy, grey-brown plumage; black spots on upperwings. *Black-billed Cuckoo* (p. 258): all-black bill; lacks rufous tinge to primaries; less prominent, white undertail spots; red rather than yellow eye ring; juvenile has buff eye ring and may have buff wash on throat and undertail coverts.

BLACK-BILLED CUCKOO

Coccyzus erythropthalmus

The Black-billed Cuckoo does not sound anything like its famous European relative, nor does it sound quite like any other bird in Canada, except for the equally elusive Least Bittern. Black-billed Cuckoos vocalize in loud bursts from shrubby thickets, repeating deep *coo-coo* and *cow* notes in tangled melodies. Only when vegetation is in full bloom will males issue their loud, long, irregular calls, advertising to females that it is time to nest. After a brief courtship, which may last for only a week, newly joined Black-billed Cuckoo pairs construct a makeshift nest, incubate their eggs and raise their young, and then return promptly to their covert lives. •Although this bird is uncommon or occasionally fairly common throughout much of southern Canada from Alberta eastward, it remains an enigma to many would-be observers. Arriving in late May, this cuckoo quietly hops, flits and skulks through low, dense deciduous vegetation in its ultra-secretive search for sustenance. • The Black-billed Cuckoo is one of few birds that thrive on hairy caterpillars, particularly tent caterpillars, and there is evidence to suggest that populations of this bird increase when a caterpillar infestation occurs in their area. • This cuckoo migrates as far as South America to avoid North American winters.

ID: sexes similar; brown upperparts; white underparts; long tail with white spots on undertail; dark, downcurved bill; reddish eye ring. *Juvenile:* buff eye ring; may have buff tinge on throat and undertail coverts. *In flight:* whitish underwing with brownish trailing edge and tip.
Size: *L* 28–33 cm; *W* 44 cm.
Habitat: dense, second-growth woodlands, shrubby areas and thickets; often in tangled riparian areas and abandoned farmlands with low, deciduous vegetation and adjacent open areas.
Nesting: in a shrub or small deciduous tree; flimsy nest of twigs is lined with vegetation; pair incubates 2–5 blue-green eggs for 10–14 days; occasionally lays eggs in other birds' nests.
Feeding: gleans hairy caterpillars from leaves, branches and trunks; also eats other insects and berries.
Voice: fast, repeated *cu-cu-cu* or *cu-cu-cu-cu-cu*; also a series of *ca, cow* and *coo* notes.
Similar Species: *Mourning Dove* (p. 256): short, straight bill; pointed, triangular tail; buffy, grey-brown plumage; black spots on upperwings. *Yellow-billed Cuckoo* (p. 257): yellow bill; rufous tinge to primaries; larger, more prominent, white undertail spots; lacks red eye ring.

BARN OWL

Tyto alba

The haunting look of this night hunter has inspired many superstitions—its white, heart-shaped face and black, piercing eyes give the Barn Owl an eerie look. Likewise, its call consists of a variety of hisses, screams, cries and other strange noises. • True to their name, Barn Owls often nests and roost in barns, but will also use silos, tree cavities, crevices in cliffs and old mine shafts. • Almost any open area can serve as a foraging site for this species, and Barn Owls are surprisingly tolerant of human activities. Nest boxes have stabilized some populations, but others are threatened by modern agricultural practices, particularly the loss of hedgerows and old wooden buildings. • Barn Owls forage at dusk and at night, often along the shoulders of busy highways. As a result, they experience frequent mortality from vehicles. • Barn Owls are found on every continent except Antarctica. In Canada, regularly occurring populations are found only in southern British Columbia and southern Ontario. Severe winter weather limits the ability of Barn Owls to survive in much of the country. The grassland habitats preferred by foraging Barn Owls are disappearing as a result of urbanization and changing farm practices. The eastern population in Ontario is classified as endangered, with the breeding population estimated at less than 10 pairs.

ID: sexes similar; medium-sized owl; large, round head; white, heart-shaped facial disc; dark eyes; pale bill; golden brown upperparts spotted with black and gray; creamy white, black-spotted underparts; long legs; white undertail and underwings; females are darker and have more spotting on underparts. *In flight:* broad wings; buoyant, somewhat uneven flight; deep, slow wingbeats.
Size: *L* 33–38 cm; *W* 1.0–1.2 m.
Habitat: roosts and nests in cliffs, hollow trees, unoccupied buildings, caves, bridges, tree groves and riverbanks; requires open areas for hunting.
Nesting: in a tree, cliff crevice, artificial platform or nest box; no actual nest is built but pellets and dead voles often present; female incubates 3–8 whitish eggs for 29–34 days; young hatch asynchronously.
Feeding: mostly nocturnal; forages more by sound than sight; eats mostly small rodents, especially voles.
Voice: harsh, raspy screeches and hisses; also metallic clicking sounds.
Similar Species: *Short-eared Owl* (p. 272): boldly streaked upperparts; yellow eyes; black "wrist" patches. *Snowy Owl* (p. 264): winter only; larger; stockier; mostly white, often with dark barring and spots; yellow eyes. *Barred Owl* (p. 269): barred chest; streaked belly; darker facial disc.

FLAMMULATED OWL

Otus flammeolus

As the campfire settles into glowing embers, from beyond the edges of the clearing comes an odd, low-pitched hooting—a Flammulated Owl has shaken off the torpor of summer daytime to begin the evening's hunt. • Flammulated Owls breed from southern British Columbia and the western United States to central Mexico. In BC, these small, dark-eyed owls favour valley sides in mountainous areas dominated by open, mature Douglas-firs or ponderosa pines. • Increased attention by birders and raptor researchers has revealed these birds, once considered very rare and local, to be much more widespread and numerous than previously believed. In the past, their distribution was restricted to the Okanagan, Thompson and Nicola valleys, but they have recently been found breeding in the Cariboo region of BC. • Flammulated Owls rarely feed on small rodents, preferring large insects. • This owl is highly migratory, arriving in early May and departing in October to spend the winter in Mexico and Guatemala. • The Flammulated Owl is the only small North American owl with dark eyes.

grey adult

ID: sexes similar; small owl; minute "ear" tufts; dark eyes; variable grey or rufous plumage; white eyebrows indent top of rusty or grey facial disc; greyish bill; dark-centred white spots on shoulders and wing coverts; dark-streaked, whitish breast. *In flight:* long, pointed wings are usually brown and grey with white barring.

Size: *L* 15–18 cm; *W* 38 cm.

Habitat: *Breeding:* mid-elevation, open coniferous forests, especially mature ponderosa pine and Douglas-fir, with thick, shrubby understorey.

Nesting: in an unlined natural cavity or artificial nest box; female incubates 1–4 creamy white eggs for 21–24 days.

Feeding: strictly nocturnal; catches mostly moths, beetles and katydids in flight or picks them from the ground or foliage.

Voice: muffled, low-pitched, mellow hoot given about every 3 seconds; heard from twilight to dawn.

Similar Species: *Western Screech-Owl* (p. 261): larger; yellow eyes; generally brown or grey; face usually has less pronounced white markings and less rusty coloration. *Northern Saw-whet Owl* (p. 274) and *Northern Pygmy-Owl* (p. 266): heavier; yellow eyes; no prominent "ear" tufts.

WESTERN SCREECH-OWL

Megascops kennicottii

Like many of the smaller owls, the Western Screech-Owl is a fierce and adaptable hunter, occasionally adding birds and even mammals larger than itself to its usual diet of insects, amphibians and rodents. This chunky, open-woodland owl passes the daylight hours concealed in dense shrubs or roosting in a hollow tree, waiting until late evening to commence its hunting activities. The habitat requirements of the Western Screech-Owl include a secluded roosting site, a tree hollow for nesting, semi-open ground for hunting and a stretch of contiguous woods for dispersal. • Between March and June, distinctive "bouncing-ball" whistles indicate the presence of a nesting pair of Western Screech-Owls. • When threatened, this owl will often stand erect, draw its body feathers in and hold its "ear" tufts up, a posture that allows this cryptically coloured bird to blend in with its tree branch or tree trunk surroundings. • The Western Screech-Owl is a year-round resident of open deciduous or mixed forests from Alaska south through British Columbia, east to the Rockies in the U.S. and into Mexico. Two subspecies are found in Canada; the *macfarlanei* subspecies is classified as endangered because of the loss of its mature riparian woodland habitat to urban development and agriculture.

ID: sexes similar (female is slightly larger); small owl; feathered "ear" tufts; yellow eyes; dark bill; streaky, grey-brown overall; facial disc has partial dark border; narrow, dark, vertical breast stripes. *In flight:* barred, greyish or brownish wings with pale linings; dark "wrist" patch; white-spotted upperwings.

Size: *Male: L* 21–23 cm; *W* 46–51 cm. *Female: L* 25–28 cm; *W* 56–61 cm.

Habitat: open deciduous or mixed forests, often near water; also large woodlots in rural areas.

Nesting: in a natural cavity or artificial nest box; little if any nest material added; female incubates 2–5 white eggs for about 26 days.

Feeding: nocturnal; swoops from a perch to capture invertebrates such as earthworms and insects; also takes rodents, amphibians and small songbirds.

Voice: courtship song is a distinctive series of soft, accelerating, even-pitched whistles and clear notes, with a rhythm like that of a bouncing ball coming to a stop; also gives a short trill followed by a longer trill; pairs often harmonize.

Similar Species: *Flammulated Owl* (p. 260): smaller; dark eyes; some rusty coloration on face. *Long-eared Owl* (p. 271): long, narrow body; tall "ear" tufts set very close together. *Northern Saw-whet Owl* (p. 274): rufous brown streaks on white underparts; lacks "ear" tufts.

EASTERN SCREECH-OWL

Otus asio

grey adult

red adult

The diminutive Eastern Screech-Owl is a year-round resident of deciduous woodlands south of the Canadian Shield, but its presence is rarely detected. • Most screech-owls sleep away the daylight hours snuggled safely inside a tree cavity or an artificial nest box. An encounter with one of these owls is usually the result of a sound cue—the noise of a mobbing horde of chickadees or squawking gang of Blue Jays will occasionally alert you to an owl's presence during daylight hours. More commonly, you will find this owl by listening for the male's eerie, horse-whinny courtship calls and loud, spooky trills at night. • Despite its small size, the Eastern Screech-Owl is an adaptable and formidable hunter. It has a varied diet that ranges from insects, small rodents, earthworms and fish to birds larger than itself. • Eastern Screech-Owls are polychromatic, showing red or grey colour phases. The red birds are less common in Canada because they are less able to withstand our cold winters. Mixed-coloured pairs may produce intermediate-coloured young that are buffy brown.

ID: sexes similar (female is slightly larger); small owl; short "ear" tufts; reddish or greyish overall; dark breast streaking; yellow eyes; pale greyish bill. *Juvenile:* buffy red or grey facial disc. *In flight:* barred, rounded wings; fairly prominent shoulder spots; stubby tail; usually flies close to the ground; rapid wingbeats with short glides; sometimes hovers.

Size: *L* 20–23 cm; *W* 50–55 cm.

Habitat: mature deciduous forests, open deciduous woodlands, riparian woodlands, orchards and shade trees with natural cavities.

Nesting: in a natural cavity or artificial nest box; no lining is added; female incubates 3–8 white eggs for about 26 days.

Feeding: small mammals, earthworms, fish, birds and insects, including moths in flight; feeds at dusk and at night.

Voice: horselike "whinny" that descends in pitch and speeds up to a tremolo.

Similar Species: *Boreal Owl* (p. 273) and *Northern Saw-whet Owl* (p. 274): long, reddish streaks on white underparts; lack "ear" tufts. *Long-eared Owl* (p. 271): much longer, slimmer body; longer, closer-set "ear" tufts; rusty facial disc; white-blotched, greyish brown upperparts.

GREAT HORNED OWL

Bubo virginianus

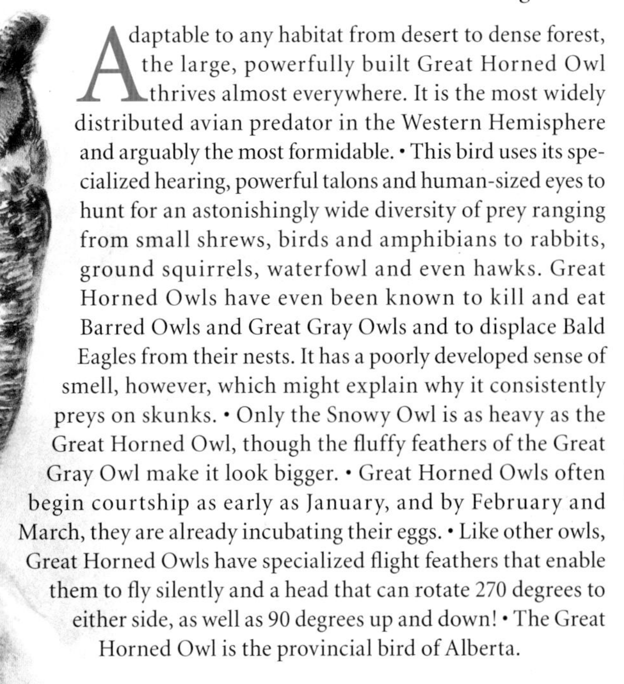

Adaptable to any habitat from desert to dense forest, the large, powerfully built Great Horned Owl thrives almost everywhere. It is the most widely distributed avian predator in the Western Hemisphere and arguably the most formidable. • This bird uses its specialized hearing, powerful talons and human-sized eyes to hunt for an astonishingly wide diversity of prey ranging from small shrews, birds and amphibians to rabbits, ground squirrels, waterfowl and even hawks. Great Horned Owls have even been known to kill and eat Barred Owls and Great Gray Owls and to displace Bald Eagles from their nests. It has a poorly developed sense of smell, however, which might explain why it consistently preys on skunks. • Only the Snowy Owl is as heavy as the Great Horned Owl, though the fluffy feathers of the Great Gray Owl make it look bigger. • Great Horned Owls often begin courtship as early as January, and by February and March, they are already incubating their eggs. • Like other owls, Great Horned Owls have specialized flight feathers that enable them to fly silently and a head that can rotate 270 degrees to either side, as well as 90 degrees up and down! • The Great Horned Owl is the provincial bird of Alberta.

ID: sexes similar (female is slightly larger); very large, stocky owl; prominent, wide-set "ear" tufts; yellow eyes; heavily mottled, brown body; densely barred underparts; rufous facial disc outlined in black; white throat. *In flight:* broad, rounded wings; steady, rather stiff wingbeats.
Size: *L* 46–64 cm; *W* 1.0–1.3 m.
Habitat: open, mixed forests, agricultural fields, sagebrush grasslands, riparian woodlands, marshes, suburban parks and wooded edges of landfills.
Nesting: in the abandoned stick nest of another large bird; also in a tree stub; rarely in an abandoned building, crevice or cave; female incubates 2–6 dull white eggs for 30–37 days.
Feeding: opportunistic; takes small mammals, birds, snakes and amphibians; hunts mainly at dawn and dusk, usually from a perch.

Voice: breeding season call is 4–6 deep hoots: *hoo-hoo-hoooo hoo-hoo* or *eat-my-food, I'll-eat you*; female gives higher-pitched hoots when pairs duet.
Similar Species: *Long-eared Owl* (p. 271): much smaller; very slim; narrower head; closer-set "ear" tufts; mostly vertical breast markings; darker "wrist" patch in flight. *Barred Owl* (p. 269): smaller; white-speckled, greyish brown upperparts; bolder facial disc; coarse, vertical belly streaks; no "ear" tufts. *Spotted Owl* (p. 268) and *Great Gray Owl* (p. 270): lack prominent "ear" tufts.

SNOWY OWL

Bubo scandiacus

When the mercury dips and the landscape hardens in winter's icy grip, Snowy Owls move south from their Arctic breeding grounds in search of food. During particularly hard winters, when lemmings and voles are scarce in northern Canada and Alaska, these owls can be especially numerous in southern Canada. At this time, Snowies, mostly immatures, can be seen perched on drift logs, lamp posts, power poles, trees, buildings and even city skyscrapers. On very windy days, these owls prefer to remain on the ground, where they blend in perfectly against the snow-covered landscape. • While their diet mostly consists of small mammals in northern regions, these opportunistic owls feed on a variety of small to medium-sized mammals and various birds while on their wintering territories. Feathered to the toes, a Snowy Owl can remain active at cold temperatures that send other owls to the woods for shelter. • As Snowy Owls age, their plumage becomes lighter. Males are also lighter, so pure white Snowies are almost certain to be older males. • Banding studies have revealed that females often return to the same winter territory year after year, even to the same perches, but males are less predictable. • The Snowy Owl is the provincial bird of Québec.

♀

ID: large owl; yellow eyes; black bill; no "ear" tufts. *Male:* mostly white; very little dark flecking. *Female:* generally white; dark barring on upperparts and breast. *Immature:* heavier, brownish grey barring on body. *In flight:* somewhat pointed wings; underwings mostly or all white.

Size: *L* 52–71 cm; *W* 1.2–1.5 m.

Habitat: *Breeding:* open tundra. *In migration* and *winter:* open country including log-littered seashores, farmlands, grasslands, marshes, rocky islands and airport fields; often perches on fence posts, buildings and power poles.

Nesting: on a raised ridge, outcrop or hummock; slight depression is sparsely lined with grass and feathers; mostly the female incubates 2–6 dull whitish eggs for about 33 days.

Feeding: opportunistic; swoops from a perch, often punching through the snow to take mice, voles, grouse, hares, weasels and and small waterfowl and marine birds.

Voice: generally silent on wintering grounds; barking *krow-ow* or repeated *rick* call on nesting grounds.

Similar Species: no other owl is predominantly white and lacks "ear" tufts.

NORTHERN HAWK OWL

Surnia ulula

The Northern Hawk Owl has a circumpolar distribution and resides in the northern boreal forest, where muskeg and mosquitoes dominate the summer landscape and icy temperatures prevail in winter. Like the Great Gray Owl, the Northern Hawk Owl is an irruptive winter visitor in semi-open boreal or coniferous-deciduous forests of southern Canada, meaning that it may be common in some winters and rare in others. When a "hawk owl year" comes around, be sure to make the best of the event—there might not be a repeat performance for a decade or more. • Northern Hawk Owls summer where the days are long, so they are comfortable hunting in daylight. Hawk owls are by no means the only owls that hunt diurnally, but they are the ones most likely to rest on exposed perches during the day. • With its long, slender, hawklike features, and direct, accipter-like flight pattern, it's easy to understand how this bird got its name. No matter how much it behaves like a hawk, however, there is no mistaking the face of an owl or this bird's obvious use of sound to hunt prey.

ID: sexes similar; medium-sized owl; rounded head without "ear" tufts; fine, horizontal barring on underparts; white facial disc bordered with black; pale bill; yellow eyes; white-spotted forehead. *In flight:* short, pointed wings; long, tapered tail.
Size: *L* 38–43 cm; *W* 80–90 cm.
Habitat: black spruce bogs and muskegs, old burns and tree-bordered clearings.
Nesting: in an abandoned woodpecker cavity or on a broken treetop; nest is unlined; female incubates 5–7 whitish eggs for 25–30 days.
Feeding: swoops from a perch; eats mostly voles, mice and birds; also takes some large insects in summer.

Voice: usually quiet; whistled breeding trill; call is an accipiter-like *kee-kee-kee.*
Similar Species: *Short-eared* (p. 272), *Boreal* (p. 273) and *Northern Saw-whet* (p. 274) *owls:* much shorter tails; vertical breast streaks. *Northern Pygmy-Owl* (p. 266): much smaller; 2 "false eyes" on back of head.

NORTHERN PYGMY-OWL

Glaucidium gnoma

Even when the Northern Pygmy-Owl is looking the other way, two dark "false eyes" on the back of its head stare blankly toward you. Because predators are less likely to attack a bird that is apparently looking in their direction, the pygmy-owl is usually effective in "watching" its own back. • Canada's smallest owl, the Northern Pygmy-Owl regularly catches prey that outweighs it. On such occasions, the owl is often dragged some distance before it kills its quarry. Pygmy-owls occasionally venture into townsites in the mountains and foothills to hunt sparrows and finches at bird feeders. • The Northern Pygmy-Owl hunts mainly at dawn and dusk, when it can sometimes be spotted as a blob at the top of a tall tree. With a poorly developed facial disc, it must depend more on vision than hearing to locate prey. • By day, the secretive Northern Pygmy-Owl tends to perch in thickets, where it is protected from predators. To find one of these owls, you can rely either on luck or on chickadees. Small bands of these songbirds delight in mobbing this little owl, and the sound can lead attentive naturalists right up to the pint-sized predator. By imitating this owl's whistled call, birders might themselves be mobbed by chickadees and other small forest birds.

red adult

ID: sexes similar; small, compact owl; 2 black "false eyes" on back of large, rounded head; white-spotted crown, neck and back; yellow eyes; short wings; reddish brown upperparts; pale underparts with bold, dark streaking; indistinct facial disc; grey to yellowish bill. *In flight:* rounded, mottled underwings; long, barred tail.
Size: *L* 18 cm; *W* 38 cm.
Habitat: open coniferous, deciduous or mixed forests, often in riparian areas; occasionally townsites in winter.

Nesting: in an unlined tree cavity or old woodpecker hole; female incubates 3–7 glossy white eggs for about 28 days.
Feeding: eats large insects and small rodents; small birds form a major part of the diet in some areas.
Voice: main song is an easily imitated series of monotonous, low-pitched *toot* notes, uttered once every 2–3 seconds, remarkably like a wooden recorder; series of 20–30 rapid-fire, decelerating *toot* notes when agitated; all calls the same pitch.
Similar Species: *Boreal Owl* (p. 273) and *Northern Saw-whet Owl* (p. 274): prominent facial discs; lack "false eyes." *Northern Hawk Owl* (p. 265): much larger; lacks "false eyes." *Western Screech-Owl* (p. 261): "ear" tufts; larger eyes. *Flammulated Owl* (p. 260): "ear" tufts; dark eyes; active only at night.

BURROWING OWL

Athene cunicularia

The Burrowing Owl is a delightful bird that, true to its name, nests underground in burrows! Once common in the grasslands of the Canadian prairies, its numbers have dwindled to less than one percent of what they once were, and the species is listed as endangered in Canada. Collisions with vehicles, increased numbers of predators, the effects of agricultural chemicals and the conversion of native grasslands to cropland and rangeland or to urban areas are some of the challenges facing this bird throughout its nesting, migration and wintering ranges. Fortunately, many prairie landowners have agreed to preserve Burrowing Owl habitat, and some have provided artificial nesting burrows to attract these birds to grasslands where suitable burrows are lacking. Band recoveries have shown that at least some Burrowing Owls that breed in British Columbia spend their winters south of San Francisco, while prairie birds winter from Oklahoma and Texas south to Mexico. • During the day, these ground-dwelling birds can be seen atop fence posts or on dirt mounds near their nests. When perched at the entrance to its burrow, this owl looks very similar to the ground squirrels with which it closely associates.

ID: sexes similar; small, ground-dwelling owl; rounded head without "ear" tufts; yellow eyes; long legs; white eyebrows; light to dark brownish body with white spotting on upperparts; dark brown barring on white underparts; short tail; often bobs. *In flight:* weak "wrist" markings.
Size: *L* 23–28 cm; *W* 50–60 cm.
Habitat: *Breeding:* open, short-grass terrain and sagebrush grasslands; occasionally on lawns and golf courses. *In migration* and *winter:* dry, open grasslands, agricultural lands and irrigation canals.
Nesting: in an underground burrow; nest chamber is lined with bits of dry manure, food debris, feathers and fine grass; female incubates 6–11 white eggs for 28–30 days.

Feeding: hunts mostly at dusk and dawn; eats larger insects and rodents; occasionally takes frogs, snakes and, when available, rodents and small grassland songbirds.
Voice: harsh *chuk* call; chattering *quick-quick-quick*; rattlesnake-like warning call from within burrow; male's courtship call is a mournful, whistled *coo-coo*.
Similar Species: *Short-eared Owl* (p. 272): larger; long, narrow wings; shorter legs and tail; buffier plumage. *Other small owls* (pp. 260–62, 266, 273–74): shorter legs; found in treed environments.

SPOTTED OWL

Strix occidentalis

Late in the 20th century, the Spotted Owl became the focal point of discussions about timberland management across its range in the Pacific Northwest. The northern *caurina* subspecies, which just ranges into extreme southwestern British Columbia, needs large areas of old-growth rainforest with minimal fragmentation in which to nest and forage. A pair of Spotted Owls requires over 800 hectares of contiguous habitat as a home range. Urbanization, recreational activities and logging interests have depleted much of this bird's habitat. Another pressure on the species is the recent range expansion of its slightly larger relative, the Barred Owl, into suitable habitat. The two species are so closely related that they occasionally hybridize. • The Spotted Owl hunts at night, waiting patiently for a meal to scamper within range, usually an arboreal or terrestrial rodent. • This owl tolerates remarkably close human approach, though females will attack intruders at the nest site. • In Canada, Spotted Owls have undergone catastrophic population declines and are classified as endangered. Currently, there are probably fewer than 25 of these owls remaining in southwestern BC. Without drastic intervention, it is thought that the species will be extirpated from Canada by 2012.

ID: sexes similar (female is slightly larger); large owl; white-mottled, brown overall; large, rounded head without "ear" tufts; large, dark eyes; pale yellowish bill; white-tipped, brown feathers give spotted look to chest; whitish throat. *In flight:* nearly rectangular, barred, brown wings; paler underparts; white-spotted upperparts.

Size: *Male: L* 43–46 cm; *W* 1.0 m. *Female: L* 46–48 cm; *W* 1.1 m.

Habitat: late mature to old-growth rainforest; also fir and mixed forests of ponderosa pine, white fir and Douglas-fir to 1000 m elevation; forages in unlogged areas and tends to avoid crossing bushy areas and clear-cuts.

Nesting: in an unlined cavity in a stump or tree or (rarely) on a tree platform; pairs use the same nest site for years or alternate between 2 or more sites; female incubates 2–3 white or pearly grey eggs for 28–32 days.

Feeding: generally nocturnal; perches and waits to pounce on small mammals, especially voles, mice and squirrels; also eats insects, amphibians and birds (occasionally other owls).

Voice: *whoo-whoo, hoo hoo* call; female gives similar 3-note hoot that is hollower and higher pitched.

Similar Species: *Barred Owl* (p. 269): slightly larger; often paler; horizontal breast bars; thick, vertical belly streaks; more likely in moist bottomlands. *Western Screech-Owl* (p. 261): much smaller; less white in plumage; prominent "ear" tufts; yellow eyes; vertically streaked underparts. *Great Horned Owl* (p. 263): larger; conspicuous "ear" tufts; large, yellow eyes. *Great Gray Owl* (p. 270): much larger; large facial disc; yellow eyes; white "bow tie."

BARRED OWL

Strix varia

The maniacal cacophony produced by calling Barred Owls is unforgettable. The escalating laughs, hoots and howls reinforce the bond between pairs but is enough to scare the uninitiated out of the woods. This owl's more typical call is a series of hoots that sounds like *Who cooks for you? Who cooks for you all?* Barred Owls tend to be more vocal during late evening and early morning when the moon is full, the air is calm and the sky is clear. At the height of courtship and when raising young, pairs may continue their calls well into daylight hours, and they may hunt actively day and night. • Barred Owls do best near water, preferring mature, mixed second-growth and old-growth forests. They are often found in swampy woods, densely forested ravines and even heavily treed suburbia, a sign of their adaptability. Their range has expanded westward from eastern North America, and the species is now resident across southern British Columbia, including southeastern Vancouver Island. • Although this owl's eyesight in darkness is 100 times keener than that of humans, it can also locate prey using sound alone. • Barred Owls have relatively weak talons, and they mainly prey on smaller animals, such as voles. Nonetheless, they sometimes take small birds and even smaller owls.

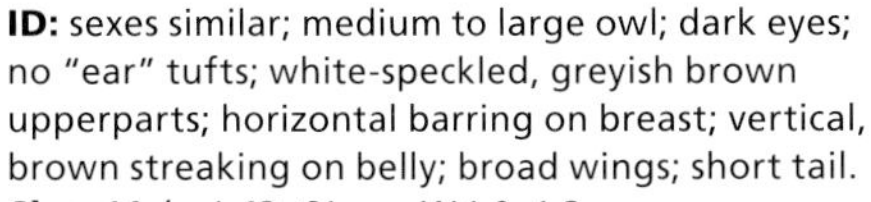

ID: sexes similar; medium to large owl; dark eyes; no "ear" tufts; white-speckled, greyish brown upperparts; horizontal barring on breast; vertical, brown streaking on belly; broad wings; short tail.

Size: *Male: L* 43–61 cm; *W* 1.0–1.3 m. *Female: L* 56–61 cm; *W* 1.1–1.3 m.

Habitat: mature, mixed second-growth and old-growth forests, especially in dense stands near swamps, streams and lakes; also wooded city and residential parks and undisturbed large woodlots.

Nesting: in an unlined cavity, hollow or broken top of a large, often dead tree; female incubates 2–4 pure white eggs for 28–33 days.

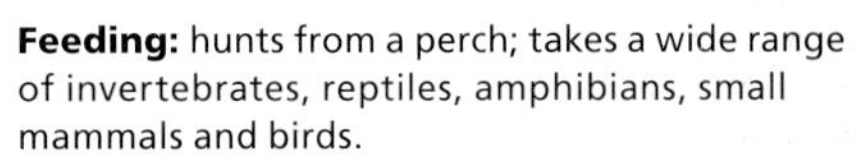

Feeding: hunts from a perch; takes a wide range of invertebrates, reptiles, amphibians, small mammals and birds.

Voice: often calls during the day but generally at night; typical call sounds like *Who cooks for you? Who cooks for you all?*; also a variety of "laughs," hoots and howls.

Similar Species: *Spotted Owl* (p. 268): slightly smaller; paler bill; white-spotted, brown breast and underparts; limited to old-growth forests. *Great Horned Owl* (p. 263): larger; conspicuous "ear" tufts; large, yellow eyes; barred underparts. *Great Gray Owl* (p. 270): very large; all-grey plumage; large facial disc; yellow eyes; white "bow tie."

GREAT GRAY OWL

Strix nebulosa

With a face like a satellite dish, the Great Gray Owl's large head swivels smoothly, focusing instantly on the slightest sound or movement beneath up to 60 centimetres of snow or within a summer's growth of grass and wildflowers. When prey is detected, the owl launches itself in noiseless flight, gliding downward on fixed wings and punching through the deep snow or dense vegetation to capture its meal. • Like many owls, the Great Gray's facial discs funnel sound into its asymmetrically placed ears, enabling it to pinpoint the precise location of its prey using triangulation. • Although it is North America's largest owl, the Great Gray Owl owes its bulk largely to a mass of fluffy insulation. In reality, it weighs 15 percent less than the smaller Snowy Owl or Great Horned Owl. • Great Gray Owls are widespread across the boreal forests of the Northern Hemisphere, and it is the only *Strix* owl to breed on both sides of the Atlantic. • These owls occasionally appear in good numbers south of their boreal breeding range, usually in years when northern small mammal populations have crashed. • The Great Gray Owl is Manitoba's provincial bird.

ID: sexes similar; very large owl; white-marbled, greyish brown plumage; prominent, concentrically barred facial disc; rounded head without "ear" tufts; yellow eyes; black "chin"; white "bow tie" collar; small, yellow bill; long tail. *In flight:* large, nearly rectangular, barred, brown wings; long, faintly banded tail.

Size: *Male: L* 61–84 cm; *W* 1.3–1.5 m. *Female: L* 74–84 cm; *W* 1.4–1.5 m.

Habitat: *Breeding:* undisturbed, dense, pure or mixed spruce or poplar stands adjacent to open muskeg, fens, bogs or meadows. *Nonbreeding:* boreal forest clearings; also roadsides, bogs and open meadows.

Nesting: usually near a spruce bog or muskeg; may use an old raptor or crow nest, broken-topped snag or artificial platform; female incubates 3–5 dull white eggs for 28–36 days.

Feeding: sit-and-wait predator; hunts mostly at dawn and dusk; eats mostly small rodents, especially voles.

Voice: deep, booming *whoo-hoo-hoo*; also a series of widely spaced, deep *whooo* notes.

Similar Species: *Barred Owl* (p. 269): smaller; less pronounced facial disc; dark eyes; horizontally barred breast. *Great Horned Owl* (p. 263): conspicuous "ear" tufts; brown body; rufous facial disc outlined in black; dark bill; dense, horizontal streaking on underparts.

LONG-EARED OWL

Asio otus

A master of disguise and illusion, the Long-eared Owl blends into its shady, wooded background by flattening its feathers and assuming a thin, vertical form. If that tactic fails, this medium-sized owl may puff up its feathers, spread its long wings to double its size, and then hiss defiantly in a threat display. • The "ears" on top of the Long-eared Owl's head are only ornamental feathers. Its real ears are hidden under the facial disc, and their asymmetrical size, shape and placement enhance the owl's judgement of distance and direction. This allows the Long-eared Owl to hunt by sound in total darkness. This nocturnal predator hunts in open areas, but returns to dense stands of trees to roost during the day. • All owls, as well as many other birds such as herons, gulls, crows and hawks, regurgitate "pellets"—the indigestible parts of their prey compressed into an elongated ball. The feathers, fur and bones that make up the pellets are interesting to analyze, because they reveal what the animal has eaten. Owl pellets can often be found under frequently used roost sites. • Long-eared Owls are most noticeable during the winter months, when they roost in woodlots, hedgerows or isolated tree groves. They usually roost in groups, sometimes with Short-eared Owls.

ID: sexes similar; slim, medium-sized owl; long "ear" tufts; yellow eyes; rusty brown facial disc; white-blotched, greyish brown upperparts; dark crisscross pattern on underparts; long, white feathers between eyes. *In flight:* long wings; mostly grey upperwing with orangey patch near tip; dark-barred wing tips and trailing edges; prominent, black "wrist" patch; pale buff underwings.
Size: *L* 35–41 cm; *W* 91–119 cm.
Habitat: *Breeding:* riparian, mixed woodlands, shrublands and wooded tangles near open country and marshes. *In migration* and *winter:* open coniferous and mixed forests and tangles adjacent to open habitats.
Nesting: in an abandoned stick nest; occasionally in a cliff cavity or on the ground; female incubates 5–7 slightly glossy white eggs for 26–28 days.

Feeding: nocturnal; flies low, pouncing on prey from the air; eats mostly voles and mice.
Voice: low, ghostly *quoo-quoo* breeding call; *weck-weck-weck* alarm call; also issues various shrieks, hisses, whistles, barks, hoots and dovelike coos.
Similar Species: *Short-eared Owl* (p. 272): buffier plumage; tiny, rarely seen "ear" tufts; light underpart streaking is denser on breast; lacks rusty facial disc; generally seen on the ground or in open areas. *Great Horned Owl* (p. 263): much larger; wider-set "ear" tufts; darker disc edge; horizontally streaked underparts; plainer wings. *Western Screech-Owl* (p. 261) and *Eastern Screech-Owl* (p. 262): smaller; wider-set "ear" tufts; less compressed body.

SHORT-EARED OWL

Asio flammeus

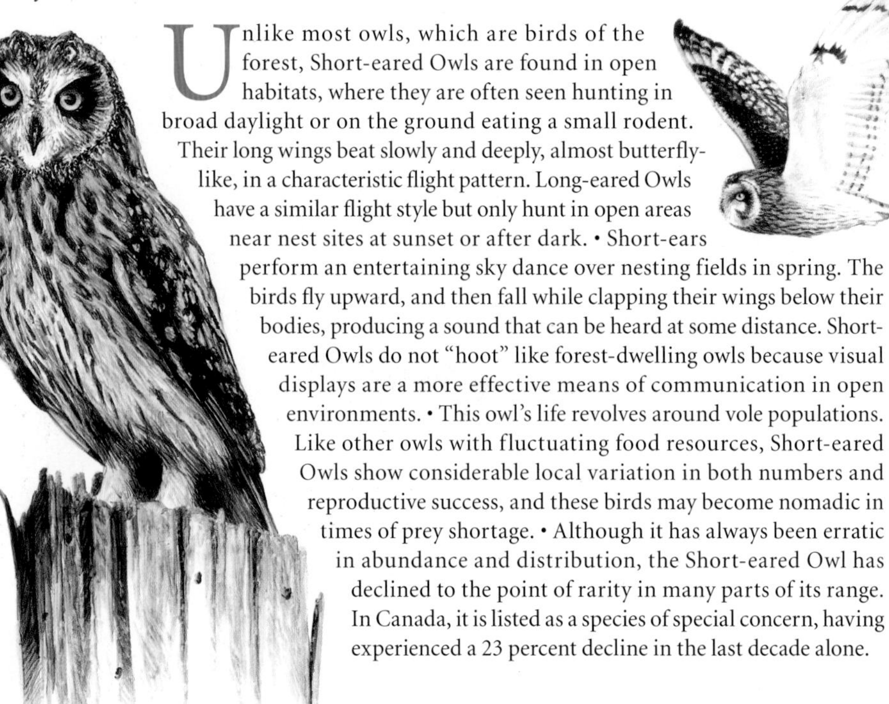

Unlike most owls, which are birds of the forest, Short-eared Owls are found in open habitats, where they are often seen hunting in broad daylight or on the ground eating a small rodent. Their long wings beat slowly and deeply, almost butterfly-like, in a characteristic flight pattern. Long-eared Owls have a similar flight style but only hunt in open areas near nest sites at sunset or after dark. • Short-ears perform an entertaining sky dance over nesting fields in spring. The birds fly upward, and then fall while clapping their wings below their bodies, producing a sound that can be heard at some distance. Short-eared Owls do not "hoot" like forest-dwelling owls because visual displays are a more effective means of communication in open environments. • This owl's life revolves around vole populations. Like other owls with fluctuating food resources, Short-eared Owls show considerable local variation in both numbers and reproductive success, and these birds may become nomadic in times of prey shortage. • Although it has always been erratic in abundance and distribution, the Short-eared Owl has declined to the point of rarity in many parts of its range. In Canada, it is listed as a species of special concern, having experienced a 23 percent decline in the last decade alone.

ID: sexes similar (female is slightly larger); medium-sized owl; short, inconspicuous "ear" tufts; pale facial disc; yellow eyes set in black sockets; brown upperparts heavily marked with dark brown and white; heavy, dark vertical streaking on pale buff belly. *In flight:* deep, slow wingbeats; long wings; dark "wrist" crescents; pale patch near base of primaries.

Size: *L* 33–43 cm; *W* 1.0–1.2 m.

Habitat: *Breeding:* wet shrublands, dry marshes, grasslands, long-grass agricultural fields and open power line corridors. *In migration* and *winter:* alpine mountain meadows, sagebrush grasslands, marshes, agricultural fields, log-littered shores, and rangeland.

Nesting: on wet ground in an open area; slight depression is sparsely lined with grasses; female incubates 4–12 creamy white eggs for 24–37 days.

Feeding: mostly eats voles, mice and shrews; sometimes takes small birds.

Voice: generally quiet; loud *keee-ow* in winter; breeding male utters a soft *toot-toot-toot*.

Similar Species: *Long-eared Owl* (p. 271): greyer, more mottled plumage; rusty brown face; prominent, long "ear" tufts; crisscross markings on underparts; darker orangey patch at base of upper primaries. *Burrowing Owl* (p. 267): smaller; much longer legs; brown, horizontal barring on white underparts; white eyebrows; white-spotted upperparts. *Other owls* (pp. 259–66, 268–70, 273–74): rarely seen in open areas during daylight.

BOREAL OWL

Aegolius funereus

The Boreal Owl routinely ranks in the top five of the most-desired species to see, according to birder surveys throughout North America. Known in the Old World as "Tengmalm's Owl," it lives primarily in the boreal forests of Canada and Alaska, as well as from Scandinavia to Siberia. Scattered populations also occur in subalpine forests in the western mountain ranges south to New Mexico. • In winter, the secretive Boreal Owl often departs from its nocturnal summertime activities in dense forest to visit more open woodlands and even suburban backyards. Occasionally, one may be found roosting near a bird feeder or in a sheltered nest box. This small owl is known to be well adapted to snowy forest environments—it is quite capable of locating and catching prey that lives beneath the snow. • When rodent populations are high, a male may mate with several females; when prey is scarce, pairs remain monogamous. • Because of the Boreal Owl's remote habitat and nocturnal habits, many aspects of its ecology and behaviour are unknown. • This approachable owl was named "Blind One" by Native peoples because it was easily captured by hand.

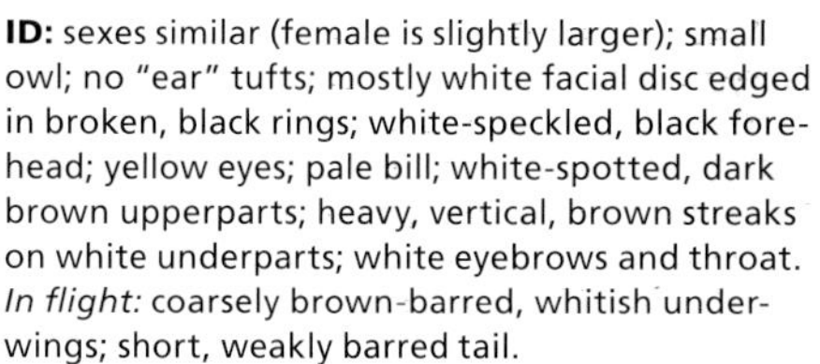

ID: sexes similar (female is slightly larger); small owl; no "ear" tufts; mostly white facial disc edged in broken, black rings; white-speckled, black forehead; yellow eyes; pale bill; white-spotted, dark brown upperparts; heavy, vertical, brown streaks on white underparts; white eyebrows and throat. *In flight:* coarsely brown-barred, whitish underwings; short, weakly barred tail.

Size: *L* 23–31 cm; *W* 55–74 cm.

Habitat: *Breeding:* mature coniferous and mixed forests. *In migration* and *winter:* more open woodlands, suburban gardens and parklands.

Nesting: in an unlined natural tree hollow, woodpecker cavity or nest box; female incubates 2–5 dull white eggs for 26–32 days.

Feeding: hunts from a low perch for small rodents (especially voles and mice) and birds; also takes insects; often plunges through the snow to catch prey in winter; may cache food.

Voice: accelerating, whistled *whew-whew-whew-whew-whew-whew*; loud, hoarse *hooo-aaak*; grating, hissing threat call; male's main spring song is a low, trilled *toot* series, increasing in volume.

Similar Species: *Northern Saw-whet Owl* (p. 274): slightly smaller; warmer brown overall; redder streaks on underparts; streaked forehead; dark bill. *Northern Hawk Owl* (p. 265): much longer tail; fine, horizontal barring on underparts.

NORTHERN SAW-WHET OWL

Aegolius acadicus

By day, Northern Saw-whet Owls roost quietly under the cover of dense, low branches and brush to avoid attracting mobbing forest songbirds. • Although saw-whets feed primarily on forest mice, these opportunistic hunters take whatever they can, whenever they can and often store what they cannot eat. When temperatures are below freezing, these small owls may catch more than they can eat at a single sitting. The extra food is usually stored in trees, where it quickly freezes. When new hunting efforts fail, a hungry owl can return to thaw out its frozen cache by "incubating" it as it would a clutch of eggs. • Fledgling Northern Saw-whet Owls differ more in colour from adults than any other North American owl except the Boreal Owl. • The common name "saw-whet" comes from this owl's continuous, whistling call, which is thought to be similar to the sound of a large saw being sharpened or "whetted." Saw-whets usually only call in the dark of night. • The *brooksi* subspecies, which lives on Haida Gwaii (Queen Charlotte Islands), BC, is unique not only in structure but in the food it eats. Unlike other owls, it feeds mostly on marine invertebrates when the tide is low.

ID: sexes similar; small owl; no "ear" tufts; unbordered facial disc; light forehead streaking; wide, white eyebrows form *Y*; yellow eyes; dark bill; white-spotted, brown upperparts; vertical, rufous brown streaks on white underparts. *In flight:* short tail; coarsely brown-barred, pale buff underwings.

Size: *Male: L* 18–23 cm; *W* 43–55 cm. *Female: L* 18–21 cm; *W* 43–46 cm.

Habitat: *Breeding:* pure and mixed coniferous and deciduous forests; wooded city parks and ravines. *In migration* and *winter:* riparian mixed woodlands, suburban parks and patches of dense conifers.

Nesting: in an unlined woodpecker cavity or nest box; female incubates 5–6 white eggs for 27–29 days.

Feeding: mainly nocturnal; hunts from a perch for small rodents; also eats insects, amphibians, intertidal fish and small birds.

Voice: evenly spaced, whistled notes repeated about 100 times per minute: *whew-whew-whew-whew*, continuous and easily imitated.

Similar Species: *Boreal Owl* (p. 273): black-bordered, pale face; white-speckled forehead; pale bill. *Northern Hawk Owl* (p. 265): much longer tail; fine, horizontal barring on underparts; white spotting on black forehead.

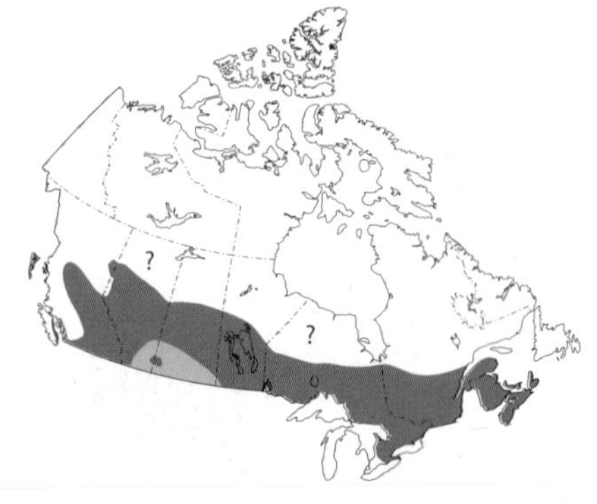

COMMON NIGHTHAWK

Chordeiles minor

To many people, the sound of a nighthawk is the sound of summer evenings. Both male and female nighthawks fly high above the ground, uttering nasal *peent* calls, and displaying males put on dramatic aerial courtship displays. From a great height, the male dives swiftly toward the ground, thrusting his wings forward at the bottom of the dive. Wind rushing through his primary feathers produces a hollow, booming sound that advertises his presence to rival males and prospective females. • Mild-mannered by day, the Common Nighthawk rests on the ground or perches lengthwise on a horizontal tree branch, its colour and shape blending perfectly into the texture of the bark. If approached, a nighthawk will remain motionless, relying on its cryptic plumage for protection. • Very much a bird of urban and suburban places, nighthawks mostly nest on gravel rooftops. • Like other members of the nightjar family, the Common Nighthawk is adapted for catching insects in midair—its bill is surrounded by feather shafts that help to funnel insects into its mouth. In recent decades, however, reduced food supplies have contributed to the decline of this species and several other aerial insectivores. Based on a 49 percent decline in Common Nighthawk numbers over three generations (10 years), the species was recently listed as threatened in Canada.

♂

♂

ID: cryptic, mottled light and dark plumage; large head; tiny bill; heavily barred underparts; prominent bristles around mouth. *Male:* white throat and undertail stripe. *Female:* buff throat. *In flight:* bold, white patches on long, pointed wings; shallowly forked, barred tail; erratic flight.
Size: *L* 22–25 cm; *W* 61 cm.
Habitat: *Breeding:* open and semi-open habitats such as grasslands, agricultural fields, marshes, forest openings, weedy meadows, barren islands, lakeshores, gravel pits, rooftops and abandoned parking lots. *In migration:* any area with large numbers of flying insects, often near water.
Nesting: in sand, gravel or rocks; occasionally on a building roof; lays eggs directly on the ground; mostly the female incubates 2 heavily speckled, white or pale grey eggs for 19–20 days.
Feeding: hunts primarily at dawn and dusk; catches insects in flight; takes mosquitoes, moths and other flying insects.
Voice: repeated, nasal *peent peent*; male makes a deep, hollow *vroom* with wings during courtship flight.
Similar Species: *Common Poorwill* (p. 276): smaller; generally paler; rounded wings without white patches; shorter, broader, rounded tail with white-tipped outer feathers. *Whip-poor-will* (p. 278) and *Chuck-will's-widow* (p. 277): less common; shorter, rounder wings; rounded tail; lack white wing patches.

COMMON POORWILL

Phalaenoptilus nuttallii

Back in 1946, the discovery of a Common Poorwill that appeared to be hibernating through winter in a rock crevice surprised the scientific community. Cold to the touch, it had no detectable breathing or heartbeat. Although most poorwills migrate to warmer climates for winter, a few choose to enter a short-term torpor in which their body temperature drops as low as 6°C and their oxygen intake is reduced by more than 90 percent. The 1946 discovery was not the first suggestion of this strange habit in poorwills: the Hopi call this bird *holchoko*, "the sleeping one," and in 1804, Meriwether Lewis found a mysterious "goatsucker...to be passing into the dormant state." • Poorwills are named after the sound of their calls, a distinctive *poor-will* or *poor-will-up*. These birds are more often heard than seen, and on warm evenings, they are sometimes seen roosting on backcountry roads. They are readily identified by their reddish eye shine. • In Canada, Common Poorwills are found in the Cypress Hills of southeastern Alberta and southwestern Saskatchewan and in the Okanagan region of British Columbia.

ID: cryptic, mottled light to dark brown plumage; very large head and eyes; tiny bill; pale throat; finely barred underparts; prominent bristles around mouth. *Male:* white tail corners. *Female:* buff corners on tail feathers. *In flight:* short, rounded wings; broad, rounded tail; moth- or batlike fluttering.

Size: *L* 19–22 cm; *W* 41–44 cm.

Habitat: *Breeding:* dry habitats, including sagebrush-covered hills, grasslands and open Douglas-fir and ponderosa pine woodlands.

Nesting: on bare ground, often by a fallen log or rock; pair incubates 2 white, sometimes pink-tinged eggs for 20–21 days; often raises 2 broods per season.

Feeding: sallies from a low perch or the ground, mostly at night, for large, flying insects.

Voice: mostly the male utters the familiar *poor-will* call at dusk and through the night; hiccuplike sound can be heard at the end of the phrase at close range.

Similar Species: *Common Nighthawk* (p. 275): larger; long, pointed wings with white patches near base of primaries; dark underpart barring continues onto longer, narrower, notched tail.

CHUCK-WILL'S-WIDOW

Caprimulgus carolinensis

During daylight, you would be lucky to see this perfectly camouflaged bird roosting on the furrowed bark of a horizontal tree limb or sitting among scattered leaves on the forest floor. Even during nesting, the Chuck-will's-widow is virtually invisible, incubating its eggs and raising its young on the forest floor. At dusk, however, it is easily detected as it calls its own name while patrolling the evening skies for flying insects. • This bird's core range is in the hot, humid southeastern U.S., so it is definitely pushing its luck at the northern limit of its range in extreme southern Ontario. Fewer than 10 known pairs of Chuck-will's-widows currently nest in Ontario, confined mostly to protected stands of scarce Carolinian forest and pure pine plantations. It is thought to be an irregular breeder in Canada—only one nest has ever been found, in 1977, at Point Pelee, Ontario. Elsewhere in eastern Canada, the Chuck-will's-widow is a vagrant in migration.

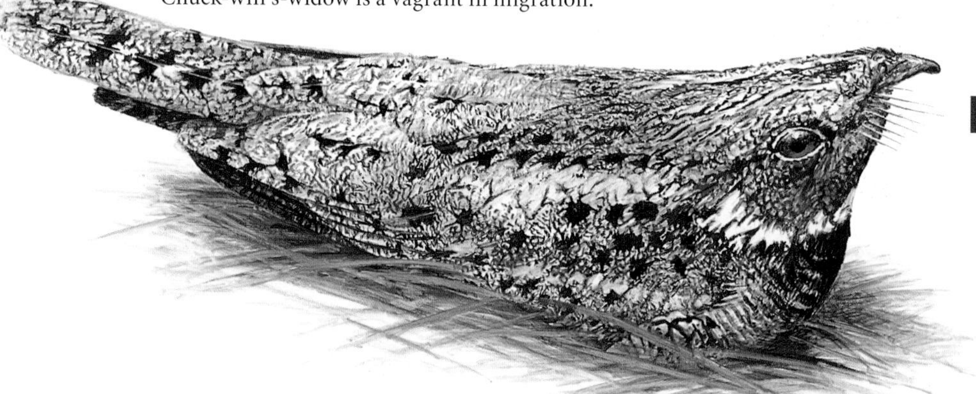

ID: mottled brown-and-buff body with overall reddish tinge; pale brown to buff throat; whitish "necklace"; dark breast. *Male:* white inner edges on outer tail feathers. *In flight:* long, rounded wings and tail.

Size: *L* 28–33 cm; *W* 62–65 cm.

Habitat: sandy, open oak and pine woodlands and savannahs.

Nesting: on bare ground; no nest is built; female incubates 2 heavily blotched, creamy white eggs for about 21 days.

Feeding: catches insects on the wing or by hawking; eats beetles, moths and other large, flying insects.

Voice: 3 loud, whistled notes often paraphrased as *chuck-will's-widow.*

Similar Species: *Whip-poor-will* (p. 278): smaller; white or buff "necklace" contrasts with black throat; greyer overall; male shows much more white in tail feathers; female's dark tail feathers bordered with buff on outer tips. *Common Nighthawk* (p. 275): forked tail; white wing patches; male has white throat; female has buff throat.

WHIP-POOR-WILL

Caprimulgus vociferus

This nighttime hunter fills the late evening and early-morning hours with repeated calls of its own name: *whip-poor-will.* One bird was documented singing more than 28,000 songs in one night! Although it is heard throughout many of the open woodlands within its range, this cryptic bird is rarely seen. Only occasionally is a Whip-poor-will observed roosting on an exposed tree branch or alongside a quiet road. • Ground-nesting Whip-poor-wills time their egg laying to the lunar cycle so that hatching occurs during the full moon, when their moth prey is readily captured. • The Whip-poor-will is a member of the nightjar, or "goatsucker," family. During the days of Aristotle, there was a widely believed superstition that these birds would suck milk from the udders of female goats, causing the goats to go blind! • Within days of hatching, young Whip-poor-wills scurry away from their nest in search of protective cover. Until the young are able to fly, about 20 days after hatching, the parents feed them regurgitated insects. • In 2009, Whip-poor-wills were listed as threatened in Canada because of a 30 percent decline on breeding bird surveys in the last decade. Like other aerial insectivores, habitat loss and changes to the insect prey base has seriously depleted this bird's populations.

ID: mottled, brown-grey overall with black flecking; reddish tinge on rounded wings; black throat. *Male:* white "necklace." *Female:* buff "necklace." *In flight:* long, rounded wings and tail; male has distinctive, white tail corners; female has buff-tipped tail corners.
Size: *L* 23–25 cm; *W* 41–49 cm.
Habitat: woodland edges and ungrazed woodlands with open glades; small woodlots in migration.

Nesting: on the ground, sometimes in dry leaf litter; mostly the female incubates 2 brown-and grey-blotched, dull white eggs for 19–21 days.
Feeding: almost entirely nocturnal; catches insects in flight, often high in the air; takes mosquitoes, blackflies, midges, beetles, flying ants, and other insects but prefers moths.
Voice: far-carrying, frequently repeated *whip-poor-will,* accented on last syllable; most often uttered at dusk and through the night.
Similar Species: *Common Nighthawk* (p. 275): less rufous in plumage; longer, pointed wings with conspicuous, white patch; squared tail with white bar.

BLACK SWIFT

Cypseloides niger

The Black Swift, North America's largest swift, is a fast-flying aerial species that is strongly localized in its mountain breeding range. Like other swifts, it performs virtually every activity, even mating, on the wing. • Often called "Cloud Swift," this swift forages high in the air when skies are clear, following insect abundance, but periods of rain or low-pressure centres bring feeding flocks closer to the ground where they can be seen. Black Swifts hunt insects on the wing for much of the day, but as the sun sets, they rocket back to the canyons to spend the night atop their nests. • Black Swifts are migratory, visiting Canada from late spring to early autumn. They occur in British Columbia and southwestern Alberta, where they breed singly or in small colonies. Only a handful of nests have been found in Canada because these birds attach their nests to steep, vertical walls behind sheets of cascading spray of inland waterfalls. • Swifts cast a characteristic boomerang silhouette in flight. They are shaped much like swallows, with tapering wings, small bills and sleek bodies, but they are only distantly related.

ID: sexes similar: black overall; slender, sleek body; very small legs; white-tipped feathers around forehead. *In flight:* long, tapering wings angle backward; broad and fanned tail, notched in male; underwings appear lighter; rapid wingbeats; often glides in broad circles.

Size: *L* 18 cm; *W* 46 cm.

Habitat: *Breeding:* forested habitats near rivers and streams; steep, usually wet cliffs with ledges for nesting. *Foraging* and *in migration:* high in the air on thermals and updrafts; lower over forests, marshes, canyons, grasslands and even cities.

Nesting: semicolonial; on a ledge on a cliff face, often near or behind a waterfall; nest is made of mosses; pair incubates 1 white egg for 24–27 days.

Feeding: captures flying insects, especially stoneflies, caddisflies and mayflies, on the wing.

Voice: high-pitched *plik-plik-plik-plik* near the nest site.

Similar Species: *Vaux's Swift* (p. 281): much smaller; paler overall; pale throat and rump; short, stubby tail; flaps more and glides less. *White-throated Swift* (p. 282): white underparts from "chin" to vent; white flank patches; white tips on secondaries. *Swallows* (pp. 342–48): lack boomerang-shaped flight silhouette.

CHIMNEY SWIFT

Chaetura pelagica

Resembling flying cigars, Chimney Swifts are more at home on the wing than perched. The frequent flyers of the bird world, they feed, drink, bathe, collect nest material and even mate while in flight! Indeed, they cannot perch like most birds and cling to cracks in cliffs, chimneys and the sides of buildings with their sharp claws, using their stiff tails for support. Swifts are distinguished by their rapid, "twinkling" flight style and their narrow wings, which are often stiffly bowed in the shape of a boomerang. • Roosting Chimney Swifts once relied on natural tree cavities for nest sites, but in recent times, almost all have adapted to living in brick chimneys and other artificial structures. • In the past, it was not uncommon to see hundreds of swifts entering into and roosting in large chimneys during fall migration. However, its numbers in Canada have declined significantly during the last 30 years, so much so that the species has recently been listed as threatened throughout its Canadian range. The cause of these widespread declines, which have also affected swallows, nightjars and other aerial insectivores, is unknown, but is likely related to reduced insect populations resulting from pesticide use and habitat loss.

ID: sexes similar; brown overall; slim body; long, thin, pointed wings; squared tail. *In flight:* rapid wingbeats; boomerang-shaped profile; erratic flight pattern.
Size: *L* 11–14 cm; *W* 30–32 cm.
Habitat: forages over cities and towns; roosts and nests in chimneys or other artificial structures; rarely nests in tree cavities or cliffs.
Nesting: usually nests singly; in the interior of a chimney or tree cavity, or in the attic of an abandoned building; half-saucer nest of dead twigs is attached to a vertical wall with saliva; pair incubates 4–5 white eggs for 19–21 days; often one or more helpers assist in feeding the young and incubating the eggs.
Feeding: insects captured in flight are swallowed whole during continuous flight.
Voice: rapid, chattering flight call: *chitter-chitter-chitter*; also a rapid series of *chip* notes.
Similar Species: *Swallows* (pp. 342–48): broader, shorter wings; smoother flight pattern; most have a forked or notched tail; lack boomerang-shaped flight silhouette. *Other swifts* (pp. 279, 281–82): not found in eastern or central Canada.

VAUX'S SWIFT

Chaetura vauxi

The migratory Vaux's Swift, the smallest swift in North America, arrives in British Columbia in late April or early May and departs in September. Like other swifts, it is often described as a "cigar with wings" and is rarely seen other than in the air. • While it is likely that the Vaux's Swift breeds throughout its summer range in the province, only a handful of actual nests have ever been found. Most of these have been in human-made structures or buildings, away from this species' typical natural sites in tree cavities. • Roost sites are very important for swifts, and unlike most birds, they cling vertically to surfaces with their small, strong claws. • On cold days, roosting swifts can lower their body temperature and remain in a state of torpor. • Like many of the aerial insect-foraging birds in Canada, Vaux's Swift populations are in decline. • This swift was named for William Sansom Vaux, an eminent mineralogist and friend of John Kirk Townsend, who first described the species in 1839 from specimens collected on the Columbia River in Oregon.

ID: sexes similar; brownish grey overall; paler throat and rump; small bill. *In flight:* long, pointed, swept-back wings; short, squared tail; fast flight with shallow, rapid wingbeats.
Size: *L* 13 cm; *W* 31 cm.
Habitat: *Breeding:* mixed woods and forests with rivers and tree cavities; also cities and towns. *In migration:* open sky over forests, rivers, marshes and lakes where flying insects are available; may roost in chimneys or crevices in buildings.
Nesting: open half-circle of loosely woven twigs is glued with sticky saliva to the inside of a hollow tree or chimney; pair incubates 6–7 white eggs for 18–19 days.
Feeding: diet almost entirely flying insects; often seen feeding high in the sky in the company of Violet-green Swallows.
Voice: several high-pitched *chip* notes followed by an insectlike trill in flight; makes booming sound with wings in midair and at nest, perhaps to discourage predators.
Similar Species: *White-throated Swift* (p. 282): larger; black with white areas, mainly on underparts; long, forked tail. *Black Swift* (p. 279): much larger; black overall; tail often forked; variable wingbeats; more commonly soars.

WHITE-THROATED SWIFT

Aeronautes saxatalis

This bird's genus name *Aeronautes* means "sky sailor," and this avian marvel certainly earns its wings as a true aeronaut. During its lifetime, the average White-throated Swift will likely fly more than 1.5 million kilometres, a distance equal to travelling around the world 40 times! Only brief, cliff-clinging rest periods, extremely bad weather and annual nesting duties keep this bird out of the air. • This scimitar-winged aerial insectivore lives in western North America and visits British Columbia every summer to breed at about two dozen widely scattered but traditional rocky cliff sites in the southern Interior. • The White-throated Swift is easily recognized by its loud, sharp, scraping call notes, black-and-white coloration and rapid, rather erratic flight. It is also considered the fastest North American swift, clocked at up to 320 kilometres per hour. At this speed, it certainly ranks among the fastest birds in the world. • White-throated Swifts reach the northern limit of their range in British Columbia.

ID: sexes similar; dark upperparts; white on throat tapers along underparts to undertail coverts; black flanks; white sides of rump. *In flight:* long, tapering, swept-back wings; long, narrow, slightly forked tail often held in a point.

Size: *L* 15–18 cm; *W* 38 cm.

Habitat: forages widely over a variety of open habitats. *Breeding:* high, vertical cliffs, canyons and rock bluffs in open country; rarely on tall buildings, freeway overpasses and under bridges.

Nesting: colonial; in a crevice or on a ledge; nest cup of plant material and some feathers is stuck together with gluelike saliva; pair incubates 4–5 creamy white eggs for 24 days.

Feeding: snatches flying insects in midair.

Voice: loud, shrill, descending *skee-jee-ee-ee-ee-ee-ee*.

Similar Species: *Vaux's Swift* (p. 281): smaller; brown upperparts; paler underparts; short, squared tail. *Black Swift* (p. 279): slightly larger; black overall; tail often appears notched. *Violet-green Swallow* (p. 344) and *Tree Swallow* (p. 343): smaller; blue, green or olive brown upperparts; all-white underparts.

RUBY-THROATED HUMMINGBIRD

Archilochus colubris

While hovering, the Ruby-throated Hummingbird probes its long, thin bill deep into the heart of a flower and uses its extendible tongue to lap up the nectar. Hummingbirds span the ecological gap between birds and bees—they feeds on the sweet, energy-rich nectar that flowers provide in exchange for pollinating the blossoms. • At full tilt, these nickel-weight speedsters are capable of speeds up to 100 kilometres per hour. When hovering or in flight, they beat their wings 55 to 75 times per second, and their hearts can beat up to 1200 times a minute. • This bird's bottlecap-sized nests are shingled with lichen for camouflage. Spiderwebs used in nest construction allow the nest to expand as the youngsters grow. • Ruby-throats migrate to Central America each winter, flying 800 nonstop kilometres over the Gulf of Mexico. • Except for a few accidental records of the Rufous Hummingbird, the Ruby-throat is the only hummingbird found east of the Rockies in Canada. There are over 300 hummingbird species in the world, all restricted to the Western Hemisphere.

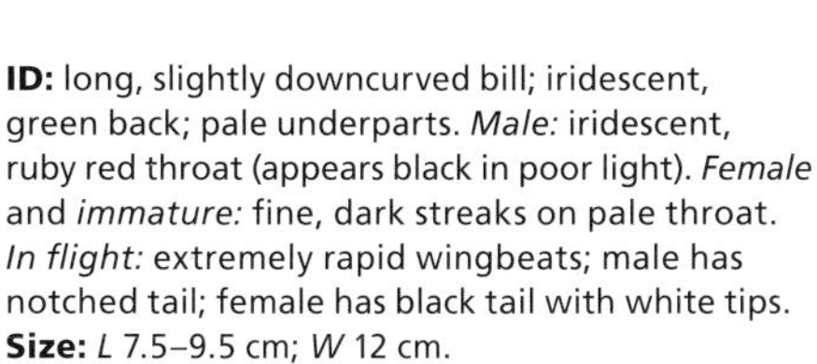

ID: long, slightly downcurved bill; iridescent, green back; pale underparts. *Male:* iridescent, ruby red throat (appears black in poor light). *Female* and *immature:* fine, dark streaks on pale throat. *In flight:* extremely rapid wingbeats; male has notched tail; female has black tail with white tips.
Size: *L* 7.5–9.5 cm; *W* 12 cm.
Habitat: open deciduous or mixed woodlands; also orchards, parks and gardens with flowers.
Nesting: on a horizontal tree limb; tiny, deep cup nest of lichen, moss and plant fibres is bound with spider silk; female incubates 2 white, pea-sized eggs for 13–16 days; often 2–3 broods per season.

Feeding: drinks nectar from flowers and sugar-sweetened water from feeders; also eats small insects and spiders and feeds at sapsucker wells.
Voice: soft buzzing of wings in flight; loud *chick* and other high squeaks in courtship or territorial battles.
Similar Species: *Rufous Hummingbird* (p. 287): male has rufous back and flanks; female has rufous patches on tail; rare east of Alberta.

BLACK-CHINNED HUMMINGBIRD

Archilochus alexandri

The Black-chinned Hummingbird is a western species that reaches the northern limit of its breeding range in south-central British Columbia. Males arrive in BC from their winter grounds in Mexico in late April or early May, and females arrive one to two weeks later. By mid-August, most of these hummers have left the province. • Throughout May, males court perched females with impressive, diving swoops that create a loud, whistling sound as wind rushes through the birds' wing and tail feathers. • A Black-chinned Hummingbird may eat up to three times its body weight in nectar in one day when the weather is cold. When insects are abundant, it can survive without nectar. • The Black-chinned Hummingbird is the western counterpart of the Ruby-throated Hummingbird of the East—the females of these two species are virtually indistinguishable in the field. • Deeply influenced by Greek mythology, naturalist and hummingbird taxonomist H.G.L. Reichenbach named several hummingbird genera after famous Greeks, including the notable poet Archilochus. The species name *alexandri* honours a doctor who collected the first specimens in Mexico.

ID: long, slender, slightly downcurved bill; small head; greyish green nape and crown; iridescent, green back; dirty white underparts. *Male:* black "chin"; gorget appears purple in good light. *Female* and *juvenile:* faintly streaked, white throat; white-tipped tail feathers.

Size: *L* 9.5 cm; *W* 11.5–12.5 cm.

Habitat: open ponderosa pine woods and deciduous copses with shrubs; riparian shrubs bordering marshes and sloughs; also urban and residential gardens and feeders.

Nesting: on a branch; female moulds a nest cup of plant down mixed with spiderwebs and insect cocoon fibres; female incubates 2 white eggs for 16 days.

Feeding: eats small insects, spiders and nectar; also takes sugar water from feeders.

Voice: buzzy and chipping alarm calls; male's soft, high-pitched, warbling courtship song is accompanied by wing buzzing.

Similar Species: *Anna's Hummingbird* (p. 285): range does not overlap. *Rufous Hummingbird* (p. 287) and *Calliope Hummingbird* (p. 286): female and immature have rufous or peach colour on sides and flanks.

ANNA'S HUMMINGBIRD

Calypte anna

Anna's Hummingbird is a western species that reaches the northern edge of its breeding range in British Columbia. Of the four regular hummingbird species that could arrive at a well-stocked sugarwater feeder in British Columbia, none can compare with the male Anna's Hummingbird for sartorial splendour. He may not have the rich reddish brown coat of a Rufous Hummingbird male, but a male Anna's stands out in the crowd with a rose red "bib" that covers his head and throat. • Once restricted as a nesting species to the Pacific slope of northern Baja California and southern California, Anna's Hummingbird has been extending its breeding range since the 1930s, an expansion partially attributed to feeders. The species was first observed in BC in the mid-1940s and first observed nesting on Vancouver Island in 1958. It has since continued to increase and expand its nesting range into interior locations. What is really surprising is that this hummingbird still maintains its February nesting cycle even in northern locations and raises two families a year on southern Vancouver Island.

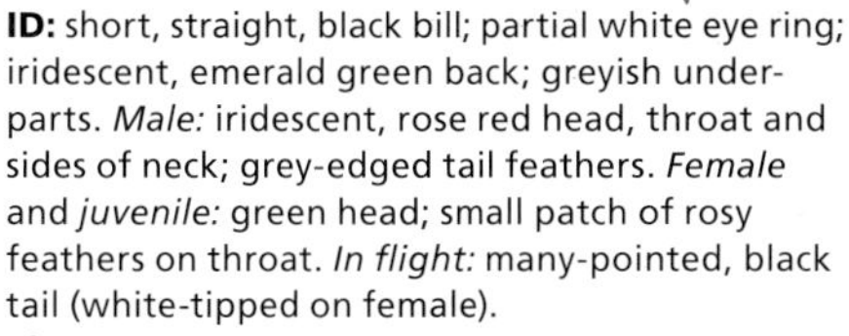

ID: short, straight, black bill; partial white eye ring; iridescent, emerald green back; greyish underparts. *Male:* iridescent, rose red head, throat and sides of neck; grey-edged tail feathers. *Female* and *juvenile:* green head; small patch of rosy feathers on throat. *In flight:* many-pointed, black tail (white-tipped on female).

Size: *L* 8–10 cm; *W* 12.5–14 cm.

Habitat: mostly urban and residential gardens and parks; also open, mixed woods.

Nesting: straddled on the branch of a tree or shrub; cup nest of plant down and spiderwebs is lined with downy material; female incubates 2 white eggs for 16–17 days.

Feeding: gleans small insects and spiders from vegetation; sips flower nectar and sugar water; males are highly territorial to other males at feeders.

Voice: high, short call note; excited, chattering chase call; male's complex songs include a varied series of high, scratchy, dry notes from a perch and sharp squeaks at the end of a diving display.

Similar Species: male is distinctive. *Rufous Hummingbird* (p. 287): range differs; female has rufous flanks. *Black-chinned Hummingbird* (p. 284): female is smaller with whiter underparts and unmarked or finely streaked throat.

CALLIOPE HUMMINGBIRD

Stellula calliope

Glistening in the slanting rays of the early-morning sun, the iridescent, streaked gorget of the male Calliope Hummingbird suggests nectar dripping from the bird's needlelike bill. Although it feeds among clusters of wildflowers, breeding male Calliopes often perch high atop shrubs or small trees to flash their unusual, pinkish red gorgets and make dazzling, high-speed dips and dives to entice the females to mate. • The Calliope Hummingbird is not only North America's smallest breeding bird, but it is also the world's smallest long-distance migrant, travelling up to 8000 kilometres annually to and from its winter home in Mexico. Calliopes breed from south-central British Columbia and southwestern Alberta, south to Baja California and east to Colorado. • To attract pollinators such as insects and hummingbirds, plants produce colourful flowers with sweet, energy-rich nectar. As hummingbirds visit flowers for the nectar, they spread pollen from one flower to another like bees, fertilizing the plants. • Novice birders often ponder the pronunciation of this bird's name. It is generally accepted as *kuh-LYE-o-pee.*

ID: iridescent, green upperparts; square tail with greenish base; short, green bill. *Male:* light green flanks; buffy grey underparts; whitish breast and belly; narrow, pinkish red gorget streaks extend down white throat from bill. *Female:* peach-washed flanks; white underparts; dark green throat spots. *In flight:* blackish tail (white-tipped on female).

Size: *L* 8 cm; *W* 10–11.5 cm.

Habitat: *Breeding:* open forests, woodland edges, rangeland copses, shrublands, gardens and orchards.

Nesting: in a tree or tall shrub; tiny cup nest is made of plant down, mosses, lichens and spiderwebs; nest may be reused; female incubates 2 white eggs for 15–17 days.

Feeding: eats flower nectar, small flying insects and sugar water at feeders; often feeds at sapsucker wells.

Voice: high-pitched *chip* calls; male utters high-pitched, chattering *tsew* notes during relatively short courtship display.

Similar Species: *Rufous Hummingbird* (p. 287): larger; rufous tail base; at rest, wing tips are shorter than tail tip. *Black-chinned Hummingbird* (p. 284) and *Anna's Hummingbird* (p. 285): females lack peach-coloured flanks.

RUFOUS HUMMINGBIRD

Selasphorus rufus

The male Rufous Hummingbird, like most of its kin, is a delicate avian jewel, but its beauty hides a relentlessly aggressive nature. Sit patiently in a flower-filled meadow or alongside a sugarwater feeder, and you'll soon notice the territoriality and feistiness of most hummingbirds—constantly buzzing past one another and chasing rivals for considerable distances. • Hummingbirds have to be "power smart," because they are raging metabolic furnaces. The males must defend their feeding territories as insurance policies for their high-energy needs. • The tiny Rufous Hummingbird may spend part of its year in tropical regions, but it migrates each summer as far north as southern Alaska. Its spring journey coincides with the flowering times of plants, and in Canada, salmonberry and currant flowers are important food sources. • At the onset of nest-building and egg-laying, males abandon the females and head inland or upslope to higher elevations to take advantage of early-blooming flowers. Eventually, the females and immatures follow, with the entire population ultimately moving southward through late-blooming mountain meadows and brushlands all the way to their wintering range in Mexico.

ID: long, straight, black bill. *Male:* orange-brown back, tail and flanks; iridescent, scarlet throat; white breast and belly; green crown; rufous tail. *Female:* green back; red-spotted throat; rufous sides and flanks; white underparts; rufous base and white tips to most tail feathers. *In flight:* multi-pointed tail.

Size: *L* 8–9 cm; *W* 11.5 cm.

Habitat: *Breeding:* mixed forest edges, residential gardens, riparian thickets and orchards. *In migration:* nearly any habitat with abundant flowers including gardens, shrubby edges of coniferous and deciduous forests, clearcuts and burns, brushy slopes and alpine meadows.

Nesting: saddled on the branch of a tree or shrub; tiny cup nest is lined with downy plant materials and covered with lichens or mosses; female incubates 2 white eggs for 15–17 days.

Feeding: darts and hovers to catch insects and sip flower nectar; readily visits sugarwater feeders.

Voice: call is a low *chewp chewp*; confrontation call is a rapid, exuberant *ZEE-chuppity-chup!*

Similar Species: *Calliope Hummingbird* (p. 286): female has pale, peach-coloured sides. *Anna's Hummingbird* (p. 285): green back and flanks; male is slightly larger with rosy crown and gorget.

BELTED KINGFISHER

Megaceryle alcyon

The Belted Kingfisher is easily identified by its large head and bill, shaggy blue crest and distinctive, rattling call. Never far from a river, lake or fish-bearing waterway, this bird is generally found perched on a bare branch that extends over a productive patch of water. • As its name suggests, the kingfisher preys primarily on fish. With a precise headfirst dive, this bird can catch fish up to 60 centimetres beneath the water's surface or snag a frog immersed in shallow water. • With an extra reddish band across her belly, the female Belted Kingfisher is more colourful than her mate. • When nesting, pairs take turns excavating a long tunnel into an earth or sandstone bank using their bills to chip away at the soil and then kicking out loose material with their tiny feet. • Most kingfishers are migratory, but a few, often males, may overwinter in southern Canada where open water can be found. Kingfishers are found around the world, but the Belted Kingfisher is the only one found in Canada. • Alcyon (Halcyone) was the daughter of the wind god in Greek mythology; she and her husband were transformed into kingfishers.

ID: shaggy crest; bluish head and back; large, heavy bill; blue-grey breast band; white collar and underparts; short legs; small, white spot in front of eyes; female has rust-coloured "belt" (occasionally incomplete). *In flight:* large head; short tail; flies low over water; often hovers.
Size: *L* 28–36 cm; *W* 51 cm.
Habitat: *Breeding:* gravel pits, road cuts and vertical banks along rivers, lakes, marshes and seashores. *In migration* and *winter:* lakes, sloughs, oxbows, reservoirs, rivers, brackish lagoons and sheltered marine waters.
Nesting: near water; in a cavity at the end of an excavated burrow 2–4 m long; pair incubates 6–8 white eggs for 22–24 days.
Feeding: dives for small fishes; also takes frogs, salamanders and aquatic invertebrates such as tadpoles; may eat small rodents and birds.
Voice: fast, repetitive, rattling cackle resembles a heavy teacup shaking on a saucer.
Similar Species: *Steller's Jay* (p. 333) and *Blue Jay* (p. 334): more intense blue colour; smaller bills and heads; backswept crests.

LEWIS'S WOODPECKER

Melanerpes lewis

This odd-looking western woodpecker is unique not only for its green-and-pink coloration but also because it does much of its foraging as flycatchers do, catching insects on the wing. In flight, the Lewis's Woodpecker is often mistaken for a crow because of its slow wingbeats and buoyant flight. It is also one of the few woodpeckers that is frequently seen perched on wires. • The Lewis's Woodpecker prefers open forest and deciduous bottomlands throughout its range in Canada, but in winter, it is restricted to residential areas and orchards. It often forages in recently burned areas, and in open forests in general. Breeding from the Interior of British Columbia eastward to southwestern Alberta, it occurs occasionally as a vagrant in areas east of its breeding range. • Competition with European Starlings and loss of snag habitat through modern wildfire suppression has greatly diminished this woodpecker's numbers. It is classified as a species of special concern, with only an estimated 600 pairs breeding in Canada. • The Lewis's Woodpecker is named for explorer Meriwether Lewis. Although not a formal naturalist, Lewis recorded a great many concise and original natural history observations.

ID: sexes similar; medium-sized woodpecker; sharp, stout bill; dark greenish black upperparts; pinkish belly; dark red face; light grey breast and collar; glossy, dark undertail coverts. *Juvenile:* dark brown head; mottled brown underparts; lacks red face and grey collar. *In flight:* appears mostly dark, especially from above; long, dark wings and tail; crowlike flight with flaps and glides.

Size: *L* 28 cm; *W* 53 cm.

Habitat: *Breeding:* open ponderosa pine and Douglas-fir woodlands; riparian black cottonwood stands with snags for nesting. *In migration* and *winter:* open woodlands, orchards, urban and residential areas with shrubs and farmland.

Nesting: male selects a stub of a live or dead tree in which to excavate a cavity; pair incubates 6–7 white eggs for 13–14 days.

Feeding: sallies from a perch to catch flying insects in spring and summer; takes nuts, pine seeds and berries locally in winter.

Voice: quiet away from the nest site; utters a wheezy contact call and harsh *churr* notes.

Similar Species: no other woodpecker is dark green; all other woodpeckers fly with pronounced undulation.

RED-HEADED WOODPECKER

Melanerpes erythrocephalus

Once common throughout its range, Red-headed Woodpecker numbers have declined dramatically over the past century as a result of intensive agricultural practices, pesticide use and loss of nesting habitat. The introduction of the European Starling, which outcompetes many native woodpecker species for nesting cavities, and the removal of dead standing trees in woodlots have also been detrimental to the species. Red-heads sometimes use utility poles treated with creosote for nesting—with an often deadly outcome for the young. • Red-headed Woodpeckers are opportunistic foragers, consuming a variety of adult and larval invertebrates, seeds, nuts, fruit and even young birds, eggs or small mammals. They are not only expert at flycatching but are also one of the few woodpecker species that stores food. Although these woodpeckers rarely overwinter in southern Manitoba or Ontario, bird feeders that offer nuts, sunflower seeds, corn and suet will occasionally entice a few to remain through the chilly weather. • With significant declines throughout its whole range, the Red-headed Woodpecker is now classified as threatened in Canada.

juvenile

ID: sexes similar; crimson red head, "chin," throat and "bib"; black back, wings and tail; white rump and underparts; large, white wing patches. *Juvenile:* brownish grey head, breast, back, wings and tail; white rump, underparts and wing patches. *In flight:* white on secondaries and rump contrasts with black back, wings and tail.

Size: *L* 21–24 cm; *W* 43 cm.

Habitat: open deciduous forests (especially oak woodlands), urban parks, river edges and open areas with scattered trees.

Nesting: in a natural hole or in a cavity excavated in a dead tree or wooden pole; nest cavity is lined with wood chips; pair incubates 4–7 white eggs for about 14 days.

Feeding: omnivorous diet includes insects, worms, grubs, nuts, seeds and fruit; may eat mice, young birds and eggs.

Voice: loud series of *kweer* or *queer* notes; occasionally a chattering *kerr-r-ruck*.

Similar Species: adult is distinctive; large, square wing patch near speculum and white rump distinguishes juvenile from other young woodpeckers. *Red-bellied Woodpecker* (p. 291): whitish face and underparts; black-and-white-barred back. *Yellow-bellied Sapsucker* (p. 293): juvenile has white patch on hind wing.

RED-BELLIED WOODPECKER

Melanerpes carolinus

Deep in southwestern Ontario, remnant stands of Carolinian forest serve as a stronghold for the Red-bellied Woodpecker. Widely distributed throughout the eastern U.S. and west to the Great Plains, this bird reaches the extreme limit of its North American range in these forests. Although it is found year-round, its numbers fluctuate depending on the availability of habitat and the mildness of winter conditions. In recent years, mild winter weather has enabled the Red-bellied Woodpecker to increase its numbers in Ontario. Aided by maturing forests in the northeast, an increase in backyard feeders and the maturation of urban tree plantings, the species expanded its range northward and westward during the latter half of the 20th century, and the species is now regularly observed as a vagrant in southern extremes of Saskatchewan, Manitoba, Québec, New Brunswick and Nova Scotia. • One of the earlier nesting birds, Red-bellies' eggs usually hatch by mid-April. • These birds often issue noisy, rolling *churr* calls as they poke about wooded landscapes in search of food. Red-bellied Woodpeckers seldom extract insects from trees or wood, instead foraging opportunistically on fruit, nuts, seeds, flying insects and even small mammals or young birds.

ID: red forehead, crown and nape (female has red on nape only); greyish white face; black and white barring on back; white patches on rump and topside base of primaries; reddish tinge on yellowish buff belly. *Juvenile:* dark grey crown; streaked breast. *In flight:* short, broad wings; narrow, white wing patch contrasts with barred upperparts.
Size: *L* 23–27 cm; *W* 41 cm.
Habitat: mature deciduous woodlands, primarily Carolinian forest; occasionally in wooded residential areas.
Nesting: in a natural cavity or the abandoned cavity of another woodpecker; female selects one of several nest sites excavated by male; pair incubates 4–5 white eggs for 12–14 days.

Feeding: eats mostly insects, seeds, nuts and fruit; also tree sap, small amphibians, bird eggs or small fish; forages in trees, on the ground or occasionally on the wing.
Voice: soft, rolling *churr* call; drums in second-long bursts.
Similar Species: *Northern Flicker* (p. 301): grey crown; dark barring on brown back; black "bib"; large, dark spots on underparts. *Red-headed Woodpecker* (p. 290): all-red head; unbarred, black back and wings; white patch on trailing edge of wings.

WILLIAMSON'S SAPSUCKER

Sphyrapicus thyroideus

Male and female Williamson's Sapsuckers are so radically different in appearance that naturalists long believed them to be separate species. Twenty-one years after a female Williamson's specimen was misidentified as a male "Black-breasted Woodpecker," a nesting pair of Williamson's was discovered and the mistake was corrected! The male is boldly patterned in black and white, whereas the female, with her barred back and wings, brown head and dark breast, resembles the juveniles of other sapsuckers. • Williamson's Sapsuckers often return to the same nest tree each year and excavate a new cavity. Some trees may contain 30 or more old nests. • Like other sapsuckers, Williamson's drills holes, called "sap wells," into tree trunks. • Within Canada, this sapsucker breeds only in the Interior of south-central British Columbia. Often found in mature larch forests, which are being extensively harvested, and with an estimated population of only 430 breeding individuals, Williamson's Sapsucker has been listed as endangered in Canada. It has been recorded as a vagrant eastward to Saskatchewan. • Robert S. Williamson, after whom this sapsucker is named, was a topographical engineer and United States Army lieutenant who led the Pacific Railroad Survey across Oregon during the mid-1800s.

ID: medium-sized woodpecker; white rump; yellowish belly. *Male:* generally black; red "chin"; white "moustache" and eyebrow; broad, white wing patch; white-streaked , blackish brown flanks; black tail. *Female:* brown head; dark brown and whitish bars on upperparts; blackish "bib" and upper breast; dark-barred flanks. *In flight:* white-barred, dark underwings; white rump.

Size: *L* 23 cm; *W* 44 cm.

Habitat: *Breeding:* mixed, open coniferous forests, especially Douglas-fir and western larch; also trembling aspen stands.

Nesting: in a tree cavity; tree and occasionally nest cavity often reused in consecutive years; cavity is lined with wood chips; pair incubates 5–6 white eggs for 12–14 days.

Feeding: eats sap from wells drilled in trees; also gleans for ants and other insects.

Voice: loud, shrill *chur-cheeur-cheeur*; initially fast drumming slows, with lengthening pauses between taps.

Similar Species: *Red-naped Sapsucker* (p. 294): red forehead and crown; pale yellow belly; white shoulder patches. *Northern Flicker* (p. 301): red underwings; large, dark spots on underparts. *Yellow-bellied Sapsucker* (p. 293): red forecrown; wider white facial stripes; different ranges.

YELLOW-BELLIED SAPSUCKER

Sphyrapicus varius

Widely known for their amusing name, Yellow-bellied Sapsuckers are conspicuous in spring, when the males hammer out their Morse code–like territorial rituals on trees, metal poles, buildings and almost any object that generates plenty of volume. The drumming consists of an "introductory roll" of several rapidly repeated taps followed by a brief pause, and then several much slower and irregularly spaced taps. This sapsucker will often reply to and eventually fly over to investigate loud tapping sequences of a similar nature made by humans. • Lines of parallel wells freshly drilled in tree bark is a sure sign that a sapsucker is in the neighbourhood. Throughout spring and summer, Yellow-bellies make routine visits to each well site to feed on the sweet, sticky sap that pools in the wells and the insects that the sap attracts. Sapsuckers don't actually suck sap; they lap it up with a fringed tongue that resembles a paintbrush. • Like all woodpeckers, a stiff tail helps brace the birds against tree trunks when foraging or drumming. The catlike calls of sapsuckers are also distinctive. • Unlike most woodpeckers, Yellow-bellied Sapsuckers are highly migratory, spending their winters as far south as Panama.

ID: medium-sized woodpecker; black and white barring on back; conspicuous, white shoulder patch; red forecrown; black and white facial stripes; broad, black "bib"; yellow wash on lower breast and belly. *Male:* red "chin" and throat. *Female:* white "chin" and throat. *Juvenile:* brownish overall; washed-out head markings.
Size: *L* 20–23 cm; *W* 35–40 cm.
Habitat: *Breeding:* mainly trembling aspen and balsam poplar forests; also mixed forests with white spruce.
Nesting: in a cavity, usually in a live aspen, poplar or birch tree with heart rot; often reuses nest trees; cavity is lined with wood chips; pair incubates 5–6 white eggs for 12–13 days.
Feeding: drills neat rows of "sap wells" in live trees to collect sap and trapped insects; also eats wild fruit; flycatches for insects.

Voice: catlike *meow* or nasal squeal; territorial/courtship hammering has a Morse-code quality and rhythm; hammering is rapid, then slows near the end.
Similar Species: *Red-naped Sapsucker* (p. 294): small, red patch on back of head. *Williamson's Sapsucker* (p. 292) and *Red-breasted Sapsucker* (p. 295): ranges do not overlap in BC. *Red-headed Woodpecker* (p. 290): juvenile lacks white wing patch.

RED-NAPED SAPSUCKER

Sphyrapicus nuchalis

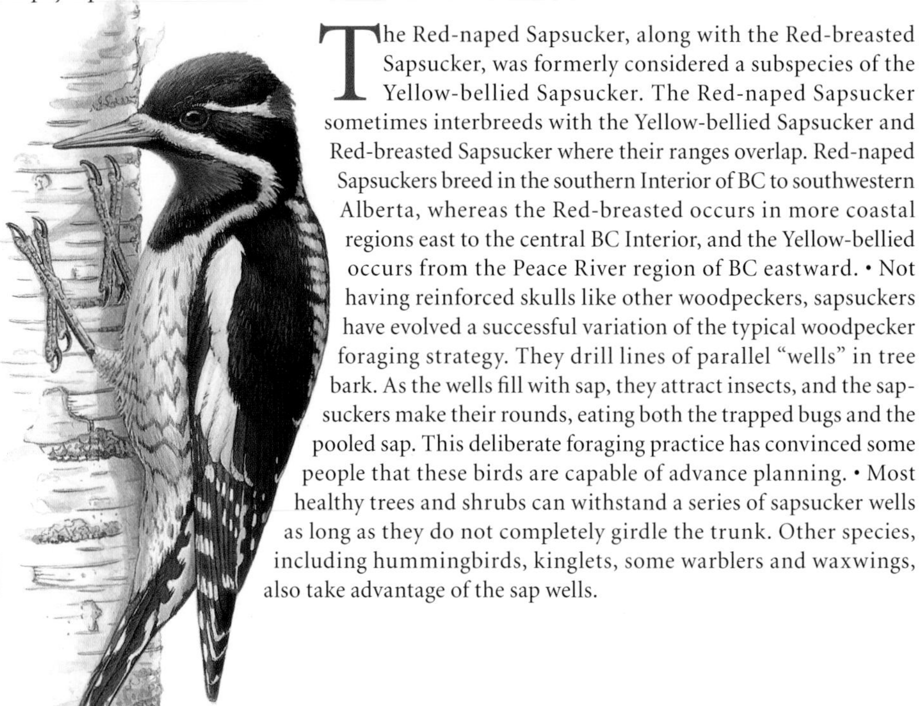

The Red-naped Sapsucker, along with the Red-breasted Sapsucker, was formerly considered a subspecies of the Yellow-bellied Sapsucker. The Red-naped Sapsucker sometimes interbreeds with the Yellow-bellied Sapsucker and Red-breasted Sapsucker where their ranges overlap. Red-naped Sapsuckers breed in the southern Interior of BC to southwestern Alberta, whereas the Red-breasted occurs in more coastal regions east to the central BC Interior, and the Yellow-bellied occurs from the Peace River region of BC eastward. • Not having reinforced skulls like other woodpeckers, sapsuckers have evolved a successful variation of the typical woodpecker foraging strategy. They drill lines of parallel "wells" in tree bark. As the wells fill with sap, they attract insects, and the sapsuckers make their rounds, eating both the trapped bugs and the pooled sap. This deliberate foraging practice has convinced some people that these birds are capable of advance planning. • Most healthy trees and shrubs can withstand a series of sapsucker wells as long as they do not completely girdle the trunk. Other species, including hummingbirds, kinglets, some warblers and waxwings, also take advantage of the sap wells.

ID: medium-sized woodpecker; striped, black-and-white head; red forehead and nape spot; messy, black-and-white back; white wing stripe; brown chevrons on pale underparts; yellow-washed upper breast. *Male:* all-red throat. *Female:* red throat with white "chin." *Juvenile:* mottled brown; prominent, white wing stripe. *In flight:* white upperwing patch; short, rapid flights.

Size: *L* 22 cm; *W* 41 cm.

Habitat: *Breeding:* mostly trembling aspen and black cottonwood forests and willow and birch thickets; also forages over clear-cuts and burns.

Nesting: excavates a cavity in a live birch, cottonwood or aspen, often near water; pair incubates 4–5 white eggs for 12–13 days.

Feeding: eats sap and insects in sap wells; takes insects in flight; eats fruit; also feeds on cambium bark layer.

Voice: catlike *neeah* call; also a loud *kweear* on its territory and a series of wavering *wika* notes; several fast drumming taps, then slower single and double taps.

Similar Species: *Red-breasted Sapsucker* (p. 295): all-red head and breast; generally blacker back. *Williamson's Sapsucker* (p. 292): male has all-dark back and red only on "chin"; female has black breast and yellow belly.

RED-BREASTED SAPSUCKER

Sphyrapicus ruber

Red-breasted Sapsuckers, together with Red-naped Sapsuckers and Yellow-bellied Sapsuckers, form what ornithologists consider a superspecies. These three species have largely separate distributions, but until 1983, they were collectively considered subspecies of the "Yellow-bellied Sapsucker." In Canada, the Red-breasted Sapsucker is found only in British Columbia. • Of the four sapsucker species in BC, the Red-breasted is the most widely distributed, the least migratory and most frequently seen in winter. It is known to interbreed with the Red-naped Sapsucker in a few spots within Interior valleys where their ranges overlap. • Bold in appearance, yet shy and reclusive by nature, the Red-breasted Sapsucker is the only sapsucker found in dense coniferous forests, especially coastal rainforests. • Like most woodpeckers, sapsuckers feed their young almost constantly, returning to the nest with food every five to ten minutes. During nesting, insects are taken more regularly and fed to the nestlings. • The wells that Red-breasted Sapsuckers drill are used extensively by Rufous Hummingbirds and Calliope Hummingbirds. • Birders from the East who meet a Red-breasted Sapsucker in British Columbia are often reminded of the similar-looking Red-headed Woodpecker, which is only found east of the Rocky Mountains.

ID: sexes similar; medium-sized woodpecker; red head, "chin," throat and breast; small, white patch above bill; darkly streaked, yellow-washed, pale belly and sides; black-and-white back. *Juvenile:* mottled brown with prominent white wing stripe. *In flight:* white patch on black wings.

Size: *L* 21 cm; *W* 41 cm.

Habitat: *Coast:* mainly dense mixed and coniferous mature and old-growth forests for breeding; mixed woodlands (especially red alder) and residential parks and woodlots in winter. *Interior:* mixed deciduous and coniferous forests, black cottonwood bottomlands and riparian woodlands for breeding; urban parklands, orchards and gardens in winter.

Nesting: excavates a cavity in a live tree; occasionally uses a snag; pair incubates 3–6 white eggs for 12–13 days.

Feeding: eats primarily tree sap; also takes insects, especially ants, and fruit; occasionally flycatches for insects.

Voice: loud, hoarse, descending *queeoh* spring call is often repeated several times; also a softer *cheer*; drumming starts with several fast taps, then becomes an irregular series of single and double taps.

Similar Species: *Red-naped Sapsucker* (p. 294) and *Yellow-bellied Sapsucker* (p. 293): little range overlap; Red-naped is restricted mainly to southern Interior of BC and Yellow-bellied to the Peace River region of BC and eastward; differences in pattern of black, red and white on head, throat and "bib"; usually more white on back. *Williamson's Sapsucker* (p. 292): white stripes on black head; black breast.

DOWNY WOODPECKER

Picoides pubescens

The Downy Woodpecker is the smallest, most widely distributed and most familiar North American woodpecker and the most likely woodpecker to be found in neighbourhood ravines and wooded city parks. It closely resembles the less common Hairy Woodpecker but is much smaller, with a comparatively tiny bill and black spots on the white outer tail feathers (as opposed to all-white outer tail feathers on the Hairy). It also has a softer call and a gentler tap than its larger cousin. Only in British Columbia are Hairy Woodpeckers more common than Downies, though Downies are still the most common woodpeckers at backyard suet feeders. • Male and female Downies may maintain a loose territory year-round, with each sex dividing up the resources in parts of the trees on which they feed. • The Downy's small bill is amazingly effective at poking into tiny crevices, even stems on flowering plants, and it uses its long, barbed tongue to extract dormant invertebrates and wood-boring grubs that larger woodpeckers might miss. • Woodpeckers have feathered nostrils that filter out the sawdust produced by their excavations and long, barbed tongues that can reach far into crevices to extract grubs and other morsels.

ID: small, sparrow-sized woodpecker; white head with black crown, eye stripe and "moustache" stripe; short, stubby bill; white back; black wings with bold, white spots or bars; black-spotted, white outer tail feathers. *Male:* red patch on back of head. *Female:* black patch on back of head.
Size: *L* 15–18 cm; *W* 30 cm.
Habitat: open deciduous or mixed forests, parks, orchards and riparian woodlots; residential yards, urban parks and orchards; avoids extensive mixed or coniferous stands.

Nesting: in a dying or decaying tree trunk or limb; pair excavates a cavity with entrance hole 2.5 cm in diameter; pair incubates 3–6 white eggs for 11–13 days.
Feeding: forages mostly for spiders and insects such as beetles and ants; also eats berries and seeds; attracted to feeders with sunflower seeds and suet in winter.
Voice: long, descending trill or "whinny"; calls are a sharp *pik*, *ki-ki-ki* or whiny *queek queek*; drumming is shorter but more frequent than that of Hairy Woodpecker.
Similar Species: *Hairy Woodpecker* (p. 297): larger; bill is as long as head is wide; wholly white outer tail feathers; louder, sharper call.

HAIRY WOODPECKER

Picoides villosus

A second or third look is often required to confirm the identity of the Hairy Woodpecker because it is often confused with its smaller cousin, the Downy Woodpecker. However, Hairies forage on the larger trunks and limbs of trees, and at feeders, Hairies don't stay long, while Downies tend to linger. As well, vocalizations differ between the two species. • Like other members of the woodpecker family, the Hairy Woodpecker has evolved features that help to cushion the repeated shocks of a lifetime of hammering. These include a sturdy bill, strong neck muscles, a flexible, reinforced skull and a brain that is tightly packed in its protective cranium. Other woodpecker adaptations include a stiff tail to prop the bird up while it hammers on tree trunks and a remarkably long, sticky, barb-tipped tongue for extracting insects from small, tight spaces. The tongue of some woodpeckers can be more than four times the length of the bill—made possible by twin structures that wrap around the perimeter of the skull and store the tongue in much the same way that a measuring tape is stored in its case. The long, manoeuverable tip is sticky with saliva and finely barbed to help seize reluctant wood-boring insects. • Across their range in North America, the plumage of Hairies varies significantly, with at least 17 known subspecies.

ID: medium-sized woodpecker; black-and-white head with black crown; white eyebrow and "moustache" stripe; thick bill is about as long as head is wide; white back; black wings with bold, white spots or bars; white outer tail feathers. *Male:* red patch on back of head. *Female:* no red patch.
Size: *L* 19–24 cm; *W* 39 cm.
Habitat: mature, mixed coniferous and hardwood forests; usually in denser, more mature stands with conifers; also in fire-burned woods and parks with larger trees.
Nesting: pair excavates a cavity in a live or decaying tree trunk or limb; cavity is lined with wood chips; pair incubates 4–5 white eggs for 12–14 days.
Feeding: chips, drills or debarks trees mainly for bark- and wood-boring beetles and their larvae; also eats berries; visits suet feeders.

Voice: loud, sharp *peek* or "whinny"; series of notes lasting about 2 seconds; drumming is longer but less frequent than that of Downy Woodpecker.
Similar Species: *Downy Woodpecker* (p. 296): smaller; tiny bill; black spots on white outer tail feathers; usually found on smaller branches and twigs. *American Three-toed Woodpecker* (p. 299): black and white barring on sides and back (mostly white patch on interior birds); 3-toed feet; yellow crown patch on male and juvenile.

WHITE-HEADED WOODPECKER

Picoides albolarvatus

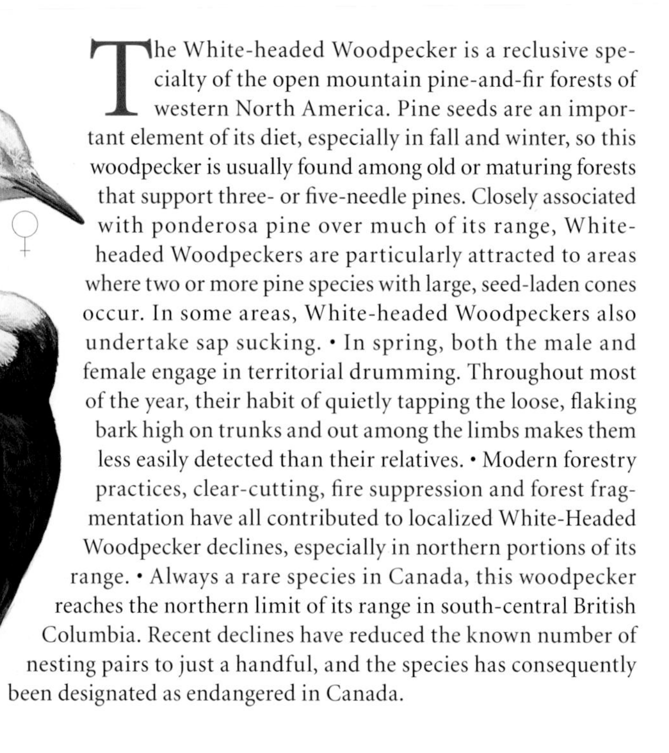

The White-headed Woodpecker is a reclusive specialty of the open mountain pine-and-fir forests of western North America. Pine seeds are an important element of its diet, especially in fall and winter, so this woodpecker is usually found among old or maturing forests that support three- or five-needle pines. Closely associated with ponderosa pine over much of its range, White-headed Woodpeckers are particularly attracted to areas where two or more pine species with large, seed-laden cones occur. In some areas, White-headed Woodpeckers also undertake sap sucking. • In spring, both the male and female engage in territorial drumming. Throughout most of the year, their habit of quietly tapping the loose, flaking bark high on trunks and out among the limbs makes them less easily detected than their relatives. • Modern forestry practices, clear-cutting, fire suppression and forest fragmentation have all contributed to localized White-Headed Woodpecker declines, especially in northern portions of its range. • Always a rare species in Canada, this woodpecker reaches the northern limit of its range in south-central British Columbia. Recent declines have reduced the known number of nesting pairs to just a handful, and the species has consequently been designated as endangered in Canada.

ID: all-black plumage except for white head, throat and wing patch; dark eyes; *Male:* red patch at back of head. *Female:* no red patch. *Juvenile:* red crown patch; duller black plumage.
Size: *L* 21–23 cm; *W* 38–40 cm.
Habitat: open mountain pine forests, especially ponderosa pine; may also forage among red and white fir, incense-cedar, Douglas-fir and sequoia.

Nesting: in a standing dead tree, typically a conifer; pairs take turns excavating nest cavity; may create new holes in the same tree for many years; pair incubates 4–5 white eggs for about 14 days.
Feeding: pries open cones for seeds; also chips large flakes of bark for insects; gleans from trunks and needle clusters; occasionally eats sap.
Voice: call is a sharp, rattling *tea-deek* or *tea-dee-deek*, easily overlooked as voice of Hairy Woodpecker.
Similar Species: no other woodpecker in Canada has a white head.

AMERICAN THREE-TOED WOODPECKER

Picoides dorsalis

Evidence of its foraging activities often betrays the American Three-toed Woodpecker long before it is seen. In seeking insects and their eggs, this resourceful bird flakes off bits of bark from old and dying conifers, exposing the bare trunk. After months or years, the tree takes on a distinctive reddish color and is skirted with bark chips. A feeding bird's tapping and flaking noises can often be heard at short range, and its contact calls are often quite musical. • While foraging, Three-toed Woodpeckers often listen for grubs moving beneath the bark and in the wood. These woodpeckers help to control infestations of the spruce bark beetle, a major forest pest. • Although it may appear to be rare throughout much of its range, the American Three-toed Woodpecker is often fairly common in areas where standing dead trees are infested with wood-boring beetle larvae. Burned-over areas and soggy muskeg bogs are also favourite haunts of this elusive bird. Major insect infestations within montane and boreal forests have greatly expanded foraging opportunities for this uncommon woodpecker. • Both the Three-toed Woodpecker and the Black-backed Woodpecker have three toes instead of four, the usual number for woodpeckers.

ID: medium-sized woodpecker; blackish head; white "moustache"; thin, white eye line; black and white barring on back; white underparts; black bars on white sides; 3-toed feet; black tail with white outer feathers. *Male:* yellow crown patch. *Female:* darkish crown. *In flight:* barred, black-and-white wings; short, fluttering flights.
Size: *L* 21–24 cm; *W* 38 cm.
Habitat: mature spruce and fir forests, especially around concentrations of fire-killed or diseased trees.
Nesting: usually in a living or dead conifer; also a tree with heart rot; pair excavates a cavity and incubates 3–7 white eggs for 11–14 days.

Feeding: gleans under bark flakes for larval and adult wood-boring insects; also eats other insects and berries; occasionally feeds at sap wells.
Voice: low *pik* or *teek* call; prolonged series of short drumming bursts.
Similar Species: *Black-backed Woodpecker* (p. 300): all-black back and neck; stronger call and drumming. *Hairy Woodpecker* (p. 297): more white on head; large, white back patch; more white on wings; no barring on flanks.

BLACK-BACKED WOODPECKER

Picoides arcticus

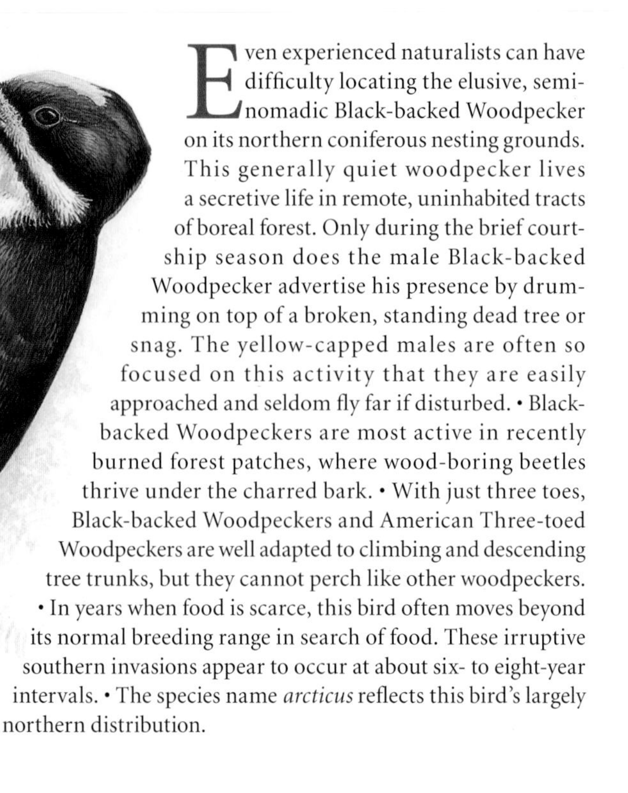

Even experienced naturalists can have difficulty locating the elusive, semi-nomadic Black-backed Woodpecker on its northern coniferous nesting grounds. This generally quiet woodpecker lives a secretive life in remote, uninhabited tracts of boreal forest. Only during the brief courtship season does the male Black-backed Woodpecker advertise his presence by drumming on top of a broken, standing dead tree or snag. The yellow-capped males are often so focused on this activity that they are easily approached and seldom fly far if disturbed. • Black-backed Woodpeckers are most active in recently burned forest patches, where wood-boring beetles thrive under the charred bark. • With just three toes, Black-backed Woodpeckers and American Three-toed Woodpeckers are well adapted to climbing and descending tree trunks, but they cannot perch like other woodpeckers. • In years when food is scarce, this bird often moves beyond its normal breeding range in search of food. These irruptive southern invasions appear to occur at about six- to eight-year intervals. • The species name *arcticus* reflects this bird's largely northern distribution.

ID: medium-sized woodpecker; all-black back; black head with white line below eye; black "moustache" stripe; white underparts; black-barred sides; black tail with unmarked, white outer feathers; 3-toed feet. *Male:* yellow crown. *Female:* black crown. *In flight:* black-and-white-barred wings.

Size: *L* 23–25 cm; *W* 41 cm.

Habitat: coniferous forests, especially burned-over sites with many standing dead trees and blackened stumps.

Nesting: usually in a dead or dying conifer; pair excavates a cavity and incubates 3–6 white eggs for 12–14 days.

Feeding: chisels away bark flakes to expose larval and adult wood-boring insects; may eat some nuts and fruits.

Voice: low *kik* call; prolonged drumming in a series of short bursts.

Similar Species: *American Three-toed Woodpecker* (p. 299): black-and-white-barred back; weaker call and drumming. *Hairy Woodpecker* (p. 297): more white on head and wing; large, white back patch; no barring on flanks.

NORTHERN FLICKER

Colaptes auratus

Although woodpeckers are typically found in trees, the Northern Flicker is more apt to spend its time on the ground. It appears almost robinlike as it hops about on grassy meadows, fields and forest clearings searching for ants and other insects. On the wing, however, the flicker's powerful, undulating flight easily distinguishes it as a woodpecker. • After bathing in a dusty depression to remove oils and bacteria, a flicker will pick up ants and preen rigorously. Ants produce formic acid, which kills small parasites on the birds' skin and feathers. • Two subspecies of flickers occur in Canada: the "Red-shafted" race is found in southern and coastal areas of British Columbia, and the "Yellow-shafted" race is found in northeastern BC, north-central Alberta and southeast to the Maritimes. Hybridization occurs in the region that runs from southern Alaska through the Rocky Mountains to southwestern Saskatchewan.

"Yellow-shafted"

ID: large woodpecker; darkly barred, brown back and wings; black "bib"; white rump patch. *"Red-shafted":* grey face; brown crown and nape; male has red "moustache." *"Yellow-shafted":* mostly brown face; greyish crown; red nape patch; male has black "moustache." *In flight:* undertail and underwings pinkish red on "Red-shafted" and yellow on "Yellow-shafted."

Size: *L* 32–33 cm; *W* 51 cm.

Habitat: *Breeding:* open, mixed woodlands and forest edges. *In migration* and *winter:* open, mixed forests, farmlands, wooded pastures, suburban gardens and city parks.

Nesting: pair excavates a cavity in a dead or dying tree, telephone pole or fence post; may use a nest box; pair incubates 5–8 white eggs for 11–16 days.

Feeding: eats mainly ground-dwelling ants; also takes berries, fruits, seeds and suet; may flycatch.

Voice: loud, laughlike, rapid *kick-kick-kick-kick-kick-kick*; rapidly repeated *woika* or *wicka* courtship call; also drums in soft, muffled volleys.

Similar Species: *Red-bellied Woodpecker* (p. 291): black-and-white pattern on back; white cheeks; more red on head; dark underwings.

PILEATED WOODPECKER

Dryocopus pileatus

This crow-sized bird, the sixth largest woodpecker in the world, is an unforgettable sight. With its flaming red crest, breathtaking flight and maniacal call, this impressive deep-forest dweller is successful at stopping most hikers in their tracks. • A pair of Pileated Woodpeckers requires at least 40 hectares of mature forest as a breeding territory, and the pair may stay together in the vicinity of its nest site year-round. • This bird's large, distinctively oval-shaped nest cavities and roost holes also reveal its presence, as do favoured feeding trees, which sometimes have huge rectangular holes, up to one metre long, bored into their sides. Trees must be of sufficient girth to support a Pileated Woodpecker nest or roost cavity, prompting concern in areas where extensive old-growth forests have been converted to younger stands. Nevertheless, Pileated Woodpeckers appear to be increasing throughout much of their range. • A variety of secondary cavity-nesting birds, including Wood Ducks, goldeneyes, Buffleheads, mergansers, American Kestrels and several owl species, all rely extensively on abandoned Pileated nests or roosting holes for their own nests. • There is no real consensus on whether this bird's name is pronounced *pie-lee-ated* or *pill-e-ated*—it is generally a matter of preference and good-natured debate.

ID: large woodpecker; bold, red crest; long, black bill; predominantly black upperparts; white head; black eye stripe; white throat. *Male:* red "moustache" stripe. *Female:* black "moustache" stripe. *In flight:* black upperwings with white patch at base of primaries; mostly white underwings with black trailing edge and primary tips; strong, direct flight.

Size: *L* 40–50 cm; *W* 68–76 cm.

Habitat: extensive tracts of mature and old-growth deciduous forest or dense, mature deciduous stands; urban and suburban habitats in winter.

Nesting: pair excavates a cavity in a dead tree; nest cavity is usually lined with wood chips; pair incubates 3–5 white eggs for 15–18 days.

Feeding: chisels rotting wood in search of wood-boring insects, especially carpenter ants; also eats fruit and seeds; visits suet feeders.

Voice: irregular *kik-kik-kikkik-kik-kik* or fast, rolling *woika-woika-woika-woika*; also a long series of *kuk* notes; loud, resonant drumming.

Similar Species: other woodpeckers are much smaller and lack head crest. *American Crow* (p. 337) and *Common Raven* (p. 339): lack white underwings in flight.

PASSERINES

Passerines are also commonly known as "songbirds" or "perching birds." Although these terms may be easier to comprehend, they are not as strictly accurate, because some passerines neither sing nor perch, and a number of nonpasserines do sing and perch. In a general sense, however, these terms represent passerines adequately: they are among the best singers, and they are typically seen perched on a branch or wire.

It is believed that passerines, which all belong to the order Passeriformes, make up the most recent evolutionary group of birds. Theirs is the most numerous of all the orders, representing about 40 percent of the bird species in Canada and nearly three-fifths of all living birds worldwide.

Passerines are grouped together based on the sum total of many similarities in form, structure and molecular details, including such things as the number of tail and flight feathers and reproductive characteristics. All passerines share the same foot shape, with three toes facing forward and one facing backward, and none have webbed toes. Also, all passerines have a tendon that runs along the back side of the knee; tightening it gives the bird a firm grip when perching.

Some of our most common and easily identified birds, such as the Black-capped Chickadee, American Robin and House Sparrow, are passerines, but so are some of the most challenging and frustrating birds to identify—until their distinctive songs and calls are learned.

Flycatchers

Shrikes & Vireos

Jays & Crows

Larks & Swallows

Chickadees, Nuthatches & Wrens

Kinglets, Bluebirds & Thrushes

Mimics, Starlings & Waxwings

Wood-warblers

Sparrows, Tanagers, Grosbeaks & Buntings

Blackbirds & Allies

Finchlike Birds

OLIVE-SIDED FLYCATCHER

Contopus cooperi

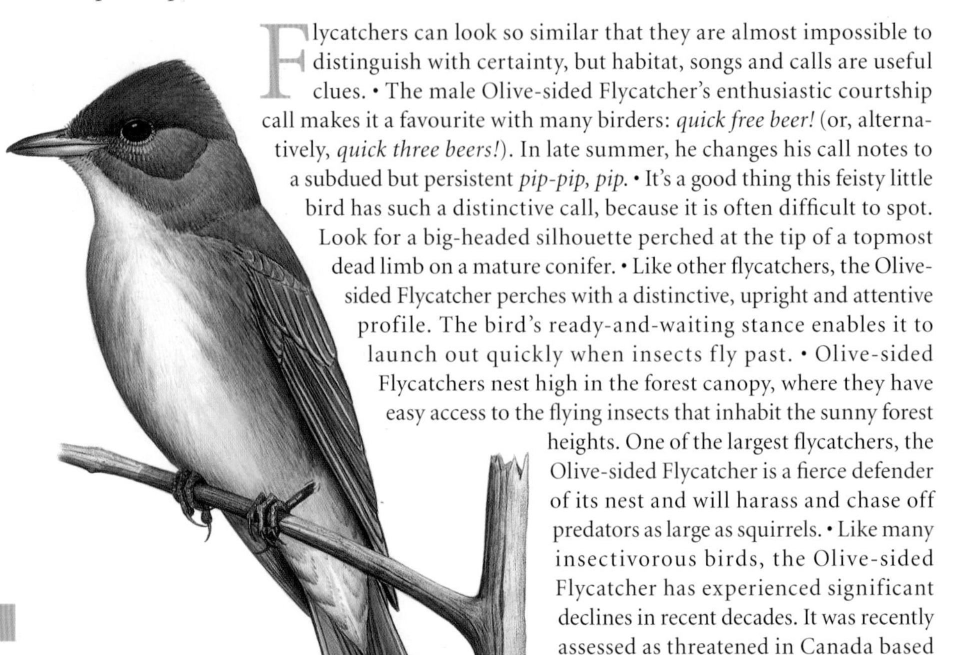

Flycatchers can look so similar that they are almost impossible to distinguish with certainty, but habitat, songs and calls are useful clues. • The male Olive-sided Flycatcher's enthusiastic courtship call makes it a favourite with many birders: *quick free beer!* (or, alternatively, *quick three beers!*). In late summer, he changes his call notes to a subdued but persistent *pip-pip, pip.* • It's a good thing this feisty little bird has such a distinctive call, because it is often difficult to spot. Look for a big-headed silhouette perched at the tip of a topmost dead limb on a mature conifer. • Like other flycatchers, the Olive-sided Flycatcher perches with a distinctive, upright and attentive profile. The bird's ready-and-waiting stance enables it to launch out quickly when insects fly past. • Olive-sided Flycatchers nest high in the forest canopy, where they have easy access to the flying insects that inhabit the sunny forest heights. One of the largest flycatchers, the Olive-sided Flycatcher is a fierce defender of its nest and will harass and chase off predators as large as squirrels. • Like many insectivorous birds, the Olive-sided Flycatcher has experienced significant declines in recent decades. It was recently assessed as threatened in Canada based on a 29 percent population decline in the last 10 years.

ID: sexes similar; large flycatcher; proportionately large head; short, squared tail; sturdy bill is dark above and light below; olive grey to olive brown upperparts; whitish throat and belly; dark olive "vest"; small, white rump patches (often hard to see).

Size: *L* 19 cm; *W* 33 cm.

Habitat: *Breeding:* edges of coniferous or mixed forests with snags or tall trees with dead branches, often bordering lakes, meadows, swamps and marshes; also forest edges near logged areas and recent burns. *In migration:* woodlands of all types, including juniper and oak, riparian areas and desert oases.

Nesting: usually in a conifer; saddled on a horizontal branch; compact nest of grasses and twigs is lined with soft plant material; female incubates 3–4 brown-marked, pale eggs for 14 days.

Feeding: sallies from a high perch to catch large flying insects.

Voice: utters a descending *pip-pip-pip* when excited; male's song is a chipper and lively *quick free beer!*

Similar Species: *Western Wood-Pewee* (p. 305) and *Eastern Wood-Pewee* (p. 306): smaller; more uniform coloration below; 2 faint, grey wing bars; call is a downslurred, nasal *peeer. Eastern Phoebe* (p. 317): smaller; all-dark bill; lacks white rump patches; often wags its tail. *Eastern Kingbird* (p. 321): much darker back and tail; all-white underparts; white-tipped tail.

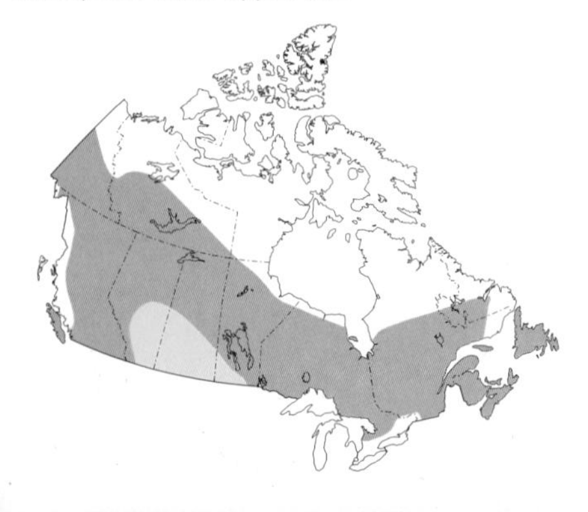

WESTERN WOOD-PEWEE

Contopus sordidulus

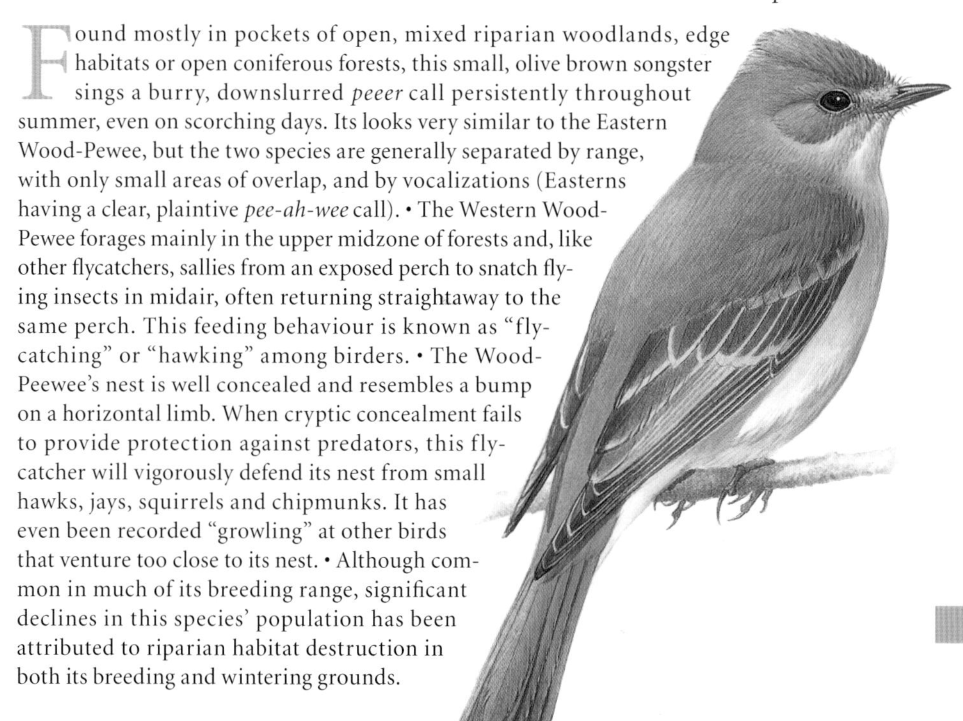

Found mostly in pockets of open, mixed riparian woodlands, edge habitats or open coniferous forests, this small, olive brown songster sings a burry, downslurred *peeer* call persistently throughout summer, even on scorching days. Its looks very similar to the Eastern Wood-Pewee, but the two species are generally separated by range, with only small areas of overlap, and by vocalizations (Easterns having a clear, plaintive *pee-ah-wee* call). • The Western Wood-Pewee forages mainly in the upper midzone of forests and, like other flycatchers, sallies from an exposed perch to snatch flying insects in midair, often returning straightaway to the same perch. This feeding behaviour is known as "flycatching" or "hawking" among birders. • The Wood-Peewee's nest is well concealed and resembles a bump on a horizontal limb. When cryptic concealment fails to provide protection against predators, this flycatcher will vigorously defend its nest from small hawks, jays, squirrels and chipmunks. It has even been recorded "growling" at other birds that venture too close to its nest. • Although common in much of its breeding range, significant declines in this species' population has been attributed to riparian habitat destruction in both its breeding and wintering grounds.

ID: sexes similar; sparrow-sized flycatcher; dark olive brown upperparts; 2 faint grey wing bars; slightly peaked hind crown; dark bill with yellow-orange base; no eye ring; pale grey throat; dingy underparts with partial "vest"; pale undertail coverts; long, slightly notched tail; upright posture.
Size: *L* 15–16 cm; *W* 27 cm.
Habitat: *Breeding:* open coniferous, trembling aspen and other hardwood forests or forest edges; also orchards and riparian woodlands. *In migration:* almost any woodland habitat.
Nesting: in a fork or directly on a horizontal branch; compact, open cup is made of grasses, plant fibres and lichens; female incubates 3 darkly blotched, whitish eggs for 12–13 days.

Feeding: sallies from a perch for flying insects, often returning to the same perch many times; may also glean insects from foliage while hovering.
Voice: whistles downslurred, nasal *peeer;* also utters other short, whistled and sneezy notes.
Similar Species: *Eastern Wood-Pewee* (p. 306): songs differ; little range overlap. *Olive-sided Flycatcher* (p. 304): larger; more distinct "vest"; white rump patches may be visible. Empidonax *flycatchers* (pp. 307–16): smaller; more conspicuous wing bars; some have conspicuous eye rings; often grey or yellow-tinged underparts.

EASTERN WOOD-PEWEE

Contopus virens

Perched on an exposed tree branch in a suburban park, woodlot edge or neighbourhood yard, the male Eastern Wood-Pewee whistles his plaintive *pee-ah-wee pee-oh* all day long throughout the summer. Some even sing their charms late into the evening, long after most birds have silenced their weary courtship songs. Pewees sing incessantly, and it is far easier to detect them by song than it is to spot them in the dense shade of their wooded haunts. • Eastern wood-pewees engage in "yo-yo" flights, darting from a perch in the forest understorey, grabbing an insect in midair and returning to the same perch. • The Eastern Wood-Pewee is less aggressive and showy than most of its close relatives. It hides its hummingbird-sized nest on a tree limb and decorates the nest with pieces of lichen. • The Eastern Wood-Pewee is one of the last spring migrants to return from its wintering range in South America. Although it is one of the most common breeding birds of eastern Canada's deciduous forests, this species has shown significant population declines in breeding bird surveys since the 1960s.

ID: sexes similar; sparrow-sized flycatcher; olive grey to olive brown upperparts; 2 whitish wing bars; slightly peaked hind crown; whitish throat; grey breast and sides; whitish or pale yellow belly, flanks and undertail coverts; dark bill with dull yellow-orange base to lower mandible; no eye ring.
Size: *L* 15–16 cm; *W* 27 cm.
Habitat: open mixed and deciduous woodlands with a sparse understorey, especially woodland openings and edges; rarely in open coniferous woodlands.
Nesting: in the fork of horizontal deciduous branch, well away from the trunk; open cup of grass, plant fibres and lichens is bound with spiderwebs; female incubates 3 darkly botched, whitish eggs for 12–13 days.

Feeding: flycatches insects from a perch; may also glean insects from foliage, especially while hovering.
Voice: *chip* call; male's song is a clear, slow, plaintive *pee-ah-wee*, with the 2nd note lower, followed by a downslurred *pee-oh*, given with or without intermittent pauses.
Similar Species: *Olive-sided Flycatcher* (p. 304): larger; white rump patches; olive grey "vest"; lacks conspicuous, white wing bars. *Eastern Phoebe* (p. 317): lacks conspicuous, white wing bars; all-dark bill; often pumps its tail. *Eastern Kingbird* (p. 321): larger; white-tipped tail; brighter white underparts; all-dark bill. Empidonax *flycatchers* (pp. 307–16): smaller; more conspicuous wing bars; eye rings. *Western Wood-Pewee* (p. 305): song differs; little range overlap.

YELLOW-BELLIED FLYCATCHER

Empidonax flaviventris

Canada boasts a large number of almost-indistinguishable *Empidonax* flycatchers. "Empids" are small, drab flycatchers with pale eye rings and wing bars, and they often flip up their tails. They are best identified by where they live and what they sound like. The Yellow-bellied Flycatcher is the most secretive of the clan, but when viewed in good light, its yellow underparts make it easy to identify. • The reclusive Yellow-bellied Flycatcher lives deep within wooded boreal wetlands and dry, young forests. It does not habitually perch in the open, but it distinguishes itself from other empids by nesting on the ground. In late spring and early summer, the male spends much of its time singing plain, soft, liquid *chelek* songs and occasionally zipping out from inconspicuous perches to help reduce the insect population. Once nesting has begun, the male changes his tune to a slow, rising *per-wee* and focuses his attention on defending his nesting territory and supplying food to his growing young. • One of the last songbird migrants to return in the spring and one of the first to leave, the Yellow-bellied Flycatcher's stay on its nesting grounds is one of the shortest of any neotropical migrant.

ID: sexes similar; small flycatcher; large, rounded head; short tail; uniform olive green upperparts; yellow throat and underparts; 2 whitish wing bars; complete yellowish eye ring.
Size: *L* 13–15 cm; *W* 20 cm.
Habitat: *Breeding:* young, mixed, dry lodgepole pine and trembling aspen forests; also forested edges of bogs and beaver ponds. *In migration:* variety of moist and dry woodlands including pole-stage trembling aspen stands.
Nesting: on the ground in dense sphagnum moss or leaf litter; small cup nest of mosses, rootlets and plant stems is lined with dry grasses; female incubates 3–4 lightly brown-spotted, whitish eggs for 12–14 days.

Feeding: flycatches for insects at low to middle levels of the forest; also gleans vegetation for larval and adult invertebrates while hovering.
Voice: male's song is a soft *cheluck* or *chelek* (2nd syllable is lower pitched); calls include a chipper *pe-wheep, preee, pur-wee* or *killik*.
Similar Species: flycatchers are best distinguished by songs and calls. *Acadian* (p. 308), *Willow* (p. 310), *Alder* (p. 309) and *Least* (p. 311) *flycatchers:* different songs; white eye rings; all lack extensive yellow wash from throat to belly; all but Acadian have browner upperparts. *Pacific-slope Flycatcher* (p. 315): longer tail; duller wing; and teardrop-shaped eye ring; found only in western provinces.

ACADIAN FLYCATCHER

Empidonax virescens

The Acadian Flycatcher is virtually indistinguishable from other *Empidonax* flycatchers until it utters its quick, forceful *peet-sah* song. Although it is widely distributed in mature deciduous forests of the eastern U.S., in Canada, this summer resident is rare and occurs only in extreme southwestern Ontario, so male singers feel little pressure to issue their hallmark song in defence of a breeding territory. • Learning to identify this bird is only half the fun. Its speedy aerial courtship chases and the male's hovering flight displays are sights to behold—if you can survive the swarming hordes of blood-sucking mosquitoes inhabiting the swampy woodlands of extreme southwestern Ontario. In Canada, mature maple and beech trees provide preferred nesting sites for the Acadian Flycatcher. The nest, built on a horizontal branch up to seven metres from the ground, can be quite conspicuous because loose material often dangles from it, giving it a sloppy appearance. • The Acadian Flycatcher is classified as endangered in Canada because of its small population and fragmented habitat.

ID: sexes similar; small flycatcher; olive upperparts; black wings with 2 white (adult) or buff (juvenile) wing bars; pale yellow eye ring; dark bill with pinkish yellow lower mandible; faint olive yellow breast; whitish throat; yellow belly and undertail coverts; long primary feathers.

Size: *L* 14–15 cm; *W* 23 cm.

Habitat: *Breeding:* deciduous woodlands, riparian woodlands and swamps. *In migration:* any wooded habitat.

Nesting: in a deciduous tree, usually 4–7 m above ground, and near water; female builds a sloppy-looking cup nest from weed stems, twigs and grasses held together with spider silk; female incubates 2–4 lightly spotted, creamy white eggs for 13–15 days.

Feeding: Feeding: feeds almost exclusively on insects and spiders captured by hawking or gleaned from foliage.

Voice: male's song is an emphatic *peet-sah*; call is a similar *peet*.

Similar Species: flycatchers are best distinguished by songs and calls. *Alder Flycatcher* (p. 309): song is *free-beer*; inconspicuous, narrower, white eye ring; browner overall; smaller head relative to body. *Willow Flycatcher* (p. 310): song is an explosive *fitz-be-yew*; browner overall; smaller head; very faint eye ring. *Least Flycatcher* (p. 311): song is a liquid *che-bek*; prominent, white eye ring; rounded head; shorter wings. *Yellow-bellied Flycatcher* (p. 307): song is *chelek*; yellow wash from throat to belly.

ALDER FLYCATCHER

Empidonax alnorum

The Alder Flycatcher is often indistinguishable from other *Empidonax* flycatchers until it opens its small, bicoloured beak, revealing its identity with a hearty *fee-bee-o* or *free beer*. This nondescript bird is well named, because it is often found in alder and willow shrubs—a fact that can help in its identification. In southern regions of Canada, the Alder Flycatcher frequently competes with the Willow Flycatcher for control over dense riparian alder and willow stands. Actual nests can be hard to find, because these thickets are often located in water or deep muck. • Once this aggressive bird has been spotted, its feisty behaviour can be observed without distraction as it drives away rivals and pursues flying insects. • Many birds have to learn their songs and calls, but Alder Flycatchers instinctively know the simple phrase of their species. Even when isolated from the calls of other Alders, a young Alder Flycatcher can produce a perfectly acceptable *free beer* call when it matures. • The nearly identical Willow Flycatcher is a close relative of the Alder, and until 1973, they were considered a single species—the Traill's Flycatcher.

ID: sexes similar; small flycatcher; dull greenish brown upperparts; usually darker crown; 2 white to buff wing bars; faint, whitish eye ring; pale throat; yellow-green underparts; dark bill with orange lower mandible; longish tail.
Size: *L* 15–17 cm; *W* 21 cm.
Habitat: alder or willow thickets bordering lakes, streams or muskeg.
Nesting: in a fork in a shrub or small tree; small cup nest is loosely woven of grasses and other plant materials; female incubates 3–4 darkly spotted, white eggs for 12–14 days.
Feeding: flycatches from a perch for beetles, bees, wasps and other flying insects; occasionally takes berries or seeds.
Voice: call is a *wheep* or *peep*; male's song is a snappy *free beer!* or *fee-bee-o*.
Similar Species: flycatchers are best distinguished by songs and calls. *Willow Flycatcher* (p. 310): prefers drier and more open shrub habitats; quick, sneezy *fitz-bew* song; inconspicuous eye ring. *Least Flycatcher* (p. 311): liquid *che-bek* song; bolder eye ring; greener upperparts; different habitat. *Acadian Flycatcher* (p. 308) and *Yellow-bellied Flycatcher* (p. 307): Acadian song is *free-beer*; Yellow-bellied song is *chelek*; yellowish eye rings; greener upperparts; yellower underparts; different habitats.

WILLOW FLYCATCHER

Empidonax traillii

Toward the end of May, as the spring movement of songbirds wanes, Willow Flycatchers begin to appear in southern Canada's lowland areas. In many areas, they are the latest of all migrants to return in spring. These flycatchers forage and nest in brushy areas, usually near water, but can often be found in clear-cuts and old fields. • Similar plumage makes the various *Empidonax* flycatchers the most difficult of any genus to distinguish. Subtle differences in habitat preferences, calls and such difficult-to-observe details as mandible colour and length of outer flight feathers provide clues to identity. • The Willow Flycatcher, like many small passerines, is a frequent host of the Brown-headed Cowbird and tries to avoid raising adopted young by building a new nest over the cowbird eggs. • Only since 1973 has the Willow Flycatcher been recognized as a separate species from the Alder Flycatcher, based mainly on song and call differences. Early spring is the only time these birds can be safely distinguished from Alder Flycatchers because it is the only time that the two species sing very different songs.

ID: sexes similar; small flycatcher; olive brown upperparts; flat forehead; distinct peak on rear crown; indistinct, white eye ring; pale olive breast; yellowish belly; dark bill with pink lower mandible; whitish throat; 2 whitish or greyish wing bars; white undertail coverts.

Size: *L* 15 cm; *W* 21 cm.

Habitat: *Breeding* and *in migration:* riparian willow thickets, brushy woodland edges, young alder stands and shrubby old fields.

Nesting: in a hardwood sapling or shrub; compact cup nest is lined with grasses, hair, plant down and feathers; female incubates 3–4 brown-spotted, buff eggs for 12–13 days.

Feeding: flies from a perch to capture flying insects; also gleans insects from vegetation while hovering.

Voice: mellow *whit* call; male's song is a quick, sneezy *fitz-bew.*

Similar Species: flycatchers are best distinguished by songs and calls. *Alder Flycatcher* (p. 309): *free beer* or *fee-bee-o* song; usually more olive green back and rump; whiter wing bars. *Least Flycatcher* (p. 311): liquid *che-bek* song; bolder eye ring; greener upperparts; different habitat. *Acadian Flycatcher* (p. 308) and *Yellow-bellied Flycatcher* (p. 307): Acadian song is *free-beer*; Yellow-bellied song is *chelek*; yellowish eye rings; greener upperparts; yellower underparts; different habitats.

LEAST FLYCATCHER

Empidonax minimus

At first glance, this small bird might not look like a bully, but the Least Flycatcher is one of the boldest and most pugnacious songbirds of Canada's deciduous woodlands, readily chasing away intruders much larger than itself. This aggression may explain why Least Flycatcher nests are infrequently parasitized by Brown-headed Cowbirds. • On territory, the male Least Flycatcher is a pint-sized fluff of terror. In territorial battles, the victor wins the right to select and chase a female for considerable distances in the hope of mating with her. The courtship might not be romantic, but afterward the male does its best to defend its mate. During the nesting season, males repeat their simple *che-bek* calls throughout the day to ensure that their territorial boundaries are respected. • In suitable forests, the species often has very small territories and they are found in a rather clumped distribution, suggesting that breeding Least Flycatchers may be socially attracted to other breeding pairs. • Some female Least Flycatchers are extremely attached to their nest and eggs. A toppled tree was once found with a female still incubating. • Because the Least Flycatcher is less specific in its choice of habitat than most other flycatchers, it is more widely distributed and thus more likely to be seen.

ID: sexes similar; smallest flycatcher; proportionately large head; olive brown upperparts; bold, white eye ring; black wings with 2 white wing bars; whitish throat; subtle, olive brown vest; pale underparts, usually without obvious yellow; dark bill with yellowish orange lower mandible.

Size: *L* 13–15 cm; *W* 20 cm.

Habitat: open deciduous woodlands, suburbs, forest openings and forest edges; often in second-growth poplar stands.

Nesting: on a horizontal limb or in a fork of a small deciduous tree; small cup nest is lined with grasses, plant down and feathers; female incubates 3–6 unmarked, creamy white eggs for 14 days.

Feeding: hunts from perch for flying insects; also gleans foliage for ants, caterpillars and insects; may eat some fruits and seeds.

Voice: call note is a dry *whit*; male's song is a constantly repeated, whistled *che-bek che-bek.*

Similar Species: flycatchers are best distinguished by songs and calls. *Alder Flycatcher* (p. 309): song is a snappy *free beer!* or *fee-bee-o*; faint eye ring; mostly found near water. *Yellow-bellied Flycatcher* (p. 307): song is *chelek*; yellowish eye ring; greener upperparts; yellower underparts. *Willow Flycatcher* (p. 310): song is a quick, sneezy *fitz-bew*; indistinct eye ring; mostly found near water. *Hammond's Flycatcher* (p. 312): darker throat and underparts. *Dusky Flycatcher* (p. 314): whitish edges on outer tail feathers.

HAMMOND'S FLYCATCHER

Empidonax hammondii

The retiring, diminutive Hammond's Flycatcher is perhaps most renowned for being hard to identify in the field and is easily confused with the similar-looking Dusky Flycatcher. Habitat preferences and songs are helpful strategies for telling the two species apart, as are subtle differences in plumage, but their calls are similar enough to make field identification difficult for the casual observer. The Hammond's makes its home beneath the shady, dense canopy of mature coniferous forests throughout British Columbia and the Alberta Rockies, whereas the Dusky prefers more open, sun-drenched forest edges, woodlands and brushy areas. The slightly shorter bill and shorter, notched tail of the Hammond's also help distinguish it from the Dusky. • The Hammond's Flycatcher lives at higher elevations than other *Empidonax* flycatchers, commonly nesting at elevations up to 3000 metres in the Rockies. It tends to prefer conifer stands that are greater than 8 hectares in size and at least 80 years old. If logging and other development projects make such areas harder to come by, this species might become increasingly dependent on our national parks and protected areas for its survival.

ID: sexes similar; small flycatcher; olive grey upperparts; slightly peaked head; distinct, white eye ring; 2 white wing bars; yellowish underparts and undertail coverts; pale grey "vest"; tiny, dark bill with orangey base to lower mandible; long flight feathers make notched tail look comparatively short when bird is perched.

Size: *L* 14 cm; *W* 21–23 cm.

Habitat: *Breeding:* dense, humid, mixed, mature coniferous forests; rarely in more open forests. *In migration:* open mixed coniferous and deciduous woodlands.

Nesting: saddled on a horizontal limb of a conifer; cup nest is lined with feathers, grasses and hair; female incubates 3–4 unmarked, creamy white eggs for 12–15 days.

Feeding: flycatches and hover-gleans for insects usually in the mid-canopy; perches less in the open than similar flycatchers.

Voice: usual calls a *pip* or short *peek*; male's song usually consists of a low, rapid, 3-part *chi-pit*, a low-pitched *brrrk* and a rising *griip*, but any part may be dropped.

Similar Species: *Dusky Flycatcher* (p. 314): longer bill; longer tail has pale-edged outer feathers; clearer whistles and *whit* call notes; usually 2-phrased song. *Least Flycatcher* (p. 311): song is a repeated, whistled *che-bek*; greyer upperparts; flicks wings and tail.

GRAY FLYCATCHER

Empidonax wrightii

Distinctions between *Empidonax* flycatchers usually rely on subtle differences in plumage, vocalization and habitat, but the Gray Flycatcher can easily be identified by its habit of slowly dipping its pale-edged tail downward. Further clues are that it prefers low vegetation, tends to drop to the ground to pursue insects and makes its summer home in open ponderosa pine forests of the southern Okanagan Valley. All other small flycatchers flick their tails quickly upward, flycatch from higher perches (except in migration) and tend to prefer less open habitat. • Open ponderosa pine forests consisting of young trees and a grassy understorey are favoured for breeding, with small, grassy openings in the forests being the centre of each pair's breeding territory. Habitat that Gray Flycatchers favour is also home to rare species such as Lewis's Woodpecker, White-headed Woodpecker, badgers, spotted bats and western rattlesnakes. • The Gray Flycatcher has recently expanded its range northward into Canada. It was first observed in 1984, with confirmation of breeding occurring in 1986. Currently, this flycatcher's population is on the increase. For a flycatcher, the Gray Flycatcher has a relatively short migration range, with many birds overwintering in southern Arizona.

ID: drab, greyish upperparts; 2 pale wing bars; whitish underparts; faint, white eye ring; long, dark-tipped bill with pale lower mandible; long tail with thin, white border.

Size: *L* 13–15 cm; *W* 22 cm.

Habitat: open ponderosa and lodgepole pine forests with grassy understorey.

Nesting: in a crotch of a juniper or sage, or near the base of a thorny shrub; cup of bark, plant down, weed stems and grass is lined with feathers and hair; female incubates 3–4 unmarked, creamy white eggs for 14 days.

Feeding: sallies from a perch for flying insects; gleans foliage and the ground for insects and larvae.

Voice: 2-syllable *chuwip* song is often followed by a dry *wit* or *pit*.

Similar Species: *Dusky Flycatcher* (p. 314): usually darker with more contrast in plumage; darker lower mandible; does not dip tail. *Hammond's Flycatcher* (p. 312): darker overall; darker "vest"; smaller bill with duskier lower mandible; often flicks wings and short, thin tail. *Willow Flycatcher* (p. 310): browner upperparts; does not dip tail; prefers wetter habitats.

DUSKY FLYCATCHER

Empidonax oberholseri

Many novice birders might wonder why three virtually identical-looking birds, the Dusky, Hammond's and Gray flycatchers, are not just lumped together as a single species—after all, why make life so confusing? In fact, at one time, the Dusky Flycatcher and the Gray Flycatcher, which is extremely localized in Canada, were considered to be one species, but closer inspection of collected specimens, DNA analysis and detailed studies of bird behaviour and habitat requirements confirmed separate species status. • In the breeding season, the Dusky Flycatcher favors mid- and high-elevation forests and other montane habitats that are more open than those preferred by the Hammond's but less open, dry and shrubby than those of the Gray Flycatcher. • As with other small temperate-zone flycatchers, snow and freezing rain in May and June sometimes can have disastrous consequences, killing huge numbers of Dusky Flycatchers. Fortunately, such weather is normally a localized event; the following year, other Dusky Flycatchers return to fill the void created by the unseasonable storms.

ID: sexes similar; small flycatcher; rounded head; pale, elongated eye ring; 2 faint, white wing bars; olive brown upperparts; whitish throat; pale grey "vest"; pale yellow belly and undertail coverts; small, dark bill with orangey lower mandible; long, dark tail with lighter edges.

Size: *L* 13–15 cm; *W* 21 cm.

Habitat: *Breeding:* dry, open coniferous woods with shrubby areas. *In migration:* riparian, hardwood and coniferous woodlands and shrublands at lower elevations.

Nesting: in a crotch of a deciduous tree or shrub; cup nest of weed stems and grasses is lined with feathers, fine grasses and hair; female incubates 3–4 unmarked, creamy white eggs for 12–15 days.

Feeding: flycatches and hover-gleans vegetation for insects.

Voice: call is a flat *wit*; male's song is a quick, whistled *chrip ggrrreep pweet,* rising at the end; breeding period song is sometimes a repeated *du...DU-hic.*

Similar Species: *Hammond's Flycatcher* (p. 312): smaller bill; shorter tail; prefers denser shade. *Gray Flycatcher* (p. 313): found only in southern Okanagan Valley; paler grey overall; dips tail persistently.

PACIFIC-SLOPE FLYCATCHER

Empidonax difficilis

Fortunately for birders, the song of the Pacific-slope Flycatcher is much more distinctive than its plumage. When you enter any moist woodland along the West Coast during spring, this small flycatcher's snappy, rising *suweeet* call is one of the first sounds you'll hear. Nonetheless, *difficilis* is an appropriate designation, given the difficulties of flycatcher identification. Until the late 20th century, this common songbird and the Cordilleran Flycatcher were collectively known as "Western Flycatcher." In the British Columbia Interior, where both species are thought to occur, looking for each species' preferred habitat can help determine their identities. Pacific-slopes are generally found in mature, humid coniferous forests and mixed second-growth forests near the coast, whereas Cordillerans are found in open to dense coniferous and deciduous riparian woodlands in the eastern Rockies and foothills. • Like other flycatchers, the Pacific-slope Flycatcher hunts primarily by hawking. It launches from an exposed perch to seize a flying insect in midair then loops back to alight on the same perch, ready for its next sally. *Empidonax*—"king of the gnats"—is a wonderful name for this confusing, but endearing, group of flycatchers and is a reflection of their amazing insect-catching abilities.

ID: sexes similar; small flycatcher; olive-brown upperparts; yellowish "teardrop" eye ring; small head crest; dark bill with orange-yellow lower mandible; yellowish throat and wing bars; pale yellow underparts; brownish "vest"; brownish rump.
Size: *L* 13 cm; *W* 20 cm.
Habitat: *Breeding:* mature, humid coniferous forests, mixed pine-oak forests and mixed second-growth forests in valleys, often near water. *In migration:* more varied wooded habitats.
Nesting: in a cavity in a streambank, among the roots of an upturned tree, under building eaves or in a small tree; cup nest of mosses, lichen, rootlets and grasses is lined with shredded bark, hair and feathers; female incubates 3–4 brown-spotted, creamy white eggs for 14–15 days.
Feeding: sallies from a perch to hover-catch flying insects; also gleans crawling insects.
Voice: male's call is a single, upslurred *suweeet* or *fe-oo-eeet!*; female's call is a brief, high-pitched *tink*; male's song is a series of high-pitched, repeated phrases: *siLEEK...tup...P'SEET!*
Similar Species: none on the coast; in the BC Interior, perhaps *Gray Flycatcher* (p. 313) and *Dusky Flycatcher* (p. 314): pale grey throats; inhabit more open sites; Gray Flycatcher is found only in southern Okanagan Valley.

CORDILLERAN FLYCATCHER

Empidonax occidentalis

The Cordilleran Flycatcher and the Pacific-slope Flycatcher were formerly lumped into one species, the "Western Flycatcher." Although they are now regarded as distinct species, their similar field characteristics and vocalizations makes them difficult to distinguish and perpetuates their uncertain status. Cordilleran Flycatchers are thought to arrive in Canada about the same time as their Pacific-slope relatives, but much remains to be learned about this species' distribution and migration. Generally, Cordilleran Flycatchers prefer drier forest habitats than Pacific-slope Flycatchers. • In Canada, Cordilleran Flycatchers are found along the Rocky Mountains and foothills of Alberta and eastern British Columbia. Rocky outcroppings near water are favoured nesting locations. The Cordilleran Flycatcher has also been recorded nesting on mountain cabins and even in old American Robin nests. • Cordilleran Flycatcher populations currently appear to be stable and possibly increasing. • The scientific name *occidentalis* is Latin for "western."

ID: sexes similar; olive green upperparts; 2 pale yellowish wing bars; yellowish throat and underparts; pale yellowish, almond-shaped eye ring; dark bill with orange lower mandible.
Size: *L* 14 cm; *W* 20 cm.
Habitat: mid to high elevations in open to dense coniferous and deciduous riparian woodlands, often near seepages and springs.
Nesting: in a cavity in a small tree, bank, bridge or cliff face; cavity is lined with moss, lichens, plant fibres, bark, fur and feathers; female incubates 3–4 brown-speckled, creamy white eggs for 15 days.
Feeding: takes insects by hawking, hovering or occasionally gleaning from branches or foliage.
Voice: call is a chipper, whistled *swee-deet*; male's song is a high-pitched, 3-part *p'SEET...p'sik...seet!*
Similar Species: *Willow Flycatcher* (p. 310) and *Western Wood-Pewee* (p. 305): no eye rings. *Least* (p. 311), *Hammond's* (p. 312) and *Dusky* (p. 314) *flycatchers:* lack almond-shaped eye ring and completely orange lower mandible; songs are useful in field identification.

EASTERN PHOEBE

Sayornis phoebe

Once limited to nesting on natural cliffs and fallen trees, this adaptive flycatcher now nests almost exclusively on buildings and bridges and is consequently close to people. Its loud, emphatic *fee-bee!* is a familiar call around summer cottages, campground picnic shelters or backyard sheds east of the Rocky Mountains. • Eastern Phoebes sometimes reuse nest sites, even the same nest, for many years. Unfortunately, people often unnecessarily remove phoebes' mud nests from outbuildings. Because of their insectivorous diet, it can be an advantage to have these birds around. • Other birds pump their tails while perched, but few species can match the zest and frequency of the Eastern Phoebe's tail pumping and wagging. This habit is a valuable identification point, since the Eastern Phoebe is a drab, greyish bird with few conspicuous field marks. • Most flycatchers migrate to the tropics in the winter, but the tough Eastern Phoebe routinely remains in southern Ontario and Québec until December, and migrants may return by March. • The Eastern Phoebe became the first banded bird in North America, when John J. Audubon tied a silver string to a bird's leg in 1804 to see if it would return to the same nest site.

ID: sexes similar; medium-sized flycatcher; grey-brown upperparts; pale yellow underparts with faint yellow-green belly; no eye ring or obvious wing bars; all-black bill; dark legs; long, dark tail.
Size: *L* 17–18 cm; *W* 26 cm.
Habitat: *Breeding:* human-made structures and overhanging embankments, often near water.
Nesting: under the ledge of a building, picnic shelter, culvert or bridge; also a riverbank; cup-shaped, mud nest is lined with mosses, grasses, fur and feathers; female incubates 4–5 eggs for 16 days; 2 broods per year.

Feeding: sallies from perch for flying insects; occasionally plucks aquatic invertebrates from the water's surface.
Voice: hearty, snappy *fee-bee*, delivered frequently; call is a sharp *chip*.
Similar Species: *Western Wood-Pewee* (p. 305) and *Eastern Wood-Pewee* (p. 306): smaller; greyish underparts; pale wing bars; bicoloured bills. *Olive-sided Flycatcher* (p. 304): dark sides form open "vest." *Eastern Kingbird* (p. 321): white-tipped tail; blackish upperparts. Empidonax *flycatchers* (pp. 307–16): smaller; most have pronounced eye rings and wing bars.

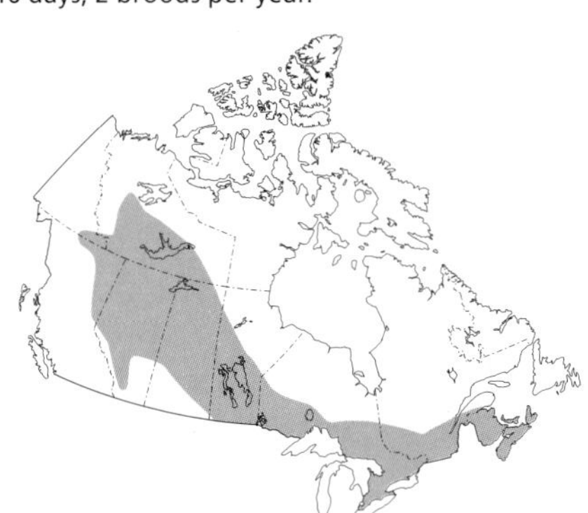

SAY'S PHOEBE

Sayornis saya

This large, handsome flycatcher, with its soft, brown plumage and apricot belly, has a subtle beauty, but its habit of flicking its tail while perched might be its most diagnostic trait. Within Canada, the Say's Phoebe lives in two distinct areas; populations in the southern Prairie provinces and southern BC are separated by about 600 kilometres from a northern population that nests primarily in Alaska, the Yukon and northern BC. In the south, it is partial to open, dry environments and thrives in sun-parched grassy valleys and hot, dry canyons. In the northwest, the Say's Phoebe is a bird of the mountains and even the tundra. It is often found around abandoned buildings that provide safe, sheltered nest sites. • Waiting patiently on a fence post or other low perch, a Say's Phoebe can confidently sally forth to hawk an easy meal. It is an early spring migrant, contending with the cold weather and scarcity of flying prey by hover-gleaning early flying insects including bees and wasps from the ground. • This bird is remarkably tolerant of arid conditions; it is capable of extracting all the water it requires from the insects it eats.

ID: sexes similar; medium-sized flycatcher; pale grey-brown head, throat, breast and upperparts; stout, all-dark bill; dark eye line; apricot-buff belly and undertail coverts; dark legs; frequently wags all-black tail.

Size: *L* 19 cm; *W* 33 cm.

Habitat: *Breeding:* open country, sagebrush steppes and shrublands; often near cliffs or abandoned buildings. *In migration:* open and semi-open habitats including rangelands, agricultural lands, logged sites, burns and rocky bluffs and canyons.

Nesting: on a ledge in or on a building, bridge, culvert or rock cliff; cup nest of soft plant materials is lined with finer materials, especially mosses and hair; female incubates 4–5 mostly unmarked, white eggs for 14 days.

Feeding: eats a variety of flying and terrestrial insects.

Voice: calls are low, mellow *pee-ter* or *pee-ur*; male's song is a melancholy *pitseedar.*

Similar Species: *Other flycatchers* (pp. 304–16, 319, 321): all lack apricot belly. *Eastern Phoebe* (p. 317): grey to white underparts. *Western Kingbird* (p. 320): yellowish belly; darker wings. *American Robin* (p. 383): larger; dark head; yellow-orange bill; streaked throat; dark red-orange underparts.

GREAT CRESTED FLYCATCHER

Myiarchus crinitus

A loud, emphatic *wheep!* alerts birders in southern Canada to the presence of a Great Crested Flycatcher. The Great Crested Flycatcher's nesting habits are unusual for a flycatcher—it is a cavity nester. Natural tree cavities and abandoned woodpecker nests are preferred nesting places, but this bird also occasionally uses nest boxes intended for bluebirds. The Great Crested Flycatcher will even decorate the entrance of its nest with a shed snakeskin. Although the purpose of this practice is not fully understood, it's likely to make any predators think twice! In some instances, this versatile bird has even been known to substitute translucent plastic wrap for genuine reptile skin. During migration, the Great Crested Flycatcher usually arrives in Canada by mid-May and departs southward by early September. Primarily an eastern bird, populations in Alberta and Saskatchewan are thought to migrate through Manitoba before heading southward, reversing this route in spring. • Songbirds such as the Great Crested Flycatcher are often thought of as birds that fly south for the winter. In reality, this flycatcher, and many other migrants, are subtropical or tropical birds of Central and South America that visit our country briefly to raise their young before returning home.

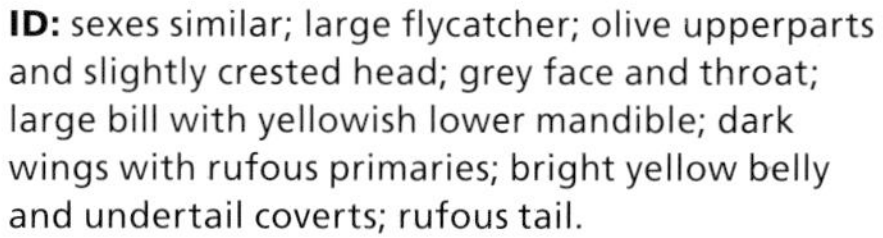

ID: sexes similar; large flycatcher; olive upperparts and slightly crested head; grey face and throat; large bill with yellowish lower mandible; dark wings with rufous primaries; bright yellow belly and undertail coverts; rufous tail.
Size: *L* 20–22 cm; *W* 33 cm.
Habitat: deciduous and mixedwood forests, near clearings or edges; occasionally in parks, orchards and cemeteries.
Nesting: in a natural cavity or woodpecker hole; occasionally in a nest box; cavity is lined with grass, bark strips, feathers and shed snakeskin; female incubates 4–5 variably marked, creamy white to pale buff eggs for 13–15 days.

Feeding: usually forages high in foliage; sallies for flying insects; also takes some fruit and the occasional small snake or lizard.
Voice: loud, burry whistled *wheep!* and rolling *prrrrreet!*
Similar Species: *Yellow-bellied Flycatcher* (p. 307): much smaller; lacks reddish brown tail and large, all-black bill. *Western Kingbird* (p. 320): typically perches on wires in the open; light grey head; black mask; black tail with white outer tail feathers.

WESTERN KINGBIRD

Tyrannus verticalis

The Western Kingbird's bickering call and aggressive nature make it a difficult bird to miss. This kingbird will fearlessly defend its breeding territory, attacking large birds such as crows and hawks, and even dive-bombing house cats that venture nearby. Witnesses to these brave attacks will find it easy to understand why this brawler was awarded the name "kingbird." Although their territories are quite large on arrival in late spring, by the time eggs are laid, the "no-trespassing" area is restricted to the immediate vicinity of each bird's nest. • Western Kingbirds are adept at catching insects in midair, and they are commonly seen hunting from fence posts, fence lines and power lines. • The male Western Kingbird's tumbling aerial courtship display is sure to enliven a tranquil spring scene. Twisting and turning all the way, he flies about 20 metres into the air, stalls and then tumbles, flips and twists as he plummets to earth. • The word *verticalis* refers to the small, hidden, red crown patch that this bird flares in courtship and territorial displays. However, it is not a good identification mark because it is rarely seen outside the breeding season. • In the fall, this bird is a regular vagrant eastward to Newfoundland.

ID: sexes similar; medium-sized songbird; grey head; small, black mask; short, black bill; white throat; greenish grey back and wings; grey breast; yellow belly and underwings; yellow undertail coverts; all-black tail with white sides.

Size: *L* 20–23 cm; *W* 38–41 cm.

Habitat: *Breeding:* open country near farm buildings and towns; also open ponderosa pine woodlands and orchards. *In migration:* almost any open habitat with buildings, telephone poles, fence lines or scattered trees.

Nesting: on a tree branch, occasionally on a transformer or in a building; bulky cup nest of grasses, plant stems and twigs is lined with hair and plant down; female incubates 3–7 heavily mottled, white to pinkish eggs for 18–19 days.

Feeding: eats insects captured in flight or gleaned from the ground.

Voice: feisty and argumentative; chatty, twittering *whit-ker-whit*; also a short *kit* or extended *kit-kit-keetle-dot*.

Similar Species: *Great Crested Flycatcher* (p. 319): slightly crested head; reddish brown tail and wings; faint wing bars; lacks white outer tail feathers. *Eastern Kingbird* (p. 321): grey-black upperparts; white underparts.

EASTERN KINGBIRD

Tyrannus tyrannus

Living up to its scientific name *Tyrannus tyrannus*, the Eastern Kingbird is a bold brawler, fearlessly attacking crows, hawks, cats and even humans that pass through its territory, pursuing and pecking at them until the threat has passed. It is a bird of open landscapes and often perches prominently on roadside wires. • The kingbird's fluttery courtship flight, which is characterized by short, quivering wingbeats reveals its gentler side, but its frequent, loud, bickering calls reveal its tyrant nature. • Unlike most other songbirds, Eastern Kingbird parents feed their young for up to seven weeks after fledging. • Sometimes referred to as the "Jekyll and Hyde" bird, the Eastern Kingbird is a gregarious fruit eater while wintering in the Amazon region of South America and an antisocial, aggressive insect eater while nesting in Canada. • Eastern Kingbirds rarely walk or hop on the ground. They prefer to fly, even for very short distances. • The red crown, for which kingbirds are named, is rarely seen outside of the courtship season.

ID: sexes similar; medium-sized songbird; grey-black upperparts; black head with slight crest; white underparts; black tail with conspicuous white tip; orange-red crown patch (rarely seen); black legs; quivering flight.
Size: *L* 22 cm; *W* 38 cm.
Habitat: *Breeding:* rural fields with scattered shrubs, trees or hedgerows; forest fringes and clearings; shrubby roadsides and farmyards; open, riparian woodlands. *In migration:* woodland edges and riparian habitats.
Nesting: on a horizontal limb of an isolated tree or shrub; cup nest of twigs, grasses and plant stems is lined with rootlets, finer grasses and hair; female incubates 3–4 darkly blotched, white to pinkish white eggs for 14–18 days.
Feeding: sallies for flying insects or picks insects from the ground.
Voice: call is a buzzy *dzee-dzee-dzee*; also a quick, loud, chattering *kit-kit-kitter-kitter*; also a nasal *dzeep*.
Similar Species: *Eastern Phoebe* (p. 317): smaller; darker head than back; lacks white-tipped tail. *Olive-sided Flycatcher* (p. 304): dark sides form open "vest"; lacks white-tipped tail. *Eastern Wood-Pewee* (p. 306): smaller; bicoloured bill; lacks white-tipped tail.

LOGGERHEAD SHRIKE

Lanius ludovicianus

Shrikes are medium-sized, predatory songbirds that are commonly seen perched, hawklike, atop shrubs or small trees, using their keen vision to search for prey. Once prey is caught, it is quickly impaled on a thorn or barbed wire fence or deposited in a fork of a shrub. These trophies demonstrate a male's hunting competence to female shrikes, and serve as a means of storing excess food items during times of plenty. This habit has earned it the nicknames "Butcher Bird" and "Thorn Bird." Loggerhead Shrikes have an uncanny memory for the location of their food stores and have been known to return to their caches as much as eight months later. • Shrikes are found in a variety of open grassland and shrubby habitats but seem unable to survive in developed areas. Populations continent wide are declining by about seven percent per year. The Loggerhead Shrike is largely resident in the central and southern areas of the Prairie provinces and is classified as threatened. Eastern populations found in Ontario and Québec are classified as endangered. Extensive recovery plans have resulted in mixed results for this population. • This bird is called "loggerhead" because of its proportionally large head.

ID: sexes similar; medium-sized songbird; large head; hooked, black bill; grey back and crown; wide, black mask; white throat and underparts; black wings with white patch; long, black tail with white corners. *Juvenile:* brownish grey, barred underparts. *In flight:* fast wingbeats; white patches in wings and tail.

Size: *L* 22–23 cm; *W* 30–31 cm.

Habitat: open habitats such as pastures, fields and prairies with scattered hawthorn shrubs, fence posts, and barbed wire; also sports fields and cemeteries.

Nesting: in the crotch of a shrub or tree; bulky cup nest of twigs and grass is lined with animal hair and feathers; female incubates 5–6 darkly spotted, pale buff to greyish white eggs for 15–17 days.

Feeding: swoops down on prey from a perch or attacks in pursuit; takes mostly large insects; regularly eats small birds and other vertebrates.

Voice: call a harsh *shack-shack*; high-pitched *bird-ee bird-ee*; song a series of warbles, trills and other notes.

Similar Species: *Northern Shrike* (p. 323): little seasonal overlap; narrower mask; longer bill; paler underparts; immature bird is brownish overall with "scaly" underparts. *Northern Mockingbird* (p. 386): much thinner bill; paler wings; lacks black mask; slower wingbeats; buoyant flight.

NORTHERN SHRIKE

Lanius excubitor

The adult Northern Shrike resembles a grey robin with the bill of a small hawk. These predatory songbirds are the winter replacement for the Loggerhead Shrike, which makes distinguishing between these very similar-looking birds less onerous for the average birder. At the end of the breeding season, Northern Shrikes retreat from the Subarctic to over-winter in southern Canada and northern United States. Like Loggerheads, they perch like hawks on exposed treetops to survey semi-open hunting grounds. • Despite its robinlike size, the Northern Shrike can quickly kill small birds, rodents, reptiles and amphibians. It consumes a higher percentage of vertebrates than the Loggerhead, especially in winter. Victims are quickly dispatched in a swift swoop and added to a storage cache of food impaled on thorns or barbs, a behaviour that has earned the shrike the nickname of "Butcher Bird" throughout its range in the Northern Hemisphere. • The juvenile is buffy brown and more heavily barred below than the adult. • *Lanius* is Latin for "butcher," and *excubitor* means "watchman" or "sentinel." "Watchful butcher" does seem a good description of the Northern Shrike's tactics.

ID: sexes similar; medium-sized songbird; grey back and crown; narrow, black mask; strongly hooked bill; white throat and underparts; black wings with white primary patch; long, black tail with white corners. *Juvenile:* brownish grey, "scaly" underparts; less distinct mask. *In flight:* prominent white patches in wings and tail; often hovers.
Size: *L* 25 cm; *W* 38 cm.
Habitat: *Breeding:* sparse coniferous woodlands and muskeg near treeline; open, shrubby braided rivers. *In migration* and *winter:* open country, including fields, shrubby riparian and terrestrial areas, forest clearings, farmyards, towns and roadsides.
Nesting: in a shrub or small tree; bulky nest is made of sticks, bark and mosses; female incubates 4–7 darkly spotted, greenish white to pale grey eggs for 15–17 days.
Feeding: swoops down on prey from a perch or attacks in pursuit; takes mostly small rodents and birds; also eats large insects.
Voice: usually silent; calls are typically harsh and nasal; song is a series of warbles, trills and other notes.
Similar Species: *Loggerhead Shrike* (p. 322): little seasonal overlap; broader "mask"; shorter bill; greyer underparts; immature bird has barred crown and back and less heavily barred underparts. *Northern Mockingbird* (p. 386): narrow eye line; paler eyes; slimmer bill; longer legs.

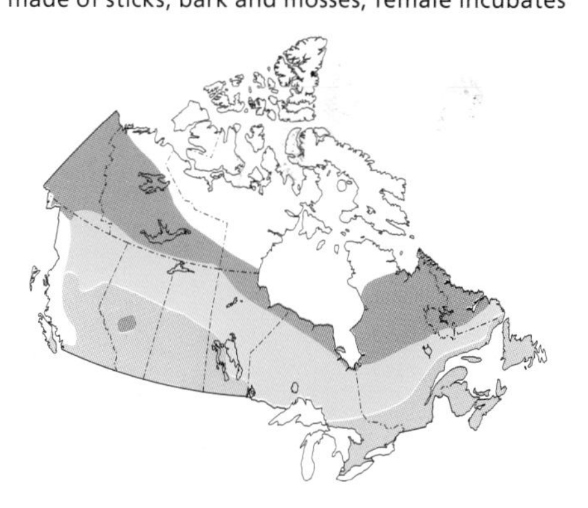

WHITE-EYED VIREO

Vireo griseus

Small, drab land birds of mixed forests, vireos have large bills and sing constantly. White-eyed Vireos can be a challenge to spot as they sneak through dense tangles of branches and foliage in search of insects but can be readily identified by their variable songs, which start and end with an emphatic *chick* note. Expert mimics, White-eyed Vireos often incorporate the vocalizations of other birds such as the Downy Woodpecker, Summer Tanager or Eastern Towhee into their songs. • Even more secretive than the bird itself is the location of its precious nest. Intricately woven with grass, twigs, bark, lichens, moss, plant down, leaves and the fibrous paper of wasp nests, the vireo's nest is hung between the forking branches of a tree or shrub. • Eight vireo species regularly occur in Canada, three of which have restricted ranges. • The White-eyed Vireo is a relatively new addition to Canada's avifauna—the first confirmed breeding record was reported in 1971. Within Canada, breeding is confined to the Carolinian forest zone of southern Ontario, an area that is home to several species of plants and animals not seen elsewhere in Canada.

ID: sexes similar; small, elusive songbird; greenish back and crown; grey face; bold, yellow "spectacles"; white throat; mostly yellow underparts; 2 white wing bars; white eyes (darker in juveniles).
Size: *L* 13 cm; *W* 19 cm.
Habitat: dense, shrubby undergrowth, scrub, abandoned pastures and woodland edges.
Nesting: in a shrub or small tree; nest hangs from a horizontal fork; intricately woven cup is lined with fine grasses and hair; pair incubates 4 lightly speckled, white eggs for 13–15 days.
Feeding: gleans insects from the ground or foliage; rarely eats fruits or seeds.
Voice: loud, snappy, 4–7-note song, usually beginning and ending with *chick* notes; often mimics other birds.
Similar Species: *Yellow-throated Vireo* (p. 325) and *Pine Warbler* (p. 410): yellow throats. *Blue-headed Vireo* (p. 327): white "spectacles"; dark eyes; yellow highlights on wings and tail.

YELLOW-THROATED VIREO

Vireo flavifrons

The Yellow-throated Vireo is usually found in mature deciduous woodlands with little or no understorey, and it takes a particular liking to tall trees, especially oaks and maples, that have a spreading canopy. Like its treetop neighbour, the Cerulean Warbler, the Yellow-throated Vireo forages high above the forest floor, making it a difficult bird to observe, despite its bright plumage. • Unmated males sing tirelessly as they search for nest sites, a process that usually includes placing a few pieces of nest material in several locations. When a female appears, the male dazzles her with its displays and leads her on a tour of potential nesting sites within his large territory. If a bond is established, the pair will mate and build an intricately woven hanging nest in the tree canopy. The male is a devoted helper, assisting the female with building the nest, incubating the eggs and rearing thc young. • Male Yellow-throated Vireos continue singing into August or September.

ID: sexes similar; small, secretive songbird; olive green back and head; bright yellow throat, breast and "spectacles"; grey rump; dark grey wings; 2 white wing bars; white belly and undertail coverts.
Size: *L* 14 cm; *W* 20 cm.
Habitat: mature deciduous woodlands with minimal understorey.
Nesting: in a tree, usually more than 5 m high; pair builds a hanging cup nest in the fork of a horizontal branch; pair incubates 4 darkly spotted, creamy white to pinkish eggs for 14–15 days; each parent takes on guardianship of separate fledglings.

Feeding: forages by gleaning upper canopy branches and foliage; eats mostly insects; also feeds on seasonally available berries.
Voice: song is a slowly repeated series of slurred, 2–3-note phrases, often including a rising *three-eight*; call is a harsh *shu-shu-shu*.
Similar Species: *Pine Warbler* (p. 410): olive yellow rump; thinner bill; faint, darkish streaking along sides; yellow belly; faint "spectacles." *White-eyed Vireo* (p. 324): white "chin" and throat; greyer head and back; white eyes. *Blue-headed Vireo* (p. 327): white "spectacles" and throat; yellow highlights in wings and tail.

CASSIN'S VIREO

Vireo cassinii

Distinct white "spectacles" set against a grey head readily distinguish this compact yet nimble songbird from other forest inhabitants. • In 1997, the former "Solitary Vireo" was split into three separate species: the Cassin's Vireo, which breeds exclusively along the West Coast, the brighter coloured Blue-headed Vireo of central and eastern North America, and the duller Plumbeous Vireo (*V. plumbeus*) of the southwestern United States and Mexico. • Some birders report that the Cassin's Vireo is a fearless nester, often swooping at human intruders near the nest, although incubating birds often allow brief close-range observations. As with other vireos, the Cassin's scolding calls often attract a host of other passerine species to the scene. • Vireos are tireless singers, and male Cassin's Vireos will even sing while on the nest. • Its breeding population is increasing both in western Canada and range-wide. • The Cassin's Vireo is named for John Cassin, a Philadelphia ornithologist who published descriptive accounts of many Pacific Coast birds.

ID: sexes similar; small, noisy songbird; olive grey back and crown; bold, white "spectacles"; short, fairly heavy, slightly hooked bill; white throat and underparts; yellowish sides and flanks; 2 white wing bars; short tail.
Size: *L* 14 cm; *W* 23 cm.
Habitat: open mixed and coniferous woodlands, frequently with dense understorey shrubs and often in drier situations.
Nesting: in a coniferous or deciduous tree; nest is suspended from a fork near the tip of a branch; cup of grasses, spiderwebs, mosses and bark strips is lined with fine grasses and hair; pair incubates 4 darkly spotted, white eggs for 14–15 days.

Feeding: mostly gleans branches, trunk and foliage for insects; sometimes hawks or hovers.
Voice: harsh *ship* and *shep* call notes are often followed by a rising *zink* note; male's song is a series of slow, high-pitched phrases: *See me... Detroit...Surreal.*
Similar Species: *Hutton's Vireo* (p. 328): more uniform, yellowish olive-brown overall; incomplete "spectacles"; smaller bill. *Warbling Vireo* (p. 329): more uniform coloration; creamy white eyebrows and lores; plain upperwings; warbling song. *Ruby-crowned Kinglet* (p. 370): thinner bill; buffy yellow underparts; lacks bold white "spectacles"; different song; constantly flicks wings.

BLUE-HEADED VIREO

Vireo solitarius

From the canopies of mature coniferous and mixedwood forests, the deliberate notes of the Blue-headed Vireo penetrate the dense foliage. The Blue-eyed Vireo prefers a different habitat than many of its relatives and is the only vireo that uses coniferous forests extensively, although it may be found nesting in deciduous situations. The distinctive white "spectacles" that frame this bird's eyes provide a good field mark, and they are among the boldest eye rings belonging to our songbirds. • During courtship, male Blue-headed Vireos fluff out their yellowish flanks and bob ceremoniously to their prospective mates. Once mating is complete and the eggs are in the nest, the parents become extremely quiet. If you are lucky, you might be able to approach close enough to look at the eggs. Once the young hatch, however, Blue-headed parents will readily scold an intruder long before its approach. They have even been seen successfully discouraging marauding Blue Jays. • Breeding bird surveys have revealed highly significant and substantial increases in Blue-headed Vireo numbers in Canada and throughout its range. These increases may be related to formerly logged forests in its nesting range regenerating into suitable habitat.

ID: sexes similar; medium-sized songbird; blue-grey head; green upperparts; white underparts; bold, white "spectacles"; yellow flanks; 2 white wing bars; dark tail; stout bill; dark legs.
Size: *L* 14 cm; *W* 9.5 cm.
Habitat: mixed coniferous-deciduous forests; also pure coniferous forests and pine plantations; rarely pure trembling aspen stands.
Nesting: in a fork of a coniferous tree; hanging, basketlike cup nest is made of grasses, rootlets, plant down, spider silk and cocoons; pair incubates 3–5 whitish eggs for 12–14 days.
Feeding: gleans foliage for invertebrates; occasionally hover-gleans to pluck insects from vegetation.
Voice: song is a slow, purposeful, robinlike *look up...see me...here I am.*
Similar Species: *White-eyed Vireo* (p. 324): yellow "spectacles"; white eyes. *Yellow-throated Vireo* (p. 325): yellow "spectacles" and throat. *Cassin's Vireo* (p. 326): greenish head. *Warbling Vireo* (p. 329), *Red-eyed Vireo* (p. 331) and *Tennessee Warbler* (p. 396): all lack white "spectacles."

HUTTON'S VIREO

Vireo huttoni

Canada's only year-round vireo is found in southwestern British Columbia. During late winter and early spring, the male Hutton's Vireo may wage continuous vocal battles throughout the day to defend his nesting territory. • Secretive and easily overlooked, this species can usually only be located by its persistent singing. But because it is an early nester, many are no longer singing when breeding bird surveys are done during mid- to late June, the peak singing period for most songbirds. This species is also easily confused with kinglets, *Empidonax* flycatchers and even a few warblers. In southwestern BC, it is far more numerous than many people believe. • A persistent "pishing" or a convincing rendition of a Northern Pygmy-Owl call (a series of low-pitched *toot* notes) will attract this year-round sprite from deep woodlands. Its lazier feeding style helps distinguish it from most small flycatchers and fall-plumaged warblers that resemble it. • John Cassin was persuaded by Spencer F. Baird to name this bird for his friend William Hutton, a field collector who first obtained the first specimen of this bird for scientific study in the 1840s.

ID: sexes similar; small, active songbird; olive brown upperparts; olive-tan underparts; incomplete, pale "spectacles"; short, slim, hook-tipped bill; 2 white wing bars.
Size: *L* 10–12 cm; *W* 20 cm.
Habitat: lower-elevation, mixed coniferous forests with openings; also conifer-dominated woodlands with deciduous trees.
Nesting: in a tree or shrub, suspended in a twig fork; deep, round cup of tree lichens is bound with spiderwebs and lined with fine, dry grasses; pair incubates 4 brown-spotted, white eggs for 14 days.

Feeding: gleans foliage and twigs for insects and spiders; may eat seasonal berries.
Voice: call is a rising, chickadee-like *reeee-dee-ree*; male's monotonous song is a nasal, buzzy series of tirelessly repeated, 2-syllable notes: *zuWhEEM, zuWhEEM, zuWhEEM*.
Similar Species: *Cassin's Vireo* (p. 326): grey head; complete, white "spectacles"; whiter throat and underparts. *Ruby-crowned Kinglet* (p. 370): slightly smaller; thinner bill; dark area between wing bars; shorter, notched tail; thinner, darker legs with yellowish feet; very active feeder. *Warbling Vireo* (p. 329) and *Orange-crowned Warbler* (p. 397): lack wing bars. Empidonax *flycatchers* (pp. 307–16): longer tails and bodies; longer, flatter bills; erect posture.

WARBLING VIREO

Vireo gilvus

This vireo lives up to its name—its velvety voice has a warbling quality and lacks the pauses that are distinctive in the songs of other vireos. In eastern North America, the varied phrases end on an upbeat note, as if asking a question, but songs heard in the West frequently have a dropping, fuzzy ending. • Fairly common and widespread in Canada, the Warbling Vireo prefers the upper canopy of forests to local parks and backyards. Because this vireo often settles close to urban areas, its bubbly, warbling songs should be familiar to many residents of southern Canada. A lot of patience is required to see this bird, because the Warbling Vireo's rather drab plumage and slow, deliberate foraging make it difficult to spot among the dappled shadows of its wooded background. • Both males and females incubate the eggs, and incubating males often sing from the nest. The hanging nests of vireos are usually much harder to find than the birds themselves, but in winter, nests are revealed as they swing precariously from bare deciduous branches.

western race

eastern race

ID: sexes similar; small, drab songbird; no "spectacles" or wing bars; olive grey upperparts; dull whitish or pale yellowish underparts; yellowish wash on flanks; creamy white eyebrows and lores; dark eyes; short, hook-tipped bill.
Size: *L* 13 cm; *W* 21 cm.
Habitat: open deciduous woodlands, urban parks and gardens, riparian forests, wooded coulees, shelterbelts and forest fragments; favours poplars and aspens for nesting.
Nesting: usually in a hardwood tree or tall shrub, suspended in a forked twig; compact, deep cup nest consists of bark strips, leaves, plant fibres and grasses; pair incubates 3–4 darkly spotted, white eggs for 13–14 days.

Feeding: gleans tree foliage, often from underneath, for insects; also eats spiders and berries in migration.
Voice: *eah* and *vit* calls; male's song is a squeaky, warbling *receiver receiver receiver receipt!*
Similar Species: *Philadelphia Vireo* (p. 330): dark eye line and lores; yellowish throat and breast. *Red-eyed Vireo* (p. 331): black eye line extends to bill; blue-grey crown contrasts with olive back; red eyes. *Tennessee Warbler* (p. 396): slender bill; pale eyebrow line; dark lores; white undertail. *Orange-crowned Warbler* (p. 397): yellow overall; slimmer bill.

PHILADELPHIA VIREO

Vireo philadelphicus

You will have to search trembling aspen or balsam poplar woodlands to see the Philadelphia Vireo. Like other vireos, Phillies are difficult to observe because they prefer perching and singing near the tops of exceptionally leafy trees. • While many similar-looking birds sound quite different, the Philadelphia Vireo and Red-eyed Vireo are two species that sound very similar but are easy to tell apart once you get them in the binoculars. Most forest songbirds are initially identified by voice, however, so the Philadelphia Vireo is often neglected because its song is nearly identical to that of the more abundant Red-eyed Vireo. • The Philadelphia Vireo has a more northerly breeding range than any other vireo, nesting in areas of the Canadian Shield that are dominated by mixed boreal forest, where it fills a niche left unoccupied by the strictly deciduous-dwelling Warbling Vireo. As with many of Canada's eastern vireos, Philadelphia Vireo populations have increased significantly throughout their range since breeding bird surveys began in 1968, with Québec having the highest densities. • This bird bears the name of the city in which the first scientific specimen was collected.

ID: sexes similar; small, nondescript songbird; greenish grey upperparts including crown; pale yellowish underparts; white eyebrow, dark eye line and lores; pale yellow throat; short, rather thick bill; short tail.
Size: *L* 13 cm; *W* 20 cm.
Habitat: open broadleaf and mixed woodlands with aspen, willow and alder components; also second-growth in burns and cutovers.
Nesting: in a deciduous tree; basketlike nest of grasses, lichens, plant down and bark strips hangs from a horizontal fork; pair incubates 4 darkly spotted, white eggs for up to 13 days.

Feeding: gleans foliage for insects; frequently hovers to search for food in foliage.
Voice: male's song is a continuous, robinlike *look-up way-up tree-top see-me here-I-am*, slightly higher pitched than Red-eyed Vireo.
Similar Species: *Red-eyed Vireo* (p. 331): black-bordered, blue-grey cap; red eyes; lacks yellow wash on belly; song is slightly lower pitched. *Warbling Vireo* (p. 329): duller upperparts; paler underparts; pale lores; partial, dark eye line (mostly behind eye); lacks yellow breast. *Tennessee Warbler* (p. 396): slimmer bill; blue-grey cap and nape; olive green back; white underparts; lacks yellow breast; more active.

RED-EYED VIREO

Vireo olivaceus

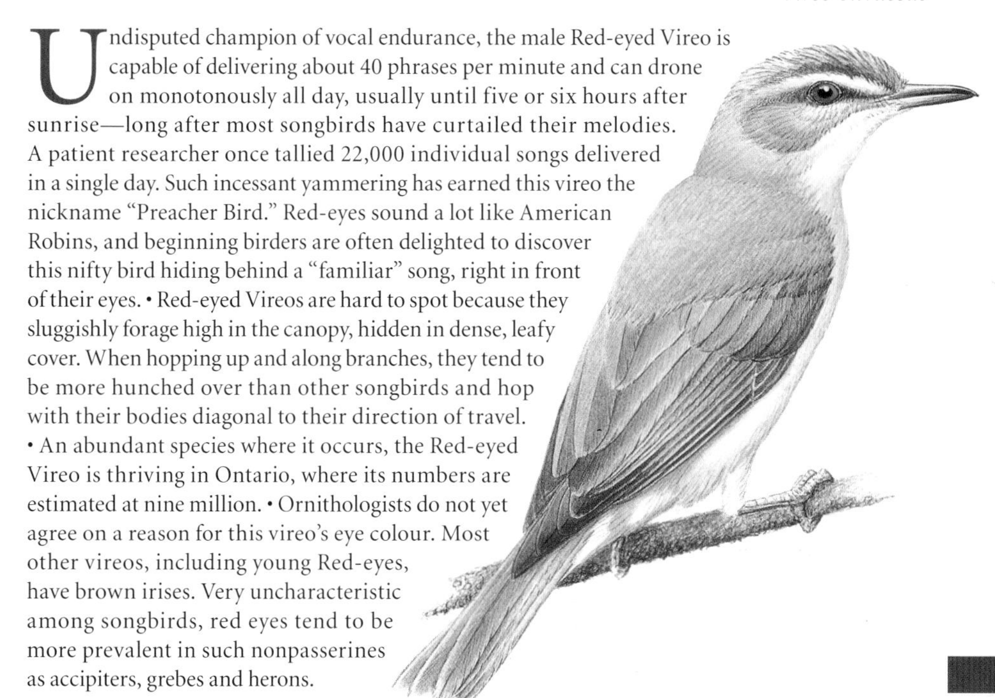

Undisputed champion of vocal endurance, the male Red-eyed Vireo is capable of delivering about 40 phrases per minute and can drone on monotonously all day, usually until five or six hours after sunrise—long after most songbirds have curtailed their melodies. A patient researcher once tallied 22,000 individual songs delivered in a single day. Such incessant yammering has earned this vireo the nickname "Preacher Bird." Red-eyes sound a lot like American Robins, and beginning birders are often delighted to discover this nifty bird hiding behind a "familiar" song, right in front of their eyes. • Red-eyed Vireos are hard to spot because they sluggishly forage high in the canopy, hidden in dense, leafy cover. When hopping up and along branches, they tend to be more hunched over than other songbirds and hop with their bodies diagonal to their direction of travel. • An abundant species where it occurs, the Red-eyed Vireo is thriving in Ontario, where its numbers are estimated at nine million. • Ornithologists do not yet agree on a reason for this vireo's eye colour. Most other vireos, including young Red-eyes, have brown irises. Very uncharacteristic among songbirds, red eyes tend to be more prevalent in such nonpasserines as accipiters, grebes and herons.

ID: sexes similar; large vireo; no wing bars; wholly olive green upperparts; blue-grey crown with black border; white eyebrow; black eye line; red eyes; white throat and underparts; yellow wash on flanks and undertail coverts.

Size: *L* 15 cm; *W* 25 cm.

Habitat: *Breeding:* favours mixed, young to mature deciduous forests, especially with trembling aspen; prefers semi-open canopy and shrubby understorey. *In migration:* prefers hardwood stands, often near water.

Nesting: in a deciduous tree or tall shrub, hanging from a forked twig; dainty, deep cup of bark strips, grasses, plant fibres and rootlets is lined with finer materials; female incubates 2–4 chestnut-spotted, white eggs for 11–15 days.

Feeding: gleans insects, especially larvae and caterpillars, from foliage; also eats berries in fall migration.

Voice: short, scolding *neeah* call; male's song is a continuous, variable run of quick, short phrases separated by short pauses: *Look-up, way-up, in-tree-top, see-me, here-I-am, there-you-are.*

Similar Species: *Philadelphia Vireo* (p. 330): dark eye line and lores; yellowish throat and breast. *Warbling Vireo* (p. 329): paler overall; dark eyes; shorter bill; warbling song. *Tennessee Warbler* (p. 396): blue-grey cap and nape; white eyebrow; olive green back; slimmer bill; dark eyes.

GRAY JAY

Perisoreus canadensis

Few birds in Canada rival Gray Jays for curiosity and boldness. Inquisitive and gregarious by nature, they are always ready to investigate any opportunity. Small family groups glide slowly and unexpectedly out of forests, attracted by the slightest commotion or movement, and are willing to show themselves to passersby, especially if food is available. Gray Jays are well known to anyone who camps or picnics in the coniferous forests—they will make themselves at home at your campsite and will not hesitate to steal scraps of food from an unattended plate. Although they can be bothersome at times, Gray Jays are likable birds and are probably the most mild-mannered of the jays. • Nesting earlier than any other Canadian songbird, Gray Jays build their well-insulated nests, lay their eggs and begin incubation as early as late February, allowing them to supply their nestlings with the first foods of spring. • Gray Jays often store food for future use. By coating it with sticky mucus from specialized salivary glands, they both preserve it and render it unappetizing to other birds and forest mammals. • The nickname "Whiskey Jack" is derived from the Algonquin name for this bird, *wis-kat-jon*; other names include "Canada Jay" and "Camp Robber."

ID: sexes similar; medium-sized corvid; fluffy, light grey plumage is darker and browner above and paler below; dark hood (reduced on inland birds); pale forehead; white cheek and nape; long, rounded, white-tipped tail; short, black bill. *Immature:* dark grey body; light grey "moustache." *In flight:* distinctive, bouncy flight with alternating fast flaps and short glides, usually close to the ground.

Size: *L* 29 cm; *W* 46 cm.

Habitat: coniferous and mixedwood forests, riparian thickets, bogs, subalpine and alpine areas; also picnic sites and campgrounds.

Nesting: in a conifer; saddled on a branch or in a crotch; bulky cup of sticks, bark strips, mosses and grasses is lined with softer materials; female incubates 3–5 finely spotted, greyish white eggs for 16–18 days.

Feeding: opportunistic omnivore; eats berries, fungi, insects, small mammals, bird eggs and carrion; stores food at scattered cache sites.

Voice: complex vocal repertoire includes soft, whistled *quee-oo*, chuckled *cla-cla-cla* and *churr*; also imitates other birds.

Similar Species: *Clark's Nutcracker* (p. 335): mostly pale grey; much longer bill; black-and-white wings and tail. *Loggerhead Shrike* (p. 322) and *Northern Shrike* (p. 323): black mask; larger, hooked bill; black-and-white wings and tail; favours open country. *Northern Mockingbird* (p. 386): white wing patch and outer tail feathers; longer, slimmer bill.

STELLER'S JAY

Cyanocitta stelleri

With its dark crest and velvet blue feathers, the stunning Steller's Jay is a resident jewel of western forests. Most common in dense coniferous forests, Steller's Jays frequently irrupt away from traditional ranges; this is most noticeable each fall on southern Vancouver Island. • The Steller's Jay and the Blue Jay are the only North American jays sporting crests. Four races of Steller's Jay occur in BC, and plumage differences are subtle. Over the past two decades, the eastern Blue Jay has expanded its range into Steller's territory, resulting in some oddball hybrids. • During the coldest months, these remarkably beautiful birds frequently visit townsites to feed greedily and clumsily on feeders that feature peanuts and sunflower seeds. Steller's Jays are bold, and, like all corvids, they will not hesitate to steal food scraps from inattentive picnickers. • Georg Wilhelm Steller, the first European naturalist to visit Alaska, collected the "type" specimen of this species in 1741. • The Steller's Jay is British Columbia's provincial bird.

ID: sexes similar; medium-sized corvid; glossy, deep blue upperparts; shaggy black crest; blackish head, chest and back; medium blue underparts; finely brown-barred, deep blue wings and tail; stout, black bill; forehead markings and eye crescents vary with race. *In flight:* short, round-tipped, blue tail; leisurely flight with short glides and little upward lift.
Size: *L* 29 cm; *W* 48 cm.
Habitat: *Breeding:* mixed coniferous forests. *In migration* and *winter:* variety of forested habitats; often in campgrounds, picnic areas and townsites.
Nesting: on a conifer branch; bulky cup of twigs, mosses and dry leaves is cemented with mud and lined with rootlets, pine needles and grasses; female incubates 4–5 brown-marked, pale greenish blue eggs for 16 days.
Feeding: omnivorous; eats mainly vegetable matter; also takes small mammals, bird eggs and young and carrion; visits feeders for sunflower seeds, peanuts and suet.
Voice: varied calls; harsh, far-carrying *shack-shack-shack* and grating *kresh kresh.*
Similar Species: *Blue Jay* (p. 334): light blue crest; greyish white face; sky blue upperparts with white wing and tail markings; thin, black collar.

BLUE JAY

Cyanocitta cristata

Beautiful, resourceful and vocally diverse, the Blue Jay embodies all the admirable traits of the corvid family but at times can become too noisy, overly bossy and mischievous. This loud, striking bird can be quite aggressive when competing for sunflower seeds and peanuts at backyard feeders. It rarely hesitates to drive away smaller birds, squirrels and even cats. It seems there is no predator, not even a Great Horned Owl, too formidable for this bird to cajole or harass. Conversely, when nesting, Blue Jays are so quiet and secretive that they usually go unnoticed by all but the most observant of birders. • Jays prolifically cache nuts and are very important to the forest ecosystem. In autumn, one Blue Jay might bury several thousand acorns, forgetting where most are hidden and thus planting many oak trees. • The Blue Jay has expanded its range westward over the years, partly as a result of the spread of human development. Feeders and landfills enable wintering birds to cope with harsh winters, and forest fragmentation has invited jays deeper into once-impenetrable forests. • If a Blue Jay regularly visits your yard, sketch or photograph the bird—individuals can be recognized by their characteristic head patterns. Then you can follow its visits to your feeder over the years.

ID: sexes similar; medium-sized corvid; mostly blue upperparts; conspicuous crest; white face and upper breast boldly outlined with black; pale grey and white underparts; white wing and tail patches; long, tapering tail with black bars and white corners.
Size: *L* 28 cm; *W* 41 cm.
Habitat: open deciduous and mixed woodlands, scrubby fields, suburban and urban parks and backyards.

Nesting: usually in a conifer; bulky stick nest is lined with mud and fine rootlets; pair incubates 4–7 variably marked, greenish, buff or pale blue eggs for 16–18 days.
Feeding: forages on the ground and among vegetation for a wide range of food; diet includes insects, nuts, seeds, berries, carrion and bird eggs and nestlings; frequents feeders for nuts, suet and sunflower seeds; also caches acorns in soil.
Voice: calls include *jay-jay-jay* and *queedle-queedle queedle-queedle*; expert mimic of various sounds, including the calls of some hawks.
Similar Species: *Steller's Jay* (p. 333): occurs farther west, ranges do not overlap; dark crest and head; dark underparts. *Belted Kingfisher* (p. 288): larger; unpatterned face; breast band; very different habits.

CLARK'S NUTCRACKER

Nucifraga columbiana

Occurring only in the mountains and foothills of British Columbia and Alberta, the Clark's Nutcracker provides the perfect complement to the snow-capped peaks of the Canadian Rockies. These birds have loud, dominant personalities, and they make themselves known at many viewpoints and picnic areas in our mountain parks. • During snow-free months, the Clark's Nutcracker busily stores food for winter. Ground caches, which can contain more than 33,000 pine seeds, might be miles apart, but the Clark's Nutcracker can remember where they are located for as long as nine months. These food stores allow the species to remain at high altitudes even when deep snow covers high mountain ridges. Because many of the cached seeds remain unrecovered, the species is a significant agent of whitebark pine dispersal. • When explorer Captain William Clark of the Lewis and Clark expedition saw the large, straight bill, he thought this raucous, gregarious bird was a woodpecker, so he placed it in the new genus *Picicorvus*, meaning "woodpecker-crow." The species has since been reclassified. • Despite its name, this bird actually "cracks" more conifer cones than nuts.

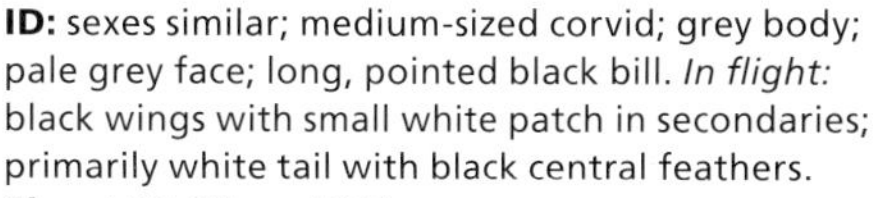

ID: sexes similar; medium-sized corvid; grey body; pale grey face; long, pointed black bill. *In flight:* black wings with small white patch in secondaries; primarily white tail with black central feathers.
Size: *L* 30–33 cm; *W* 61 cm.
Habitat: open coniferous and mixed forests, scenic overlooks, picnic sites and towns; moves to lower elevations in winter; mixed woods during irruptions.
Nesting: saddled on a conifer limb; twig and stick platform nest is lined with grass and strips of bark; pair incubates 2–4 darkly marked, pale green eggs for 16–18 days.

Feeding: omnivorous; eats mainly pine seeds; also other seeds, fruit, insects, bird eggs and nestlings and carrion; stores food for winter.
Voice: usual call is a grating *skraaaaaaa*, usually from high perch; higher-pitched and yelping calls.
Similar Species: *Gray Jay* (p. 332): smaller; fluffy, light grey plumage is darker and browner above and paler below; shorter bill; uniformly grey tail; lacks white in wings.

BLACK-BILLED MAGPIE

Pica hudsonia

Magpies are too often maligned because of their raucous demeanor and occasional nest robbing forays, but visitors who have never seen a Black-billed Magpie before are often captivated by this bird's exceptional beauty and approachability. In the past, contests were held and bounties were offered in some prairie communities to control magpie numbers. Fortunately, these adaptable birds persevered, and many people now recognize the valuable role they play in consuming large numbers of grasshoppers and other harmful insects. Its constant search for food leads it from plucking ticks off the backs of large mammals to turning over stones for grubs. • The Black-billed Magpie is an exceptional architect. Pairs construct large, domed stick nests that might take 40 days to complete. Abandoned nests remain in trees for years and are often reused by other birds such as Merlins and Long-eared Owls. • Until recently, the Black-billed Magpie, which is largely confined to western North America, was considered to be the same species as the widespread Magpie (*P. pica*) of Europe, North Africa and Asia. • Albino magpies, with pale grey and white plumage, are extremely rare but do occur in Canada.

ID: sexes similar; large corvid; iridescent, greenish blue upperparts, breast and undertail coverts; white shoulders and belly; stout black bill; long, iridescent, greenish blue tail; long, black legs. *In flight:* rounded wings with large, white areas near tips.

Size: *L* 46–56 cm; *W* 61–66 cm.

Habitat: open country, especially grasslands and agricultural areas with scattered trees and tall shrubs; subalpine with mature spruce patches; also cities and towns.

Nesting: in a tall, dense shrub or tree; large, conspicuous, domed stick nest contains a bowl of mud lined with rootlets, fine plant stems and hair; female incubates 5–13 brown-marked, greenish grey eggs for 16–21 days.

Feeding: omnivorous; eats mostly insects, but searches the ground, gleans foliage and hawks for a variety of invertebrates, small vertebrates, fruit and seeds; also eats carrion, especially roadkill, and eggs and nestlings of other birds.

Voice: many vocalizations including loud, nasal, frequently repeated *yeck-yeck-yeck* or *wah-wah-wah-wah*; also a nasal *aag-aag*.

Similar Species: none.

AMERICAN CROW

Corvus brachyrhynchos

Probably the most intelligent bird, the American Crow is inquisitive, cunning and social, adapting to an amazing variety of habitats, food resources and environmental conditions. They have learned, for instance, to feed on roadkills along highways, calmly walking onto the shoulder of the road when oncoming traffic approaches too closely. They have also been observed dropping clams or chestnuts from great heights onto a hard surface to crack the shells, one of the few examples of birds using objects to manipulate food. • Crows use a variety of calls to communicate and can expertly mimic human speech and some wildlife sounds. • The American Crow was once highly migratory in the northern part of its range, but landfills and waste grain have become sources of winter food and have contributed to this bird's year-round occurrence in parts of southern Canada. • Flocks can be composed of hundreds or even thousands of individuals—these impressive gatherings, known as "murders," are merely get-togethers in preparation for evening roosts or preambles for the fall exodus. In southwestern Ontario, as many as 90,000 crows might roost together on any given winter night. • The American Crow's cumbersome-sounding scientific name, *Corvus brachyrhynchos*, means "raven with the small nose."

ID: sexes similar; large corvid; glossy black plumage; stout, black bill; broad wings; short, fan-shaped tail. *In flight:* rarely glides.
Size: *L* 43–53 cm; *W* 94 cm.
Habitat: almost any semi-open habitat, especially agricultural areas, hedgerows and farm shelterbelts; also urban areas; avoids dense coniferous forests.
Nesting: in the canopy of a tree or tall shrub; bulky stick nest is lined with shredded bark, mosses, grasses and hair; pair incubates 4–9 variably marked, bluish or olive green eggs for 18 days.
Feeding: omnivorous and opportunistic; often feeds on the ground in small groups; eats almost anything, including small animals, insects, bird eggs and nestlings, carrion and garbage; also eats acorns, fruit and berries.
Voice: distinctive, far-carrying *caw-caw-caw.*
Similar Species: *Common Raven* (p. 339): larger and much heavier; heftier bill; shaggy throat; wedge-shaped tail; glides with wings held flat. *Northwestern Crow* (p. 338): restricted to coastal BC; slightly smaller; hoarser, lower-pitched calls.

NORTHWESTERN CROW

Corvus caurinus

The Northwestern Crow is a year-round coastal species found only from southeastern Alaska to northwestern Washington. It has adapted well to both natural and developed areas but is most closely associated with marine environments, and Northwestern Crows can be seen foraging along tidal pools and reefs exposed at low tide. Highly adaptable in its feeding habits, it survives by digging clams and catching small crabs and fish on beaches, snatching eggs from a variety of seabirds and pilfering food from picnic areas or garbage cans. During spawning runs, small flocks may follow fish upstream for up to 120 kilometres. • Northwestern Crows do not migrate, but local movements do occur. The clearing of coastal forests and this bird's adaptable nature has led to considerable expansion of its historic range in coastal British Columbia. • Like other corvids, this species is monogamous, and pairs form long-term bonds. • During nonbreeding seasons, spectacular flocks containing thousands of individuals roost at favourite sites in British Columbia. • Some experts argue that the Northwestern Crow is really a subspecies of the American Crow, differing only in size and voice. The common call of the Northwestern Crow is hoarser and slightly lower pitched than the call of an American Crow.

ID: sexes similar; large corvid; glossy, black plumage; stout, black bill; broad wings; short, fan-shaped tail; slightly smaller than American Crow.
Size: *L* 41 cm; *W* 86 cm.
Habitat: prefers marine shorelines with rocky headlands, rocky beaches and tidal mudflats; also found in agricultural areas, garbage dumps, residential areas, playgrounds and golf courses.

Nesting: in a tree or tall, dense shrub; occasionally nests on the ground in seabird colonies; bulky stick nest contains mud cup lined with bark strips, plant stems and hair; female incubates 1–6 brown-marked, bluish green or olive green eggs for 18–20 days.
Feeding: omnivorous and opportunistic; often feeds on the ground in small groups; eats mostly marine invertebrates and small fish; also takes seabird eggs, carrion, garbage and insects.
Voice: call is hoarser and slightly lower pitched than American Crow.
Similar Species: *American Crow* (p. 337): slightly larger; found in interior regions. *Common Raven* (p. 339): larger and much heavier; heftier bill; shaggy throat; wedge-shaped tail.

COMMON RAVEN

Corvus corax

The Common Raven is has been likened to a crow that has convinced itself that it's a raptor. Whether stealing food from a flock of gulls, harassing a soaring hawk in midair, dining from a roadside carcass or confidently strutting among campers at a park, this species is worthy of its reputation as a bold and clever bird. Ravens exhibit behaviours that many think of as exclusively human, executing tumbling aerobatic feats, performing complex vocalizations and playfully sliding down snowbanks on their backs. They perform actions with an individuality and practised complexity that goes far beyond the instinctive behaviour of most other birds. • The Common Raven's vocal repertoire includes an astonishing variety of sounds that are enhanced by local dialects. • Ravens can live as long as 50 years, and pairs maintain loyal, lifelong pair bonds. • Few other birds occupy such diverse ecological and geographical ranges as the raven, which survives along coastlines, in arid deserts, atop the tallest mountains, in the Arctic tundra and in human settlements.

ID: sexes similar; very large corvid; iridescent, all-black plumage; large head; heavy, black bill; shaggy throat. *In flight:* wedge-shaped tail; much soaring.
Size: *L* 61 cm; *W* 1.2–1.3 m.
Habitat: almost anywhere, including alpine areas, mountains, grasslands, parklands, lakes, marshes, canyons, seashores and cities.
Nesting: on a cliff ledge, train trestle, bridge, transmission tower, building or in a tree; bulky stick nest is lined with shreds of bark and hair; female incubates 4–6 brown- or olive-marked, greenish eggs for 18–21 days.
Feeding: omnivorous and varied diet; searches for food from the air; eats mostly meat, including carrion, small mammals and bird eggs and nestlings (especially at seabird colonies); patrols highways for roadkill; also feeds at landfills.
Voice: deep, loud, guttural, repetitive *craww-craww* or *quork quork*; also croaking *cr-r-ruck*, metallic *tok*, and many other vocalizations, including remarkably varied songs heard only at close range.
Similar Species: *American Crow* (p. 337) and *Northwestern Crow* (p. 338): smaller and slimmer; smaller bills; rounded tails; smooth throats; higher-pitched calls.

SKY LARK

Alauda arvensis

Few birds have inspired more people with their songs than the Sky Lark. In 1820, poet Percy Bysshe Shelley wrote "...That from heaven, or near it / Pourest thy full heart / In profuse strains of unpremeditated art." The Sky Lark's continuous warble, delivered high in the air and lasting up to 30 minutes for a completed song, is the longest of any North American bird. • The Sky Lark was introduced to southwestern British Columbia from Great Britain in 1903. The Fraser River delta population failed, but birds on southern Vancouver Island fared better. After a few more birds were released, small numbers became well established, and by the 1960s, their heyday, Sky Larks were a common sight around Victoria and on the Saanich Peninsula to Sidney. By the mid-1990s, there were fewer than 100 birds remaining on Vancouver Island and the San Juan Islands in Washington. Loss of breeding and foraging habitat to increased urbanization and land use changes appears to be the main reason for ongoing declines of the Sky Lark. Birders from around the world visit southern Vancouver Island to add the species to their North American list, but the Sky Lark could be extirpated from Canada in our lifetime.

ID: sexes similar; cryptically coloured, ground-dwelling songbird; brown upperparts; buff white underparts; short, brownish streaks on breast and sides; small crest; short, dark tail with pale centre. *Immature:* spotted and speckled underparts. *In flight:* white trailing edges on wings.
Size: *L* 18 cm; *W* 33 cm.
Habitat: *Breeding:* dry, open, short- to moderately high-grass fields and pastures in managed agricultural lands and airport fields. *Winter:* short-grass fields, cultivated land, grassy beaches, playing fields and airfields.
Nesting: on the ground among grasses and forbs; compact nest of coarse grasses, plant stems, leaves and rootlets is lined with finer dry grasses; female incubates 3–4 heavily brown-marked, greyish white or buff eggs for 11 days.
Feeding: omnivorous; forages on the ground mainly for weed seeds and grain; also eats insects, especially beetles, and sprouting plant leaves in summer.
Voice: long, complex warble of gurgles, whistles and liquid notes given from high in the air while circling or hovering; flight call is *cherrup* or *drirdrirk*.
Similar Species: *American Pipit* (p. 390): buff underparts with brown streaks; black tail with white outer tail feathers; lacks crest; constantly bobs tail.

HORNED LARK

Eremophila alpestris

Almost always encountered in open country, the Horned Lark nests from lowland grasslands and alpine meadows to High Arctic tundra. Its tinkling, aerial song and high, swooping courtship flight are a springtime treat. • Horned Larks migrate north at the first hint of warmth, while snow still covers the fields. In much of Canada, small flocks of Horned Larks are among the earliest birds to return to nesting areas, settling on small openings in fields and along roadsides as early as February. • Horned Larks rely on their disruptive coloration, mouselike foraging technique and low profile among scattered grass tufts to avoid detection. They might flush from open rural roadsides as cars pass by—watch for the blackish tail that contrasts with the sandy-coloured body. A good look at a perched lark reveals a handsome bird sporting a black mask, twin tiny horns and pale yellow underparts smudged with a dark crescent across the chest. • One way to distinguish a Horned Lark from a sparrow is by its method of travel: Horned Larks walk, whereas sparrows hop. • Carl Linnaeus, who encountered the species in Europe, named the Horned Lark *Alauda alpestris,* "lark of the mountains."

♂

ID: sparrow-sized songbird; mostly light brown upperparts; yellowish face and throat; black mask and breast band; white underparts; dark legs; dark tail with white outer tail feathers. *Male:* small, black "horns"; tapered, black mark from bill to cheek; black breast band; rufous tints are common. *Female:* duller brown replaces rufous and black; paler yellow areas; no "horns."

Size: *L* 18 cm; *W* 30 cm.

Habitat: *Breeding:* open areas, including pastures, rangeland, airfields, alpine meadows and tundra. *In migration* and *winter:* croplands, roadside ditches and fields.

Nesting: on the ground; in a shallow scrape among short, sparse vegetation; nest of grasses, plant stems and dead leaves is lined with finer grasses and plant down; female incubates 3–4 cinnamon-spotted, grey eggs for 11–12 days.

Feeding: forages mainly on the ground for insects and seeds.

Voice: call is a tinkling *tsee-titi* or *zoot*; flight song is a long series of tinkling, twittered whistles.

Similar Species: *Sparrows* (pp. 435–56), *Longspurs* (pp. 458–61), *American Pipit* (p. 390) and *Sprague's Pipit* (p. 391): all lack "horns," the distinctive facial pattern and solid black breast band.

PURPLE MARTIN

Progne subis

Purple Martins have been associated with humans for centuries. Native North Americans placed hollow gourds around their villages to lure these handsome swallows. Historically, Purple Martins nested in tree cavities, old woodpecker holes and natural crevices, but today, this adaptable species prefers nest boxes. If only Purple Martins were as common as their apartment-style houses! House Sparrows and European Starlings are often blamed for the failure of a Purple Martin house to attract birds, but they might not be the problem. For best results, place the house high on a pole in the middle of a large, open area near water and paint it white, so the birds do not overheat. A Purple Martin colony can provide an endlessly entertaining summer spectacle as the adults spiral around the house, and the young perch clumsily at the opening of their apartment cavity. Over winter, the cavity openings should be covered to prevent House Sparrows or starlings from moving in. • Breeding bird surveys have shown that Purple Martins are in significant decline in Canada, except in Alberta and Saskatchewan. • Despite their reputation as superb mosquito eaters, Purple Martins feed on a variety of flying insects but rarely eat mosquitoes.

ID: large swallow; dark blue upperparts; large head and bill; broad, pointed wings; shallowly notched tail. *Male:* wholly dark bluish black. *Female:* grey forehead; scaly-looking, dark and whitish underparts; chest is darker than belly. *Immature:* greyish throat; brownish wing and tail feathers.
Size: *L* 18–20 cm; *W* 46 cm.
Habitat: *Breeding:* open country near salt and fresh water, including seashores, harbours and rivers; attracted to martin condos in towns, farmyards and semi-open areas, often near water.
Nesting: loosely colonial; usually in a nest box; nest materials include feathers and grasses; female incubates 4–8 white eggs for 15–18 days.
Feeding: aerial forager; usually eats flies, ants, dragonflies, mosquitoes and other flying insects; rarely takes ants, insects or berries on the ground.
Voice: rich, pleasant, chirping *pew-pew* or *tchew-wew*; gurgling song often heard in flight.
Similar Species: *Tree Swallow* (p. 343): smaller; bright white underparts. *European Starling* (p. 389): longer bill (yellow in summer); lacks notched tail.

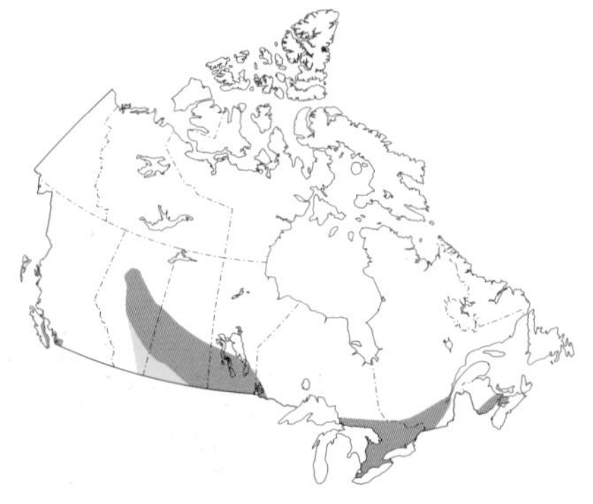

TREE SWALLOW

Tachycineta bicolor

Nesting pairs of Tree Swallows are often seen perched along rural fence lines adorned with nest boxes or flying above the tree-lined shores of lakes and wetlands. Although these swallows are often not the intended occupants of the bluebird nest boxes, they nonetheless prove to be delightful tenants. Anyone with Tree Swallows nesting nearby appreciates this bird's enormous appetite for flying insects, including mosquitoes. When feeding nestlings, parents are known to return to their young with a mouthful of insects 10 to 20 times per hour—about 140 to 300 times a day! • These hardy swallows appear in Canada as early as mid-March, and they are back in force by the end of April. Early returns often get a leg up in the fierce competition for nesting cavities, but the occasional late-spring storm can result in high mortality for these and other early returning insect-eaters. Tree Swallows and other cavity nesters benefit greatly when land-owners and foresters allow some dead trees to remain standing, especially near water. • The back of a male Tree Swallow appears steely blue in bright spring sunshine but looks much greener in late summer and fall as moult progresses.

ID: sexes similar; iridescent, bluish black upperparts; dark crown includes eyes; white underparts; dark rump; long, pointed wings; shallowly forked tail. *Immature:* dusky brown upperparts.
Size: *L* 14 cm; *W* 35–38 cm.
Habitat: *Breeding:* open areas, usually near water, in natural cavities or nest boxes. *In migration:* mostly ice-free lakes, rivers, marshes and sloughs.
Nesting: primarily singly, occasionally in loose colonies; in a natural cavity, old woodpecker hole or nest box; nest of plant stems, grasses and rootlets is lined with fine grasses and feathers; female incubates 4–7 white eggs for 14–15 days.
Feeding: plucks flying insects from the air; in cold spells may eat berries and plant parts.
Voice: call is a soft, twitterlike *silip* or *chi-veet*; song is a liquid, twittering *weet trit weet*.
Similar Species: *Violet-green Swallow* (p. 344): white patches on sides of rump; white on face extends above eyes; multicoloured upperparts; female is duller with smudgy head pattern. *Bank Swallow* (p. 346): brown upperparts. *Northern Rough-winged Swallow* (p. 345): brown throat. *Purple Martin* (p. 342): male is dark overall; female has sooty grey underparts.

VIOLET-GREEN SWALLOW

Tachycineta thalassina

Demonstrating a greater aptitude than the Tree Swallow for taking advantage of diverse habitats, the Violet-green Swallow is less reliant on riparian habitats and is found around cliffs, in towns and cities and in treeless, open areas. Violet-greens often forage with swifts at elevations of 300 to 600 metres, far higher than other swallows normally feed. • At first sight, the Violet-green Swallow looks a lot like a Tree Swallow, which is to be expected, given their close relationship. The Violet-green, a western species, nests in cliff crevices, crannies in buildings, nest boxes and, occasionally, in old Cliff Swallow or Barn Swallow nests. • This swallow occasionally eats mineral-rich soil, eggshells and shellfish fragments, possibly to renew the minerals lost during egg formation. • Swallows are swift and graceful flyers, routinely travelling at speeds of 50 kilometers per hour. • The genus name *Tachycineta* is from a Greek word meaning "I move fast," and *thalassina* is Latin for "sea green," the latter a reference to the male's upper plumage, which is reminiscent of shallow inshore ocean waters.

♂

ID: white underparts; white-sided rump; long, pointed wings extend well beyond short, notched tail on perched bird. *Male:* iridescent, green crown, nape and back; white on face extends above and behind eyes; purplish wings, rump and tail. *Female:* duller; browner head, sides of face and throat.

Size: *L* 13 cm; *W* 36 cm.

Habitat: *Breeding:* open woodlands, wooded canyons, towns and open areas near water. *In migration:* any open area, including agricultural fields.

Nesting: singly or loosely colonial; in a cavity in a tree, cliff, bank or building, or in a nest box or old swallow nest; cup of grasses and plant stems is lined with feathers; female incubates 4–6 white eggs for 14–15 days.

Feeding: feeds mostly on flying insects.

Voice: exuberant, irregular chatter: *ch-ch-ch-ch-chairTEE, chairTEE-ch-ch*; male's song is a squeaky series of single notes and rapid *tsip* repetitions.

Similar Species: *Tree Swallow* (p. 343): iridescent, bluish black upperparts; white on face stops abruptly below eyes; folded wings do not project beyond tail. *Bank Swallow* (p. 346) and *Northern Rough-winged Swallow* (p. 345): brown upperparts; lack white cheeks.

NORTHERN ROUGH-WINGED SWALLOW

Stelgidopteryx serripennis

The Northern Rough-winged Swallow is the cool-temperate member of a group of swallows that wear almost the same colours as the earthen banks in which they live. These low-flying aerialists nest in burrows alongside streams and road cuts. Occasionally, a pair might nest among a large colony of Bank Swallows, and they will also make use of the abandoned diggings of other birds and rodents. • Usually seen in small numbers, Rough-wings frequently join mixed flocks of swallows hawking insects over rivers or lakes. The Rough-wings are often completely overlooked among their similar-looking cousins. In flight, they look very much like a Bank Swallow, except that Rough-wings lack a distinct breast band, their wings are slightly longer and their wingbeats are deeper and slower. • Unlike other swallows, male Northern Rough-wings have curved barbs along the outer edges of their primary wing feathers. Although the purpose of this sawtoothed edge is a mystery, it's thought it might be used to produce sound during courtship displays. The Rough-winged Swallow's common and scientific names relate to this structure: *Stelgidopteryx* means "scraper wing," and *serripennis* means "saw feather."

ID: sexes similar; small swallow; dull brownish grey upperparts, including the head; indistinct brown wash on throat and breast; remainder of underparts whitish; broad wings; square tail. *Juvenile:* wide, rufous wing bars. *In flight:* plain brown body.

Size: *L* 14 cm; *W* 36 cm.

Habitat: *Breeding:* open areas with earthen banks, such as marine shores, road cuts, gravel pits and river margins; also drain pipes in seepage cliffs and bridges. *In migration:* forages over most habitats, especially water and fields.

Nesting: usually solitary but sometimes in a small colony; pair reuses an old burrow or excavates a new one in an earthen bank; nest chamber is lined with leaves and dry grasses; mostly the female incubates 4–8 white eggs for 12–16 days.

Feeding: catches flying insects on the wing; occasionally eats insects from the ground; drinks on the wing; generally forages at low heights.

Voice: short, squeaky *zzzrrit*, ordinarily given in a loose series of 3–6 calls.

Similar Species: *Bank Swallow* (p. 346): dark breast band; colder brown upperparts; shorter wings; shallower wingbeats. *Tree Swallow* (p. 343): female has green upperparts and clean white breast. *Violet-green Swallow* (p. 344): female has green upperparts, white cheek and rump patches and white breast. *Cliff Swallow* (p. 347): dark blue-grey head and back; buff forehead and rump patch; square tail.

BANK SWALLOW

Riparia riparia

The Bank Swallow, known as "Sand Martin" in Europe, is North America's smallest swallow. Although this bird forages in large flocks, the most spectacular sight is at breeding colonies, which can range anywhere from a few pairs to thousands of birds. Adults create a constant flurry of activity as they fly back and forth, delivering insects to their insatiable young. • Bank Swallows usually excavate their own nest burrows, first using their small bills and later digging with their feet. Most nestlings are safe from predators within their nest chamber, which can be up to 1.5 metres long but is typically 60 to 90 centimetres in length. However, because the nest is usually close to the top of a bank or cliff face, persistent predators can dig down to the burrow from above. • Bank Swallow populations throughout Canada are in decline, with some colonies now absent from traditional breeding locations. • In medieval Europe, it was believed that swallows overwintered in mud at the bottom of swamps because the huge flocks seemed to disappear overnight after breeding, and it seemed beyond imagining that these birds flew south for the winter!

ID: sexes similar; small, slender swallow; pale brown upperparts; brown crown and cheeks; white forehead, throat and ear patches; white underparts with dark, well-defined breast band; long, thin wings; long, notched tail.
Size: *L* 13 cm; *W* 33 cm.
Habitat: *Breeding:* stabilized, vertical, sandy banks, usually near water; feeds mostly over wetlands and shrublands. *In migration:* open areas near water.

Nesting: colonial; in a burrow up to 1.5 m long, near the top of a cliff face or bank; nest chamber is lined with plant materials and feathers; pair incubates 3–5 white eggs for 13–15 days.
Feeding: catches flying insects and drinks on the wing.
Voice: call is a series of buzzy twitters: *zzzrt, zzzrt.*
Similar Species: *Northern Rough-winged Swallow* (p. 345): lacks well-defined breast band. *Violet-green Swallow* (p. 344) and *Tree Swallow* (p. 343): generally glossy above; some birds drab, appearing brown above, but with pure white underparts.

CLIFF SWALLOW

Petrochelidon pyrrhonota

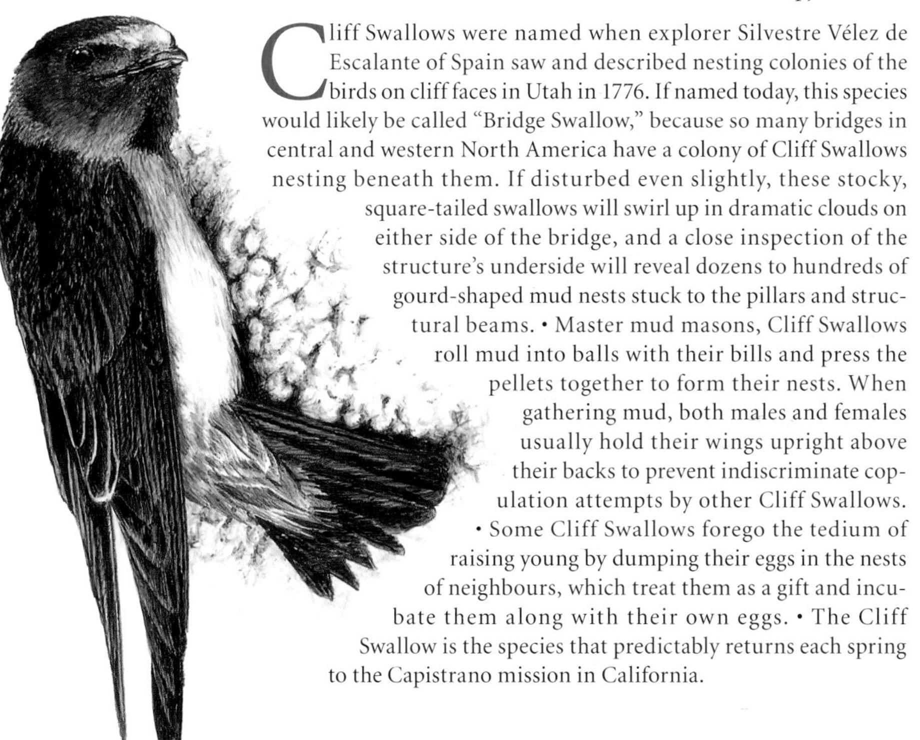

Cliff Swallows were named when explorer Silvestre Vélez de Escalante of Spain saw and described nesting colonies of the birds on cliff faces in Utah in 1776. If named today, this species would likely be called "Bridge Swallow," because so many bridges in central and western North America have a colony of Cliff Swallows nesting beneath them. If disturbed even slightly, these stocky, square-tailed swallows will swirl up in dramatic clouds on either side of the bridge, and a close inspection of the structure's underside will reveal dozens to hundreds of gourd-shaped mud nests stuck to the pillars and structural beams. • Master mud masons, Cliff Swallows roll mud into balls with their bills and press the pellets together to form their nests. When gathering mud, both males and females usually hold their wings upright above their backs to prevent indiscriminate copulation attempts by other Cliff Swallows. • Some Cliff Swallows forego the tedium of raising young by dumping their eggs in the nests of neighbours, which treat them as a gift and incubate them along with their own eggs. • The Cliff Swallow is the species that predictably returns each spring to the Capistrano mission in California.

ID: sexes similar; small swallow; dark blue-grey head and back; rusty cheek, nape and throat; pale collar; buff forehead and breast; black throat; pale underparts; orange rump; blackish wings; square, black tail.

Size: *L* 14 cm; *W* 33 cm.

Habitat: *Breeding:* mostly riparian, especially canyons and sites with suitable artificial structures such as bridges and buildings. *Foraging:* rivers, marshes, grasslands and other open country; also towns and, less often, cities.

Nesting: colonial; gourd-shaped mud nest is attached to a rock face or rough surface of an artificial structure; nest is lined with dry grasses; pair incubates 3–4 finely marked, whitish eggs for 11–16 days.

Feeding: catches flying insects and drinks on the wing.

Voice: call is a twittering *churrr-churrr;* also a *nyew* alarm call.

Similar Species: *Barn Swallow* (p. 348): brighter, more extensively blue upperparts; orangey red to white underparts; longer, deeply forked tail. *Tree* (p. 343), *Bank* (p. 346) and *Northern Rough-winged* (p. 345) *swallows:* all lack buff forehead and orange rump patch. *Cave Swallow* (p. 504): vagrant; rufous forehead; tawny rump; pale throat.

BARN SWALLOW

Hirundo rustica

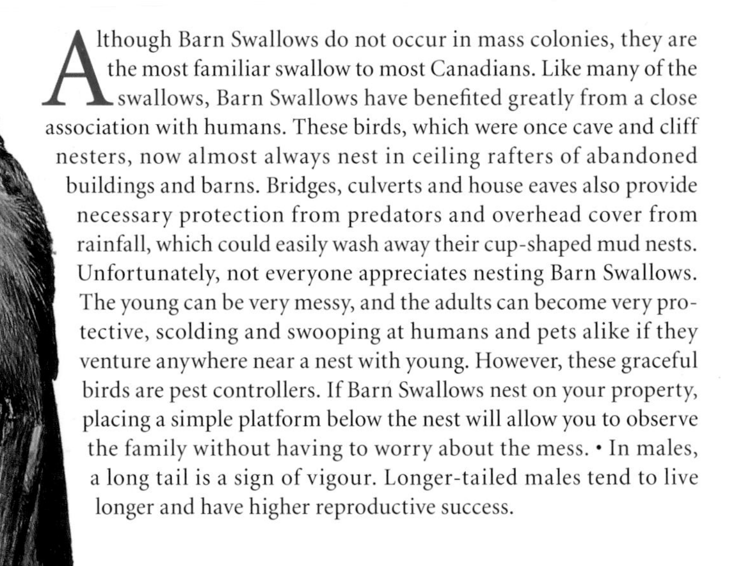

Although Barn Swallows do not occur in mass colonies, they are the most familiar swallow to most Canadians. Like many of the swallows, Barn Swallows have benefited greatly from a close association with humans. These birds, which were once cave and cliff nesters, now almost always nest in ceiling rafters of abandoned buildings and barns. Bridges, culverts and house eaves also provide necessary protection from predators and overhead cover from rainfall, which could easily wash away their cup-shaped mud nests. Unfortunately, not everyone appreciates nesting Barn Swallows. The young can be very messy, and the adults can become very protective, scolding and swooping at humans and pets alike if they venture anywhere near a nest with young. However, these graceful birds are pest controllers. If Barn Swallows nest on your property, placing a simple platform below the nest will allow you to observe the family without having to worry about the mess. • In males, a long tail is a sign of vigour. Longer-tailed males tend to live longer and have higher reproductive success.

ID: sexes similar; blue-black upperparts; rufous underparts (much paler in female and juvenile); long, deeply forked tail; blue crown, nape and ear patches; rust-coloured forehead, cheeks and throat.
Size: *L* 15–18 cm; *W* 32–34 cm.
Habitat: open rural and urban areas where bridges, culverts and buildings are found near rivers, lakes, marshes or ponds.
Nesting: singly or in a small, loose colony; on a human-made structure under a suitable overhang; cup nest of mud, grasses and saliva is lined with feathers and grasses; pair incubates 4–7 brown-spotted, white eggs for 13–17 days.
Feeding: forages exclusively on the wing for flying insects.
Voice: continuous, twittering *zip-zip-zip*; also *kvick-kvick* call.
Similar Species: *Cliff Swallow* (p. 347): square tail; buff rump and forehead; pale underparts. *Purple Martin* (p. 342): shallowly forked tail; male is completely blue-black; female has sooty grey underparts. *Tree Swallow* (p. 343): clean white underparts; notched tail.

BLACK-CAPPED CHICKADEE

Poecile atricapillus

Curious and inquisitive, Black-capped Chickadees will readily land on an outstretched hand offering sunflower seeds! They are common and familiar visitors to backyard feeders, and often swing upside down from the tips of twigs while foraging, gleaning insects or plucking berries. Chickadees cache seeds and are able to relocate hidden food up to a month later. • Black-caps travel in family groups throughout summer, then band together in fall and winter in flocks comprising both regular members and "floaters" that move between flocks. These multispecies flocks can be a birder's delight, because they often include chickadees, small woodpeckers, nuthatches, creepers, kinglets, vireos and warblers. • Most songbirds, including Black-capped Chickadees, have both songs and calls. The chickadee's *fee-bee* song is delivered primarily during courtship, to attract a mate and to defend its territory. The *chick-a-dee-dee-dee* call, which can be heard year-round, is used to keep flocks together and to maintain contact among the birds. • Competition for a tree cavity nesting site is not just limited to birds; bumblebees intent on establishing a new hive have been known to invade a chickadee cavity and chase the small birds from their nest.

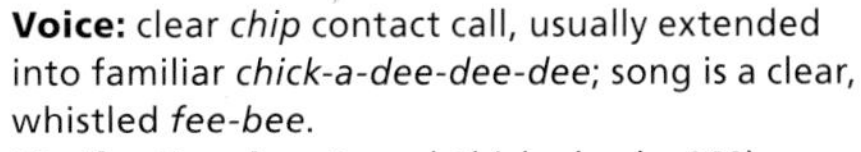

ID: sexes similar; tiny songbird; black cap and "bib"; white cheeks; long tail; olive grey back and rump; stubby, dark bill; grey wings with pale edging on secondary feathers; whitish underparts with buff or pale rufous flanks.
Size: *L* 13–15 cm; *W* 20 cm.
Habitat: deciduous and mixed forests and woodlands, riparian woodlands, wooded urban parks and backyards with feeders.
Nesting: in a cavity, typically in a snag or rotten branch; also in an old woodpecker hole or nest box; moss foundation is lined with hair; female incubates 6–8 brown-dotted, white eggs for 12–13 days.
Feeding: gleans insects, spiders and some fruit from vegetation; also visits feeders for suet and sunflower seeds.

Voice: clear *chip* contact call, usually extended into familiar *chick-a-dee-dee-dee*; song is a clear, whistled *fee-bee*.
Similar Species: *Boreal Chickadee* (p. 352): grey-brown cap and back; dusky cheeks; rich brown flanks; wing feathers lack white edgings; associated with coniferous forests. *Mountain Chickadee* (p. 350): white eyebrows; pale grey underparts with less buff; harsher calls; longer song. *Chestnut-backed Chickadee* (p. 351): dark brown crown; rufous brown back, flanks and sides; shorter tail; buzzier calls; song is a *chip* series.

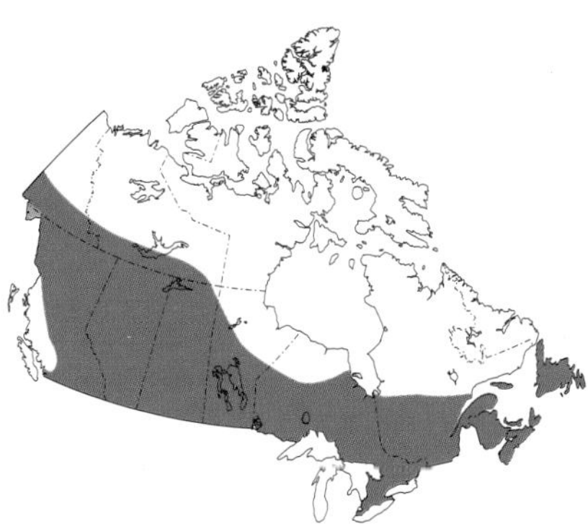

MOUNTAIN CHICKADEE

Poecile gambeli

Breeding at higher elevations than other chickadees, the Mountain Chickadee routinely nests in dry coniferous forests and may be seen foraging as far north as treeline. More commonly, they move down into the montane forests for winter. • Feeders in mountain townsites offer excellent viewing opportunities in winter—scan the flocks of chickadees, and you will often find Mountain, Black-capped and Boreal chickadees foraging together. • Harsh winter weather can force Mountain Chickadees to move to lower elevations in search of food. They sometimes depend on food stores stashed during the summer, and they stay warm by snuggling together in crevices or woodpecker cavities. • When an intruder approaches a chickadee nest, both parents flutter their wings and hiss loudly at the perceived threat. • This year-round resident spends much of its time feeding on seeds and insects on twigs, branches and trunks high in the conifer canopy. Breeding bird surveys have shown Mountain Chickadee populations to be increasing. • The name "chickadee" is an onomatopoeic description of the call most typical of the Black-capped species.

ID: sexes similar; tiny songbird; white eyebrow extends through black cap; black "bib"; white cheeks; grey upperparts; paler grey underparts; grey tail.
Size: *L* 13 cm; *W* 21 cm.
Habitat: dry montane coniferous forests, especially pine and Douglas-fir; higher elevations in summer; visits feeders.
Nesting: in a natural cavity, abandoned woodpecker nest or nest box; may excavate a cavity in soft, rotting wood; nest is made of fur, feathers, mosses and grasses; female incubates 5–9 chestnut-speckled, white eggs for up to 14 days.
Feeding: gleans vegetation for insects and spiders and their eggs; also eats conifer seeds and sunflower seeds at feeders.
Voice: drawling *chick a-day, day, day* call; song is a sweet, clear, whistled *fee-bee-bay.*
Similar Species: *Black-capped Chickadee* (p. 349): buff flanks; lacks white eyebrow. *Boreal Chickadee* (p. 352): grey-brown cap and flanks. *Chestnut-backed Chickadee* (p. 351): dark brown cap; rusty back and flanks.

CHESTNUT-BACKED CHICKADEE

Poecile rufescens

The smallest of the North American chickadees, the colourful, energetic Chestnut-backed Chickadee could fit in the palm of your hand but would hardly sit still long enough for you to hold it. Truly tiny birds, their minute size allows them to nest in the tight quarters of nearly any tree cavity. • In Canada, the Chestnut-backed Chickadee occurs throughout Pacific coastal regions and from the southern Interior of British Columbia to southwestern Alberta, becoming far less common and local from west to east. • Like the Black-capped Chickadee, the Chestnut-backed joins multispecies winter flocks of kinglets, vireos, nuthatches, creepers and lingering warblers. To view these friendly birds up close, mount a platform feeder or hang a suet block near your window. • Chestnut-backed Chickadees line their nests with plenty of hair from rabbits, coyotes, deer, moose and bear in natural areas and cats and dogs in urban yards. • With its dark cap and rusty back and flanks, the Chestnut-backed Chickadee is the most colourful member of its family. The scientific name *rufescens* is Latin for "to become reddish."

ID: sexes similar; tiny songbird; dark cap and "bib"; white cheeks; stubby, dark bill; rufous brown back and sides; whitish underparts; dark brownish grey wings and tail; short tail; darkish legs.

Size: *L* 12 cm; *W* 18 cm.

Habitat: moist coniferous, hardwood and mixed forests; also urban and residential areas.

Nesting: in a soft, rotting trunk or stub, natural cavity, abandoned woodpecker hole or nest box; excavated or natural cavity is lined with fur, feathers, mosses and plant down; mostly the female incubates 6–7 rufous-marked, white eggs for up to 15 days.

Feeding: gleans mainly insects, spiders; also seeds and fruits seasonally; readily visits backyard feeders.

Voice: higher, buzzier call notes than other chickadees; song is a *chip* series.

Similar Species: *Black-capped Chickadee* (p. 349): black crown and "bib"; greyish back; paler, buffier flanks; whistled song. *Mountain Chickadee* (p. 350): white eyebrows; black eye line. *Boreal Chickadee* (p. 352): grey-brown cap and flanks.

BOREAL CHICKADEE

Poecile hudsonicus

With five species and ample numbers, Canada is blessed with chickadees. Birders generally love these energetic little "tits," and, as the most northern representative of this endearing troupe, the Boreal Chickadee is especially sought out. As its name suggests, the Boreal Chickadee resides primarily in Canada's expansive boreal forests. Unlike the more common and familiar Black-capped Chickadee, the Boreal prefers the seclusion of coniferous forests, and it tends to be softer-spoken. During the nesting season, Boreal Chickadees are so quiet, you would never know they were there at all. • Chickadees burn so much energy that they must replenish their stores daily to survive winter; they have adapted to harsh weather by storing food in holes and bark crevices. During cold winter nights, all chickadees enter a state of torpor, in which the bird's metabolism slows so it uses less energy. • Although breeding bird surveys may not accurately reflect species trends throughout the boreal forest region, the Boreal Chickadee has experienced significant population declines on these surveys, especially in the East. • The scientific name *hudsonicus* refers to the northern (Hudsonian) region of Canada.

ID: sexes similar; small songbird; greyish brown cap; black "bib"; white cheeks; light brownish back; grey-brown flanks; light grey underparts; grey wings and tail. *In flight:* appears browner than other chickadees.
Size: *L* 13–14 cm; *W* 20 cm.
Habitat: mature and young Engelmann spruce and subalpine fir forests at higher elevations; mixed coniferous forests at lower elevations in winter.

Nesting: in a conifer; excavates a cavity in soft, rotting wood or uses a natural cavity or abandoned woodpecker nest; nest is lined with fur, feathers, mosses and grasses; female incubates 5–8 finely chestnut-dotted, white eggs for about 15 days.
Feeding: gleans vegetation for small, tree-infesting insects (adults, pupae and eggs) and spiders; also eats conifer seeds.
Voice: soft, nasal, whistled *scick-a day day day.*
Similar Species: *Black-capped Chickadee* (p. 349): black crown; buff flanks. *Mountain Chickadee* (p. 350): white eyebrow extends through black cap.

GRAY-HEADED CHICKADEE

Poecile cinctus

The Gray-headed Chickadee, also known as "Siberian Tit," breeds throughout northern Europe and Asia, from Norway eastward to northwestern North America. In Canada, the Gray-headed Chickadee is primarily found north of the Arctic Circle in northern Yukon and northwestern Northwest Territories. • Gray-headed Chickadees are generally sedentary, though some populations become nomadic during winter months. Like other chickadees, they form mixed-species flocks with other passerines outside the breeding season. • The female excavates and constructs the nest while the male guards against rival males and predators. During incubation, the male feeds the female, which remains on the nest throughout the incubation period. • Each bird stores as much as 3 kilograms of food in cavities, holes and under bark to help it through the frigid, dark winters. • The Gray-headed Chickadee can be distinguished from the similar Boreal Chickadee by its greyer cap, extensive white cheeks and dusty flanks. Juvenile Boreal Chickadees have a greyer crown and hoarser voice than adults and are sometimes misidentified as Gray-headed Chickadees. • Although the Gray-headed Chickadee is rare but apparently stable in Canada, populations in Eurasia have declined considerably because of human disturbance and especially logging.

ID: sexes similar; small songbird; grey-brown cap; white cheeks; black "bib"; brown back; greyish white underparts; pale brown flanks; blackish wing feathers with pale edges. *Juvenile:* duller plumage than adult.

Size: *L* 13–14 cm; *W* 20 cm.

Habitat: coniferous taiga; spruce forests with pockets of willow.

Nesting: in a conifer; in an excavated or natural cavity in soft, rotting wood; may use an abandoned woodpecker hole or nest box; nest is lined with fur, dry mosses and feathers; female incubates 6–10 greyish white eggs for about 15 days.

Feeding: gleans vegetation for insects and spiders, also eats conifer seeds; caches food.

Voice: *pitit peer peer peer* call.

Similar Species: *Boreal Chickadee* (p. 352): grey-brown cap and back; smaller white cheek patches; rich brown flanks; wing feathers lack white edgings; slightly shorter tail. *Black-capped Chickadee* (p. 349): black crown; buff flanks; leucistic individuals may have greyish instead of black crown.

TUFTED TITMOUSE

Baeolophus bicolor

This bird's insatiable appetite and amusing antics at feeders keep observers entertained. Grasping an acorn or sunflower seed with its tiny feet, the dexterous Tufted Titmouse strikes its dainty bill repeatedly against the hard outer coating to expose the inner core. • Titmice are small, "friendly" songbirds belonging to the same family as chickadees. A common species of deciduous forests of eastern North America, the Tufted Titmouse reaches the northern edge of its range in the Carolinian forest zone of southern Ontario, where it is a rare local resident in oak forests. Its northward expansion has been partially attributed to climate change and increased bird feeding. • Titmice maintain pair bonds throughout the year, even when joining small, multispecies flocks for the cold winter months. Young often remain with the parents for more than a year to help raise the next brood. • In late winter, mating pairs break from their flocks to search for nesting cavities and soft lining material. If you are fortunate enough to have Titmice living in your area, you can attract them by setting out strands of your own hair that have accumulated in a hairbrush.

ID: sexes similar; small, lively songbird; prominent crest; grey upperparts; white face; bold, black eyes; white underparts; orange flanks.
Size: *L* 15–16 cm; *W* 25 cm.
Habitat: deciduous woodlands (mostly oak), groves and suburban parks with large, mature trees.

Nesting: in a natural cavity or bird house lined with soft vegetation and animal hair; female incubates 5–6 finely dotted, white eggs for 12–14 days.
Feeding: forages on branches and occasionally on the ground for insects, spiders and seeds; also visits feeders.
Voice: harsh, scolding calls; song is a whistled *peter-peter-peter.*
Similar Species: none; grey plumage and crest are unique.

BUSHTIT

Psaltriparus minimus

A unique, dishevelled-looking resident of southwestern British Columbia, the Bushtit is slowly expanding its range on Vancouver Island and the Sunshine Coast. A tiny, hyperactive, grey cotton ball with a long, narrow tail, it seems to be constantly on the move, bouncing from one shrubby perch to another, looking for something to keep its hungry little engine running. • The Bushtit is a fastidious perfectionist when it comes to nest building and will test every fibre to ensure that the hanging, socklike structure fulfills all tenets of Bushtit architecture. Surprisingly, it may desert both nest and mate if the sanctity of its nest is violated. • When they are not fully engrossed in the business of raising young, Bushtits travel in flocks of up to 40 birds or in mixed-species flocks, examining everything of interest and filling the neighbourhood with charming, bell-like, tinkling calls. In cold weather, flocks huddle together in a tight mass to reduce heat loss. • This bird's musical calls give rise to its genus name *Psaltriparus*, from the Greek word *psaltris*, meaning "player of the lute" or "a harpist."

ID: tiny, fluffy songbird; dull grey back and wings; brown crown; light brown cheek patch; pale buff wash on greyish underparts; long, dull grey tail; black legs. *Male:* dark eyes. *Female:* pale yellow eyes.
Size: *L* 11 cm; *W* 15 cm.
Habitat: shrubby areas in mixed coniferous and deciduous woodlands; also urban and residential areas.
Nesting: in a shrub or near the tip of a tree branch; pair builds a socklike hanging nest, intricately woven with mosses, lichens, cocoons, spider silk, fur and feathers; pair incubates 5–7 white eggs for 12 days.

Feeding: gleans lower vegetation for tiny insects and spiders; also eats some small fruits, berries and seeds; visits suet feeders regularly.
Voice: excited lisping notes; trilled alarm call; short, high, buzzy contact notes, similar to the calls of kinglets and titmice; high, falling series of notes warns of an aerial predator.
Similar Species: *Chickadees* (pp. 349–53): slightly larger; dark caps and "bibs"; mostly white cheeks; heavier bills; different calls.

RED-BREASTED NUTHATCH

Sitta canadensis

The nasal *yank-yank-yank* call of the Red-breasted Nuthatch is a familiar sound in forested regions throughout much of the country. This bird has a more northerly range than the White-breasted Nuthatch, occurring throughout boreal forests in the northern U.S. and Canada. • Some winters, Red-breasted Nuthatches stage periodic southward invasions, called irruptions, which make them common at feeders during some winters and absent from them the next. Irruptions are triggered by food shortages, not weather. • Like woodpeckers and creepers, nuthatches are specialized tree gleaners, working tree trunks and limbs in search of bark-dwelling insects. But unlike other gleaners, nuthatches typically move headfirst down trunks rather than upward, allowing them to spot prey in nooks and crannies that upwardly mobile birds would miss. Nuthatches are especially attracted to backyard feeders filled with suet or peanut butter. • While building its nest, the Red-breasted Nuthatch smears the entrance of its nesting cavity with sap—the sticky doormat likely serves to prevent ants and other animals from entering the nest chamber. Invertebrate parasites can be a serious threat to nesting success, because they can transmit diseases and parasitize nestlings.

ID: small, tree-climbing bird; white face with dark crown and thick, dark eye line; short, straight bill; blue-grey upperparts; short tail. *Male:* black head; rich peach underparts. *Female:* blue-grey head; pale peach underparts; narrower eye line.
Size: *L* 11 cm; *W* 21 cm.
Habitat: *Breeding:* mixed coniferous and deciduous forests; also suburban areas with mature trees. *In migration* and *winter:* mixed, open, mature woodlands; also feeders.
Nesting: in a natural crevice, excavated cavity, old woodpecker hole or nest box; nest is lined with bark, grasses and fur; female incubates 5–7 rufous-marked, white to pinkish eggs for 12–14 days.

Feeding: forages down tree trunks for insects (adults and larvae) and spiders; also eats conifer seeds and feeder foods.
Voice: nasal *yank-yank-yank* or *rah-rah-rah-rah* calls, slower and more nasal than the White-breasted Nuthatch.
Similar Species: *White-breasted Nuthatch* (p. 357): larger; black cap; white face and breast; no black eye line; rufous undertail coverts; prefers hardwoods. *Pygmy Nuthatch* (p. 358): southern BC only; stockier; brown crown descends to black eye line; stouter bill; paler underparts; more grey on flanks.

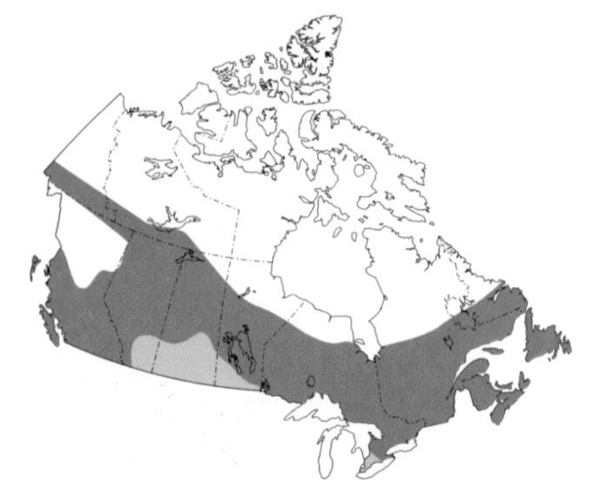

WHITE-BREASTED NUTHATCH

Sitta carolinensis

The White-breasted Nuthatch has a somewhat dizzying view of the world, proficiently moving headfirst down tree trunks and pausing mid-descent to utter its noisy, nasal call. Nuthatches are unique among tree-climbing songbirds because they use one foot to brace themselves while the other holds onto the bark—in contrast, woodpeckers and the Brown Creepers use their tails for stability. • Pairs of White-breasted Nuthatches remain in permanent, year-round territories in deciduous and mixed forests. They store their gathered food items, with each cache site containing just one type of food. Stored foods are supplemented by regular visits to feeders within the territory. A typical feeder visit lasts just long enough for the bird to select a seed, pick it up and then flutter off. • Nuthatches are named for their habit of lodging nuts and seeds in crevices, and then hammering them with their powerful bills—thus "hatching" the nut to get at the meat within. • According to recent breeding bird surveys, nuthatch populations are on the rise throughout Canada and the U.S.

ID: small songbird; bright white face and breast; blue-grey back and wings; rusty undertail coverts; slightly upturned bill; short tail. *Male:* black cap. *Female:* greyer cap.

Size: *L* 15 cm; *W* 28 cm.

Habitat: deciduous and mixed deciduous-coniferous forests; treed residential areas; also feeders.

Nesting: in a natural crevice, old woodpecker cavity or nest box; nest is built of mosses, bark strips and feathers; female incubates 5–9 chestnut-spotted, whitish eggs for 12–14 days.

Feeding: mainly eats larval and adult insects and spiders; regularly visits feeders for nuts, seeds and suet.

Voice: loud, nasal *yank* call, often as a series; also rapid series of similar notes: *wer wer wer wer wer* or *whi whi whi whi.*

Similar Species: *Red-breasted Nuthatch* (p. 356): smaller; black eye stripe; buff or orangey underparts and underwings; prefers conifers. *Pygmy Nuthatch* (p. 358): southern BC only; smaller; brown crown descends to black eye line; warm buff breast and undertail coverts.

PYGMY NUTHATCH

Sitta pygmaea

Maintaining a year-round existence in interior British Columbia means surviving the winter, so the Pygmy Nuthatch spends a lot of time finding food. During daylight hours, this energetic tree climber moves along limbs in ponderosa pine forests, probing and calling incessantly. With a body designed mainly for foraging among clumps of needles, the Pygmy Nuthatch seems barely capable of keeping itself airborne as it flutters awkwardly between adjacent trees. • Like the larger nuthatches, Pygmy Nuthatches are gregarious by nature, usually forming flocks in fall and winter. At night, when the temperature drops, this resourceful bird seeks the shelter and warmth of communal roosts in tree cavities, where numbers can reach 80 birds. These little birds can drop their body temperature, using controlled hypothermia to save energy and survive cold winter nights. • Pygmy Nuthatches are even social during the nesting season—a breeding pair might have as many as three unmated male "helpers" looking after its nestlings. • Because this bird's nests are located in dead pines, forestry practices that leave these snags are beneficial to this species and all cavity-nesting birds. • The Pygmy's piping, high-pitched voice is quite unlike the nasal, rhythmic calls of other nuthatches.

ID: sexes similar; small songbird; blue-grey back and wings; brown crown; black eye line; white cheeks, throat and nape patch; dark-and-white wing edges; pale buff underparts; greyish blue flanks; short tail.
Size: *L* 10 cm; *W* 18–19 cm.
Habitat: open, mature to old-growth ponderosa pine forests.
Nesting: in an excavated cavity, old woodpecker hole or nest box; lining is soft plant material, fur and feathers; female incubates 6–8 sparsely red-spotted, white eggs for 15–16 days; unmated males often assist with nesting duties.
Feeding: mainly adult and larval insects; also eats pine seeds and suet at feeders.
Voice: high-pitched, piping *te-dee te-dee*; also varied, persistent, loud chipping and squeaking notes.
Similar Species: *Red-breasted Nuthatch* (p. 356): larger; white eyebrow stripe; smaller bill; orangey buff or rusty underparts; nasal calls. *White-breasted Nuthatch* (p. 357): larger; white face with dark crown; greyish and white underparts; rufous undertail coverts; nasal calls.

BROWN CREEPER

Certhia americana

The Brown Creeper is probably one of the most inconspicuous birds in Canada. Favouring sizable stands of large trees much of the year, it often goes unnoticed until what looks like a flake of bark suddenly starts ascending a tree trunk, often in a spiral fashion. The camouflage is so effective that even when a creeper is detected, it can be easily lost again against the background until it moves once more. • Completely at home amid the peeling bark, the Brown Creeper even builds its nest there. Each nest has one entrance and exit hole. • In contrast to nuthatches, which forage down tree trunks, the Brown Creeper works its way up. When it reaches the upper branches during a feeding foray, the creeper flies down to the base of a neighbouring tree to begin another foraging ascent. Its long, stiff tail feathers prop it up against vertical tree trunks as it inches its way skyward. • The thin whistle of the Brown Creeper is so high-pitched that birders frequently fail to hear it. • Like nuthatches and chickadees, groups of creepers occasionally roost together in tree cavities—an activity that effectively reduces heat loss.

ID: sexes similar; small bird that creeps up tree trunks; mottled, brown and white upperparts; brown head; pale eyebrow; white throat and breast; thin, downcurved bill; white underparts; buffy undertail coverts.
Size: *L* 13 cm; *W* 19 cm.
Habitat: *Breeding:* old-growth and mature coniferous or mixed coniferous-deciduous forests. *In migration* and *winter:* mixed mature woodlands.
Nesting: wedged under loose bark; nest of grasses, mosses, bark strips and feathers; female incubates 5–6 faintly marked, white eggs for 14–17 days.

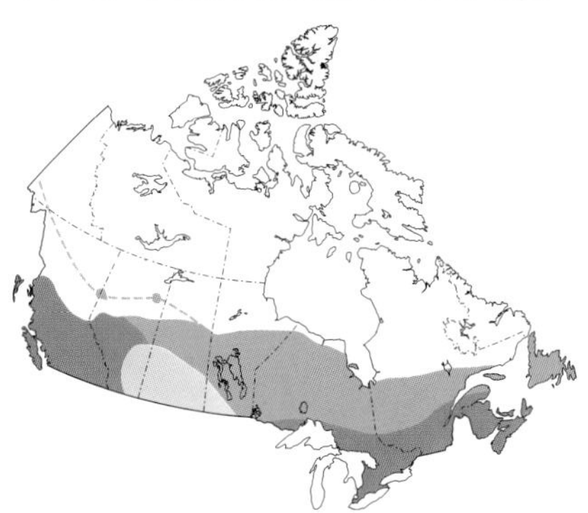

Feeding: creeps up tree trunks, probing under loose bark for adult and larval invertebrates.
Voice: call is a single, high *tseee*; male's song is a faint, high-pitched *trees-trees these-trees, see the trees*.
Similar Species: plumage is unique. *Wrens* (pp. 360–67) and *nuthatches* (pp. 356–58): forage on trunks but do not ascend trunks in spiral fashion.

ROCK WREN

Salpinctes obsoletus

Wrens are small, brown songbirds with loud "bubbly" songs and tails that are often held up at an angle. There are 74 species in the New World, eight of these occurring regularly in Canada. • The mysterious Rock Wren has dull, cryptic markings and can be hard to find in its rocky habitat. The male expertly bounces his songs off surrounding rocks to maximize their effectiveness. One of western North America's best songsters, the male Rock Wren might use 100 or more different song types. It may perch out in plain sight, profiled against the sky, or frustrate birders by singing from a hidden location. • Hidden in cracks and crevices, the entrance of a Rock Wren nest is typically "paved" with as many as 1500 small pebbles, bones, shells and other flat items. This "welcome mat" might be intended to protect the nest from moisture, make it easier to find in confusing terrain or reduce the risk of marauding ants. • Surprisingly, this species is not known to drink water, even in captivity; it derives enough liquid from its food.

ID: sexes similar; large wren; white-speckled, pale brownish grey upperparts; pale eyebrows; long, slightly downcurved bill; finely grey-streaked, white throat and breast; light buff belly and flanks; buffy tips on tail feathers. *In flight:* cinnamon rump; black-barred, white undertail coverts.

Size: *L* 15 cm; *W* 23 cm.

Habitat: open country with rock outcroppings, canyons, talus slopes, boulder piles and large gravel quarries; also boulder-strewn clear-cuts.

Nesting: in a deep cavity or crevice among rocks; nest of grass and rootlets is lined with hair or fur; female incubates 5–6 brown-spotted, white eggs for 14–16 days.

Feeding: gleans from the surface or probes in cracks and crevices for insects and spiders.

Voice: buzzy calls; male's song consists of loud, buzzy, trilled phrases, each repeated 3–6 times: *tra-lee tra-lee tra-lee*; *tick-EAR* alarm call.

Similar Species: *Canyon Wren* (p. 361): only in extreme south-central Interior BC; largely brown-barred, cinnamon-rufous back, wings, tail and underparts; longer bill; white throat; downward-cascading song. *House Wren* (p. 364): smaller; shorter bill; plainer, browner upperparts; duskier underparts; bubblier song; prefers woodland edges.

CANYON WREN

Catherpes mexicanus

Visitors to western North America's broad, steep-sided, arid canyons have probably heard the song of the male Canyon Wren without catching sight of the bird itself. In fact, most people are surprised to discover that the songster is a small bird. The song, which echoes off the canyon walls, ripples and cascades downward in pitch as if it were recounting the action of tumbling boulders. Few landbirds are as restricted to rock cliffs and outcrops as the Canyon Wren, which inhabits territories year-round and nests in protected rock crevices. • These small birds forage tirelessly, even during the hottest parts of the day, searching nooks and crevices with great vigilance for hidden insects and spiders. Scurrying among the rocks, it may look like a small rodent darting about in quick, gliding movements and passing easily through narrow crevices. But its habit of regularly raising and lowering its tail and hindquarters every few seconds confirms the Canyon Wren's identity. Near the nest, this Wren becomes secretive and moves stealthily to avoid betraying the nest's location. • Parents continue to care for young up to two weeks after fledging.

ID: sexes similar; large wren; brown-barred, cinnamon-rufous back, wings, tail and underparts; grey-streaked crown; clean white throat and upper breast; rufous-barred tail; long, downcurved bill.
Size: *L* 14 cm; *W* 19 cm.
Habitat: steep rocky canyons, cliffs and tumbled boulder patches.
Nesting: in a crevice or ledge among rocks or on a cave shelf; cup nest of mosses, twigs and lichens is lined with fur and feathers; female incubates 5–6 chestnut-flecked, white eggs for up to 18 days.
Feeding: gleans rocks, exposed ground and vegetation for insects and spiders.

Voice: high-pitched, far-carrying calls; male's song is a startling cascade of descending 1- and 2-note whistles: *dee-ah dee-ah dee-ah dah-dah-dah.*
Similar Species: *Rock Wren* (p. 360): shorter bill; brownish grey upperparts; unbarred underparts; lightly streaked throat and breast; buzzy, trilling song. *House Wren* (p. 364): much shorter bill; plainer, browner upperparts; more bubbly song; prefers woodland edges.

CAROLINA WREN

Thryothorus ludovicianus

A little, brown bird with a big, booming voice, the Carolina Wren's year-round, clear, ringing *tea-kettle tea-kettle tea-kettle tea* song can't be missed. Males and females perform lively "duets" at any time of day and in any season. • Carolina Wrens are southerners but have expanded northward over the past century into southern Ontario, Québec and Atlantic Canada. With significantly increasing densities found in the adjacent northeastern United States, we can expect more colonizing wrens to immigrate into eastern Canada. • In years of mild winter weather, pioneering colonies of Carolina Wrens become fairly large and stable, but a winter of frigid temperatures and ice-rain can completely decimate an otherwise healthy population. Fortunately, such disasters represent only a temporary loss of these stubborn feathered delights. • Carolina Wrens occur around homes and have an odd penchant for nesting in flowerpots, old shoes, mailboxes and even overturned canoes. If conditions are favourable, two broods can be raised in a single season.

ID: sexes similar; small, active wren; rich brown upperparts including nape, cheek and crown; longish, slightly downcurved bill; bold, white eyebrow; rusty cheeks; white throat; rich buffy underparts; longish tail.
Size: *L* 14 cm; *W* 19 cm.
Habitat: dense forest undergrowth, especially shrubby tangles and thickets; often near rural residences.
Nesting: in a nest box or natural cavity; pair fills the cavity with twigs and vegetation and lines it with fine materials; nest cup can be domed and may include a snakeskin; female incubates 4–5 brown-blotched, white eggs for 12–16 days.
Feeding: usually forages in pairs on the ground and among vegetation; eats mostly insects and other invertebrates; also takes berries, fruits and seeds; will visit feeders for peanuts and suet.
Voice: loud, repetitious *tea-kettle tea-kettle tea-kettle*; female often chatters while male sings.
Similar Species: *House Wren* (p. 364): lacks prominent, white eyebrow. *Red-breasted Nuthatch* (p. 356): blue-grey upperparts; black (male) or dark grey (female) crown; lacks loud, ringing calls.

BEWICK'S WREN

Thryomanes bewickii

The Bewick's Wren investigates all the nooks and crannies of its territory with endless curiosity and exuberant animation. As this charming bird briefly perches to scan its surroundings for food, its long, narrow tail flits and waves from side to side, occasionally flashing with added verve as the bird scolds an approaching intruder. • The songs of Bewick's males in western North America are simpler than those in eastern regions. Nevertheless, western Bewick's males can still produce a bewildering variety of songs and calls at different times of the year. • A former breeding species in the far southwestern corner of Ontario, the Bewick's Wren is now confined in Canada to southwestern British Columbia, where breeding bird surveys show a 10 percent annual population decrease since the late 1980s. Although declines have been attributed to habitat loss, severe winters, pesticide use and competition with various other cavity-nesting species, widespread range expansion of the aggressive, nest-destroying House Wren appears to have been accompanied by the disappearance of Bewick's.

west-central race

eastern & Pacific race

ID: sexes similar; small, slender wren; brown or greyish brown upperparts; bold, white eyebrow; downcurved, grey bill; white throat; pale underparts; dark-barred undertail; often flicks long, banded tail from side to side.
Size: *L* 13 cm; *W* 18 cm.
Habitat: shrubby or open, mixed, wooded residential, agricultural, clear-cut, riparian and park areas.
Nesting: in a cavity or crevice, old woodpecker hole, building, nest box or wood pile; cup of twigs and grasses is lined with feathers; female incubates 3–8 darkly marked, white eggs for 14–16 days.

Feeding: probes crannies for adult and larval insects and spiders; also visits suet feeders.
Voice: extremely variable calls, including harsh *dzheer*; alarm call is a peevish *dzeeeb* or *knee-deep*; male's song is a bold, clear *chick-click, for me-eh, for you*.
Similar Species: *House Wren* (p. 364): lacks white eyebrow. *Winter Wren* (p. 365): more compact; darker brown; more mottling and barring; smaller bill; shorter tail.

HOUSE WREN

Troglodytes aedon

The House Wren and its loud, bubbly warble are a familiar suburban sight and sound. They typically skulk in dense brush but won't hesitate to express displeasure with intruders by delivering harsh, scolding notes. The breeding male will perch within some brushy foliage and belt out its song for much of the day and is often quite approachable. • All it usually takes to attract these feathered charmers is some shrubby cover and a small cavity in a dead tree or nest box. Even an empty flowerpot, vacant drainpipe, abandoned vehicle or forlorn hanging glove can serve as a nest site. • Don't despair if a nest box in your backyard is packed full of sticks and left abandoned without any nesting birds in sight. Male House Wrens, like several of their relatives, build a number of "dummy" nests for potential partners to inspect and complete. An offering rejected by one female might become another's "dream home." • House wrens can be aggressive and highly territorial and are known to puncture the eggs of other birds that nest nearby. • This wren ranges all the way from Canada to southern South America—the broadest longitudinal distribution of any wren.

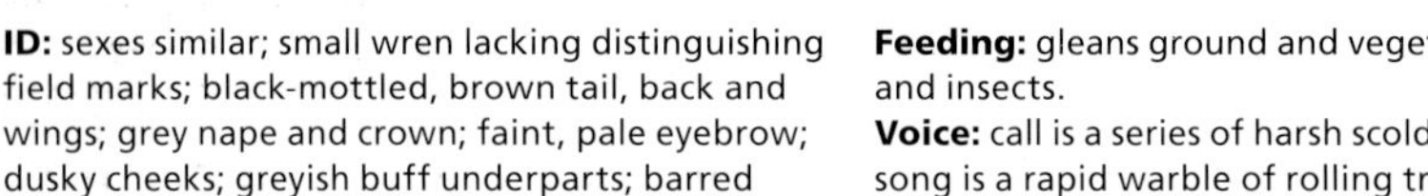

ID: sexes similar; small wren lacking distinguishing field marks; black-mottled, brown tail, back and wings; grey nape and crown; faint, pale eyebrow; dusky cheeks; greyish buff underparts; barred undertail coverts.
Size: *L* 12 cm; *W* 14 cm.
Habitat: *Breeding:* shrubby fields, thickets or open woodland edges; towns and farmyards, often near buildings.
Nesting: in a tree crevice, old woodpecker cavity or nest box; nest of twigs and grass is lined with feathers, fur and soft materials; female completes nest started by male; female incubates 4–12 chestnut-marked, whitish eggs for 12–13 days.
Feeding: gleans ground and vegetation for spiders and insects.
Voice: call is a series of harsh scolding notes; male's song is a rapid warble of rolling trills and rattles lasting 2–3 seconds, rising in a musical burst and falling at the end.
Similar Species: *Marsh Wren* (p. 367): prominent, white eyebrow; rustier plumage; white-streaked, black back; shorter tail; found in marshy areas. *Winter Wren* (p. 365): smaller; darker plumage; prominent, black barring on belly and flanks; shorter tail; long, tumbling, warbling song; prefers moist woodlands. *Sedge Wren* (p. 366): heavily streaked crown and back; short tail; found in sedgy marsh areas.

WINTER WREN

Troglodytes troglodytes

The loudest singer of its size, the male Winter Wren has one of the most vibrant songs of any species. Long and melodious, the bubbly song stands out in both length (10–15 seconds) and sheer exuberance, and none of the many phrases are repeated. The male's distinctive song makes the bird's presence known, but heavy forest growth often hides the singer. Because of the Winter Wren's tiny size, secretive behaviour and preference for relatively inaccessible habitats, our knowledge of its life is incomplete. • While the female raises the young, the male wren brings food to the nest and defends his territory through song. At night, the male sleeps away from his family in an unfinished nest. • Outside of the breeding season, several Winter Wrens might huddle together in a nest or sheltered crevice, but usually each bird lays claim to its own patch of moist coniferous forest, defending it against other wrens. Although a migratory species in most of its Canadian range, in coastal and southern British Columbia, it remains a year-round resident. • The Winter Wren is the only wren also found in the Old World, where it is known simply as a "Wren."

ID: sexes similar; tiny wren; dark brown upperparts, including wings; distinct, pale eyebrow; pale brown throat and breast; remainder of underparts heavily barred with black; short tail is usually held upraised.
Size: *L* 10 cm; *W* 13 cm.
Habitat: wet, mixed coniferous forest with closed canopy, dense understorey shrubs and fallen woody debris.
Nesting: in a natural cavity, old woodpecker hole, under bark, among tree branches or in a building; bulky nest consists of twigs, mosses, grasses, feathers and fur; female incubates 5–8 chestnut-flecked, creamy to pinkish eggs for 12–14 days.
Feeding: gleans ground or vegetation for insects and spiders.
Voice: call is a sharp *chip-chip*; male's song is a melodious series of quick trills and twitters, often lasting more than 10 seconds.
Similar Species: *House Wren* (p. 364): paler overall; pale eyebrow is weak or absent; lacks barring on flanks and undertail coverts; shorter song; prefers drier, more open habitats.

SEDGE WREN

Cistothorus platensis

Like most wrens, the Sedge Wren is secretive and difficult to observe, keeping itself well concealed in dense stands of sedges and tall, damp grasses. Birders not familiar with the Sedge Wren's distinctive *chap-chap-chap-chap, chap, churr-r-r-r-r* call will consider these birds much less numerous than they really are. These wrens usually do not perch in full view and are reluctant to flush until they are almost stepped upon. • Sedge Wrens are feverish nest builders, with construction beginning immediately after the male settles on a nesting territory. Each energetic male might build several incomplete nests throughout his territory before the females arrive. The decoy or "dummy" nests are not wasted and often serve as dormitories for young and adult birds later in the season, or they may be used for renesting or second nesting attempts. • Since their habitat is susceptible to droughts and flooding, Sedge Wren populations tend to be highly mobile from one year to the next. • This species was once called "Short-billed Marsh Wren."

ID: sexes similar; tiny wren; dark crown and back faintly streaked with white; faint, pale eyebrow; blackish eye line; dusky nape; brown barring on wing coverts and primaries; whitish throat, breast and belly; buffy orange sides, flanks and undertail coverts; short, narrow tail is often held upraised.
Size: *L* 10–11 cm; *W* 14 cm.
Habitat: damp grassy fields, dense sedge stands, dry prairie, pond or lake edges, salt marshes; avoids nesting in areas of standing water and cattails.
Nesting: near the ground in sedges or grasses; well-built globe nest with side entrance is woven from sedges and grasses and lined with fine grass, feathers and hair; female incubates 4–8 unmarked, white eggs for 12–16 days.
Feeding: forages low in dense vegetation; picks and probes for adult and larval insects and spiders; occasionally catches flying insects.
Voice: call is a sharp, staccato *chat* or *chep*; male's song is a few short staccato notes followed by a rattling trill: *chap-chap-chap-chap, chap, churr-r-r-r-r.*
Similar Species: *Marsh Wren* (p. 367): white-streaked, black back; unstreaked crown; bold, white eyebrow; rusty, unstreaked shoulders; prefers wetter habitats. *House Wren* (p. 364): brown, unstreaked upperparts; pale eye ring; lacks pale eyebrow. *Winter Wren* (p. 365): much darker upperparts; unstreaked crown; heavily barred underparts; found in treed habitats.

MARSH WREN

Cistothorus palustris

The energetic and reclusive Marsh Wren is almost always associated with cattail marshes or dense, wet meadows. Although it prefers to stay hidden in the deep vegetation, the male's noisy, chattering call each nesting season can alert you to his presence. John J. Audubon dismissed the song as "the grating of a rusty hinge," but many birders appreciate the reedy, gurgling outpouring of emotion. For other birders, the frustration of being able to hear but not see this vociferous songster is almost more than they can bear. Patient observers are usually eventually rewarded with a brief glimpse of this bird perched atop a cattail as it briefly evaluates its territory. • During the breeding season, male Marsh Wrens build numerous globe-shaped nests within their territory but will only choose one as the actual nesting site. The remaining dummy nests are not wasted—they serve to divert predators from the real nest and are used as dormitories later in the season. • Marsh Wrens occasionally destroy the nests and eggs of other Marsh Wrens and marsh-nesting songbirds. It is believed that this behaviour reduces competition for limited food resources, especially when it comes time to feed their young. • Until recently, this bird was known as "Long-billed Marsh Wren."

ID: sexes similar; small wren; prominent, white eyebrow; rufous brown upperparts and wings; black shoulder patch streaked with white; brown nape; dark crown; pale throat; greyish brown or rusty underparts; brown tail with black bands is usually held upright.
Size: *L* 13 cm; *W* 15 cm.
Habitat: freshwater and brackish wetlands; cattail and bulrush marshes; occasionally in wetter tall-grass or sedge marshes.
Nesting: in cattails, bulrushes, sedges or shrubs; dome-shaped nest of aquatic plants with side entrance is lined with fine materials; female incubates 4–10 heavily dotted, pale brown eggs for 12–16 days; occasionally polygamous; usually has 2 broods per year.
Feeding: gleans vegetation for insects and spiders; infrequently, small crustaceans and snails.
Voice: song is a rapid but weak series of gurgling notes followed by a trill: *cut-cut-turrrrrrr-ur*; call is a harsh *chek* or *tsuck*, often doubled or tripled.
Similar Species: *Bewick's Wren* (p. 363): plain brown back; longer, banded tail; simpler, trilled song; prefers drier habitats. *House Wren* (p. 364): unstreaked, grey-brown upperparts; pale buffy underparts; lacks bold, white eyebrow; bubbly, warbling song; prefers drier habitats. *Sedge Wren* (p. 366): smaller; streaked back and crown.

AMERICAN DIPPER

Cinclus mexicanus

Should you encounter a small, dark bird standing on an exposed boulder next to a fast-flowing mountain stream, it will likely be an American Dipper. Formerly known as "Water Ouzel," this unique, aquatic songbird bends its legs incessantly, bobbing to the roar of the torrent, then dives suddenly into the water in search of aquatic insects. Fitted with scaly nose plugs, strong claws, dense plumage, inner eyelids to protect against water spray and an oil gland to waterproof its feathers, the American Dipper survives a lifetime of these ice-cold forays. • The American Dipper's loud, ringing song, which it sings even in the depth of winter, can be heard above the roar of rivers and the babbling of brooks. The melody is wrenlike, a clue that the two birds are related. No other songbird has the American Dipper's combination of plumage, foraging technique and song. • Small, low bridges with steel support girders provide good nest sites for this bird and have aided the nesting success and distribution of dippers in some areas of Canada. The species' presence is regarded as an indicator of good stream health.

ID: sexes similar; medium-sized aquatic songbird; slate grey overall; brownish grey head; white eyelids; straight, dark bill; long, pale legs; short, wrenlike tail is often raised; constantly bobs body.
Size: *L* 19 cm; *W* 28 cm.
Habitat: clear, fast-flowing streams and rivers with rocky bottoms; may also use the margins of clear, cold lakes, ponds and subalpine tarns; backwaters in winter.
Nesting: usually on a cliff ledge, near a waterfall, on a midstream boulder or under a bridge; domed or ball-like nest with side entrance is made of mosses and grasses and lined with leaves; female incubates 4–5 white eggs for 13–17 days.
Feeding: forages in streams by walking, swimming and diving for aquatic insects and other invertebrates and their larvae; also eats tiny fishes and their eggs.
Voice: vocal year-round; high-pitched, buzzy call notes; warbled song is clear and melodious; alarm call is a harsh *tzeet*.
Similar Species: none.

GOLDEN-CROWNED KINGLET

Regulus satrapa

Barely larger than a hummingbird, the Golden-crowned Kinglet is the smallest songbird in North America. The birds are rather tame and can often be coaxed closer by making squeaking or "pishing" sounds. Identify kinglets from afar by their perpetual motion and chronic, nervous wing flicking. These diminutive wonders forage by hanging upside down, or hovering, and picking at the undersides of leaves. • Surprisingly hardy, these birds nest throughout the boreal forest and overwinter at northern latitudes. In winter, Golden-crowned Kinglets are commonly seen and heard among multispecies flocks that often include Black-capped Chickadees, Boreal Chickadees, Red-breasted Nuthatches and Brown Creepers. These small flocks move through forests, decorating tall spruces, pines, firs and naked deciduous hardwoods like Christmas ornaments. Like chickadees, kinglets can lower their body temperature at night to conserve energy. Even if some individuals die over the winter, each pair usually raises two broods each summer, helping maintain population sizes. • With a voice beyond the hearing of many people, the Golden-crowned Kinglet often goes unnoticed, but once spotted, this little jewel of the coniferous forest canopy is friendly and approachable.

ID: tiny, very active songbird; greyish green upperparts; tiny bill; grey underparts; black-bordered, yellow crown patch; white eyebrow; black eye line; white face with faint, black "moustache" stripe; black wings; 1 white wing bar. *Male:* orange-red in centre of crown patch. *Female:* all-yellow crown patch.
Size: *L* 10 cm; *W* 18 cm.
Habitat: *Breeding:* mixed and pure, mature coniferous forests and plantations, especially those dominated by spruce; also conifer plantations. *In migration* and *winter:* coniferous, deciduous and mixed forests and woodlands.

Nesting: usually high in a conifer, well out on a branch; squarish nest of mosses, lichens, bark strips and fine grasses is lined with fine plant materials and hair; female incubates 5–11 brown-and-grey-speckled, whitish eggs for 15 days; usually has 2 broods.
Feeding: gleans or hover-gleans from vegetation for insects and spiders; also eats seeds, sap and suet.
Voice: call is a very high-pitched *see-see-see*; male's song is a faint, accelerating *tsee-tsee-tsee-tsee, why do you shilly-shally?*
Similar Species: *Ruby-crowned Kinglet* (p. 370): bold, broken, white eye ring; crown and face lack bold markings. *Hutton's Vireo* (p. 328): olive brown head and back; pale eye ring; 2 white wing bars.

RUBY-CROWNED KINGLET

Regulus calendula

Come May and June, the Ruby-crowned Kinglet's familiar loud, rolling song echoes through Canada's boreal and mountain forests. At that time of year, the forest emerges from its long winter sleep, and the sound of cold wind blowing through conifer boughs is replaced by the joyful singing of birds. • When engaged in spring courtship or chasing off a rival, the male holds its small, ruby crown erect in an effort to impress prospective mates. Throughout most of the year, however, the crown remains hidden among the dull feather-tips atop the bird's head—invisible even to the magnified view of binoculars. • The female Ruby-crown lays up to 12 eggs, the largest clutch size of any North American songbird its size. • During migration, Ruby-crowned Kinglets are regularly seen flitting about treetops, intermingling with a colourful assortment of warblers and vireos. • The purpose of the constant wing-flicking behaviour of kinglets is not known, but this activity helps distinguish them from the similar Hutton's Vireo and Least Flycatcher. • Based on breeding bird survey data, populations of this common boreal forest resident appear to be declining.

ID: tiny, active songbird; broken, white eye ring; tiny bill; greyish green upperparts, including head; black wings; 2 prominent, white wing bars; pale greenish grey underparts. *Male:* red crown patch. *Female:* no crown patch.
Size: *L* 10 cm; *W* 19 cm.
Habitat: *Breeding:* coniferous and mixed forests at higher elevations, especially those dominated by spruce. *In migration* and *winter:* variety of forests and shrublands; also suburban backyards.
Nesting: saddled on a conifer branch near the tip; nest of mosses, lichens and plant fibres is lined with feathers and plant down; female incubates 5–12 lightly brown-speckled, white or buff eggs for 13–14 days.
Feeding: gleans and hover-gleans vegetation for insects and spiders and their eggs; also eats some fruit and suet.
Voice: short *je-dit* call; male's song is a loud, spirited combination of clear notes and whistles: *see si seeseesee here-here-here ruby ruby ruby see.*
Similar Species: *Golden-crowned Kinglet* (p. 369): boldly patterned face; yellow-orange (male) or all-yellow (female) crown; lacks broken eye ring. *Hutton's Vireo* (p. 328): olive brown head and back; pale eye ring; 2 white wing bars. *Orange-crowned Warbler* (p. 397): no eye ring or wing bars. Empidonax *flycatchers* (pp. 307–16): complete eye ring or no eye ring at all; larger bills.

BLUE-GRAY GNATCATCHER

Polioptila caerulea

Most people would never know of this beautiful little bird's proximity. Blue-gray Gnatcatchers typically remain well up in the trees, and only their squeaky, fussy-sounding, banjolike notes give them away. Constantly on the move, foraging gnatcatchers flash their white outer tail feathers, an action that reflects light and may scare insects into flight. • During courtship, a male gnatcatcher accompanies his prospective mate around his territory. Once a bond is established, the pair remains close during nest building and egg laying, and the male takes an active part in helping raise the young. • Although these birds undoubtedly eat gnats, these insects do not represent a substantial portion of their diet. • Over the years, the Blue-gray Gnatcatcher has been slowly increasing in number and expanding its range in Canada, pushing north as far as south-central Ontario and southwestern Québec. • The scientific name *caerulea* is from the Latin word for "blue."

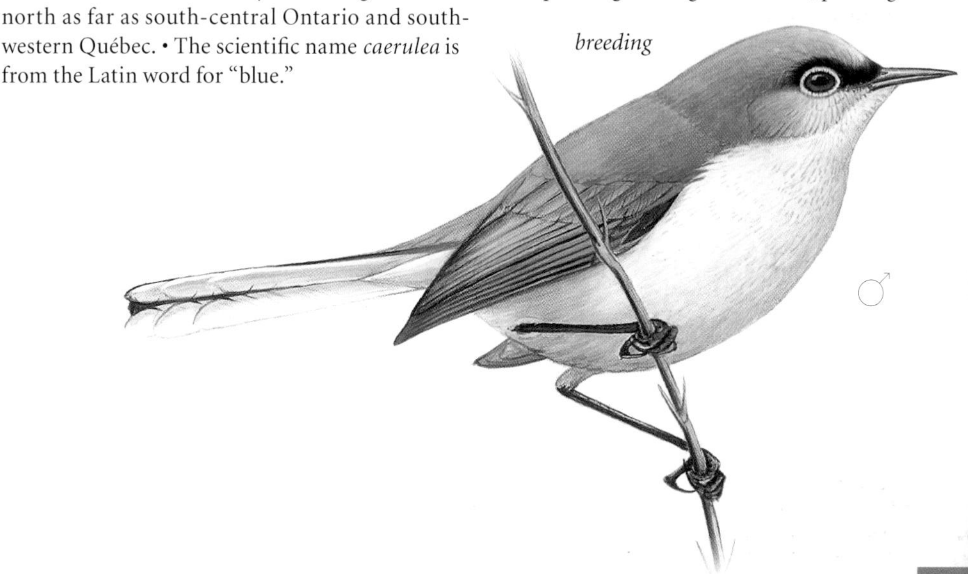

ID: tiny songbird; blue-grey upperparts; white eye ring; thin, dark bill; pale grey underparts; no wing bars; black uppertail; white outer tail feathers. *Breeding male:* darker upperparts; black forehead.
Size: *L* 11 cm; *W* 15 cm.
Habitat: deciduous woodlands and woodland edges, suburban parks and gardens, often near water.
Nesting: on a limb or a fork in a deciduous tree; lichen-covered cup nest of plant fibres and grass is bound by spider silk; pair incubates 3–4 variably marked, bluish white eggs for up to 15 days.

Feeding: gleans vegetation and flycatches for insects, spiders and other invertebrates.
Voice: call is a high-pitched, banjolike, twanging *chee*; male's song is a low warble, often beginning with *zee-u zee-u*.
Similar Species: *Golden-crowned Kinglet* (p. 369), *Ruby-crowned Kinglet* (p. 370) and *Hutton's Vireo* (p. 328): olive green overall; short tails; wing bars.

NORTHERN WHEATEAR

Oenanthe oenanthe

Most of the 20 species of wheatears in the Old World inhabit desert regions. The Northern Wheatear, by contrast, nests in the Arctic tundra of North America and Eurasia. This bird has a very unusual postbreeding strategy, migrating to wintering grounds south of the Sahara in Africa. Two races occur in Canada, each with a unique migration route: *O. o. oenanthe* breeds in northwestern Canada and Alaska and migrates across Asia and the Middle East to Africa; *O. o. leucorhoa*, found in northeastern Canada and Greenland, flies non-stop across the Atlantic to western Europe, a distance of about 2400 kilometres, before turning southward. In fall, vagrants occasionally show up in southern Canada, especially in the East. • These nervous birds restlessly forage on the ground or in foliage for insects, actively flicking their tails. They often perch on rocks and posts on coastal headlands, sometimes joining Snow Buntings and Lapland Longspurs. • Unfortunately, Canadians rarely get to see Northern Wheatears in their attractive breeding attire. When they arrive in southern Canada, most birds are in nonbreeding or immature plumage.

♀ ♂ *breeding eastern race*

♀ ♂ *breeding western race*

ID: white rump and tail base; pale cream belly. *Breeding male:* grey cap, nape and back; thick, black eye line; cinnamon-buff throat and breast; black wings and tail tip. *Female* and *nonbreeding male:* grey-brown cap, nape and back; dark brown wings and tail tip.
Size: *L* 15 cm; *W* 30 cm.
Habitat: *Breeding:* rocky tundra. *In migration:* open country, including coastal meadows, vacant lots and barren fields.
Nesting: on the ground, usually in a crevice under rocks or in an abandoned burrow; cup nest of grass, twigs and fur is lined with mosses, lichens and grass; mostly the female incubates 5–6 reddish-marked, pale blue eggs for 13–14 days.
Feeding: run-and-stop foraging on the ground; may fly out from a perch to catch insects; eats mostly insects and other hard-bodied invertebrates; occasionally feeds on berries.
Voice: calls are clicking *chack-chack* and whistled *hweet*; male's song is a jumble of warbled notes.
Similar Species: *American Pipit* (p. 390): slimmer; streaked back and underparts; narrow, white outer tail feathers; brown rump.

EASTERN BLUEBIRD

Sialia sialis

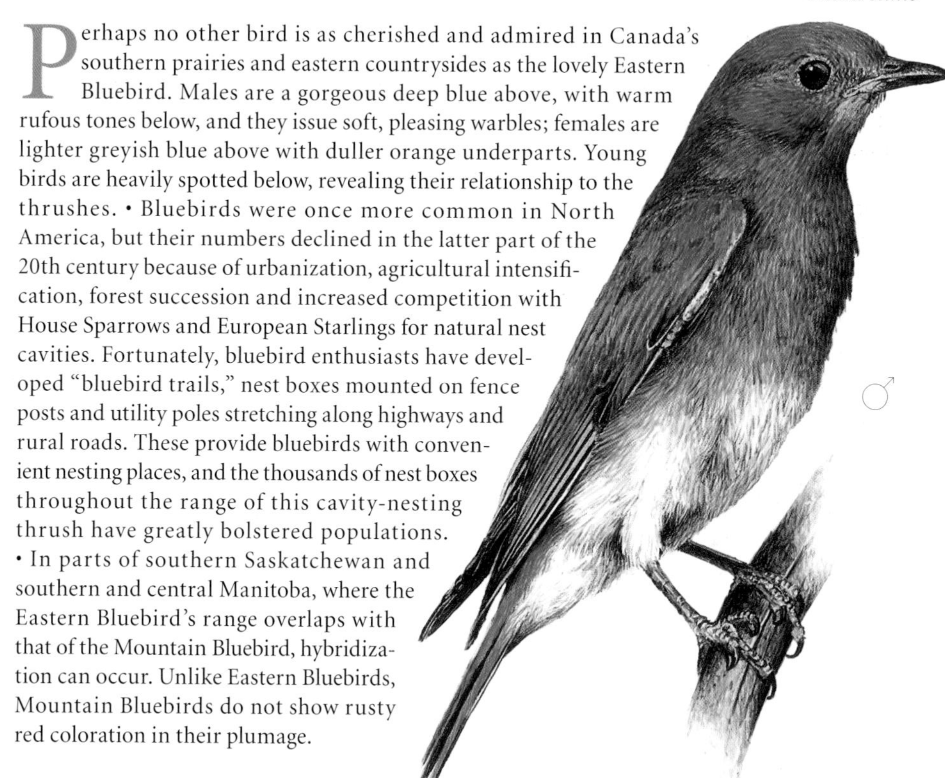

Perhaps no other bird is as cherished and admired in Canada's southern prairies and eastern countrysides as the lovely Eastern Bluebird. Males are a gorgeous deep blue above, with warm rufous tones below, and they issue soft, pleasing warbles; females are lighter greyish blue above with duller orange underparts. Young birds are heavily spotted below, revealing their relationship to the thrushes. • Bluebirds were once more common in North America, but their numbers declined in the latter part of the 20th century because of urbanization, agricultural intensification, forest succession and increased competition with House Sparrows and European Starlings for natural nest cavities. Fortunately, bluebird enthusiasts have developed "bluebird trails," nest boxes mounted on fence posts and utility poles stretching along highways and rural roads. These provide bluebirds with convenient nesting places, and the thousands of nest boxes throughout the range of this cavity-nesting thrush have greatly bolstered populations. • In parts of southern Saskatchewan and southern and central Manitoba, where the Eastern Bluebird's range overlaps with that of the Mountain Bluebird, hybridization can occur. Unlike Eastern Bluebirds, Mountain Bluebirds do not show rusty red coloration in their plumage.

ID: rusty red throat, breast and sides; white belly and undertail coverts; dark bill and legs. *Male:* deep blue upperparts. *Female:* thin, white eye ring; greyish head and back tinged with blue; bluish grey wings and tail; paler rust-coloured underparts. *Immature:* greyish overall; hints of blue in wings and tail; speckled breast.

Size: *L* 18 cm; *W* 33 cm.

Habitat: open country with scattered trees and fence lines; forest clearings and edges; golf courses and cemeteries; often near bluebird nest boxes.

Nesting: in a natural cavity, abandoned woodpecker hole or nest box; nest of grass, stems and twigs is lined with feathers and hair; female incubates 3–7 pale blue to white eggs for 12–16 days; usually 2 broods per year.

Feeding: swoops from a perch or pursues flying insects; also forages on the ground for worms, snails, other invertebrates and berries.

Voice: call is a chittering *pew;* male's song is a gurgling *turr, turr-lee, turr-lee.*

Similar Species: *Mountain Bluebird* (p. 375): lacks rusty red underparts; female has pronounced eye ring. *Lazuli Bunting* (p. 469): white belly; conical bill; darker upperparts; white wing bars.

WESTERN BLUEBIRD

Sialia mexicana

Each feather's microscopic structure, not pigmentation, creates the blue of the Western Bluebird and most other blue birds. Iridescence produces shiny blues; dull blues come from "Tyndall scatter," the same process that produces the blue of the sky. • The Western Bluebird is found from interior British Columbia to southwestern Alberta and south to Mexico. Along with the Mountain Bluebird, the Western Bluebird is a brilliant harbinger of spring in BC. Formerly common on the southwest coast, the Western Bluebird is now limited by habitat modification and competition from European Starlings mainly to the southern BC Interior. • Western Bluebirds usually manage to raise two broods of young each year. As with most birds that double clutch, the second set of eggs is often laid when young from the first set are still reliant on their parents for food. Almost every brood has uncles, aunts, nieces, nephews and cousins on hand to lessen the load. • Insectivorous during the nesting season, Western Bluebirds switch to a diet dominated by berries and fruits through the fall and winter. Western Bluebirds often flock together with Yellow-rumped Warblers around good crops of berry-producing shrubs.

ID: medium-sized, stocky songbird; blue wings; orange breast; pale belly; upright posture. *Male:* deep blue head and upperparts; deep rusty breast, shoulder patches and flanks; blue-washed, white belly and undertail coverts. *Female:* paler than male; partial whitish eye ring; brownish grey upperparts; pale blue mostly on tail and wings. *Immature:* white spotting on dark grey back, breast and flanks.

Size: *L* 18 cm; *W* 34 cm.

Habitat: open Douglas-fir and ponderosa pine forests; also farmlands and transmission corridors.

Nesting: in a natural cavity, old woodpecker hole or nest box; nest is made of grasses, plant stems and feathers; female incubates 4–6 pale blue eggs for 13–17 days.

Feeding: flycatches from a perch and gleans the ground for insects; eats berries in fall and winter.

Voice: calls are a low, chippy warble, a *chuk* and dry chatter; male's song is a harsh but upbeat *cheer cheerful charmer.*

Similar Species: *Mountain Bluebird* (p. 375): male is blue overall; female is paler and greyer. *Lazuli Bunting* (p. 469): smaller; male has sturdy bill, white wing bars, whiter underparts and shorter wings.

MOUNTAIN BLUEBIRD

Sialia currucoides

Despite its name, the Mountain Bluebird is a fairly common summer resident of the aspen parkland ecoregion eastward to Manitoba. The almost-fluorescent sky blue of the male Mountain Bluebird dazzles on sunny spring mornings, and few birds rival him for good looks, cheerful disposition and boldness. It is not surprising that bluebirds are viewed as the "birds of happiness." • The Mountain Bluebird differs from most thrushes in that it nests in cavities, prefers more open terrain, frequently hovers and eats more insects. • Bluebirds have profited from forest clearing, livestock raising and the installation of nest boxes, but they have suffered from the practice of fire suppression and the introduction of European Starlings and other aggressive nesting competitors. • Mountain Bluebirds often raise two broods per year. In such cases, fledglings from the first brood sometimes help to gather food for the second brood. • Spring and fall migrations most often consist of small groups, but on occasion, Mountain Bluebirds migrate in flocks of more than 100 birds. Mountain Bluebirds wander and explore more country than Western Bluebirds and are regular winter vagrants to eastern Canada.

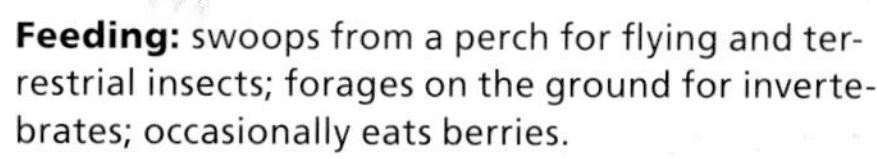

ID: slim, medium-sized songbird. *Male:* sky blue upperparts, slightly paler below; whitish undertail coverts. *Female:* bluish grey head, back, wings and tail; white eye ring; pale "chin"; grey underparts; sometimes pale rufous buff on breast. *Immature:* darker back; white-spotted flanks.
Size: *L* 18 cm; *W* 36 cm.
Habitat: *Breeding:* open parkland, rangeland and subalpine. *In migration:* open terrain, grasslands, forest edges, clear-cuts and agricultural fields.
Nesting: in a natural cavity, stump or nest box; female selects one of several sites where male displays; nest of grass, plant stems and twigs is lined with feathers; female incubates 5–6 pale blue eggs for 13 days.

Feeding: swoops from a perch for flying and terrestrial insects; forages on the ground for invertebrates; occasionally eats berries.
Voice: call is a low *turr turr*; male's song is a short warble of repeated *chur* notes.
Similar Species: *Western Bluebird* (p. 374): rusty breast; male is deeper blue; back of female is usually buffier; spotting on immature includes back. *Eastern Bluebird* (p. 373): male is extensively orange below; female is duller with pale orange extending to neck and flanks. *Blue Jay* (p. 334): prominent crest. *Lazuli Bunting* (p. 469): white belly; conical bill; darker upperparts; white wing bars. *Townsend's Solitaire* (p. 376): peach-coloured patches on wings and tail; white outer tail feathers.

TOWNSEND'S SOLITAIRE

Myadestes townsendi

Few birds characterize our upper-elevation mountain forests or steep dirt banks and road cuts better than the Townsend's Solitaire. Slim and elegant but otherwise inconspicuous, this thrush drops to the ground to snatch food in the manner of a bluebird. It favours cut-banks with overhangs for nesting but will also nest on the ground in alpine areas. • Solitaires are inconspicuous, perching for minutes at a time at the tip-top of tall trees or snags or on the upturned roots of fallen trees. • During the colder months, Townsend's Solitaires move southward and to lower elevations, defending feeding territories among junipers and mistletoe-bearing trees and shrubs. In winter, small numbers regularly appear in eastern Canada. Their winter range is highly correlated with the distribution of juniper trees, and juniper patches are actively defended through singing, perching in exposed sites, and by aggressively displacing any bird that might threaten the Townsend's winter berry supply. • During the summer months, Townsend's Solitaires are rarely seen in groups—this solitary tendency is represented in the bird's name. • Although generally quiet, the male may suddenly burst into a bout of sustained song at almost any time of the year.

ID: sexes similar; grey overall; darker wings and tail; peach-coloured wing patches (very evident in flight); bold, white eye ring; dark, shortish bill; white outer tail feathers; perches upright. *Immature:* brown body heavily spotted with buff; pale eye ring.

Size: *L* 22 cm; *W* 37 cm.

Habitat: *Breeding:* steep banks; coniferous and open mixed forests; also alpine. *In migration* and *winter:* open woodland edges and grassy slopes with patches of juniper shrubs.

Nesting: on the ground, in a vertical bank or cliff face or among upturned tree roots; cup nest of twigs and grass is well lined with conifer needles; female incubates 3–4 brown-marked, pale blue eggs for 11–14 days.

Feeding: feeds mainly on insects and spiders in summer; eats berries, especially juniper, at other times.

Voice: call note is a single, piping *heep*; whistled calls; male's rambling song is a mixture of whistled and mumbled notes, often lasting 10–30 seconds.

Similar Species: *Grey Catbird* (p. 385): black cap; red undertail coverts. *Bluebirds* (pp. 373–75): females lack peach-coloured wing patches and white outer tail feathers. *Northern Mockingbird* (p. 386): slightly larger; paler underparts; longer, often upraised tail; bold, white wing patches in flight.

VEERY

Catharus fuscescens

ssp. salicicola (*western*)

ssp. fuscescens (*eastern*)

Like a musical waterfall, the male Veery's voice descends through thick undergrowth in liquid ripples. Even more reclusive than other *Catharus* thrushes, the Veery spends most of its time on the ground, where it nests and forages in short, springy hops amid tangled vegetation. Its reddish plumage distinguishes it from most other thrushes, except perhaps the russet-backed Swainson's Thrush. • In spring and early summer, Veeries often start singing well before dawn and are prominent again near dusk after most songbirds have called it a day. • Veeries are nocturnal migrants that can fly up to 285 kilometres a night at elevations of more than 2000 metres. These birds migrate to South America each winter, so there's a very good chance that the Veery pairs nesting in your local ravine might soon be travelling to the rainforests of the Amazon. • The Veery was first described by Alexander Wilson in 1831 as "Wilson's Thrush" or "Tawny Thrush"; its present name is an interpretation of the male's song.

ID: sexes similar; medium-sized thrush; rich rufous upperparts, including head; very thin, whitish eye ring; buffy throat; faintly brown-streaked upper breast; whitish grey underparts; greyish flanks and face patch; pale, pinkish legs.

Size: *L* 18 cm; *W* 31 cm.

Habitat: cool, moist deciduous and mixed forests with a dense understorey of shrubs and ferns.

Nesting: on or near the ground, in a bush or small tree; bulky nest is made of leaves, weeds, bark strips and rootlets; female incubates 4–5 pale bluish green eggs for 10–14 days.

Feeding: gleans the ground and lower vegetation for invertebrates in summer; adds berries to diet in migration.

Voice: call is a high, whistled *feeyou*; male's song is a fluty, descending series of whistled notes: *da-vee-ur, vee-ur, vee-ur, veer, veer, veer.*

Similar Species: *Swainson's Thrush* (p. 380): pale "spectacles"; red-tinged, olive brown upperparts and flanks; rising, flutelike song. *Hermit Thrush* (p. 381): rufous tints limited to tail, back and wings; darker breast spotting; flutelike song has ascending and descending phrases. *Gray-cheeked Thrush* (p. 378) and *Bicknell's Thrush* (p. 379): grey-brown upperparts; dark breast spots; brownish grey flanks.

GRAY-CHEEKED THRUSH

Catharus minimus

Few Canadians know of the Gray-cheeked Thrush, but keen birders find this inconspicuous bird a source of great interest. This champion migrant of thrushes winters as far south as Peru and regularly summers in the Subarctic, farther north than any other North American thrush. Each spring and fall, the Gray-cheeked Thrush migrates through southern and central Canada to the northern taiga, where it nests among willows and stunted black spruce. Unfortunately, the inaccessibility of this muskeg region has prevented most birders and ornithologists from documenting more than a few nesting records for this elusive bird. • The Gray-cheeked Thrush travels primarily at night, so it is most often seen or heard rustling through shrub-covered leaf litter on early mornings. • Gray-cheeked Thrushes will settle in almost any habitat while migrating, but they do not stay for long, rarely uttering more than a simple warning note during their brief refuelling stops. • The *minimus* subspecies that is found throughout Newfoundland is greyer than the subspecies found in the rest of Canada.

ID: sexes similar; grey-brown upperparts; grey cheeks; whitish eye ring (often inconspicuous); heavily spotted breast; pale underparts; brownish grey flanks.
Size: *L* 18–20 cm; *W* 32 cm.
Habitat: *Breeding:* dwarf black spruce and willows near treeline and on coastal islands; also muskeg and coniferous forests. *In migration:* variety of forested areas, parks and backyards.
Nesting: in a tree, often a willow, usually near the ground; nest is woven from twigs, moss, grass, weeds, bark strips and rootlets; female incubates 4 brown-spotted, pale blue eggs for 12–14 days.

Feeding: hops along the ground, picking up insects and other invertebrates; may also eat berries in migration.
Voice: call is a downslurred *wee-o*; male's song is typically thrushlike in tone, ending with a clear, descending whistle: *wee-a, wee-o, wee-a, titi wheeee.*
Similar Species: *Bicknell's Thrush* (p. 379): breeding ranges overlap only in Atlantic Canada; base of lower mandible is noticeably yellow; different song. *Swainson's Thrush* (p. 380): prominent eye ring; buff cheeks and upper breast. *Hermit Thrush* (p. 381): reddish tail; olive brown upperparts; lacks grey cheeks. *Veery* (p. 377): reddish brown upperparts; very pale breast streaking.

BICKNELL'S THRUSH

Catharus bicknelli

Before 1995, the Bicknell's Thrush was considered a subspecies of the Gray-cheeked Thrush. The two species are very difficult to tell apart, especially where breeding ranges overlap in Atlantic Canada. Identification problems are compounded when browner Newfoundland Gray-cheeks appear in other parts of Atlantic Canada during migration. Both thrushes may summer in the Saguenay region of Québec. The best way to distinguish the two species is not by appearance but by song, although even this may be challenging. The Bicknell's Thrush has a higher-pitched, more nasal *ch-ch zreee p-zreeeew p-p-zreeee*, with the last note rising. • This shy bird nests primarily in dense, stunted conifers at the tops of mountains or in second-growth forests within a small breeding range that extends from Québec City to Nova Scotia and south to New York State. It regularly overwinters only on islands in the Greater Antilles in the Caribbean. • Forest fragmentation and habitat loss threaten this thrush on both its breeding and wintering grounds. With only 2500 pairs nesting within a patchy distribution in the Maritime provinces and Québec, the Bicknell's Thrush is listed as a species of special concern in Canada.

ID: sexes similar; warm grey-brown upperparts; indistinct, whitish eye ring; plain brown cheeks; straight bill; pale throat; reddish tinge to wings and tail; heavily brown-spotted, buff breast; off-white underparts; brownish grey flanks; pale pink legs.
Size: *L* 16 cm; *W* 29 cm.
Habitat: *Breeding:* open- and closed-canopy coniferous and mixed forests, mainly in darker, moister boreal forests dominated by balsam fir. *In migration:* mainly coastal scrub and tuckamore but occasionally second-growth and overgrown gardens.
Nesting: up to 6 m high in a stunted conifer; nest of grasses, plant stems, mosses, bark and small twigs is lined with grass, fine rootlets and sometimes dry leaves; female incubates 3–6 brown-spotted, pale blue eggs for 13–14 days.

Feeding: forages on the ground or gleans branches and foliage; eats mostly adult and larval insects, spiders and earthworms.
Voice: call note is a buzzy, descending *vee-ah*, similar to Gray-cheeked Thrush but higher-pitched and less slurred; male's song is a high-pitched, nasal *ch-ch zreee p-zreeew p-p-zreeee*, with the last note rising;
Similar Species: *Gray-cheeked Thrush* (p. 378): slightly larger; upperparts more grey-brown (except Newfoundland *minimus* race); darker lower mandible. *Swainson's Thrush* (p. 380): buffy lores; more buff on face and breast; greyer upperparts. *Hermit Thrush* (p. 381): distinct white eye ring; buff rump. *Wood Thrush* (p. 382): very reddish upperparts; dark spots extend onto belly. *Veery* (p. 377): reddish upperparts and breast spotting.

SWAINSON'S THRUSH

Catharus ustulatus

Perched atop the tallest tree in its territory, the male Swainson's Thrush is one of the last forest songsters to be silenced by nightfall. Continuing well into spring evenings, its phrases rise ever higher and then disappear, as if the bird has run out of air or inspiration. On its breeding grounds, the Swainson's Thrush is rarely seen, and its ethereal and flutelike song can be readily confused with the rising and falling, fluty song of the Hermit Thrush. • A spotted thrush that forages mainly on the ground for invertebrates, the Swainson's Thrush also hover-gleans from the airy heights of trees like a warbler or vireo. During migration, these thrushes occasionally visit backyards and neighbourhood parks but will give a sharp warning call and quickly vanish at the first sign of approaching danger. • Two identifiable subspecies occur: the Russet-backed Thrush (*C. u. ustulatus*) favours Pacific coastal forests; the Olive-backed Thrush (*C. u. swainsoni*) is found throughout the rest of Canada, east from BC's coastal mountains to Atlantic Canada. One of the more common passerines of the Canadian forests, an estimated 8 million Swainson's Thrushes occur in Ontario alone.

ssp. swainsoni

ssp. ustulatus

ID: sexes similar; medium-sized thrush; olive brown (ssp. *swainsoni*) to reddish brown (ssp. *ustulatus*) upperparts; distinct, buffy eye ring; dense, black spotting on buffy breast; buffy face and throat; white belly and undertail coverts; rufous-tinged, olive flanks.

Size: *L* 18 cm; *W* 30 cm.

Habitat: *Breeding:* hard- and mixedwood forests and thickets; prefers moist areas with spruce and fir; dense undergrowth not a requisite. *In migration:* low-elevation mixed woodlands, shrubby transmission corridors, riparian thickets and shrubby parks and gardens.

Nesting: in a fork of a small tree or tall shrub; nest of grasses, mosses and fine twigs lined with fine grasses; female incubates 3–5 brown-speckled, pale blue eggs for 10–14 days.

Feeding: eats berries year-round; also takes insects and spiders.

Voice: call is a sharp *whit*; male's song is a slow, rolling, rising spiral: *Oh, Aurelia will-ya, will-ya will-yeee.*

Similar Species: *Gray-cheeked Thrush* (p. 378): greyish cheeks; less or no buff wash on breast; lacks conspicuous eye ring. *Veery* (p. 377): noticeably rusty upperparts; very thin, whitish eye ring; buffier breast with faint brown streaks. *Hermit Thrush* (p. 381): rufous tints on upperparts, especially tail and primaries; descending song phrases.

HERMIT THRUSH

Catharus guttatus

ssp. faxoni

If the beauty of forest birds were gauged by sound rather than appearance, there is no doubt that the Hermit Thrush would be deemed one of the most beautiful birds in Canada. In spring and early summer, the male takes up a prominent perch and adds its lovely, airy notes to the dawn and dusk choruses, even on cloudy days. • On its breeding grounds, the Hermit Thrush lives up to its name, spending much of its time quietly skulking about in the understorey or on the forest floor. When alarmed, it flicks its wings, raises its tail and utters harsh *chuck* notes. • Like many *Catharus* thrushes, the Hermit Thrush shows noticeable regional differences in plumage and size. Coastal British Columbia birds (ssp. *guttatus*) are smaller and darker, whereas mountain birds (ssp. *auduboni*) are the largest. The *faxoni* subspecies is found in eastern Canada. • The Hermit Thrush is unique among spotted thrushes in being the only species that winters in North America, with small numbers wintering in southwestern BC and southern Ontario. • Breeding bird survey data shows populations increasing in western Canada and decreasing in eastern Canada.

ssp. auduboni

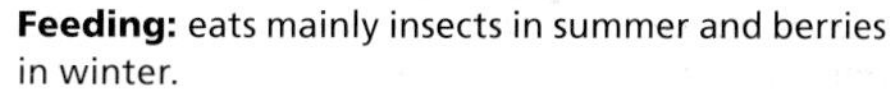

ID: sexes similar; medium-sized thrush; warm brown upperparts, including head; thin, white eye ring; streaked cheeks; reddish wings and tail; white throat; heavily black-spotted, white breast; white underparts; buffy flanks.

Size: *L* 18 cm; *W* 30 cm.

Habitat: *Breeding:* coniferous and mixed woodlands from sea level to alpine. *In migration* and *winter:* variety of hardwood and coniferous woodlands and forest edges, moist riparian areas; also shrubby parks and gardens.

Nesting: usually on the ground; occasionally in a small tree or shrub; mostly in conifers; bulky nest is made of grasses, mosses and hair; female incubates 3–4 light blue eggs for 11–13 days.

Feeding: eats mainly insects in summer and berries in winter.

Voice: call is a dry *chuck*; male's song is a series of ethereal, flutelike notes, some rising and others falling, in no set order, always preceded by a lone, thin note.

Similar Species: *Swainson's Thrush* (p. 380): olive brown upperparts, including tail, without rusty tone; warm buff face; ascending song. *Veery* (p. 377): rich rufous upperparts; buffy breast lightly marked with brown; descending song. *Gray-cheeked Thrush* (p. 378) and *Bicknell's Thrush* (p. 379): grey cheeks; lack conspicuous eye ring. *Fox Sparrow* (p. 449): stockier build; conical bill; rusty breast streaking.

WOOD THRUSH

Hylocichla mustelina

The clear, flutelike, whistled song of the Wood Thrush is one of the most characteristic melodies of eastern deciduous forests. These rich rufous brown thrushes, with heavily speckled underparts, have a split syrinx, or vocal organ, which enables them to sing two notes simultaneously and harmonize with themselves. Their hauntingly ethereal songs delight listeners. • Still common but on the decline, Wood Thrushes face loss of habitat and other threats, both here and in their Central American wintering habitats. Broken forests and diminutive woodlots have allowed the invasion of raccoons, skunks, crows, jays, cowbirds and other pests and predators that traditionally had little access to nests insulated deep within vast hardwood forests. Many tracts of woodlands that have been urbanized or developed for agriculture now host families of American Robins rather than the once-prominent Wood Thrush. Breeding bird survey data shows annual Canadian population declines of 4 percent, though annual declines in New Brunswick are as much as 20 percent. • Naturalist and author Henry David Thoreau considered the Wood Thrush's song to be the most beautiful of avian sounds.

ID: sexes similar; large thrush; plump body; reddish brown upperparts; rusty brown forehead, crown and nape; bright white eye ring; streaked cheeks; black-spotted throat and breast; white underparts with bold, black spots; brown wings, rump and tail.
Size: *L* 20 cm; *W* 32 cm.
Habitat: moist, mature and preferably undisturbed deciduous woodlands and mixed forests.
Nesting: low in a fork of a deciduous tree; bulky cup nest of grass, twigs, moss, weeds, bark strips and mud is lined with softer materials; female incubates 3–4 pale greenish blue eggs for 13–14 days.

Feeding: forages on the ground and gleans vegetation for insects and other invertebrates; also eats berries.
Voice: calls include a *pit pit* and *bweebeebeep*; male's song is 3–5-note, bell-like phrases with each note at a different pitch and followed by a trill: *Will you live with me? Way up high in a tree, I'll come right down and...seeee!*
Similar Species: Catharus *thrushes* (pp. 377–81): smaller spots on underparts; most have coloured wash on sides and flanks; all lack bold, white eye ring and rusty cap and back. *Fox Sparrow* (p. 449): stockier build; conical bill; rusty breast streaking.

AMERICAN ROBIN

Turdus migratorius

Among our most familiar and easily recognized birds, the American Robin occurs almost everywhere. • Anyone who has ever watched a foraging robin might think that it is listening for worms beneath a lawn. In fact, it is looking for movements in the grass, and tilting its head to the side helps because the bird's eyes are placed on the sides of its head. • In winter, many American Robins form flocks, move to lower latitudes or elevations and switch to a diet of berries. Although robins are widely recognized as harbingers of spring in many regions of Canada, the proliferation of non-native shrubs with copious fruit encourages many more of these birds to stay north for the winter. • Robins usually raise two broods per year, though in parts of southern BC, they may raise three. While the male cares for fledglings from the first brood, the female incubates the second clutch of eggs. Young robins are easily distinguished from their parents by their dishevelled appearance and heavily spotted undersides. • The American Robin was named by English colonists after the Robin (*Erithacus rubecula*) of their native land, though the birds are only distantly related.

ID: slate grey upperparts; broken, white eye ring; yellow bill; long, grey tail; white tips to outer tail feathers. *Male:* blackish head; brick red underparts; streaked, white throat. *Female:* dark grey head; clear white throat; paler rufous underparts. *Immature*: heavily spotted breast.
Size: *L* 25 cm; *W* 43 cm.
Habitat: *Breeding:* residential lawns, gardens, urban parks, woodlot edges and openings, burned areas, clear-cuts, bogs and fens. *Winter:* near fruit-bearing trees and springs.
Nesting: in a tree, shrub or artificial structure; bowl of twigs and grasses has inner cup of mud, grasses and mosses; female incubates 4 pale blue eggs for 12–14 days; may have 2–3 broods.
Feeding: forages on the ground for earthworms and insects in summer and berries in winter.
Voice: call is a rapid *tut-tut-tut*; male's song is an evenly spaced warble: *cheerily cheer-up cheerio*.
Similar Species: *Varied Thrush* (p. 384): occurs mainly in western Canada, casual elsewhere; orange eye line; dark breast band; orange wing stripe. Catharus *thrushes* (pp. 377–81): smaller; lack extensive red or rufous on breast; underwing bar is visible in flight. *Wood Thrush* (p. 382): slightly smaller; redder back; more heavily spotted underparts than immature American Robin.

VARIED THRUSH

Ixoreus naevius

Amid American Robins in urban parks and gardens, the Varied Thrush rarely gets a second glance, yet this bird is among the most attractive of North America's many thrushes. • The whistling tones of the male's long, drawn-out song easily penetrate the dense coniferous forests and drifting mist enshrouding its habitat, with each tone at a different pitch and seemingly coming from a different location. When finally revealed, the songster shows a plumage pattern unmatched by any other North American bird. • The Varied Thrush is typically a bird of damp coastal coniferous forests, but it extends its breeding range east into Alberta in appropriate habitat. • Heavy snowstorms often drive the Varied Thrush into the lowlands in abundance. Varied Thrushes are typically western birds, but every year, a few birds seem to become disoriented and head east, showing up for extended periods in other provinces all the way to the Maritimes. • Berries, fruits, seeds, nuts, acorns and suet are some of the offerings that might encourage a lengthy visit from a Varied Thrush—provided Blue Jays and other backyard regulars don't dominate the wealth of goodies.

ID: sexes similar; large thrush; dark grey upperparts; orangey eyebrow, throat, breast and wing bars; dark "V" on breast; white belly and undertail coverts; pinkish orange legs. *Male:* dark head; black breast band; "scaly" grey flanks. *Female:* paler than male, especially head and breast band. *Immature:* grey-blotched breast.

Size: *L* 24 cm; *W* 41 cm.

Habitat: *Breeding:* mature coniferous and mixed-wood forests, especially old-growth. *In migration* and *winter:* riparian woodlands, road edges, thickets and brushy suburban areas; also dense coniferous cover near an active feeding station.

Nesting: near the trunk or branch tip of a tree or tall shrub; loose, open cup of grasses, mosses, rootlets and small twigs is lined with grasses; female incubates 3–4 light blue eggs for 12–13 days.

Feeding: forages on the ground mainly for insects in summer and berries and seeds in winter.

Voice: *chup* call; male's song is a series of whistles of different pitches, often about 10 seconds apart.

Similar Species: *American Robin* (p. 383): plain breast; plain grey wings; immature is more heavily spotted, with no peach on wings or head.

GRAY CATBIRD

Dumetella carolinensis

The Gray Catbird is a member of the Mimidae, a family of medium-sized, long-tailed, ground-loving birds with the ability to mimic sounds and other bird's songs. A catbird in full song issues a nonstop, squeaky barrage of warbling notes, interspersed with poor imitations of other birds. Occasionally, these birds let go with loud, catlike meows that could even fool a feline. The male can even counter-sing by using each side of its syrinx separately. • The male's courtship activities involve an unusual "mooning" display in which he raises his long, slender tail to show off his striking, cinnamon red undertail coverts. As if proud of his colourful posterior, the male often looks back over his shoulder in mid-performance. • The male catbird vigorously defends his nesting territory, often to the benefit of neighbouring ground-nesting warblers, towhees and sparrows. • Although this species is one of the few that can distinguish Brown-headed Cowbird eggs from its own, confused individuals have been seen expelling their own eggs from the nest in error. • Gray Catbirds occupy thick brushy habitats and can be difficult to glimpse. *Dumetella*, Latin for "small thicket," reflects the Gray Catbird's favourite habitat.

ID: sexes similar; medium-sized songbird; slate grey overall; black cap; rufous undertail coverts; light grey underparts; long, blackish tail; black eyes, bill and legs. *In flight:* rounded wings and tail.
Size: *L* 22 cm; *W* 28 cm.
Habitat: dense thickets, brambles, shrubby or brushy areas and hedgerows, often near water; also along forest edges.
Nesting: in a dense shrub; bulky cup nest is loosely built with twigs, leaves and grasses and lined with finer grasses; female incubates 2–6 greenish blue eggs for 12–15 days; often 2 broods per year.
Feeding: forages on or near the ground for insects, spiders and other invertebrates.
Voice: common call is a catlike *mew*; male's song is a variety of warbles, squeaks and mimicked phrases repeated only once.

Similar Species: *Brewer's Blackbird* (p. 478): female has more pointed bill and lacks black cap and rufous undertail. *Northern Mockingbird* (p. 386): plainer underparts; blackish eye line; longer, darker, white-edged tail; large, white wing patches in flight. *Townsend's Solitaire* (p. 376): prominent eye ring; peach rather than white in wings.

NORTHERN MOCKINGBIRD

Mimus polyglottos

Undisputed masters of avian mimicry, male Northern Mockingbirds have a vocal repertoire of more than 150 different songs. They flawlessly imitate a wide array sounds including crows, chickadees, human wolf-whistles, barking dogs, musical instruments and even fire truck sirens and the backup beeps of garbage trucks. The sound replications are so accurate that even computer analysis of sonograms is often unable to detect differences between the original source and the mockingbird. Unmated males often sing throughout the night in summer, sometimes from the tops of chimneys, to the displeasure of homeowners trying to sleep. • During winter, Northern Mockingbirds establish and defend territories in berry-rich areas, including suburban parks and gardens. Generous offerings of suet and raisins can go a long way toward attracting mockingbirds and other feathered friends. • The Northern Mockingbird lifts its wings in a display known as "wing flashing," which may serve to flush insects and scare off predators. • Barely ranging into the southern extremes of several provinces across Canada, our harsh winters may prove to be an insurmountable barrier to the northward expansion of this typically non-migratory species. • The scientific name *polyglottos* is Greek for "many tongues" and refers to the bird's varied vocal repertoire.

ID: sexes similar; medium-sized songbird; grey back, nape and crown; pale face; yellow eyes; white throat; black wings; 2 narrow, white wing bars; whitish underparts; long, blackish tail; white outer tail feathers. *Juvenile:* paler overall; lightly spotted breast. *In flight:* large, white patch at base of primaries; rounded tail.
Size: *L* 25 cm; *W* 36 cm.
Habitat: dense tangles, shrublands, thickets, agricultural areas and riparian forests; suburban gardens and orchard margins with an abundance of available fruit.
Nesting: often in a small shrub or low in a tree; cup nest is built with twigs, grass, fur and leaves; female incubates 3–5 brown-blotched, bluish grey to greenish eggs for 12–13 days.

Feeding: gleans vegetation and forages on the ground for beetles, ants, wasps and grasshoppers; also eats berries; visits feeders for suet and raisins.
Voice: call is a harsh *chair* or *chuck*; male's song is a variable musical medley, with phrases often repeated 3 times or more; habitually imitates other songs and noises; often silent and reclusive.
Similar Species: *Loggerhead Shrike* (p. 322) and *Northern Shrike* (p. 323): thicker, hooked bills; black mask; juveniles are stockier with finely barred underparts. *Townsend's Solitaire* (p. 376): prominent eye ring; peach rather than white in wings. *Sage Thrasher* (p. 387): heavily streaked underparts; fainter white wing bars. *Clark's Nutcracker* (p. 335): much larger; larger bill; darker underparts; white patch on trailing edge of secondaries; more white in tail.

SAGE THRASHER

Oreoscoptes montanus

Thrashers are large, ground-dwelling mimic thrushes. The Sage Thrasher, which depends heavily on open sagebrush flats in the extreme south-central region of British Columbia and near the southern border of Alberta and Saskatchewan, is the smallest and plainest of the thrashers. The males are distinctive for performing both a melodious song that can last several minutes and an exaggerated, undulating courtship flight. Because of these two attributes, this species is considered a closer relative to mockingbirds than to other thrashers and was formerly known as "Mountain Mockingbird." Other mockingbird-like mannerisms include slowly raising and lowering its tail while perched and holding its tail high while running along the ground. • Typical of thrashers, the Sage Thrasher will run away rather than fly if disturbed. • This species is classified as endangered in Canada because of its small population and threats to its preferred sagebrush habitat. Most of the Canadian population of Sage Thrashers is found in the southern Okanagan region of British Columbia • The scientific name *Oreoscoptes montanus* is Greek for "mimic of the mountains," a misnomer since this bird is rarely found in mountainous habitat.

ID: sexes similar; pale greyish brown upperparts; short, slim, dark bill; yellow eyes; white to buffy underparts with heavy, dark streaking; 2 thin, white wing bars; dark legs; plumage fades by late summer. *In flight:* white tips on long, darkish tail.
Size: *L* 22 cm; *W* 30–31 cm.
Habitat: *Breeding:* open, arid shrub-steppe dominated by sagebrush and bitterbrush interspersed with grasses and forbs.
Nesting: in sagebrush, occasionally in saskatoon and hawthorn shrubs; bulky nest of coarse twigs, rootlets and bark strips is lined with grasses, hair and fine rootlets; pair incubates 3–5 boldly spotted, rich blue eggs for 11–13 days.
Feeding: gleans ground and shrubs for insects and spiders.
Voice: low *tup* call; male's song is a sustained, complex, mellow warble with repeated phrases and little change in pace or pitch; often nocturnal.
Similar Species: *Brown Thrasher* (p. 388): long, reddish tail; reddish back; heavily streaked underparts. *Northern Mockingbird* (p. 386): plainer underparts; longer, darker, white-edged tail; large, white wing patches in flight. *Swainson's Thrush* (p. 380) and *Hermit Thrush* (p. 381): stockier; dark eyes; browner upperparts; spotted breasts; unmarked bellies; pinkish legs; prefer forest habitats.

BROWN THRASHER

Toxostoma rufum

The Brown Thrasher is a relatively common nesting bird in the Prairies and southeastern Canada, and it is not uncommon to find one lurking in a shrubby woodland edge, brushy coulee or overgrown pasture. • Thrashers have an even more extensive song repertoire than the Northern Mockingbird, and some 3000 distinct phrases have been catalogued for the species. The male adds new phrases to his repertoire by reproducing sounds made by other male thrashers and neighbouring birds. The Brown Thrasher's complex song of repeated phrases distinguishes it from the similar-sounding Gray Catbird, which rarely repeats a phrase, and the Northern Mockingbird, which tends to repeat each phrase three times. • The Brown Thrasher frequently feeds in the open but is rarely found far from dense cover. At the slightest sign of danger, it scurries for safety. • Brown Thrashers nest close to the ground and aggressively defend their eggs and nestlings against predatory snakes, weasels, skunks and other animals. Breeding bird survey data reveals declines in this bird's population of 2.5 percent annually. The declines are primarily attributed to habitat destruction. Placing fences around thick patches of shrubs and wooded areas bordering wetlands, rivers and streams can prevent cattle from devastating thrasher habitat.

ID: sexes similar; large, terrestrial songbird; rich rufous upperparts and wings; 2 white wing bars; pale underparts with heavy brown spotting and streaking; greyish cheeks; yellow or orangey eyes; long, slender, downcurved bill. *In flight:* buffy wing linings; long, reddish brown tail with buffy tail corners.

Size: *L* 27–30 cm; *W* 33 cm.

Habitat: dense shrubs and thickets, overgrown pastures (especially with hawthorns), woodland edges and brushy areas; rarely close to human habitation.

Nesting: low in a shrub; rarely on the ground; cup nest of grass, twigs, leaves and bark is lined with rootlets and fine vegetation; pair incubates 3–6 reddish-dotted, pale blue eggs for up to 14 days; often 2 broods per year.

Feeding: gleans the ground and digs among debris for larval and adult invertebrates; also eats seeds and berries.

Voice: calls include a harsh *shuck*, a soft *churr* or a whistled, 3-note *pit-cher-ee*; male sings a large variety of phrases, with each phrase usually repeated: *dig-it dig-it, hoe-it hoe-it, pull-it-up, pull-it-up*.

Similar Species: *Sage Thrasher* (p. 387): greyer overall; smaller bill; inhabits sagebrush areas. *Hermit Thrush* (p. 381) and *Wood Thrush* (p. 382): smaller; no wing bars; dark eyes; shorter bills and tails.

EUROPEAN STARLING

Sturnus vulgaris

Thank the Shakespeare Society for this species, perhaps the most damaging non-native bird ever introduced to North America. About 100 European Starlings were released in New York City in 1890 and 1891 as part of an ill-fated effort to introduce all the birds mentioned in Shakespeare's writings. The species was first recorded Canada in 1914. Now abundant throughout North America, aggressive starlings often drive native species such as Eastern Bluebirds, Tree Swallows and Red-headed Woodpeckers from nest cavities. Despite many concerted efforts to control and even eradicate this species, the European Starling has continued to assert its claim in the New World. The estimated North American population is an astounding 200 million birds. Encouraging results from breeding bird surveys show a significant declining trend in starling populations in Canada and throughout North America. • European Starlings are one of our longest-lived songbirds—one survived for almost 18 years. Their success is partly a result of their association with urban buildings and other structures that provide them with nesting and roosting sites. • Courting starlings are infamous for their ability to reproduce the sounds of other birds such as Killdeers, Red-tailed Hawks, jays and Soras.

breeding

ID: sexes similar; medium-sized songbird. *Breeding:* black body with iridescent, purple head and iridescent, green back and body; yellow bill; brown-spotted undertail coverts; pink legs. *Nonbreeding:* black body heavily spotted with white; black head heavily streaked with white; black bill; rufous edges to wing feathers. *Juvenile:* drab grey-brown overall; pale throat; dark bill. *In flight:* pointed wings; short, blunt tail.

Size: *L* 22 cm; *W* 41 cm.

Habitat: generally in urban and suburban areas, farmsteads, feedlots, and a variety of open wooded habitats except deep forests.

Nesting: in a natural or artificial cavity or nest box; cavity is filled with a variety of natural or artificial (plastic, string) materials; mostly the female incubates 4–6 bluish to greenish white eggs for 12–14 days; 2–3 broods per year.

Feeding: omnivorous; mostly a ground forager; diverse diet includes invertebrates, berries, seeds and human food waste.

Voice: very variable call; male's song is a variety of whistles, squeaks and gurgles, including a harsh *tseeeer* and a whistled *whooee*; imitates other birds.

Similar Species: *Rusty Blackbird* (p. 477): male has yellow eyes, longer tail and no spotting; nonbreeding bird shows rusty tinges. *Brewer's Blackbird* (p. 478): no spotting; dark legs; longer tail; male has iridescent, bluish green body, purplish head and yellow eyes; female is brown overall. *Brown-headed Cowbird* (p. 480): smaller; stouter bill; dark legs; male has brown head; female has paler head than immature European Starling; immature has streaked underparts.

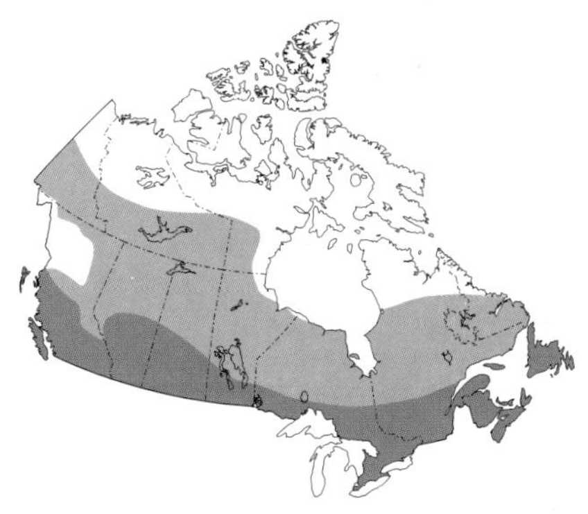

AMERICAN PIPIT

Anthus rubescens

Pipits are sparrowlike, terrestrial songbirds with long, slender bills and white outer tail feathers. The American Pipit's plain wardrobe and habit of continuously bobbing its white-sided tail makes it instantly recognizable. It forages by walking with an upright posture, rather than hopping or bent over like a sparrow. • American Pipits breed in Arctic and alpine tundra habitats. Many birds are already paired when they reach the breeding territories, having gone through the preliminaries of courtship and pair formation at lower elevations and latitudes. This strategy is thought to save valuable time, allowing pairs to rapidly proceed with nesting and raising young during a very brief Arctic summer. Harsh weather during the breeding season can cause high mortalities among nestlings. • Like Sprague's Pipit of the grasslands, American Pipits performs marvellous aerial courtship and territory advertising flights. • In fall migration, flocks of American Pipits carpet open areas, especially short-grass areas, barren fields, airports and lowland shorelines. Flocks of pipits can go unnoticed to untrained eyes, because their dull brown and buff plumage blends into the landscape. • This bird was formerly known as "Water Pipit."

nonbreeding

breeding

ID: sexes similar; slender, sparrowlike songbird; faintly streaked, greyish brown upperparts; whitish buff to pale peach-coloured underparts; 2 faint, white wing bars; pale eyebrow and throat; slim, dusky bill; black legs and feet; dark eyes; bobs tail up and down while feeding. *In flight:* brownish grey flight feathers; greyish white wing lining; blackish tail with white outer feathers.

Size: *L* 15–18 cm; *W* 26 cm.

Habitat: *Breeding:* open, alpine meadows, Arctic and alpine tundra, often on slopes near seepages. *In migration* and *winter:* agricultural fields, young clear-cuts, short-grass grasslands and shorelines.

Nesting: concealed in grasses on the ground or a hummock; base of dried grasses and sedges is lined with finer grasses and hair; female incubates 3–7 brown-spotted, pale eggs for 14–15 days.

Feeding: gleans the ground for invertebrates in summer; adds seeds in winter.

Voice: high, thin calls, often repeated; male delivers shrill series of *chwee* notes during exuberant aerial display on breeding grounds.

Similar Species: *Vesper Sparrow* (p. 440): hops rather than walks; does not bob tail; dark cheek patches; conical bill. *Brewer's Sparrow* (p. 438): unstreaked breast; conical bill; stout body; lacks white outer tail feathers. *Sprague's Pipit* (p. 390): pale back with strong streaking; paler buff breast.

SPRAGUE'S PIPIT

Anthus spragueii

This prairie vocalist delivers an uplifting melody from high above the ground, its song carrying across the open landscape. As he sings, the male often flies in a continuous circle, sometimes for an hour or more at a time. The Sprague's Pipit has few musical rivals. Its plumage, however, is quite ordinary. It wears a pattern common to many prairie passerines: camouflaged browns with white outer tail feathers visible only in flight. • With its very small breeding distribution and highly cryptic plumage and habits, the Sprague's Pipit is one of the least-known birds in North America. Nesting is restricted to the southern Prairie provinces and the northern Great Plains, and it requires extensive native grassland areas greater than 150 hectares in size. The Sprague's Pipit has adapted poorly to the conversion of grasslands to agricultural crops, tame haylands and pastures, and the species has been designated as threatened in Canada. Breeding bird survey data shows it declining at a rate of 7 percent annually. Because 75 percent of Canada's native grasslands have been lost to cultivation, this pipit is most likely to be found in rangeland areas or protected prairie.

ID: sexes similar; slender, sparrowlike songbird; heavily buff-streaked, greyish brown upperparts; lighter underparts; faint breast streaks; thin, pinkish yellow bill; pale face; dark eyes; light-coloured legs. *In flight:* brownish grey flight feathers; greyish white wing lining; white outer tail feathers.

Size: *L* 16–17 cm; *W* 25 cm.

Habitat: native short-grass prairie; prefers ungrazed or lightly grazed sites.

Nesting: in a depression on the ground; well-built cup nest is made of woven grass and often domed with grasses; female incubates 4–5 heavily spotted, white eggs for 13–14 days.

Feeding: walks along the ground foraging for grasshoppers, beetles, moths and other invertebrates; may also eat seeds.

Voice: seldom-heard flight call is a nasal *squeep*; male's song is a swirling, descending, bell-like *choodly choodly choodly choodly.*

Similar Species: *American Pipit* (p. 390): darker plumage; darker legs; bobs its tail. *Vesper Sparrow* (p. 440): heavier bill; chestnut shoulder patch. *Baird's Sparrow* (p. 445): heavier bill; whitish underparts; lacks white outer tail feathers.

BOHEMIAN WAXWING

Bombycilla garrulus

A flock of waxwings swarming over a suburban neighbourhood is guaranteed to dispel even the most severe winter blues. Faint, quavering whistles announce an approaching flock, then the birds swoop down to perch and take turns stripping berries from mountain-ashes, junipers and other trees and bushes. • In winter, nomadic Bohemian Waxwings appear in central and southern Canada from more northern breeding areas to pass the blustery season. They stay just long enough to deplete a berry source before moving on. During summer, Bohemian Waxwings retreat into the boreal forests of central and northern Canada, where they are one of the last birds to initiate nesting, timing their breeding so as to have plenty of ripening fruit to feed their young. • Unlike most songbirds, Bohemian Waxwings do not have a real song, partly because they do not defend a nesting territory. • Waxwings are named for the "waxy" red spots on their secondary feathers. Actually enlargements of the feather shafts, these spots get their colour from the berries the birds eat. • Over the last 20 years, this species' breeding range has expanded eastward. • This nomadic avian wanderer is named for Bohemia, once considered the ancestral home of the Romany people.

ID: sexes similar; slender songbird; greyish brown head, crest, breast and belly; rufous undertail coverts; cinnamon crest; black mask and throat separated by white streak; reddish forehead and cheek; grey wings with white and yellow edging; greyish tail with black band and yellow tip. *Juvenile:* dull greyish upperparts; whitish face; heavily grey-streaked, whitish underparts and throat; no mask; reddish markings restricted to undertail coverts. *In flight:* pointed, greyish brown wings; white "wrist" crescent on dark outer half of upperwing.

Size: *L* 20 cm; *W* 37 cm.

Habitat: *Breeding:* shrubby situations in open, northern boreal forests, frequently near water. *In migration* and *winter:* any habitat with berry-bearing trees and shrubs.

Nesting: in a conifer; cup nest of twigs, grasses, mosses and lichens is sometimes lined with hair; female incubates 4–6 heavily black-spotted, pale bluish grey eggs for 12–16 days.

Feeding: eats mostly berries and other fruits year-round; in summer, also flycatches insects.

Voice: call is a faint, high-pitched, quavering whistle.

Similar Species: *Cedar Waxwing* (p. 393): smaller; tawnier; red spots only on wing tip; yellowish belly; white undertail coverts; slightly higher call notes.

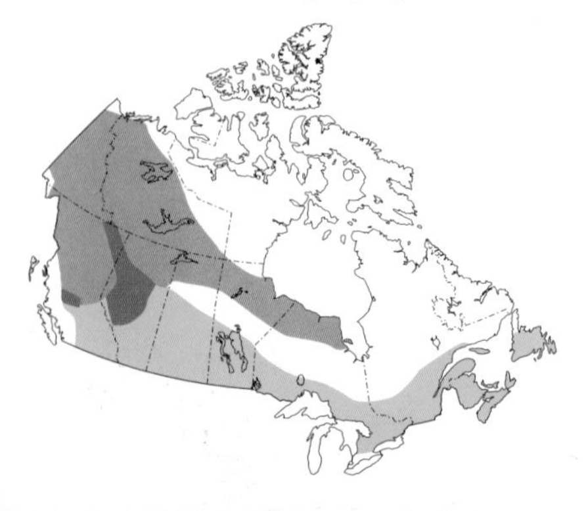

CEDAR WAXWING

Bombycilla cedrorum

Graceful and dapper, Cedar Waxwings have a decidedly suave look. Like their Bohemian cousins, they are highly gregarious, showing little hostility toward each other and sometimes engaging in communal feeding, passing fruit from bird to bird. Waxwings gorge themselves on fruit left hanging on tree and shrub branches in fall and winter, occasionally eating themselves into flightless intoxication on fermented fruit. In summer, Cedar Waxwings add insects to their diet, capturing them by gleaning foliage or by flycatching near wetlands. Favoured plants include serviceberry and mountain-ash. • Cedar Waxwings, unlike the slightly larger, more northerly Bohemian Waxwings, inhabit the south-central half of Canada. In cold winters, most breeding birds migrate southward, but if food is plentiful during warm spells, small numbers will remain in locations in southern Canada. Flocks show little site loyalty and move around looking for food. • Cedar Waxwings nest later than most songbirds, often timing the arrival of young when berries are readily available. This species is increasing over most of its range and is expanding into new locations.

ID: sexes similar; medium-sized songbird; rich brown upperparts, including head and crest; grey rump; black mask and throat separated by white streak; rich brown breast shading to pale yellow belly; white undertail coverts; small, waxy, red "drops" on grey wings; greyish tail with black band and yellow tip. *Juvenile:* dull greyish brown upperparts; pale throat; greyish brown streaking on white underparts. *In flight:* pointed, unmarked, greyish brown wings.

Size: *L* 18 cm; *W* 30 cm.

Habitat: open woodlands, shrubby forests, wetlands and road edges; also farm shelterbelts and wooded residential parks and gardens, especially near fruit trees and water.

Nesting: on a branch or in a vine; bulky, open cup nest of twigs and lichens is lined with fine grasses and hair; female incubates 3–5 sparsely dark-spotted, very pale blue eggs for 12–16 days.

Feeding: mostly eats berries and other fruit; also sallies for flying insects.

Voice: faint, high-pitched, trilled whistle, often given in flight.

Similar Species: *Bohemian Waxwing* (p. 392): larger; greyer overall; reddish face; grey belly; rufous undertail coverts; white areas and yellow stripe on wings.

BLUE-WINGED WARBLER

Vermivora pinus

During the mid-1800s, the Blue-winged Warbler began expanding its range eastward and northward from its home in the central-midwestern United States, finding new breeding territories in the overgrown fields and pastures of abandoned human settlements. Eventually, it came into contact with the Golden-winged Warbler, a bird with completely different looks but almost identical habitat requirements and breeding biology. Where both species share the same habitat, a distinctive, fertile hybrid known as "Brewster's Warbler" can be produced. This hybrid more closely resembles the Golden-winged Warbler, but it retains the thin, black eye line and the touch of yellow on the breast from its Blue-winged parent. In rare cases, when two of these hybrids are able to reproduce successfully, a second-generation hybrid known as "Lawrence's Warbler" is produced. It more closely resembles the Blue-winged Warbler but has the black mask, chin and throat of the Golden-winged Warbler. • The Blue-winged Warbler was first recorded in Ontario in 1908, with the first provincial breeding record documented in 1956. It is now well established in the most southerly parts of the province and continues a slow expansion.

♀ ♂

breeding

ID: *Male:* bright yellow head and underparts; olive yellow upperparts; bluish grey wings and tail; black eye line; thin, dark bill; 2 white wing bars; white to yellowish undertail coverts. *Female:* duller coloration with more olive on head.

Size: *L* 12 cm; *W* 19 cm.

Habitat: second-growth woodlands, willow thickets and woodland edges and openings; also shrubby, overgrown fields and pastures.

Nesting: on or near the ground, concealed by vegetation; cone-shaped nest of grasses, leaves and bark strips is lined with soft materials; female incubates 4–5 finely brown-spotted, white eggs for about 11 days.

Feeding: gleans insects and spiders from the lower branches of trees and shrubs.

Voice: buzzy, 2-note song: *beee-bzzz.*

Similar Species: *Prothonotary Warbler* (p. 419): lacks black eye line and white wing bars. *Pine Warbler* (p. 410): darker; white belly; faint streaking on sides and breast. *Yellow Warbler* (p. 400): yellow wings; lacks black eye line. *Prairie Warbler* (p. 412): rare; black streaking on sides and flanks; darker wings.

GOLDEN-WINGED WARBLER

Vermivora chrysoptera

Unlike people, who can build fences around their property, the male Golden-winged Warbler uses song to defend his nesting territory. If song fails to repel an encroaching rival, then body language and aggression calls warn the intruder to stay away. When a male's claim is seriously challenged, a warning call, a raised crown and a spread tail may be employed. The last resort is to physically remove the competitor in a high-speed chase or winged duel. • The battle to maintain breeding territory is not confined within the species—the Golden-winged Warbler is losing ground to its colonizing relative, the Blue-winged Warbler, which seems to be outcompeting the Golden-wing through hybridization. Golden-winged Warblers prefer moister sites than Blue-winged Warblers, and in this habitat, one can find mostly pure Golden-wings. • Although this warbler has increased in numbers and expanded its distribution for more than a century, the species is now declining or has disappeared from previously occupied regions. In Canada, Golden-wing numbers have declined 79 percent in the last 10 years, and it is now classified as threatened. Possible causes of this decline include hybridization with Blue-wings, nest parasitism by Brown-headed Cowbirds and habitat loss on both its summer and winter ranges.

breeding

ID: yellow forecrown and wing patch; dark "chin," throat and mask are bordered by white; bluish grey upperparts and flanks; white underparts; white spots on underside of tail. *Male:* black throat. *Female* and *immature:* duller overall; grey throat and mask.

Size: *L* 12 cm; *W* 19 cm.

Habitat: moist, shrubby fields, woodland edges and early-succession forest clearings.

Nesting: on the ground, concealed by vegetation; open cup nest of grasses, leaves and bark is lined with softer materials; female incubates 4–5 variably marked, pinkish to pale cream eggs for about 11 days.

Feeding: gleans insects and spiders from tree and shrub canopies.

Voice: call is a sweet *chip*; male's buzzy song begins with a higher note: *zee-bz-bz-bz*.

Similar Species: *Yellow-rumped Warbler* (p. 405): white throat; dark breast patches; yellow sides. *Yellow-throated Warbler* (p. 505): rare; 2 white wing bars; yellow throat; lacks yellow forecrown. *Black-throated Green Warbler* (p. 407): 2 white wing bars; black streaking on sides; lacks dark mask. *Brewster's Warbler* and *Lawrence's Warbler* (hybrids): see Blue-winged Warbler (p. 394).

TENNESSEE WARBLER

Vermivora peregrina

Tennessee Warblers lack the bold, brilliant colours of other warblers but are nevertheless difficult to overlook because of their loud, distinctive, three-part song. In Canada, the Tennessee Warbler is a migrant and summer visitor, breeding mainly in the boreal forests and taiga habitats. • Migrating Tennessee Warblers often sing their tunes and forage for insects high in the forest canopy. However, inclement weather and the need for food after a long flight often force these birds to lower levels in the forest. Females build their nests on the ground and, unlike the males, usually remain close to the forest floor when feeding. • Tennessee Warbler populations fluctuate according to the availability of spruce budworms, which are their primary food source during the breeding season. This songbird is more abundant in eastern Canada, where densities can exceed 500 males per 100 hectares when spruce budworm populations are high. • Ornithologist Alexander Wilson discovered this species along the Cumberland River in Tennessee and named it for that state. This warbler is only a migrant through Tennessee, however, and breeds almost exclusively in Canada.

breeding

ID: *Breeding male:* yellow-green upperparts; grey nape and crown; white eyebrow; black eye line; dusky cheek patches; white throat and underparts; short, thin bill. *Breeding female:* similar to male but upperparts duller; olive grey crown; yellow wash on breast and eyebrow. *Nonbreeding:* similar to breeding female; olive yellow upperparts; yellow eyebrow; yellow underparts; white undertail coverts; male may have white belly.

Size: *L* 10–13 cm; *W* 20 cm.

Habitat: *Breeding:* mature coniferous or mixed mature forests; occasionally spruce bogs and swamps. *In migration*: wide variety of wooded habitats, including farmsteads, parks and gardens.

Nesting: on the ground; nest of grasses, bark strips and mosses is lined with fur and finer grasses; female incubates 2–6 variably marked, white eggs for 11–12 days.

Feeding: gleans foliage for insects and spiders.

Voice: call is a high, sharp *chip*; male's song is a loud, sharp, accelerating *ticka-ticka-ticka swit-swit-swit-swit chew-chew-chew-chew-chew.*

Similar Species: *Warbling Vireo* (p. 329): stouter overall; thicker bill; upperparts less greenish. *Philadelphia Vireo* (p. 330): paler underparts; yellowish undertail coverts; thicker bill. *Orange-crowned Warbler* (p. 397): dull streaking on breast and flanks; yellow undertail coverts; lacks pale eyebrow.

ORANGE-CROWNED WARBLER

Vermivora celata

Don't be disappointed if you can't see the Orange-crowned Warbler's telltale orange crown, because its most distinguishing characteristic is its lack of field marks—wing bars, eye rings and colour patches are all conspicuously absent. • This warbler is frequently encountered low in shrubs and bushes as an olive-coloured bundle flitting nervously while picking insects from leaves, buds and branches. Only the closely related Tennessee Warbler rivals this species for dullness of plumage. Its plain appearance also makes it frustratingly similar to females of other warbler species. • Three subspecies occur in Canada, all with subtle differences in plumage, colour and size. Orange-crowned Warblers seen in Atlantic Canada are generally of the palest and greyest race, so are even more drab than those seen elsewhere in Canada. • Wood-warblers are strictly confined to the New World. All 109 warbler species, 56 of which occur north of Mexico, are thought to have originated in South America, which boasts the greatest diversity of these birds. Orange-crowned Warblers are most common in western Canada.

breeding

♂

♀

ID: sexes similar; olive yellow to olive grey overall; faintly greenish streaked underparts; bright yellow undertail coverts; thin, faint, dark eye line; yellow eyebrow and broken eye ring; thin bill; faint orange crown patch (rarely seen).
Size: *L* 13 cm; *W* 19 cm.
Habitat: *Breeding:* deciduous or mixed forests, shrubby slopes, woodlands and riparian thickets. *In migration:* wide variety of wooded habitats, including farmsteads, riparian deciduous woods, thickets, parks and gardens.
Nesting: on the ground or low in a shrub; well-hidden nest of coarse grasses and bark strips is lined with finer grasses; female incubates 4–5 russet-speckled, white eggs for 12–14 days.
Feeding: gleans foliage for invertebrates and berries; also takes berries, nectar and sap.
Voice: call is a bright, thin *tsip*; male's faint trill song rises slightly, then breaks downward at midpoint.
Similar Species: *Tennessee Warbler* (p. 396): greener upperparts; unstreaked underparts. *Nashville Warbler* (p. 398): greyer head; prominent, white eye ring; greener back and wings; yellower underparts. *Ruby-crowned Kinglet* (p. 370): broken eye ring; 2 wing bars. *Yellow Warbler* (p. 400): female has brighter yellow head and underparts, larger eyes and complete eye ring.

NASHVILLE WARBLER

Vermivora ruficapilla

breeding western race

The male Nashville Warbler has plenty to sing about! Its plumage is a bright mixture of green, yellow and bluish grey, with splashes of rufous and white thrown in for good measure. Fortunately for beginning birders, this species retains much the same plumage from juvenile to adult. • Compared to most other warblers, Nashville Warblers are partial to drier, more open, second-growth forest and brushy areas, often at the edge of a dry forest or burn area. Considered a rare bird in the 1800s, Nashville Warblers have benefited from the clearing of old-growth forests for timber and agriculture. Quite conspicuous in summer, in migration they are much more retiring and can easily slip by unnoticed. • This species has a most unusual distribution, with two widely separated summer populations: one eastern and the other western. This separation is believed to have occurred thousands of years ago when a single core population was split during continental glaciation. • With a population of 15 million, the Nashville Warbler is estimated to be the most abundant bird species in Ontario. Like the Tennessee, Cape May and Connecticut warblers, the Nashville Warbler bears a name that reflects where it was first collected in migration and not its breeding or wintering range.

breeding eastern race

ID: plain, olive green upperparts; bright yellow underparts; white belly; bold, white eye ring; thin, dark bill. *Male:* bluish grey head; rufous crown patch, usually hidden. *Female:* duller overall; paler grey head and nape.

Size: *L* 10 cm; *W* 19 cm.

Habitat: prefers second-growth mixed forests; also wet coniferous forests, riparian woodlands, cedar-spruce swamps and moist, shrubby, abandoned fields.

Nesting: on the ground; well-hidden nest of grasses, leaves and mosses is lined with fine grasses; female incubates 2–5 white eggs, wreathed in reddish brown, for 11–12 days.

Feeding: gleans foliage for adult and larval insects, including caterpillars, flies and aphids.

Voice: usual call is a rather sharp *twit*; male's song begins with a thin, high-pitched *see-it see-it see-it see-it*, followed by a trilling *ti-ti-ti-ti-ti*.

Similar Species: *MacGillivray's* (p. 427), *Connecticut* (p. 425) and *Mourning* (p. 426) *warblers:* larger; grey "hood" extends to upper breast; completely yellow underparts; pinkish bills. *Wilson's Warbler* (p. 430): yellow head; black or olive crown.

NORTHERN PARULA

Parula americana

The smallest of the Canadian wood-warblers, the Northern Parula is typically found in older coniferous forests, where the lichens it uses during nesting have had a chance to mature. It occurs throughout eastern North America and in southern Canada from Manitoba to the Maritime provinces. • Male Northern Parulas spend most of their time singing and foraging among the tops of tall coniferous spires. The young spend the first few weeks of their lives enclosed in a fragile, socklike nest suspended from a tree branch. Once they have grown too large for the nest and their wing feathers are strong enough to allow a short, awkward flight, the young leave their warm abode, dispersing themselves among the surrounding trees and shrubs. As warm summer nights slip away to be replaced by cooler fall temperatures, newly fledged Northern Parulas migrate to the warmer climes of Central America; mature birds winter as far north as Florida. • Breeding bird surveys have revealed significant increases in Northern Parula populations throughout eastern North America, but especially in Canada, where numbers have increased by 2 percent annually since the 1960s. • *Parula* is Latin for "little titmouse." The description refers to the bird's habit of occasionally feeding upside down from branches, like a chickadee.

♂

♀

breeding

ID: blue-grey upperparts; olive patch on back; 2 bold, white wing bars; broken, white eye ring; black eye line; yellow "chin," throat and breast; white belly and flanks. *Male:* 1 black and 1 orange breast band. *Female:* duller than male; band across breast is absent or greatly reduced.
Size: *L* 11 cm; *W* 18 cm.
Habitat: moist, mature coniferous forests, humid riparian woodlands and swampy deciduous woodlands, especially where lichens hang from branches.
Nesting: usually in a conifer; female weaves small hanging nest into strands of tree lichens; may add lichens to camouflage nest; pair incubates 4–5 brown-marked, whitish eggs for 12–14 days.
Feeding: forages for insects and other invertebrates by hovering, gleaning or hawking; feeds from tips of branches and occasionally on the ground.
Voice: song is a rising buzzy trill, ending with an abrupt, lower-pitched *zip*.
Similar Species: *Cerulean Warbler* (p. 416): streaked breast and sides; lacks white eye ring. *Yellow-rumped Warbler* (p. 405): yellow rump and crown; lacks yellow throat. *Kirtland's Warbler* (p. 411) and *Yellow-throated Warbler* (p. 505): heavy black streaking on sides. *Canada Warbler* (p. 431): more uniform grey-blue upperparts and yellow underparts; blackish breast streaking; no wing bars.

YELLOW WARBLER

Dendroica petechia

One of Canada's most common wood-warblers, the Yellow Warbler is usually the first warbler a birder learns to identify. Often mistakenly called "Wild Canary" because of its bright yellow plumage, it is the only warbler that routinely nests in human-influenced habitats such as hedgerows, farm shelterbelts, urban parks and gardens. Active and inquisitive, flitting from branch to branch in search of plant-eating insects, the Yellow Warbler is the home gardener's perfect houseguest and a welcome addition to any neighbourhood. • Different warbler species can coexist in a limited environment because they partition the food resources by foraging exclusively in certain areas. Yellow Warblers also partition among themselves, with males tending to forage in treetops and females foraging closer to ground level. In the forest heights, males take advantage of the exposed singing perches, and the females remain in the dense understorey foliage, where their nests are concealed. • The Yellow Warbler is also popular with the Brown-headed Cowbird, which deposits an egg or two of its own into this warbler's nest. Unlike most forest songbirds, the Yellow Warbler often recognizes the foreign eggs and either abandons the nest or builds over the clutch. Some persistent pairs build over and over, creating bizarre, multilayered, high-rise nests.

breeding

ID: bright yellow body; yellowish green crown, cheeks, nape, back and rump; stout, black bill; bright yellow highlights in dark olive tail and wings. *Breeding male:* reddish streaks on breast and sides. *Breeding female:* drabber; smudgy olive streaks on breast and flanks; white-edged inner flight feathers.

Size: *L* 13 cm; *W* 19 cm.

Habitat: habitat generalist; moist, open woodlands, dense scrub, shrubby meadows, second-growth woodlands, riparian woods and urban parks and gardens.

Nesting: in a fork of a hardwood tree or small shrub, especially willow and wild rose; neat, compact cup nest is made of grasses and plant down; female incubates 4–5 variably marked, off-white eggs for 11–12 days.

Feeding: gleans foliage for insects, especially caterpillars.

Voice: lacklustre chipping call, sometimes repeated; male's song is a fast, repeated *sweet-sweet-sweet summer sweet.*

Similar Species: *Wilson's Warbler* (p. 430): smaller bill; darker crown; greener upperparts; plain yellow underparts. *Orange-crowned Warbler* (p. 397): olive green to dull yellow overall; faintly streaked underparts; thin, dark eye line. *Common Yellowthroat* (p. 428): darker face and upperparts; female lacks bright yellow wing edgings; seldom seen high in trees. *American Goldfinch* (p. 494): black wings and tail; male often has black forehead.

CHESTNUT-SIDED WARBLER

Dendroica pensylvanica

When colourful waves of warbler migrants flood across the Canadian landscape each May, the Chestnut-sided Warbler is consistently ranked among the most anticipated arrivals. Boldly patterned males never fail to dazzle onlookers as they flit about at eye level, often within arm's reach. • Chestnut-sided Warblers favour early-succession forests, which have become abundant in the past century. Although clear-cut logging and prescribed forest burns have adversely affected other warbler species, they have created suitable habitat for the Chestnut-sided Warbler in many parts of its Canadian range. A good indicator of this species' success is the fact that each spring and summer, you can easily see more Chestnut-sided Warblers in a single day than John J. Audubon saw in his entire life—he saw only one! • Although other warblers lose some of their brighter colours in fall but are still recognizable, the Chestnut-sided Warbler undergoes a complete transformation and could easily masquerade as a flycatcher or kinglet in its green-and-grey coat.

♂

breeding

ID: *Breeding male:* chestnut sides and flanks; white underparts; yellow crown; black legs; yellowish wing bars; black mask; prominent, white eye ring. *Breeding female:* paler; dark streaking on yellow crown; incomplete mask; less chestnut on flanks. *Nonbreeding:* yellow-green crown, nape and back; white eye ring; grey face and sides of neck; greyish white underparts.

Size: *L* 11–14 cm; *W* 20 cm.

Habitat: shrubby, deciduous second-growth woodlands, abandoned fields and orchards; especially in areas that are regenerating after logging or fire.

Nesting: low in a shrub or sapling; small cup nest of bark strips, grass, roots and weed fibres is lined with fine grasses, plant down and fur; female incubates 4 brown-marked, whitish eggs for 11–13 days.

Feeding: gleans trees and shrubs at midlevel for insects.

Voice: musical *chip* call; male's song is a loud, clear *so pleased, pleased, pleased to MEET-CHA!*

Similar Species: *Bay-breasted Warbler* (p. 414): black face; dark chestnut hindcrown, upper breast and sides; buff belly and undertail coverts; white wing bars. *American Redstart* (p. 418): female is more greyish overall and has large, yellow patches on wings and tail. *Flycatchers* (pp. 304–21): less obvious eye rings; perch upright. *Ruby-crowned Kinglet* (p. 370): smaller; buffy underparts; more active.

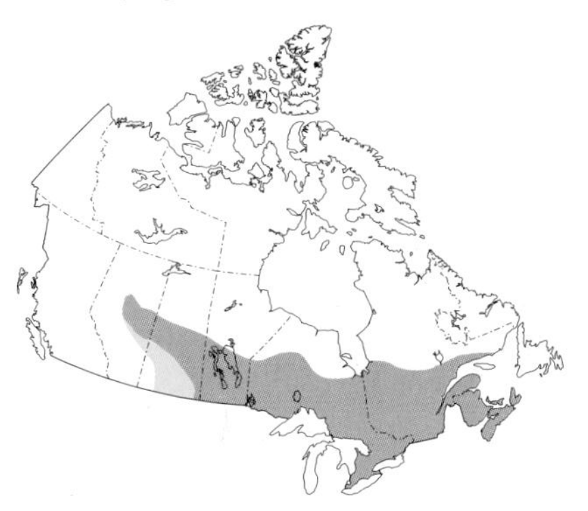

MAGNOLIA WARBLER

Dendroica magnolia

The Magnolia Warbler is widely regarded as one of the most beautiful wood-warblers. Like a customized Cadillac, the Magnolia comes fully loaded with all the fancy features—bold eyebrows, flashy wing bars and tail patches, an elegant "necklace," a bright yellow rump and breast and a dark mask. In autumn, it moults into a much duller nonbreeding plumage. • This beautiful warbler frequently forages along the lower branches of trees and among shrubs, allowing reliable, close-up observations. It is an active feeder, often fanning its tail to reveal the distinctive, white band on the upper tail and the equally distinctive undertail pattern of white basal half and black distal half. • Like many songbirds, Magnolia Warblers migrate primarily at night. Unfortunately, many night-flying birds are killed each year when they collide with buildings, radio towers and other high structures. • Originally called "Black-and-yellow Warbler" by its discoverer, Alexander Wilson, this species was renamed Magnolia Warbler because the first specimen was collected from a magnolia tree. • This warbler breeds across southern Canada and the extreme northern continental U.S. and winters from Florida, Mexico and the West Indies to Panama.

♀ ♂ *breeding*

ID: *Breeding male:* black upperparts; blue-grey crown; black mask; white eyebrow and lower eye arc; yellow underparts with bold, black streaks; white wing bars often blend into larger patch; yellow rump; unique tail pattern with white band and black tip. *Female* and *nonbreeding male:* duller overall; grey face; 2 distinct, white wing bars; streaked, olive back.
Size: *L* 10–13 cm; *W* 18 cm.
Habitat: *Breeding:* open coniferous and mixed forests, mostly in natural openings and along edges, often near water; favours young balsam fir and white spruce stands. *In migration:* wide variety of wooded habitats, including farmsteads, parks and gardens.

Nesting: on a conifer branch; loose nest is built of grasses, rootlets and plant stems; female incubates 3–5 variably marked, white eggs for 11–13 days.
Feeding: gleans vegetation and flycatches for spiders, insects and their larvae.
Voice: call is a hoarse *vink*; male's song is weak and variable, but always ends with a 2-note phrase that rises at the end: *pretty pretty lady, swee swee swee witsy* or *wheata wheata wheet-zu.*
Similar Species: *Yellow-rumped Warbler* (p. 405): white throat; yellow hindcrown; white belly. *Canada Warbler* (p. 431): lacks yellow rump and white patches on wings and tail. *Cape May Warbler* (p. 403): chestnut cheek patch on yellow face; lacks white tail band. *Prairie Warbler* (p. 412): dusky jaw stripe; faint, yellowish wing bars.

CAPE MAY WARBLER

Dendroica tigrina

Cape May Warblers require forests that are least 50 years old for secure nesting habitat and an abundance of canopy-dwelling insects. The Cape May holds the feeding rights to the very tops of the spruces, and throughout most of their almost exclusively Canadian breeding range, these small birds seem to be spruce budworm specialists. In years of budworm outbreaks, Cape Mays can successfully fledge more young. The use of pesticides to control budworms can adversely affect populations of this warbler. Short logging rotations can also contribute to declines, because this species is linked to old white spruce stands. • The Cape May's tubular tongue is unique among wood-warblers. It uses its specialized tongue to feed on nectar and fruit juices on its tropical wintering grounds. • Named for Cape May County, New Jersey, where the first scientific specimen was collected in 1811, this bird was not recorded there again for more than 100 years! • The plumage of the breeding male is recognized by the species name *tigrina*, Latin for "striped like a tiger."

♂ *breeding*

♀

ID: dark streaking on yellow underparts; yellow side collar; dark olive green upperparts; yellow rump; clean white undertail coverts. *Breeding male:* chestnut cheeks on yellow face; dark crown; large, white wing patch. *Female:* paler overall; 2 faint, thin, white wing bars; greyish cheeks and crown.

Size: *L* 12–14 cm; *W* 18 cm.

Habitat: mature coniferous and mixed forests, especially dense old-growth stands of white spruce and balsam fir.

Nesting: near the top of a spruce or fir, often near the trunk; cup nest of moss, weeds and grass is lined with feathers and fur; female incubates 6–7 rufous-spotted, whitish eggs for about 12 days.

Feeding: gleans treetop branches and foliage for spruce budworms, flies, beetles, moths, wasps and other insects; occasionally hover-gleans.

Voice: call is a very high-pitched *tsee*; male's song is a very high-pitched, weak *see see see see*.

Similar Species: *Bay-breasted Warbler* (p. 414): male has black face and chestnut throat, upper breast and sides and 2 white wing bars; buff underparts lack black streaking. *Black-throated Green Warbler* (p. 407): black throat and/or upper breast; white lower breast and belly; 2 white wing bars; lacks chestnut cheek patches. *Magnolia Warbler* (p. 402): white tail band; less streaking on underparts; lacks chestnut cheek patches and yellow side collar.

BLACK-THROATED BLUE WARBLER

Dendroica caerulescens

Dark and handsome, the male Black-throated Blue Warbler is a treasured sight to the eyes of any birder. The female looks nothing like her male counterpart, however, appearing more like a vireo or a plain-coloured Tennessee Warbler. • If you are fortunate enough to meet this bird during its migration or over the course of the nesting season, you will notice that it is more deliberate in its foraging behaviour than most other warblers. This warbler prefers to work methodically over a small area, snatching up insects among branches and foliage. It is generally shy and inconspicuous, foraging secretly within the dense confines of low shrubs and saplings. • Although Black-throated Blue Warblers prefer solitude among deciduous forests, agriculture and urban development have forced them from many of these haunts. They tend to come out into the open in migration, joining Black-and-white Warblers to glean insects from tree branches or competing with Palm Warblers for food on the ground. • Fortunately for birders, the Black-throated Blue Warbler retains its strikingly distinctive plumage throughout the year.

ID: *Male:* black face, throat, upper breast and sides; dark blue upperparts; clean white underparts and wing patch. *Female:* olive-brown upperparts; unmarked, buff underparts; faint, white eyebrow; small, buff to whitish wing patches (may not be visible).
Size: *L* 13–14 cm; *W* 19 cm.
Habitat: upland deciduous and mixed forests with dense understorey of deciduous saplings and shrubs; also in second-growth woodlands and brushy clearings.
Nesting: in a low fork of a dense shrub or sapling, usually within 1 m of the ground; open cup of weeds, bark strips and spiderwebs is lined with moss, hair and pine needles; female incubates 4 variably marked, creamy white eggs for 12–13 days.
Feeding: thoroughly gleans branches and understorey for caterpillars, moths, spiders and other insects; occasionally eats berries and seeds.
Voice: call is a short *tip*; male's song is a slow, wheezy *I am soo lay-zeee*, rising slowly throughout.
Similar Species: male is distinctive. *Tennessee Warbler* (p. 396): paler cheeks; greener back; lacks white wing patches. *Warbling Vireo* (p. 329) and *Philadelphia Vireo* (p. 330): stouter bills; paler cheeks; more yellow-white underparts; lacks white wing patches. *Cerulean Warbler* (p. 416): broader, yellowish eyebrow; 2 white wing bars.

YELLOW-RUMPED WARBLER

Dendroica coronata

"Myrtle" ♂

"Audubon's" breeding ♀ ♂

The Yellow-rumped Warbler is the most abundant and widespread wood-warbler in North America. • Other warblers seem to appreciate the Yellow-rump's company and its alertness, and they tag along with it in fast-moving foraging flocks. • This species comes in two forms: the common, yellow-throated "Audubon's Warbler" of the West, and the white-throated "Myrtle Warbler," which breeds in the North and east of the Rocky Mountains. In general, Audubon's Warblers are more abundant in southwestern Canada, and Myrtle Warblers are more abundant north of 55°N latitude. These two races of the Yellow-rumped Warbler were once considered separate species, but because of their overlapping ranges, ability to interbreed and genetic similarities, they are now considered a single species. • Adults are generally quiet when they have eggs or young to guard. When they are noisy and aggressive, it is a good sign that the young have left the nest. • Small numbers of Yellow-rumped Warblers overwinter in southern Canada. During winter, they subsist on berries, seeds and suet, instead of their usual diet of insects.

ID: yellow rump and patch on side of chest; dark cheek; thin, white eyebrow (boldest on "Myrtle"); dark streaking on breast and flanks; white underparts; 2 faint, white wing bars; white above and below eyes; grey tail. *Male:* yellow crown patch; black streaking on blue-grey upperparts; black breast; yellow ("Audubon's") or white ("Myrtle") throat. *Female:* dark streaking on grey-brown upperparts; all-grey crown.
Size: *L* 14 cm; *W* 23 cm.
Habitat: *Breeding:* coniferous or mixed forests. *In migration* and *winter:* open, mixed deciduous forests and brushy areas, often riparian; also gardens and orchards.
Nesting: in the fork of a tree branch; compact nest of weed stalks, twigs and rootlets is lined with feathers, hair and plant down; female incubates 3–5 brown-marked, creamy white eggs for 12–13 days.
Feeding: hawks, hover-gleans or gleans vegetation for insects; eats mostly berries in fall and winter.
Voice: *"Audubon's":* liquid, rising *swip* call. *"Myrtle":* sharp, dry *kep* call. *Male:* song is 6–8 repetitions of the same note followed by a rising or falling trill: *seet-seet-seet-seet trrrrrr.*
Similar Species: *Magnolia Warbler* (p. 402): black streaks on yellow underparts; may show yellow rump. *Chestnut-sided Warbler* (p. 401): chestnut sides on otherwise clean white underparts; lacks yellow rump. *Cape May Warbler* (p. 403): heavily streaked yellow throat, breast and sides; male lacks yellow crown. *Yellow-throated Warbler* (p. 505): yellow throat; bold, white eyebrow, ear patch and wing bars; lacks yellow crown and rump.

BLACK-THROATED GRAY WARBLER

Dendroica nigrescens

Once seen, the Black-throated Gray Warbler is instantly recognizable—it looks like a yellow-challenged Townsend's Warbler. A close relative of the Townsend's, which has a moderate amount of yellow in its plumage, the Black-throated Gray has just one small, yellow spot in front of the eyes. This species is considered "peripheral" in Canada because it reaches the northern limit of its North American range on the southwest coast of British Columbia. Because the Black-throated Gray Warbler is only a summer visitor, the species' breeding biology is poorly known; less than 10 nests have ever been found in BC. • The male Black-throated Gray Warbler's shrill, buzzy song is typical for a warbler and very similar to the songs of its close relatives, the Townsend's Warbler and the Hermit Warbler. • The Black-throated Gray Warbler is one of the very few western warblers that rarely crosses the Continental Divide. It is a regularly occurring vagrant east of its range, and individuals have been recorded attempting to overwinter in southern Ontario.

ID: black crown; black and white head stripes; yellow spot in front of eyes; grey nape and back; black streaks on white underparts; 2 white wing bars; blackish legs. *Male:* black throat. *Female:* whitish throat; paler back and head.
Size: *L* 13 cm; *W* 20 cm.
Habitat: low-elevation hardwood and mixed hardwood-softwood forests.
Nesting: on the branch of a tree or tall shrub; small cup nest consists of weed stalks, grasses and plant fibres; female incubates 3–5 brown-marked, white eggs for about 12 days.
Feeding: hawks, hover-gleans and gleans vegetation for insects and their caterpillars.
Voice: usual call is a low *tep*; male's song is a lazy, oscillating *weezy-weezy-weezy-wee-zee*, often rising or falling at the end.
Similar Species: *Black-and-white Warbler* (p. 417): ranges do not overlap; white-streaked back; white crown stripe; yellowish feet. *"Myrtle" Yellow-rumped Warbler* (p. 405): white throat; yellow rump and side patch; less white in tail.

BLACK-THROATED GREEN WARBLER

Dendroica virens

The Black-throated Green Warbler makes its home high up in coniferous spires. Conifer crowns can be penthouses for neotropical warblers, and many species choose to nest and forage exclusively at these great heights. Because so many warbler species occur in our conifer forests, the food supply in these areas can come under pressure. Fortunately, many species of warblers can coexist because they partition food supplies by foraging exclusively in certain parts of the trees. Black-throated Greens feed just below the crowns on the outer branches. • If it weren't for this warbler's flitting habits and its unmistakable *zoo zee zoo zoo ZEE* song, it would often escape detection. Males have two different songs, one given around the nest and the other used to mark territorial boundaries. It is a persistent singer, and one male was recorded performing as many as 450 songs in one hour. • The male Black-throated Green Warbler is accurately named—its green, black and white plumage is unique among the wood-warblers. The female and juvenile are similar but have mottled black or pale yellow throats. A diagnostic field mark in all plumages is a yellow wash on the undertail coverts.

breeding

♀

♂

ID: yellow face; plain olive crown, nape and upperparts; faint, olive eye line extends to cheek; blackish wings; 2 white wing bars; black breast and flanks; white underparts; pale yellow undertail coverts. *Male:* black "chin," throat and upper breast. *Female:* whitish throat.

Size: *L* 10–13 cm; *W* 18 cm.

Habitat: mature coniferous and mixed forests; deciduous woodlands with beech, maple or birch; also cedar swamps, hemlock ravines and conifer plantations.

Nesting: on the horizontal branch of a conifer; small cup nest is made of grasses, twigs, stems and spiderwebs; female incubates 3–5 rufous-marked, creamy white to grey eggs for about 12 days.

Feeding: gleans vegetation or flycatches for insects and spiders.

Voice: call is a clear *chip*; flight call is a high, rising *swit*; song is a fast *zoo zee zoo zoo ZEE* or *see-see-see SUZY!*

Similar Species: *Townsend's Warbler* (p. 408): yellow breast; dark ear patch. *Blackburnian Warbler* (p. 409): female has yellowish underparts and angular, dusky facial patch. *Cape May Warbler* (p. 403): heavily streaked yellow throat, breast and sides. *Pine Warbler* (p. 410): yellowish breast and upper belly.

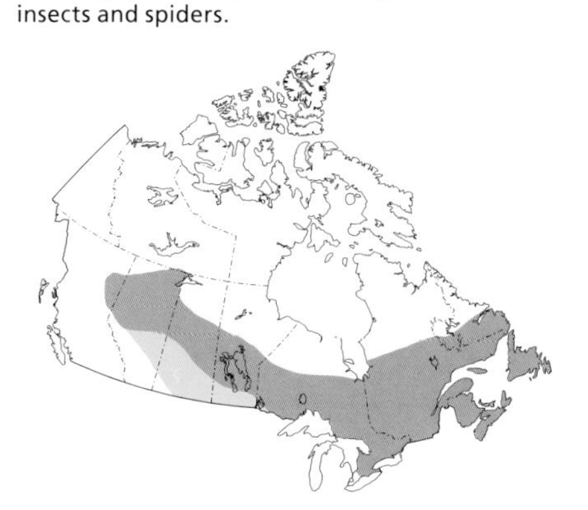

TOWNSEND'S WARBLER

Dendroica townsendi

Hardier than many of its close relatives, the stripe-headed Townsend's Warbler is found nesting throughout much of British Columbia and into southwestern Alberta, with some of the highest densities of breeding Townsend's Warblers in North America occurring along the BC coast. Although rare in winter and found only in the Strait of Georgia Lowland area of BC, the Townsend's sometimes joins other cold-hardy warblers in dense cover, where it offers a much better look at its attractive plumage as it forages at suet feeders. • The male, uttering its wheezy song from the tip of a conifer branch in the upper canopy, is difficult to locate amid the dense, dark foliage. Conifer crowns are preferred foraging sites for many wood-warblers, making warbler watching a neck-straining experience. • Some Townsend's Warblers bear evidence of hybridization with the yellow-headed and more southern Hermit Warbler, which makes identifying some Townsend's difficult. The latest evidence indicates that this hybridization is more extensive than formerly believed. • The Townsend's Warbler has been recorded as a vagrant east to Newfoundland.

♂ ♀

breeding

ID: greenish back; grey wings; yellow face and lower eye arc; dark ear patch; yellow breast; dark stripes on sides; 2 white wing bars; white belly, flanks and undertail coverts; blackish legs. *Male:* black crown, throat and cheek patch; black and yellow head stripes. *Female:* olive where male is black; yellowish white throat.

Size: *L* 13 cm; *W* 20 cm.

Habitat: *Breeding:* mature coniferous forests, especially along the BC coast and at higher elevations in the Interior. *In migration* and *winter:* mixed coniferous and deciduous forests, especially with Douglas-fir, western red-cedar and grand fir; also gardens and riparian areas.

Nesting: saddled on a horizontal branch in a conifer; shallow nest of grasses, mosses, bark strips, twigs and plant fibres is lined with mosses, feathers and hair; mostly the female incubates 4–5 brown-marked, white eggs for 11–14 days.

Feeding: gleans vegetation and hawks for insects; occasionally visits suet feeders.

Voice: chippy call; male's unevenly patterned song consists of wiry *zee* and *zoo* notes and ends with a flourish.

Similar Species: *Black-throated Green Warbler* (p. 407): unmarked back; yellow face and cheek without dark ear patch. *Hermit Warbler* (p. 505): grey back; white breast, belly and undertail coverts; all-yellow face lacks dark ear patch.

BLACKBURNIAN WARBLER

Dendroica fusca

Ablaze in spring with a fiery orange throat, the dazzling Blackburnian Warbler dwells high among the towering conifers. Widely regarded as one of the most beautiful warblers in Canada, the Blackburnian stays hidden in the upper canopy for much of the summer. • Different species of wood-warblers are able to coexist through a partitioning of foraging niches and feeding strategies. This intricate partitioning reduces competition for food sources and prevents the exhaustion of particular resources. Some warblers inhabit high treetops, a few feed and nest along outer tree branches—some at high levels and some at lower levels—and others restrict themselves to inner branches and tree trunks. Blackburnians have found their niche predominantly in the outermost branches. • Although Blackburnian Warblers are regarded as forest-interior species, making them susceptible to forest fragmentation in both their summer and winter ranges, breeding bird surveys since the 1960s suggest that its numbers have remained stable. • This bird's name is thought to honour the Blackburne family of England, which collected the type specimen and managed the museum in which it was housed.

♂

♀

breeding

ID: *Male:* fiery, reddish orange upper breast, head and throat; orange-yellow head with angular, black facial patch; 2 broad, black crown stripes; blackish upperparts; large, white wing patch; yellowish to whitish underparts; dark streaking on sides and flanks; may show some white on outer tail feathers. *Female:* more yellowish orange head, throat and breast.
Size: *L* 12–14 cm; *W* 22 cm.
Habitat: mature conifers, especially tall spruce and hemlock; also mixed forests.
Nesting: high in a mature conifer, often near a branch tip; cup nest of bark, twigs and plant fibres is lined with conifer needles, moss and fur; female incubates 3–5 rufous-blotched, white to greenish white eggs for about 13 days.
Feeding: forages on the uppermost branches, gleaning budworms, flies, beetles and other invertebrates; occasionally hover-gleans.
Voice: call is a short *tick*; male's song is a soft, faint, high-pitched *ptoo-too-too-too tititi zeee* or *see-me see-me see-me see-me*.
Similar Species: *Prairie Warbler* (p. 412): yellow underparts; faint yellowish wing bars; black facial stripes do not form solid, angular patch. *Black-throated Green Warbler* (p. 407): unstreaked, olive green back and rump. *Townsend's Warbler* (p. 408): yellow throat.

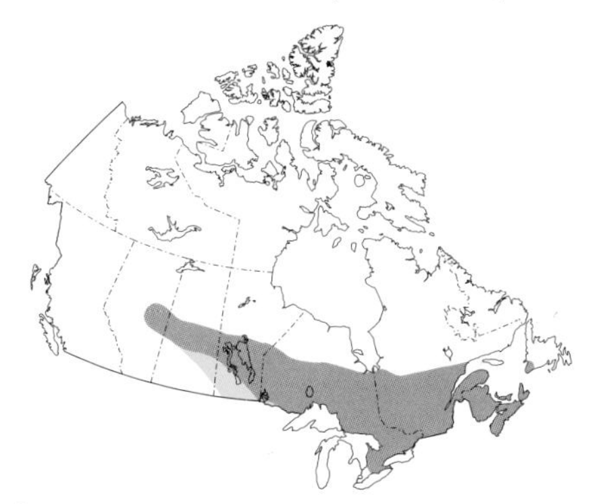

PINE WARBLER

Dendroica pinus

This unassuming bird is perfectly named because it is bound to majestic, sheltering pines of eastern North America. Pine Warblers can be difficult to find because they typically forage near the top of very tall, mature pine trees. They are particularly attracted to white and red pines, avoiding those with shorter needles. The species may have been more widespread before many tall pines were cut for shipbuilding in the 1800s and early 1900s. Populations are increasing in Canada as pine forests are replanted and mature. Occasionally, foraging Pine Warblers can be seen smeared with patches of sticky pine resin. • The Pine Warbler's modest appearance is similar to that of a number of immature and fall-plumaged vireos and warblers, forcing birders to take a good, long look before making a positive identification. This warbler is most often confused with the Bay-breasted Warbler or Blackpoll Warbler in drab fall plumage. Its song is a musical trill, which can also cause confusion with species that have similar trills, such as Chipping Sparrows, Dark-eyed Juncos and Worm-eating Warblers. • Unlike most other warblers, the Pine Warbler's wintering grounds are almost exclusively in North America.

♀

breeding

♂

ID: olive green head and back; dark greyish wings and tail; whitish to dusky wing bars; yellow throat and breast; faded dark streaking or dusky wash on sides of breast; white belly and undertail coverts; faint, dark eye line; faint, broken, yellow eye ring; female has duller plumage.

Size: *L* 13–14 cm; *W* 22 cm.

Habitat: *Breeding:* open, mature pine woodlands and mature pine plantations. *In migration:* mixed and deciduous woodlands.

Nesting: toward the end of a pine limb; deep, open cup nest of twigs, bark, grasses, pine needles and spiderwebs is lined with feathers; pair incubates 3–5 brown-speckled, whitish eggs for 10–13 days.

Feeding: gleans the ground and foliage for insects, berries and seeds; may hang upside down on branch tips.

Voice: call is a sweet *chip*; male's song is a short, musical trill.

Similar Species: *Prairie Warbler* (p. 412): distinctive, dark facial stripes; darker streaking on sides; yellowish wing bars. *Kirtland's Warbler* (p. 411): darker streaking on sides; broken, white eye ring; bluish grey upperparts. *Bay-breasted Warbler* (p. 414) and *Blackpoll Warbler* (p. 415): immature and fall birds have dark streaking on head and/or back; long, thin, yellow eyebrow. *Yellow-throated Vireo* (p. 325): bright yellow "spectacles"; grey rump; lacks streaking on sides.

KIRTLAND'S WARBLER

Dendroica kirtlandii

The Kirtland's Warbler has long been regarded as one of North America's rarest songbirds. The world's breeding population is almost totally confined to Michigan, and until recently, this globally endangered species was considered a former breeder in Canada. Kirtland's Warblers bred in central Ontario from Barrie eastward to the Petawawa area near Ottawa. In 1974, the world's population dipped to only 167 singing males on territory. With increased protection and habitat manipulation, the population of singing males had increased to 1400 by 2006, when Kirtland's Warblers returned to the Petawawa area. In 2007, Canada's second confirmed breeding was discovered in that same area. • Kirtland's Warblers require dense stands of young jack pines for nesting, a habitat specialization that is responsible for the species' rarity. As these stands age, the warblers abandon them. Recent increases in Kirtland's numbers have been attributed to controlled forest burns, which create a mosaic of ideal nesting habitat, and intensive Brown-headed Cowbird controls. • Although the Kirtland's Warbler has been recorded in southern Québec, the vast majority of sightings are from southwestern Ontario, as these warblers travel to Michigan from their wintering grounds in the Bahamas. • Kirtland's Warblers regularly bob their tails and are our largest species of warbler.

breeding

♀

♂

ID: yellow underparts; dark streaks confined to sides; greyish upperparts; white crescents above and below eyes; faint wing bars; white-spotted tail. *Male:* darker bluish grey back; dark lores. *Female:* lighter grey back; paler yellow underparts; less distinct eye crescents.

Size: *L* 14 cm; *W* 20 cm.

Habitat: large stands of young jack pines, 1–6 m in height, with ground cover of ferns, grasses, blueberries and brambles.

Nesting: in a slight depression under low vegetation; open cup of grasses, sedges and pine needles is lined with hair, moss and plant fibres; female incubates 3–6 finely brown-spotted, whitish eggs for about 14 days.

Feeding: gleans insects from ground cover up to midlevel of canopy; occasionally hover-gleans; forages on ripe blueberries.

Voice: strong, clear *chip* call; male's song is a loud, distinctive *chip-chip-che-way-o*, rising in pitch and intensity.

Similar species: *Magnolia Warbler* (p. 402): black mask; yellow rump; white in wings and tail; female and fall birds lack dark mask and have olive green backs. *Canada Warbler* (p. 431): complete eye ring; yellow lores; streaky, black "necklace"; lacks wing bars. *Yellow-rumped Warbler* (p. 405): localized yellow patches; 2 white wing bars; brighter yellow rump; white undertail coverts; does not bob tail.

PRAIRIE WARBLER

Dendroica discolor

This conspicuous, tail-wagging yellow warbler, with its flashy facial and flank markings, inhabits woodland edges and open scrublands. It prefers early successional habitats and areas throughout the rocky Canadian Shield with such poor soil conditions that vegetation remains short and scattered. A male Prairie Warbler may return each year to a favoured nesting site until the vegetation in that area grows too tall and dense. • On their nesting grounds, song wars occasionally result in physical fights between competing males. When the dust and feathers clear, the victor resumes his slow, graceful, butterfly-like courtship flight. • Although Prairie Warblers are considered quite common in many parts of the eastern U.S., they are rare in southern Ontario, the northern limit of their range. Estimates suggest that fewer than 500 breeding pairs exist in the province. • In Atlantic Canada, Prairie Warblers are seen only as rare vagrants. Most birds appear in fall as reverse migrants in the wave of southbound warblers passing through the U.S. eastern seaboard or Ontario and are brought to the East Coast by fall storms. • Although this species spread and increased in numbers until the mid-20th century, it is one of the few warblers that has experienced significant population declines since the 1960s.

♂

breeding

♀

ID: olive grey upperparts; bright yellow face and underparts; dark cheek stripe and eye line; black streaking on sides; inconspicuous chestnut streaks on back; 2 faint, yellowish wing bars; white undertail coverts; female has duller plumage.
Size: *L* 12–13 cm; *W* 18 cm.
Habitat: dry, open, scrubby sand dunes and rocky pine-oak-juniper scrublands; also woodland edges and young pine plantations with deciduous scrub.
Nesting: low in a tree; open cup nest of soft vegetation is lined with animal hair; female incubates 4 brown-spotted, whitish eggs for 11–14 days; occasionally in small, loose colonies.
Feeding: gleans, hover-gleans and occasionally hawks for insects; also eats berries and tree sap exposed by sapsuckers; caterpillars are favoured for nestlings.
Voice: sweet *chip* call; male's buzzy song is an ascending series of *zee* notes.
Similar Species: *Pine Warbler* (p. 410): lighter streaking on sides; whitish wing bars; lacks distinctive dark streaking on face. *Yellow-throated Warbler* (p. 505): white belly; bold, white wing bars, eyebrow and ear patch. *Bay-breasted Warbler* (p. 414) and *Blackpoll Warbler* (p. 415): immature and fall birds have white bellies and wing bars and paler upperparts with dark streaking.

PALM WARBLER

Dendroica palmarum

breeding
ssp. hypochrysea
♂

A habitual "tail wagger," the Palm Warbler can easily be picked out in mixed-species foraging flocks, even when its plumage details are hard to observe. In fall, this tail wag is as good a field mark as any, but in breeding plumage, the male has a handsome, if understated, appearance. Palm Warblers also differ from other *Dendroica* warblers in being much more terrestrial, almost always observed feeding on the ground or low in shrubs and trees. • The song of the Palm Warbler is one of the most distinctive sounds of Canada's black spruce bogs—not counting the buzzing of mosquitoes. This bird sings proudly from the stunted trees, adding life to an environment that is not exactly bird-rich. • Contrary to its common name, the Palm Warbler has little to do with palm trees, even on its southern wintering grounds. It was found overwintering in the vicinity of palms in the Caribbean when first collected, hence its name, but it breeds in tamarack and spruce bogs in boreal forest regions across Canada. • There are two subspecies of Palm Warblers, which can be easily separated by birders. The "Yellow Palm Warbler" (ssp. *hypochrysea*) nests from Ottawa eastward; the "Western Palm Warbler" (ssp. *palmarum*) nests west of our nation's capital.

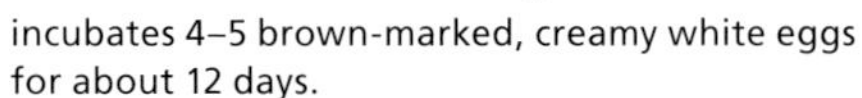

♂
breeding
ssp. palmarum

ID: constantly pumps tail at rest and when feeding; yellow undertail coverts; white tail corners. *Breeding:* rufous crown; yellow eyebrow and throat; faintly yellow breast and belly; darkly streaked, greyish brown upperparts; 2 brownish wing bars; dull yellowish rump. *Nonbreeding:* greyish brown crown; paler, smudgy streaking on underparts. *"Yellow" (eastern) subspecies:* larger; all-yellow underparts; more pronounced belly streaking.

Size: *L* 13 cm; *W* 20 cm.

Habitat: *Breeding:* open to semi-open bogs and muskeg dominated by black spruce. *In migration* and *winter:* open, shrubby areas; parks, golf courses and beaches.

Nesting: on the ground, often in sphagnum moss; small cup nest of grass, weeds and bark is lined with grasses, animal hair and feathers; female incubates 4–5 brown-marked, creamy white eggs for about 12 days.

Feeding: forages mostly on the ground for insects and seeds; also gleans foliage and hawks for insects.

Voice: clipped *chik* call; male's song is a monotonous, insect-like trill.

Similar Species: *Yellow-rumped Warbler* (p. 405): does not wag tail; localized yellow patches; 2 white wing bars; brighter yellow rump; white undertail coverts. *Blackpoll Warbler* (p. 415): female has 2 white wing bars, olive green rump and white undertail coverts. *Prairie Warbler* (p. 412): dark jaw stripe; darker eye line; lacks chestnut crown and dark streaking on breast. *Pine Warbler* (p. 410): faint, whitish wing bars; white undertail coverts; lacks chestnut cap and bold, yellow eyebrow.

BAY-BREASTED WARBLER

Dendroica castanea

You will have to search deep within stands of old-growth spruce and fir to find the handsome Bay-breasted Warbler. It typically forages midway up trees, often on the inner branches, so it's often difficult to spot. • Like all migratory birds, Bay-breasted Warblers face many dangers on their travels. Their annual trip north to the expansive boreal forest, however, seems well worth it for the abundance of summer food that is found there. • Bay-breasted Warblers are spruce budworm specialists, and their populations fluctuate from year-to-year, along with the cyclical rise and fall of this budworm. These birds are invaluable when it comes to long-term suppression of budworm outbreaks, typically moving to where the larvae are most numerous. It is estimated that in outbreak years, Bay-breasted Warblers can eat 13,000 budworms per hectare through the breeding season. With the lack of sizable outbreaks in recent years, this species has been declining at a rate of 4 percent annually. Deforestation in both its summer and winter habitats, environmental contaminants and migration hazards have all also contributed to its decline over the past few decades. This bird is truly an international resident, so its conservation requires the efforts of several nations.

♀

♂

breeding

ID: *Breeding male:* broad, black mask; chestnut crown, throat, sides and flanks; creamy yellow belly, undertail coverts and patch on side of neck; 2 white wing bars. *Breeding female:* paler overall; dusky face; whitish to creamy underparts and neck patch; faint chestnut cap; rusty wash on sides and flanks. *Nonbreeding:* yellow olive head and back; darkly streaked crown and back; whiter underparts.

Size: *L* 13–15 cm; *W* 20–22 cm.

Habitat: mature coniferous and mixed boreal forest, almost exclusively in stands of spruce and fir.

Nesting: usually on a horizontal conifer branch; open cup nest of grass, twigs, moss, roots and lichen is lined with fine bark strips and fur; female incubates 4–5 darkly marked, whitish eggs for about 13 days.

Feeding: usually forages at midlevel of trees; gleans vegetation and branches for caterpillars and adult invertebrates; eats numerous spruce budworms when available.

Voice: call is a high-pitched *see*; male's song is an extremely high-pitched *seee-seese-seese-seee.*

Similar Species: *Cape May Warbler* (p. 403): chestnut cheeks on yellow face; dark streaking on mostly yellow underparts; lacks reddish crown and flanks. *Chestnut-sided Warbler* (p. 401): yellow crown; white cheek and underparts; nonbreeding bird has white eye ring, unmarked whitish face and underparts and lacks bold streaking on lime green upperparts. *Blackpoll Warbler* (p. 415): nonbreeding and immature birds have dark streaking on breast and sides, white undertail coverts and lack chestnut on sides and flanks.

BLACKPOLL WARBLER

Dendroica striata

None of the 115 or so wood-warbler species found in the Americas migrates farther than the Blackpoll Warbler. It breeds from Alaska to New England and winters in northern South America. Spring migration is mostly inland, but fall migrants fly a much greater distance over water. Taking off from the coast of the northeastern provinces or states, Blackpoll Warblers head out over the Atlantic Ocean for a nonstop flight that averages 3000 kilometres and can take 88 hours to complete! In spring migration, Blackpolls are one of the last species to pass through on their way to their boreal and taiga breeding grounds, migrating so late that they avoid the mixed-species flocks that typify other warblers. • During the nesting season, males have been recorded singing throughout the daylight hours. • The male Blackpoll Warbler is easily identified by its chickadee-like head. The resemblance between the two species is an example of evolutionary convergence, although its cause is poorly understood. Blackpoll Warblers in fall plumage are easily confused with similar-looking Bay-breasted Warblers. Most Blackpolls, however, migrate later in fall than their Bay-breasted counterparts.

♀

♂

breeding

ID: distinctive yellow legs and feet; dark wings with 2 white bars; white belly and undertail coverts; white-spotted tail. *Breeding male:* black-streaked, olive grey upperparts; black cap extends to eyes; white cheek and throat; black "moustache" streak merges with flank streaking. *Breeding female:* streaked, yellow olive head and back; small, dark eye line; pale eyebrow. *Nonbreeding:* olive upperparts; pale yellow eyebrow, throat, breast and belly; faint streaking on flanks; may have dark legs and feet.
Size: *L* 14 cm; *W* 22 cm.
Habitat: *Breeding:* northern boreal black spruce forests, muskeg and burns; occasionally mixed forests and riparian thickets. *In migration*: almost any wooded or brushy habitat.

Nesting: on the limb of a tree; cup nest is made from grasses, mosses and twigs; female incubates 4–5 variably marked, whitish eggs for about 11 days.
Feeding: gleans vegetation or flycatches for insects and spiders.
Voice: call is a loud *chip*; male's song is a very high-pitched, uniform trill: *tsit tsit tsit*.
Similar Species: *Black-and-white Warbler* (p. 417): tree-creeping behaviour; black-and-white back; bold, white eyebrow; boldly streaked flanks; dark legs and feet. *Bay-breasted Warbler* (p. 414): nonbreeding bird has bolder eye line, more yellow underparts, faint streaking on flanks and dark legs and feet.

CERULEAN WARBLER

Dendroica cerulea

Concealed in the canopy of mature southern deciduous forests in Ontario and Québec, the Cerulean Warbler leads a mysterious life. The handsome blue-and-white male, which blends into the sunny summer sky while foraging among treetop foliage, is particularly difficult to observe. Often the only evidence of a meeting with this largely unobservable canopy dweller is the sound of the male's buzzy, trilling voice. Among the wood-warblers, the Cerulean Warbler is distinguished as one of the highest canopy nesters and its migrations are among farthest—this species winters in the mountains of northern South America. • Only in the last few decades have ornithologists been able to document this bird's breeding behaviour—courtship, mating, nesting and the rearing of young all tend to take place high in the canopy, well out of sight of most casual observers. Pairs have been recorded making up to four nesting attempts after nest failures. • Butterflies and moths make up a large proportion of the Cerulean Warbler's diet. • The loss of deciduous forests has caused a notable decline in this bird's numbers in Canada. It is now classified as a species of special concern, with a Canadian population of only 1000 breeding pairs.

♀

♂

breeding

ID: white undertail coverts and wing bars. *Male:* blue upperparts; white throat and underparts; bluish black "necklace"; streaking on sides. *Female:* blue-green crown, nape and back; dark eye line; yellow eyebrow, throat and breast; pale streaking on sides.

Size: *L* 11–12 cm; *W* 20 cm.

Habitat: mature, deciduous hardwood forests and extensive woodlands with a clear understorey; particularly drawn to riparian stands.

Nesting: on the end of a branch high in a deciduous tree; open cup nest of bark strips, weeds, grass, lichens and spider silk is lined with fur and moss; female incubates 3–5 brown-spotted, grey to creamy white eggs for 12–13 days.

Feeding: gleans or flycatches insects from upper canopy foliage and branches.

Voice: sharp *chip* call; male's song is a rapid, 3-part, accelerating sequence of buzzy notes followed by a higher trilled note.

Similar Species: *Black-throated Blue Warbler* (p. 404): lacks wing bars; male has black face, "chin" and throat; female has small, white wing patch. *Bay-breasted Warbler* (p. 414) and *Blackpoll Warbler* (p. 415): similar to female Cerulean but with more yellow than green on mantle. *Pine Warbler* (p. 410): browner; similar to female Cerulean but with more yellow on mantle.

BLACK-AND-WHITE WARBLER

Mniotilta varia

Although, in a general sense, this is a normal-looking warbler, the foraging behaviour of the Black-and-white Warbler stands in sharp contrast to that of most of its kin. Rather than flitting quickly between branches, Black-and-white Warblers behave like creepers and nuthatches, distantly related families, creeping along the trunks and larger branches of trees as they search for food. The bird's unique long hind toe and claw on each foot helps in its movements. Birders who have developed frayed nerves and tired eyes watching flitty warblers will be refreshed by the sight of the Black-and-white Warbler as it methodically creeps up and down tree trunks. Even a novice birder can easily identify this two-toned warbler, which lives up to its name in both spring and fall. A keen ear also helps: the male's gentle, oscillating song—like a wheel in need of greasing—is easily recognized and remembered. • This striking species breeds across much of southern Canada and the eastern United States and winters from Florida south to Venezuela and Columbia.

breeding

ID: mostly black upperparts with white streaks; black crown with white median stripe; white eyebrow; black wings; 2 white wing bars. *Male:* black cheek patch and throat; white "moustache" stripe; white underparts heavily streaked with black. *Female:* grey cheek patch; white throat; underparts washed with pale buff and less boldly streaked.
Size: *L* 13 cm; *W* 20 cm.
Habitat: *Breeding:* moist deciduous woodlands, willow thickets and pole-stage trembling aspen stands. *In migration:* mixed woods, woodlots and coastal tuckamore and scrub.
Nesting: usually on the ground, often at the base of a shrub; small cup nest of grass, leaves, bark strips, rootlets and pine needles is lined with fur and fine grasses; female incubates 3–5 brown-flecked, creamy white eggs for 10–12 days;
Feeding: creeps along tree trunks and larger branches; eats insects and spiders.
Voice: soft, high *seet* and sharp *pit* calls; male's song is a thin, high-pitched *wee-see wee-see wee-see wee-see wee-see wee-see*, like a squeaky wheel.
Similar Species: *Blackpoll Warbler* (p. 415): *breeding* male has solid black cap and clean white undertail coverts. *Yellow-throated Warbler* (p. 505): similar feeding style; grey-and-white upperparts; bright yellow throat and breast.

AMERICAN REDSTART

Setophaga ruticilla

Behaving more like a butterfly than a bird, the American Redstart flits from branch to branch in dizzying pursuit of prey. These birds are always moving, as if they are eager to show off their contrasting plumage, which is continually displayed in an enthusiastic series of flutters, twists, turns and feather-spreading. This constant motion extends to darting out into midair and hawking for insects, much in the manner of flycatchers, and the American Redstart is equipped with flycatcher-like bristles around its mouth, which help it to sense when prey is close enough to trap it in its ready bill. They behave the same way on their Central American wintering grounds, where they are called *candelitas*—"little candles." • Many species exhibit sexual dimorphism, but the plumages of male and female American Redstarts are so different that it seems impossible that they are the same species. • If you are surprised to hear a courtship song apparently coming from the mouth of a female redstart, what you are seeing is most likely a second-year male that has not yet acquired its striking black-and-orange adult plumage. The song of the American Redstart is so variable that even experienced birders faced with an unknown warbler song will exclaim, "It must be a redstart!"

breeding

ID: dark eyes; small, dark bill; blackish legs. *Male:* black head, throat and upperparts; orangey red (yellow on first-year male) sides of breast, wing bars and tail patch; white belly and undertail coverts. *Female:* olive grey upperparts; grey head; pale "spectacles"; yellow sides of breast, wing bars and tail patch; light grey underparts.

Size: *L* 13 cm; *W* 19 cm.

Habitat: *Breeding:* moist, mixed or deciduous forests with tall shrub undergrowth, often near water. *In migration:* mixed woodlands and riparian thickets.

Nesting: in the fork of a sapling or tall shrub; compact cup of plant fibres, grasses and rootlets is lined with plant down and grasses; female incubates 4–5 brown-wreathed, white eggs for 10–13 days.

Feeding: actively gleans foliage and hawks for insects and spiders.

Voice: call is a sharp, sweet *chip*; male's song is a highly variable series of *tseet* or *zee* notes, often at different pitches.

Similar Species: male is distinctive. *Palm Warbler* (p. 413): nonbreeding female is browner, with streaked flanks and breast, yellow undertail coverts and pale eyebrow.

PROTHONOTARY WARBLER

Protonotaria citrea

The Prothonotary Warbler is unusual among the wood-warbler clan because it nests in cavities. Standing dead trees and stumps riddled with natural cavities and woodpecker excavations provide perfect nesting habitat for this bird, especially if the site is near stagnant, swampy water. Although much of its swampy habitat is inaccessible, if you are in the right place at the right time, you might be lucky enough to come across this bird as it forages for insects along tree trunks, decaying logs and debris floating on the water's surface. • Suitable Prothonotary nesting sites are so rare in Canada that a breeding pair will often return annually to the same nest cavity. Males can be very aggressive in defending their territory and often resort to combative aerial chases when songs and warning displays fail to intimidate intruders. Even with this aggressive behaviour, one in four nests is parasitized by Brown-headed Cowbirds. House Wrens have been recorded destroying Prothonotary Warbler nests in areas with limited cavities. Within Canada, these warblers are classified as endangered because of their limited distribution and a population decline of 80 percent to the current estimated population of only about 30 birds. • The species was named for its plumage which resembles the bright yellow robes of papal clerks (prothonotaries) in the Roman Catholic Church.

breeding

♂ ♀

ID: *Male:* bright orange-yellow head and throat; bright yellow underparts; large, dark eyes; long, dark bill; olive green back; unmarked, bluish grey wings and tail; white undertail coverts. *Female:* similar to male but with olive green crown, nape and cheeks.

Size: *L* 14 cm; *W* 22 cm.

Habitat: wooded, deciduous swamps.

Nesting: in a cavity in a standing dead tree, rotten stump or abandoned woodpecker hole, from water level to 3 m above ground; often returns to same nest site; female chooses one of male's moss-filled cavities, lines it with soft plant material and incubates 4–6 brown-spotted, creamy to pinkish eggs for 12–14 days.

Feeding: forages for a variety of insects and small molluscs; gleans vegetation; may hop on floating debris or creep along tree trunks.

Voice: brisk *tink* call; male's song is a loud, ringing series of *sweet* or *zweet* notes on a single pitch; flight song is *chewee chewee chee chee.*

Similar Species: *Blue-winged Warbler* (p. 394): white wing bars; black eye line; yellowish white undertail coverts. *Yellow Warbler* (p. 400): dark wings and tail with yellow highlights; yellow undertail coverts; male has reddish breast streaking. *Hooded Warbler* (p. 429): female has yellow undertail coverts and yellow-olive upperparts.

WORM-EATING WARBLER

Helmitheros vermivorum

Worm-eating Warblers are rare in Canada, and there are only about 30 confirmed sightings per year, most from southwestern Ontario. Almost all the sightings have been reported in May, though some birds inevitably surface throughout the summer and early fall. Most Worm-eating Warblers summer in the U.S., but they seem to have set their sights on southern Ontario for the latest expansion of their breeding range. To date, there is no evidence of breeding, only a good chance that the bird might nest here in the future. • This warbler's subdued colours allow it to blend in with the decomposing twigs, roots and leaves that litter the forest floor. Most of the Worm-eating Warbler's time is spent foraging for small, terrestrial insects and spiders amid the dense undergrowth and dead leaves of deciduous forests. Worm-eating Warblers regularly forage in clusters of dead leaves hanging in the undergrowth.

ID: sexes similar; black and buff-orange head stripes; dark eyes; brownish olive upperparts; rich buff-orange breast; whitish undertail coverts.
Size: *L* 11–13 cm; *W* 21–22 cm.
Habitat: steep, deciduous woodland slopes, ravines and swampy woodlands with shrubby understorey cover.
Nesting: not known to nest in Canada.
Feeding: eats mostly small insects; forages on the ground and in the understorey.

Voice: buzzy *zeep-zeep* call; male's song is a faster, thinner version of Chipping Sparrow's chipping trill.
Similar Species: *Red-eyed Vireo* (p. 331): grey crown; white eyebrow; red eyes; yellow undertail coverts. *Northern Waterthrush* (p. 422) and *Louisiana Waterthrush* (p. 423): darker upperparts; bold, white or yellowish eyebrow; dark streaking on white breast; lacks striped head. *White-throated Sparrow* (p. 453): greyish bill; yellow lores; black and white or tan crown stripes.

OVENBIRD

Seiurus aurocapilla

The Ovenbird's loud song is among the easiest to learn and a familiar sound in May and June. Sung from the ground or a low shrub, it sounds like *TEAcher TEAcher TEAcher.* • An Ovenbird will rarely expose itself in the open forest, and it usually stops calling as soon as an intruder enters its territory. Even when they are threatened, Ovenbirds stubbornly refuse to become airborne, preferring to escape on foot through dense tangles and shrubs. • The Ovenbird gets its name from the shape of its nest, which often resembles a Dutch oven. An incubating female nestled within her woven dome is usually so confident in her nest that, unless closely approached, she will choose to sit tight rather than flee approaching danger. • The Ovenbird's plain olive upperparts and spotted breast make it appear thrushlike, but its small size, striped crown and ground-nesting habits place it among the wood-warblers. The plumage of the Ovenbird varies little between the sexes and among age classes. • In years when food is plentiful, such as during spruce budworm outbreaks, Ovenbirds usually lay more eggs. They are successful in raising most of their young and are able to produce at least two broods in a single summer.

ID: sexes similar; olive brown upperparts; rufous orange crown bordered by black stripes; olive face; white eye ring; white throat; black "moustache" stripe; white underparts with black spots forming stripes to lower belly; pink legs and feet.
Size: *L* 15 cm; *W* 23 cm.
Habitat: *Breeding:* undisturbed, mature deciduous forests or poplar stands in mixed forests; usually under a closed canopy with little understorey. *In migration:* woodlands with dense, riparian shrubs and thickets; also farmsteads, parks and gardens.
Nesting: on the ground, often covered by vegetation; domed nest is made of grasses, leaves and mosses; female incubates 4–5 brown- or grey-speckled, white eggs for 11–14 days.
Feeding: gleans the ground for invertebrates, including worms, snails and insects.
Voice: loud *chip* call; male's song is a loud, distinctive *TEAcher TEAcher TEAcher*, increasing in speed and volume.
Similar Species: *Thrushes* (pp. 376–84): larger; lack rufous crown outlined in black; most have buffy or dusky breasts, with little or no streaking on lower belly. *Northern Waterthrush* (p. 422): wholly dark crown; bold, white or yellowish eyebrow; lacks white eye ring; bobs hind end constantly.

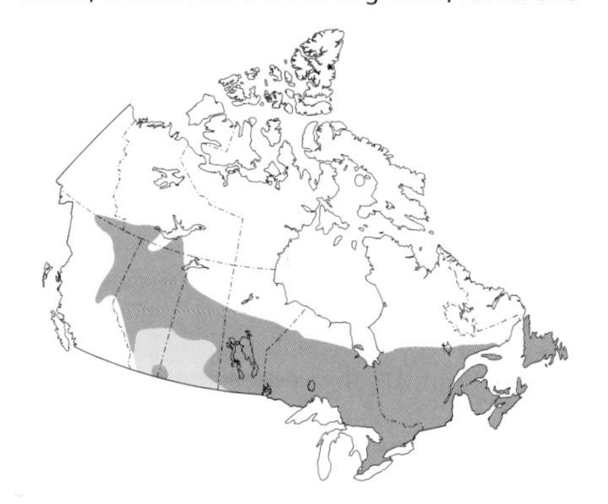

NORTHERN WATERTHRUSH

Seiurus noveboracensis

Like the Ovenbird, the Northern Waterthrush is a warbler that resembles a thrush. With a strong affinity for wet habitats, this bird mostly skulks along the shores of deciduous swamps or coniferous bogs, so birders not satisfied with simply hearing a Northern Waterthrush in its nesting territory must literally get their feet wet scrambling through shrubby tangles and over fallen logs and soggy ground if they hope to see one on its nesting territory. • Like the Ovenbird, the Northern Waterthrush is a walker and not a hopper. As it works along downed logs and branches and on the ground, its preferred gait is a tail-bobbing walk, and it teeters along like a slightly out of practice Spotted Sandpiper. • The male Northern Waterthrush uses its song to claim and defend a breeding territory of about one hectare in size. Its voice is loud and raucous for such a small bird, so it seems fitting that it was once known as "New York Warbler," a city known for its decibels. The species name *noveboracensis* means "of New York."

ID: sexes similar; plain, dark brown upperparts and tail; narrow, white or pale yellow eyebrow; brown crown and eye line; pink bill; heavily black-streaked, white or yellow-washed underparts; white or yellow undertail coverts; pink legs; bobs hind end constantly.
Size: *L* 15 cm; *W* 24 cm.
Habitat: wooded edges of swamps, lakes, beaver ponds, bogs and rivers; also moist, wooded ravines and riparian thickets.
Nesting: on the ground among grasses or fallen tree roots or trunks; small, concealed cup of mosses, leaves, twigs and inner bark is lined with finer materials; female incubates 4–6 variably marked, white eggs for about 12 days.
Feeding: feeds mostly on the ground; eats both aquatic and terrestrial invertebrates.
Voice: brisk *chip* or *chuck* call; male's distinctive, loud, 3-part song is *sweet sweet sweet, swee wee wee, chew chew chew chew,* usually down the scale and speeding up, ending with a flourish.
Similar Species: *Ovenbird* (p. 421): black-bordered, rufous crown; bold, white eye ring; warm brown upperparts; distinctive song. *Louisiana Waterthrush* (p. 423): broader, white eyebrow; unspotted, white throat; orange-buff wash on flanks.

LOUISIANA WATERTHRUSH

Seiurus motacilla

The Louisiana Waterthrush is often seen sallying along the shorelines of babbling streams and gently swirling pools in search of its next meal. This bird inhabits swamps and sluggish streams throughout much of its North American range, but in Canada, where its range overlaps that of the Northern Waterthrush, it inhabits shorelines near fast-flowing water. • Louisiana Waterthrushes have never been recorded in great numbers in Canada, with an estimated Canadian population of about 300 pairs, partly because little suitable habitat remains for them. This bird's success depends on the stewardship of private landowners to protect its nesting habitat. Only a small percentage of Canada's wild areas are officially protected by law in parks, conservation areas and ecological reserves. • Although only a confirmed breeder in southern Ontario, this bird is thought to breed occasionally in southwestern Québec. The species is a regular vagrant into Québec and the Maritime provinces. • The Louisiana Waterthrush is easily identified by its habit of bobbing its head and moving its tail up and down as it walks. The tail-wagging behaviour is so noticeable that both the genus and species names refer to "tail-wagger."

ID: sexes similar; brownish upperparts; long bill; pink legs; white underparts with orange-buff wash on flanks; long, dark streaks on breast and sides; bicoloured, buff-and-white eyebrow; clean white throat.
Size: *L* 15 cm; *W* 25 cm.
Habitat: moist, forested ravines alongside fast-flowing streams; rarely along wooded swamps.
Nesting: concealed within a rocky hollow or within a tangle of tree roots; cup-shaped nest of leaves, bark strips, twigs and moss is lined with animal hair, ferns and rootlets; female incubates 3–6 variably marked, creamy white eggs for about 14 days.
Feeding: gleans terrestrial and aquatic insects and crustaceans from rocks and debris near or in shallow water; may flip and probe dead leaves and other debris for food; occasionally catches insects by flying over water.
Voice: brisk *chick* or *chink* call; male's song begins with 3–4 distinctive, shrill, slurred *seeeew* notes followed by a warbling twitter.
Similar Species: *Northern Waterthrush* (p. 422): yellowish to buff eyebrow narrows behind eye; heavily black-streaked, white or yellow-washed underparts; lacks orange-buff flanks. *Ovenbird* (p. 421) and *thrushes* (pp. 376–84): lack broad, white eyebrow.

KENTUCKY WARBLER

Oporornis formosus

Like most of the *Oporornis* warblers, Kentucky Warblers are skulking birds that spend much of their time on the ground, overturning leaves and scurrying through dense thickets in search of insects. These birds are typically found in the moist, deciduous woodlands of the eastern U.S., but each year, a few make their way into Ontario. Most sightings are recorded in Point Pelee National Park during the last two weeks of May. • Although no nests have been confirmed in Canada, Kentucky Warblers are thought to be sporadic breeders in the deciduous woodland parks and wooded ravines of extreme southwestern Ontario. A lack of evidence might be because females are shy and elusive, and, during the breeding season, they often leave their nests before they are discovered. • Like waterthrushes and ovenbirds, the Kentucky Warbler has a habit of bobbing its tail up and down as it walks. Its song can sometimes be confused with that of the Carolina Wren or maybe even an Ovenbird, but it has a richer quality and a steadier, rolling tempo.

♂

breeding

ID: sexes similar; olive green upperparts; bright yellow "spectacles" and underparts; black crown, "sideburns" and half-mask; short tail; long, pinkish legs; no wing bars.
Size: *L* 13 cm; *W* 18–22 cm.
Habitat: moist, deciduous and mixed woodlands with dense shrubby cover and herbaceous plant growth, including wooded ravines, swamp edges and creek bottomlands.
Nesting: not known to nest in Canada.
Feeding: gleans insects while walking along the ground, flipping over leaf litter or snatching prey from undersides of low foliage.
Voice: sharp *chick*, *chuck* or *chip* call; male's musical song is a series of 2-syllable notes: *chur-ree chur-ree.*
Similar Species: *Canada Warbler* (p. 431): dark, streaky "necklace"; bluish grey upperparts. *Yellow-throated Warbler* (p. 505): bold, white eyebrow, ear patch and wing bars; white belly and undertail coverts.

CONNECTICUT WARBLER

Oporornis agilis

The Connecticut Warbler's elusive nature and primarily Canadian breeding distribution have made it one of the most sought-after birds around. A trip to any suitable mature aspen forest can reveal the Connecticut Warbler's boisterous songs, but pinpointing this ground-dwelling warbler's location is much more difficult. Connecticuts are quite secretive, and with the exception of singing males, they rarely venture out from low vegetation, preferring instead to skulk. • These warblers feed mostly on insects and spiders obtained on or near the ground. • Only a handful of nests with eggs have been found in Canada, and much of the Connecticut Warbler's breeding biology remains poorly known. • The three look-alike species of *Oporornis* warblers can best be separated in the field by their eye rings: the Connecticut has a complete eye ring; the MacGillivray's has an incomplete eye ring; and the Mourning has no eye ring at all. • Like many other North American birds, the Connecticut Warbler was named for the place in which it was first collected, even though, in this case, it is not even a common migrant there.

breeding

ID: olive green upperparts; yellow underparts; bold, complete, white eye ring; long undertail coverts make tail appear short; pink legs. *Breeding male:* blue-grey "hood." *Female:* pale grey "hood."

Size: *L* 13–15 cm; *W* 22 cm.

Habitat: young to mature trembling aspen forests with closed canopy and sparse shrub layer; open pine forests and fairly open spruce bogs and tamarack fens with well-developed understorey.

Nesting: on the ground; on a hummock or in a low shrub; messy nest of leaves, weeds, grasses, bark strips and mosses is lined with finer materials; female incubates 4–5 brown-marked, creamy white eggs for about 12 days.

Feeding: gleans invertebrates from ground leaf litter; occasionally forages among low branches.

Voice: brisk, metallic *cheep* or *peak* call; male's song is a loud, clear, explosive *chipity-chipity-chipity chuck* or *per-chipity-chipity-chipity choo.*

Similar Species: *Mourning Warbler* (p. 426): no eye ring or thin, incomplete eye ring; shorter undertail coverts; male has blackish breast patch. *Nashville Warbler* (p. 398): bright yellow throat; shorter, dark legs and bill. *MacGillivray's Warbler* (p. 427): bold, broken eye ring.

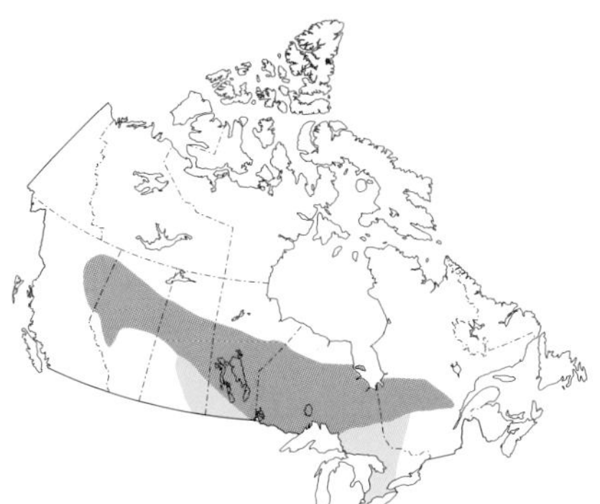

MOURNING WARBLER

Oporornis philadelphia

Mourning Warblers seldom leave the protection of their often-impenetrable habitat, which typically includes dense, shrubby thickets, raspberry brambles and nettle patches. Although Mourning Warblers can be quite common in some locations, they are seen less frequently than might be expected. Males tend to sing only on their nesting territory, and even then, they rarely sing from exposed perches. Riparian areas, regenerating cut-blocks and patches of forest that have been recently cleared by fire provide the low shrubs and sapling trees that this warbler relies on for nesting and foraging. • If invaders approach too close to a nest site, the Mourning Warbler will pretend to have a broken wing to lead predators away. • Breeding bird survey (BBS) data shows that populations have been declining for several decades in eastern Canada and increasing in western Canada. The clearing of forests in Central and South America for coffee plantations is wiping out winter habitat for neotropical migrants. • The Mourning Warbler is named for its sombre, dark hood, which reminded 19th-century ornithologist Alexander Wilson of someone in mourning. This warbler's bright yellow underparts and cheery song belie its name.

♀ ♂

breeding

ID: olive green upperparts; yellow underparts, including vent; short tail; pinkish legs. *Breeding male:* dark face, usually without eye ring; blue-grey "hood"; black upper breast patch. *Female:* lighter grey "hood"; whitish "chin" and throat; may have thin eye ring.

Size: *L* 10–15 cm; *W* 18 cm.

Habitat: shrubby, deciduous edges of roads and highways, clear-cuts and forest openings; also deciduous second growth.

Nesting: on or near the ground, often at the base of a shrub; bulky nest of leaves, weeds and grasses is lined with hair and finer grasses; female incubates 3–5 brown-blotched, creamy white eggs for about 12 days.

Feeding: forages in dense, low shrubs for invertebrates.

Voice: loud, low *check* call; male's husky, 2-part song is variable but often descends at the end: *cheery cheery cheery, chorry chorry* or *blee blee blee blee-blee choochoo.*

Similar Species: *MacGillivray's Warbler* (p. 427): bold, broken eye ring. *Connecticut Warbler* (p. 425): bold, complete eye ring; lacks black breast patch; long undertail coverts make tail look short. *Nashville Warbler* (p. 398): bright yellow throat; dark legs.

MACGILLIVRAY'S WARBLER

Oporornis tolmiei

The MacGillivray's Warbler inhabits the dense understorey shrubbery and bushy tangles along the edges of hardwood forests or residential parks. Its presence is often revealed only by its loud, rich *chip* call or the male's buzzy song. To see this warbler, birders must often crouch down and peer deep into the shadows in rapid response to this bird's faintly perceptible actions. The best time of year to observe this mountain specialty is in spring, when males sit boldly atop taller shrubs to belt out their courtship serenades. • Multiple wood-warbler species can coexist in one habitat by using different foraging niches and nest sites. The MacGillivray's Warbler has found a niche among the dense understorey shrubbery and brushy tangles. • At one time, the MacGillivray's Warbler and the Mourning Warbler were considered the same species. • John J. Audubon named this warbler for William MacGillivray, who edited and reworked the manuscript of Audubon's classic 1840 work, *Birds of America.*

♀

♂

breeding

ID: plain, olive green upperparts; bright, unmarked, lemon yellow underparts; broken, white eye ring; grey "hood" extends to breast; dark eyes; slightly downcurved, pinkish bill. *Male:* greyish blue "hood," darker on breast, forehead and eye line. *Female:* paler, more uniform "hood"; lacks dark mottling on breast.

Size: *L* 13 cm; *W* 19 cm.

Habitat: low, dense shrub thickets in clearings, transmission corridors, forest edges and clear-cuts.

Nesting: usually close to the ground in thick shrubbery; small cup nest of weed stems and grasses is lined with finer grasses; female incubates 4 brown-marked, creamy white eggs for 11–13 days.

Feeding: gleans low vegetation and the ground mostly for insects; occasionally eats seeds.

Voice: typical *chik* call; male's song is a buzzy *churr churr churr swee swee* with less-clear final notes.

Similar Species: *Mourning Warbler* (p. 426): no eye ring or thin, incomplete eye ring; shorter undertail coverts; male has blackish breast patch. *Connecticut Warbler* (p. 425): bold, complete eye ring; lacks black breast patch; long undertail coverts make tail look very short. *Nashville Warbler* (p. 398): yellow chin; complete eye ring. *Common Yellowthroat* (p. 428): no eye ring; female lacks grey throat.

COMMON YELLOWTHROAT

Geothlypis trichas

Although most warblers avoid getting their feet wet by frequenting dry, shrubby and forested areas, the Common Yellowthroat lives just above the mud and muck of wetland vegetation. Despite its bright colours and abundance in wetland areas throughout Canada, this bird can be surprisingly difficult to find. • In May and June, male Yellowthroats issue their distinctive *witchety* songs while perched atop tall cattails or shrubs. An extended look at the male in action will reveal the location of his favourite singing perches, which he visits regularly, in rotation. These strategic outposts mark the boundary of his territory and are fiercely guarded from intrusion by other males. • Common Yellowthroats do not spend all their time hidden away in marshes. When nesting is over, they can be found in almost any damp, scrubby habitat. Here, they fuel up for a short winter trip south, often no farther than California or the Carolinas. • Common Yellowthroats are often parasitized by Brown-headed Cowbirds, which are primarily open-country birds and usually target nests in less-forested habitats.

breeding

♀

♂

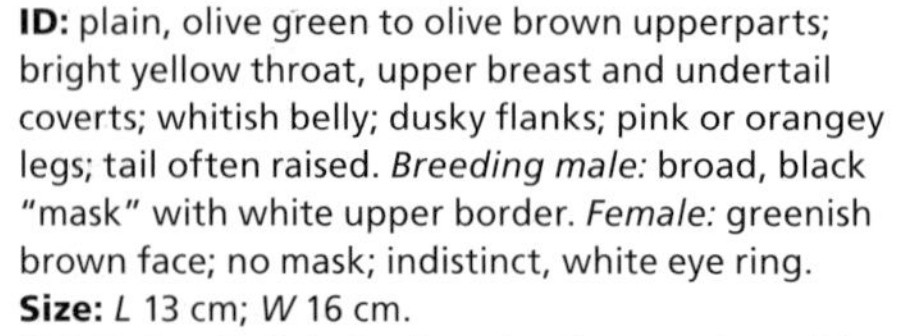

ID: plain, olive green to olive brown upperparts; bright yellow throat, upper breast and undertail coverts; whitish belly; dusky flanks; pink or orangey legs; tail often raised. *Breeding male:* broad, black "mask" with white upper border. *Female:* greenish brown face; no mask; indistinct, white eye ring.
Size: *L* 13 cm; *W* 16 cm.
Habitat: cattail, bulrush and sedge marshes with dense vegetation, including willows.
Nesting: in low vegetation over water; neat, compact cup nest of grasses and sedges is lined with finer grasses; female incubates 3–6 darkly marked, creamy white eggs for 12 days; usually raises 2 broods.
Feeding: gleans vegetation for insects and other invertebrates.

Voice: sharp *tcheck* or *tchet* call; male's song is a clear, oscillating *witchety witchety witchety-witch* or other repetitions of 3 (rarely 2) syllables.
Similar Species: male's black mask is distinctive. *Kentucky Warbler* (p. 424): yellow "spectacles"; all-yellow underparts; half-mask. *Yellow Warbler* (p. 400): brighter yellow overall; yellow highlights in wings; all-yellow underparts. *Wilson's Warbler* (p. 430): all-yellow forehead, eyebrow, cheek and underparts; may show dark cap. *Orange-crowned Warbler* (p. 397): dull yellow-olive overall; faint breast streaks. *Nashville Warbler* (p. 398): bold, complete eye ring; blue-grey crown. *MacGillivray's Warbler* (p. 427): grey "hood"; bold, broken, white eye ring; deeper olive green upperparts; yellower underparts.

HOODED WARBLER

Wilsonia citrina

With a distinct, bold, black-and-yellow "hood," this striking warbler is a favourite and much sought-after species. Hooded Warblers reach the northern limit of their range in southern Ontario, so they are rare in Canada. The first confirmed nesting was not recorded until 1940, and the species is currently classified as threatened in Canada. A forest-interior species often restricted to larger woodlots, the Hooded Warbler's habitat is becoming scarce and fragmented. However, this bird's population has doubled since the 1980s, and the increase does not show signs of ending soon. • Despite nesting low to the ground, Hooded Warblers require extensive mature forests, where fallen trees have opened gaps in the canopy, encouraging understorey growth. • Unlike their female counterparts, male Hooded Warblers may return to the same nesting territory year after year. Once the young have left the nest, each parent takes on the guardianship of half the fledged young. • Recent evidence suggests that wintering Hooded Warblers partition food resources by sex: males tend to forage in treetops, whereas the females forage near the ground. • The Hooded Warbler's eyes are larger than those of most warbler species; this is thought be related to its preference for shaded forest habitats.

♀ ♂

breeding

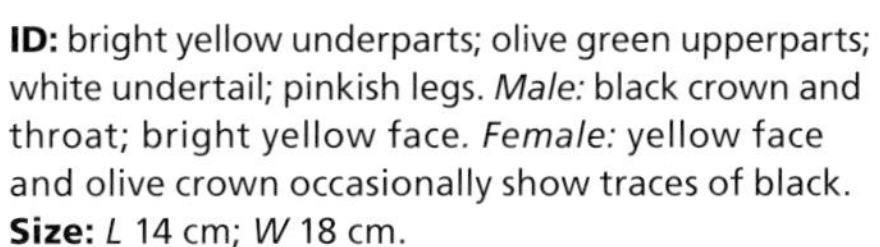

ID: bright yellow underparts; olive green upperparts; white undertail; pinkish legs. *Male:* black crown and throat; bright yellow face. *Female:* yellow face and olive crown occasionally show traces of black.
Size: *L* 14 cm; *W* 18 cm.
Habitat: clearings with dense, low shrubs in mature, upland deciduous and mixed forests; occasionally in moist ravines or mature white pine plantations with a dense understorey of deciduous shrubs.
Nesting: low in a deciduous shrub; open cup nest of fine grasses, bark strips and dead leaves is lined with animal hair, fine grasses and plant down; female incubates 4 brown-spotted, creamy white eggs for about 12 days.

Feeding: gleans insects and other forest invertebrates from the ground or shrub branches; may scramble up tree trunks or flycatch.
Voice: call note is a metallic *tink*, *chink* or *chip*; male's clear, whistling song is a variation of *whitta-witta-wit-tee-yo*.
Similar Species: *Wilson's Warbler* (p. 430), *Yellow Warbler* (p. 400) and *Common Yellowthroat* (p. 428): females lack white undertail coverts. *Kentucky Warbler* (p. 424): yellow "spectacles"; dark, triangular half-mask.

WILSON'S WARBLER

Wilsonia pusilla

Even a casual glance into a wet thicket during migration will often reveal an energetic Wilson's Warbler flickering through tangles of leaves and branches in search of food, as if every moment is precious. Birders often become exhausted while pursuing a Wilson's Warbler, but the bird never seems to tire during its lightning-fast performances. • This warbler may be found in almost any shrubby habitat during spring or fall migration, but its nesting habitat is mostly restricted to wet thickets and shrubs in montane and northern boreal forests, where it settles near bogs and shrub-lined beaver ponds. With more energy than many larger birds, the Wilson's Warbler is also quick to jump to the defence of its hidden brood, should an intruder come too close. • Often intensely golden over much of its plumage, this tiny warbler can be confused with the Yellow Warbler; a closer look usually reveals its true identity, especially if the male Wilson's flashes its black crown. • Breeding bird surveys since the 1960s have revealed significant declines in Wilson's Warbler numbers throughout North America, but especially in Canada, where overall numbers have dropped by 4 percent per year. Reasons for these rather steep declines are not readily apparent.

♀ ♂

breeding

ID: olive yellow upperparts; all-yellow underparts; yellow face; large, dark eyes; broad, yellow eyebrow. *Male:* black crown. *Female:* olive crown and forehead.
Size: *L* 10 cm; *W* 18 cm.
Habitat: *Breeding:* wet thickets and shrubs; also higher-elevation, shrubby, mixedwood habitats. *In migration:* mostly deciduous shrubs and thickets, often along edges of water bodies.
Nesting: on the ground or in a low shrub; bulky nest of dead leaves, grasses and mosses is lined with finer grasses and hair; female incubates 2–6 brown-marked, creamy white eggs for 10–13 days.

Feeding: gleans vegetation for insects and other invertebrates.
Voice: flat, low *chet* or *chuck* call; male's rapid, 2-part chattering song becomes louder and faster toward the end.
Similar Species: *Yellow Warbler* (p. 400): plainer face; heavier bill; yellow wing bars; male is bright yellow overall with reddish orange breast streaks. *Common Yellowthroat* (p. 428): female has darker face. *Kentucky Warbler* (p. 424): yellow "spectacles"; dark, angular half-mask. *Orange-crowned Warbler* (p. 397): dull yellow olive overall; faint breast streaks. *Nashville Warbler* (p. 398): bold, complete eye ring; blue-grey crown.

CANADA WARBLER

Wilsonia canadensis

The Canada Warbler is one of the few wood-warblers whose wardrobe is just as colourful in autumn as it is in spring. They are fairly inquisitive birds, and they occasionally pop up from dense shrubs in response to passing hikers. Male Canada Warblers, with their bold, white eye rings, have a wide-eyed, alert appearance. • These birds live in open defiance of winter, never staying in one place long enough to experience one! This colourful warbler arrives late and leaves early, spending three months or less in Canada before completing its 18,000-kilometre round trip to wintering grounds in South America. • Although several wood-warblers breed almost exclusively in Canada, the Canada Warbler isn't one of them. It breeds across Canada east of British Columbia, but it can also be found nesting throughout the Appalachians as far south as Georgia. • Canada Warblers are classified as threatened in Canada because of a population decline of 85 percent since 1968. Although reasons for these declines are unclear, forest succession and loss of forested wetlands in its breeding range and loss of forested habitat on its wintering grounds in the northern Andes are all potential causes.

breeding

ID: blue-grey upperparts; yellow forehead patch; complete, white eye ring; yellow throat, chest and belly; white undertail coverts; long tail is often raised. *Male:* streaky, black "necklace"; dark, angular half-mask. *Female:* blue green back; faint "necklace."

Size: *L* 12–15 cm; *W* 17–22 cm.

Habitat: moist, deciduous or mixed forests with dense shrub layer.

Nesting: on the ground, often by a fallen tree; loose, neat cup nest of leaves, grasses and weed stems is lined with hair and finer grasses; female incubates 2–6 brown-spotted, creamy white eggs for 10–14 days.

Feeding: gleans the ground and vegetation for insects and other invertebrates.

Voice: loud, quick *chick* or *chip* call; male's song begins with a sharp *chip* note and continues with a rich, variable warble.

Similar Species: *Magnolia Warbler* (p. 402): black streaks extend onto sides; yellow rump; white in wings and tail. *Kentucky Warbler* (p. 424): yellow "spectacles"; dark, triangular half-mask; greenish upperparts; yellow undertail coverts; lacks black "necklace." *Northern Parula* (p. 399): white wing bars; broken, white eye ring; white belly. *Yellow-rumped Warbler* (p. 405): yellow rump; white wing bars or patches; white belly.

YELLOW-BREASTED CHAT

Icteria virens

Seemingly too large to be a warbler, the Yellow-breasted Chat is the largest member of the wood-warbler family. Although its inclusion with the wood-warblers has been questioned, molecular studies seem to support this grouping. With its bright yellow breast and skulking behaviour, the Yellow-breasted Chat is a bizarre bird and often attracts attention to itself through strange vocalizations and noisy thrashing in dense undergrowth, more like a catbird than a warbler. When present, it seems to do its best to ensure that it is not overlooked. • The Yellow-breasted Chat breeds in riparian zones of dense scrub and brush, which are being threatened by urban sprawl, draining of wetlands and livestock grazing. In Canada, Yellow-breasted Chats have a disjointed distribution, with separate subspecies occurring in southern British Columbia, the Prairies and southern Ontario. The population in southern BC is classified as endangered because of habitat loss and small numbers. The species is also on the edge of its range in Ontario, where it has been designated as special concern, but reasonable numbers still occur in dense scrubby and bushy valleys and steep slopes in parts of Alberta and Saskatchewan.

♂

breeding

ID: white "spectacles" and "moustache" stripe; heavy, black bill; unmarked, olive green upperparts; rich yellow breast and throat; buff flanks; white belly and undertail coverts; long tail is often held upraised. *Male:* black lores. *Female:* gray lores.

Size: *L* 18 cm; *W* 25 cm.

Habitat: dense willow thickets and mixed brushy tangles; riparian shrublands and shrubby coulees, usually near water.

Nesting: low in a shrub or small, bushy tree; well-concealed, bulky nest of leaves, grasses, weed stems and bark strips is thinly lined with finer grasses; female incubates 3–4 large, brown-marked, creamy white eggs for 11–12 days.

Feeding: gleans vegetation for insects and berries; usually feeds low in thicket growth.

Voice: calls include *whoit, chack* and *kook*; male's song, which can be long, is an assorted series of whistles, squeaks, grunts, rattles and mews; often sings after dusk and before dawn.

Similar Species: *Common Yellowthroat* (p. 428): much smaller; no white "spectacles" or "moustache" stripe; finer bill; shorter tail. *Nashville Warbler* (p. 398): much smaller; white eye ring; thinner bill.

SPOTTED TOWHEE

Pipilo maculatus

Scraping back layers of dry leaves in its search for food, a Spotted Towhee puts both feet to use at the same time, doing the "towhee dance," its long claws revealing hidden morsels in the leaf litter. The resulting ruckus of ground litter being tossed about leads many people to expect a squirrel or something larger in the underbrush. What a surprise to see a colourful bird not much larger than a sparrow! • Towhees like tangled thickets and overgrown gardens, especially if they offer blackberries or other small fruits. Many pairs nest in urban neighbourhoods, where they take turns scolding the resident cats or checking out suspicious sounds. These cocky, spirited birds can often be enticed into view by "squeaking" or "pishing," noises that alert curious towhees to an intrusion. Discerning birders, however, would rather not disturb these busy birds and, instead, prefer to enjoy the sound of the towhee's clamorous exploits. • The male Spotted Towhee often selects a prominent perch from which to spit out his curious, trilled song, puffing out his chest and exposing his striking rufous flanks. At other times, the male can be curiously shy.

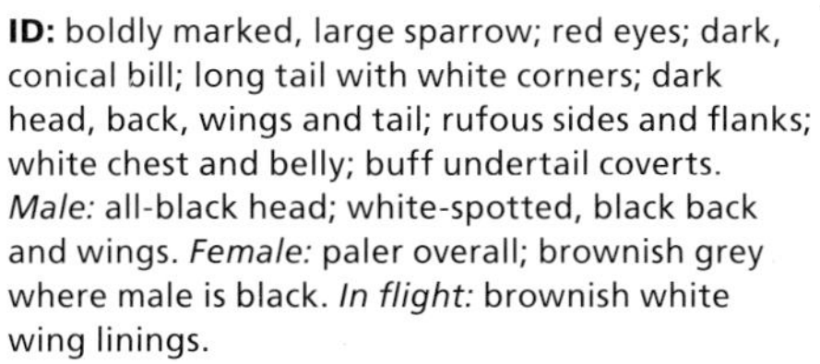

ID: boldly marked, large sparrow; red eyes; dark, conical bill; long tail with white corners; dark head, back, wings and tail; rufous sides and flanks; white chest and belly; buff undertail coverts. *Male:* all-black head; white-spotted, black back and wings. *Female:* paler overall; brownish grey where male is black. *In flight:* brownish white wing linings.

Size: *L* 18–20 cm; *W* 27 cm.

Habitat: brushy, lowland thickets and shrubby situations, often in residential areas, clear-cuts or near water.

Nesting: usually on the ground, sometimes in a low shrub; neat cup of leaves, grasses, bark strips, twigs and rootlets is lined with finer grasses and hair; pair incubates 3–5 brown-wreathed, white eggs for 12–14 days; usually 2 broods per year.

Feeding: scratches the ground vigorously to uncover seeds and insects; also eats berries, especially in winter; regularly visits feeders.

Voice: catlike *meeew* call note; male's song is 1–8 hurried notes accelerating into a fast, raspy trill: *che che che che che zheeee.*

Similar Species: *Eastern Towhee* (p. 434): lacks spotting on back. *Black-headed Grosbeak* (p. 468): male has heavy, conical, bluish grey bill, bright orange nape, breast, belly, flanks and rump and more white in wings. *Dark-eyed Junco* (p. 457): smaller; "Oregon" race has pale rufous on back and sides.

EASTERN TOWHEE

Pipilo erythrophthalmus

Towhees are cheeky birds that typically rustle about in dense undergrowth, craftily scraping back layers of dry leaves to expose the seeds, berries, earthworms and insects hidden beneath. Distinctive *drink your teeeee* or *cheweee* calls are usually the first hint that an Eastern Towhee is nearby. It might take some time to locate this wary singer, but the effort is usually rewarded with a peek at the bird's bright plumage. • Although you wouldn't guess it, this colourful bird is a member of the American Sparrow family—a group that is usually drab in colour. • The Eastern Towhee and its western relative, the Spotted Towhee, were once grouped together as a single species known as "Rufous-sided Towhee." • According to recent surveys, Eastern Towhees are declining throughout eastern North America, with habit loss the likeliest culprit. However, breeding bird survey data suggests that the Canadian population is stable to increasing. • The genus name *Pipilo* is derived from the Latin *pipo*, meaning "to chirp or peep"; the species name *erythrophthalmus* is derived from Greek words that mean "red eye."

ID: boldly marked, large sparrow; rufous sides and flanks; distinct white wing patches; white outer tail corners; white lower breast and belly; buff undertail coverts; red eyes; dark bill. *Male:* black "hood" and upperparts. *Female:* dusky brown hood and upperparts. *Immature:* dusky brown upperparts; heavily streaked, white-buff underparts; pale bill. *In flight:* short, rounded wings; long tail; prominent white "wrist" patch and tail spots.

Size: *L* 18–21 cm; *W* 25–28 cm.

Habitat: dense shrubs with leaf litter, woodland openings and edges and shrubby abandoned fields.

Nesting: on the ground or in a dense, low shrub; cup nest of twigs, bark strips, grasses and rootlets is lined with fine grasses and hair; pair incubates 3–6 brown-spotted, creamy white to pale grey eggs for 10–12 days; usually 2 broods per year.

Feeding: scratches away ground leaf litter to reveal insects, seeds and berries.

Voice: scratchy, slurred *cheweee!* or *chewink!* call; male's song is 2 high, musical, whistled notes followed by a trill: *drink your teeeee.*

Similar Species: *Spotted Towhee* (p. 433): white spots on back and wings. *Dark-eyed Junco* (p. 457): much smaller; pale bill; black eyes; white outer tail feathers.

AMERICAN TREE SPARROW

Spizella arborea

With its unassuming but doggedly regular migratory habits, the appearance of American Tree Sparrows announces the arrival of both spring and fall throughout much of southern Canada. • You know winter is approaching when you start to find more than the occasional American Tree Sparrow among the local sparrow flocks. Quietly arriving from the north, this bird is easily identified by its two-tone bill and the central breast spot on its otherwise clear underparts. The American Tree Sparrow finds its place each winter in brushy and shrubby areas alongside migrating and year-round resident species. • These birds typically nest among patches of shrubs along the Subarctic treeline northward, so the best time to see them is in late March and April, when they are in migration. As the small flocks migrate north, they offer bubbly, bright songs between bouts of foraging along the ground or in low, budding shrubs. • With adequate food supplies, the American Tree Sparrow can survive temperatures as low as –33°C. • Although both its common and scientific names (*arborea* means "tree") imply that this is a forest-dwelling bird, American Tree Sparrows are most often found in semi-open, shrubby habitats.

ID: sexes similar; small sparrow; plain grey underparts; dark central breast spot; grey head; rufous crown and eye line; white eye ring; mottled, brown-and-chestnut upperparts; dark bill with yellow lower mandible; 2 prominent, white wing bars; rufous shoulder patch; buff sides. *Nonbreeding:* grey central crown stripe. *Juvenile:* streaky breast and head; 2 buffy wing bars. *In flight:* long, rounded wings; long, grey, notched tail; grey sides of neck and rump are distinctive; pale buff underwing coverts.

Size: *L* 15 cm; *W* 24 cm.

Habitat: *Breeding:* shrubby tundra, often near water. *In migration* and *winter:* shrubby, open habitats.

Nesting: in a shrub or low vegetation; cup nest of dried grasses, mosses and sedges is heavily lined with feathers; female incubates 3–5 darkly marked, pale bluish to greenish eggs for 12–13 days.

Feeding: forages on the ground and in low shrubs for insects in summer and seeds and berries in migration and winter.

Voice: *tseet* call; male's song of late winter and spring migration is a high, whistled *tseet-tseet* followed by a short, sweet, musical series of slurred whistles.

Similar Species: *Chipping Sparrow* (p. 436): thin, blackish eye line; yellowish pink bill; lacks dark central breast spot. *Swamp Sparrow* (p. 452): white throat; lacks dark breast spot and white wing bars. *Field Sparrow* (p. 439): white eye ring; orange-pink bill; lacks dark breast spot. *White-crowned Sparrow* (p. 455): immature has rufous-and-buff-striped head, all-pinkish or all-yellow bill, unmarked breast and brownish rump.

CHIPPING SPARROW

Spizella passerina

The Chipping Sparrow and the Dark-eyed Junco obviously do not share a tailor, but they must have attended the same voice lessons. Only subtle differences distinguish the two songs: the rapid trill of the Chipping Sparrow is just slightly faster, drier and more mechanical. • Habitat preferences can sometimes help in distinguishing unseen singers, but both of these species can be found along edges of forests. • Commonly nesting at eye level or lower, the Chipping Sparrow offers the opportunity to study its courtship and nest-building rituals at close range. Some nests are so poorly constructed that the eggs can be seen through the nest materials. Chipping Sparrows are well known for their preference for conifers as nesting sites and hair as lining material for the nest. By planting conifers in your backyard and offering samples of your pet's hair—or even your own—in backyard baskets in spring, you can attract nesting Chipping Sparrows to your area and contribute to their nesting success. • A small percentage of male Chipping Sparrows are polygynous, meaning that one male may have two females sitting on separate clutches at the same time. • This bird's common name refers to its call.

nonbreeding

breeding

ID: sexes similar; small sparrow. *Breeding:* prominent, rufous cap; grey cheeks; white eyebrow and throat; black eye line; plain, grey nape, rump and underparts; all-dark bill; dark-mottled, brown back and wings; 2 white wing bars; white throat. *Nonbreeding:* paler crown with dark streaks; brown eyebrow and cheeks; pale lower mandible. *Juvenile:* brown cap; buff eyebrow; streaked chest. *In flight:* long, rounded wings; long tail; uniformly pale grey below.

Size: *L* 14 cm; *W* 21 cm.

Habitat: *Breeding:* open conifers or mixed woodland edges with young trees or shrubs to subalpine forests; often in yards and gardens with tree and shrub borders.

Nesting: on the branch of a conifer or shrub; compact cup of grasses and other plant materials is lined with hair; female incubates 3–5 darkly marked, bluish green eggs for 10–15 days.

Feeding: forages on the ground or gleans in low foliage for seeds and crawling insects.

Voice: usual call is a short, clipped *chip*; male's song is a simple, long, dull trill.

Similar Species: *Clay-colored Sparrow* (p. 437): darkly streaked, brown crown with white, central stripe; pinkish or orangey bill; brownish rump. *American Tree Sparrow* (p. 435): dark central breast spot; lacks bold, white eyebrow; rufous eye line. *Swamp Sparrow* (p. 452): lacks white eyebrow, black eye line and white wing bars.

CLAY-COLORED SPARROW

Spizella pallida

Clay-colored Sparrows go completely unnoticed, for the most part, because their plumage, habits and voice all contribute to a cryptic lifestyle. Even when the males sing at the top of their "air sacs," they are usually mistaken for buzzing insects. • Subtle in plumage, the Clay-colored Sparrow possesses an unassuming beauty. Birders looking closely at this sparrow to confirm its identity can easily appreciate its delicate shading, texture and form—features so often overlooked in birds with more colourful plumage. • Clay-colored Sparrows are common hosts for Brown-headed Cowbird eggs. These sparrows often recognize the foreign eggs, however, and many will either abandon their nest or build another in a different part of their territory. • During the late 1800s and early 1900s, Clay-colored Sparrows expanded their range, flourishing wherever brushy paths grew in forest clear-cuts and along the margins of agricultural fields. Recently, however, breeding bird surveys have revealed that these birds have declined significantly throughout their North American range. These sparrows have adapted to a variety of habitats throughout Canada: in Saskatchewan, they are particularly partial to patches of silverberry and snowberry; in British Columbia, they prefer wild rose hedgerows; and in Ontario, they inhabit shrubby, open bogs and willow scrub habitats.

breeding

ID: sexes similar; small sparrow; unstreaked, greyish white underparts; buff breast wash; grey nape; light brown cheek edged with darker brown stripes; brown crown with dark streak and pale, central stripe; pale eyebrow; white jaw stripe bordered by brown; white throat; largely pale bill. *Juvenile:* dark streaks on buff breast, sides and flanks. *In flight:* broad wings; long tail; pale buff below, including wing linings.

Size: *L* 13–14 cm; *W* 19 cm.

Habitat: *Breeding:* brushy open areas, woodland edges and openings, abandoned fields, regenerating burn sites and riparian thickets.

Nesting: in a grassy tuft or small shrub; open cup nest of twigs, weeds and rootlets is lined with fine grasses; mostly the female incubates 3–4 brown-speckled, bluish green eggs for 11–12 days; often 2 broods per year.

Feeding: forages for seeds and insects on the ground and in low vegetation.

Voice: soft *chip* call; male's song is a series of 2–5 slow, low-pitched, insectlike buzzes.

Similar Species: *Chipping Sparrow* (p. 436): breeding adult has prominent rufous cap, grey cheeks and underparts, 2 faint white wing bars and all-dark bill; juvenile lacks grey nape and buff on sides and flanks.

BREWER'S SPARROW

Spizella breweri

The Brewer's Sparrow, with its nondescript plumage, easily goes unnoticed among other nearby birds and is the ultimate challenge for birders. Fortunately, the Brewer's Sparrow favours treeless sagebrush plains and brushy hillside habitats. However, it rarely perches in the open and has a habit of making quick getaway flights. • The male's song, which often exceeds 10 seconds in length, is a remarkable outburst of rapid, buzzy trills that constantly changes in speed, pitch and quality. • This bird is found in widely separated regions in western Canada, with only casual sightings between the regions. One breeding region is in the southern Interior of BC and the prairie grasslands of southern Alberta and Saskatchewan. A subspecies known as "Timberline Sparrow" occurs in two widely separated areas—in the mountains of southwestern Alberta, and from extreme northwestern BC to southwestern Yukon. In Alberta, Brewer's Sparrows are most frequently encountered in sagebrush habitats, which, like this sparrow, have declined over time. • This bird's name honours 19th-century physician and ornithologist Dr. Thomas Brewer, who made significant contributions to our understanding of the breeding behaviour of North American birds.

breeding

ID: sexes similar; small sparrow; finely streaked, brown upperparts; light brown, unstreaked underparts; unbroken, white eye ring; brown cheek patch with darker brown border; pale eyebrow, throat and jaw stripes; light-coloured legs; long tail.
Size: *L* 12.5–14 cm; *W* 19 cm.
Habitat: *Breeding:* 2 distinct populations; sagebrush rangelands in the south; tall shrubs, especially willow and birch, in the northwest.
Nesting: in a low, dense shrub or on the ground; small, compact cup of grasses, plant stems and fibres is lined with finer grasses and hair; pair incubates 2–4 brown-marked, bluish green eggs for 11–12 days.
Feeding: forages on the ground and in low vegetation for seeds and insects, especially arthropods.
Voice: high-pitched *psst* call; male's long song of interspersed buzzes and trills varies in speed and pitch.
Similar Species: *Clay-colored Sparrow* (p. 437): white eyebrow; unstreaked, grey nape; unstreaked rump; buffier, especially on underparts; greyer and more heavily patterned in breeding plumage. *Chipping Sparrow* (p. 436): chestnut on crown and upperparts; black eye line; plainer cheeks; greyer rump.

FIELD SPARROW

Spizella pusilla

This pink-billed sparrow is a denizen of overgrown fields, pastures and forest clearings. Deserted farmland might seem "unproductive" to some people, but for Field Sparrows, it is ideal. For nesting purposes, this bird usually chooses pastures that are scattered with shrubs, herbaceous plants and plenty of tall grass. Adult Field Sparrows are extremely faithful to successful nest sites, with half returning to the same site the following year. • Field Sparrows have a very distinctive song – an accelerating series of soft whistles that increase to a trill. • Unlike most songbirds, a nestling Field Sparrow will leave its nest prematurely if disturbed. • Over time, the Field Sparrow has learned to recognize when its nest has been parasitized by the Brown-headed Cowbird. Because the unwelcome eggs are usually too large for this small sparrow to eject, the nest is simply abandoned. This sparrow has been known to make numerous nesting attempts in a single season. • Field Sparrows gather in small foraging flocks during the winter, combing the ground for nourishing seeds. Some birds remain in southern Ontario during warm winters, when they are often detected among larger flocks of American Tree Sparrows. • Since 1968, there has been a 4 percent annual decline in Canada's Field Sparrow population.

ID: sexes similar; small sparrow; orange-pink bill; grey face and throat; rusty crown with grey central stripe; rusty streak behind eye; white eye ring; 2 white wing bars; unstreaked, grey underparts with buffy red wash on breast, sides and flanks; pinkish legs. *Immature:* duller version of adult; streaked breast; faint, buff-white wing bars. *In flight:* short, rounded wings; very long tail; buff underparts, including wing linings.

Size: *L* 13–15 cm; *W* 20 cm.

Habitat: abandoned or weedy and overgrown fields and pastures, woodland edges and clearings, extensive shrubby riparian areas and young conifer plantations.

Nesting: on or near the ground, often sheltered by a grass clump, shrub or sapling; open cup nest of grasses is lined with animal hair and seed fluff; female incubates 3–5 brown-spotted, whitish to pale bluish white eggs for 10–12 days.

Feeding: forages on the ground; eats mostly insects in summer and seeds in winter.

Voice: *chip* or *tsee* call; male's song is a series of woeful, musical, downslurred whistles accelerating into a trill.

Similar Species: *American Tree Sparrow* (p. 435): dark central breast spot; dark upper mandible; lacks white eye ring. *Swamp Sparrow* (p. 452): white throat; dark upper mandible; lacks white wing bars and eye ring. *Chipping Sparrow* (p. 436): all-dark bill; white eyebrow; black eye line; lacks buffy red wash on underparts.

VESPER SPARROW

Pooecetes gramineus

For birders who live near dry grasslands and agricultural fields that teem with confusing little brown sparrows, the common Vesper Sparrow offers welcome relief. Its white-edged tail, chestnut shoulder patch and deeply undulating flight are reliable identification features. The male Vesper is also known for its bold and easily distinguished song, which is often heard in the evening (*vesper* is Latin for "evening"), usually from an elevated perch or in a brief song flight. • Vesper Sparrows spend most of their time on the ground, and they are commonly seen along roadsides, their white tail feathers flashing as they flit about. When the business of nesting begins, the male will scour the neighbourhood for a potential nesting site. The Vesper Sparrow's nest is usually located in a slight hollow in a short-grass area with enough structure to provide camouflage, a windbreak and a protective umbrella. • Changing agricultural practices and free-range domestic stock are two potential threats to the Vesper Sparrow. Breeding bird surveys have revealed that Vesper Sparrow numbers are declining throughout North America, especially in the East, where numbers have dropped by 4 percent annually since the 1960s.

ID: sexes similar; large sparrow; streaked, brown upperparts; whitish underparts and undertail; chestnut shoulder patch; dark breast streaking; white eye ring; pale cheek patch with brown border; dark "moustache" stripe; faintly streaked flanks. *In flight:* rounded wings; long tail with prominent, white outer feathers; flight is stronger than other sparrows.

Size: *L* 15 cm; *W* 25 cm.

Habitat: *Breeding:* dry grasslands and shrub steppes; edges of roads, agricultural fields, airports and pastures.

Nesting: in a slight depression on the ground; well-concealed, compact nest of grasses and plant stems is lined with finer grasses; pair incubates 3–6 brown-marked, creamy or greenish white eggs for 11–13 days; usually 2 broods per year.

Feeding: forages on the ground for insects and plant seeds.

Voice: short, hard *chip* call; male's song begins with 4 low notes, the 2nd higher in pitch, then a bubbly trill of *here-here there-there, everybody-down-the-hill.*

Similar Species: *Other sparrows* (pp. 433–39, 442–57): lack white outer tail feathers and/or chestnut shoulder patch. *Lark Sparrow* (p. 441): white-tipped tail; striking facial pattern. *American Pipit* (p. 390) and *Sprague's Pipit* (p. 391): thinner bills; lack chestnut shoulder patch; bob tails when feeding. *Lapland Longspur* (p. 459): nonbreeding bird has broad, pale eyebrow, reddish edging on wing feathers and buff wash on upper breast. *McCown's Longspur* (p. 458): more white on face and tail.

LARK SPARROW

Chondestes grammacus

Singing atop small bushes or low rock outcrops in their shrubby grassland haunts, male Lark Sparrows reminded early naturalists of the famed Sky Lark (*Alauda arvensis*) of Eurasia and North Africa. Lark Sparrow males do occasionally indulge themselves in short display flights, but they do not fly as high or as skillfully as the Sky Lark. • Typically seen in dry scrubland, open shrub-steppe and edge habitats, Lark Sparrows occasionally venture into grassy forest openings, other woodland edges and meadows. • Lark Sparrows are readily distinguished from other grassland sparrows in flight by their relatively long, rounded, white-edged and white-tipped tail. Lark Sparrows sometimes associate with Vesper Sparrows and Savannah Sparrows, but the unique pattern of white on the tail easily distinguishes the three species. Although the Lark Sparrow's head pattern is distinctively bold, most sparrows share the same basic pattern in a mixture of not-so-contrasting browns and greys. • The male Lark Sparrow fluffs his chestnut feathers, spreads his tail, droops his wings and bubbles with song in the presence of potential mates. • A formerly uncommon breeder east into southern Ontario, with the last confirmed nesting occurring in 1976, this bird is currently a regularly occurring vagrant east of Manitoba.

ID: sexes similar; large sparrow; chestnut head with white eyebrow and crown stripe; black eye line and "moustache"; heavy, grey bill; white throat; plain, white underparts with dark breast spot; brown-streaked back and wings; buffy grey sides and flanks; greyish brown rump; tail has greyish brown centre and darker outer feathers with white tips. *In flight:* long tail with white corners; strong flight with deep undulations.

Size: *L* 15 cm; *W* 28 cm.

Habitat: *Breeding:* semi-open habitats, including shrub-steppe, shrubby grasslands, roadsides, farmlands and pastures.

Nesting: in a slight depression on the ground; bulky nest of various grasses and leaves is lined with finer grasses; female incubates 4–5 dusky-marked, creamy or greyish white eggs for 11–13 days.

Feeding: walks or hops on the ground to find seeds and arthropods, especially insects.

Voice: finchlike *pik* flight call and loud, high alarm call; male's melodious and variable song, occasionally given on the wing, consists of short trills, buzzes, pauses and clear notes.

Similar Species: none; head and tail pattern are distinctive.

LARK BUNTING

Calamospiza melanocorys

A visit to the grasslands and hayfields of the southwestern Prairies provides a good chance of seeing the spectacular courtship flight of the male Lark Bunting. The male rises into the air and flutters about in circles high above the ground, beating its wings slowly and deeply. Its bell-like, tinkling song spreads over the landscape until the bird decides to fold its wings and float to the ground like a falling leaf. • The male's gorgeous, jet-black breeding plumage with large, white wing patches makes him appear to be unrelated to his drab sparrow neighbours. • The numbers and breeding range of the Lark Bunting vary markedly from year to year. In drought years, it breeds farther north than usual and becomes one of the most conspicuous and abundant birds in the southern Prairies. It breeds in native prairie and hayfields, as well as in roadside ditches. • Because the Lark Bunting's courtship behaviour evolved before the arrival of fence posts and power poles on which to perch, it developed the habit of delivering its song on the wing. • Although the Lark Bunting's highly irruptive nature and small breeding distribution makes for difficult trend analyses, breeding bird surveys have documented significant range-wide declines for the species.

♂ *breeding*

♂ ♀ *nonbreeding*

ID: large sparrow; conical, greyish blue bill; large, white wing patch (most conspicuous on male). *Breeding male:* all-black plumage. *Female:* mottled, brown-and-buff upperparts; lightly brown-streaked, whitish underparts; pale eyebrow. *Nonbreeding male:* similar to female but with darker streaking. *In flight:* mostly white tail tip.
Size: L 16 cm; *W* 28 cm.
Habitat: short-grass prairie and sagebrush, hayfields and grassy ditches.

Nesting: on the ground; sheltered by a canopy of grass or by a small bush; loosely built cup nest of grasses, roots and other plant material is lined with plant down and fur; mostly the female incubates 4–5 pale blue eggs for 11–12 days.
Feeding: walks or hops along the ground to take insects, including grasshoppers, beetles and ants; also eats seeds and waste grain.
Voice: soft *whoo-*ee call; male's song is a rich, complex, repetitious warble of clear notes interspersed with whistles and rattles.
Similar Species: *Other sparrows* (pp. 433–41, 443–62): all lack white wing patch. *Bobolink* (p. 472): male has creamy nape and white rump and back patches.

SAVANNAH SPARROW

Passerculus sandwichensis

Through spring and summer, the male Savannah Sparrow belts out his distinctive, buzzy tunes, usually while perched atop a prominent shrub, blade of grass or strategic fence post. The Savannah Sparrow is one of Canada's most common open-country birds, and its yellow lore spot and easy-to-remember song make it one of the easiest of the grassland sparrows to identify. • Savannah Sparrows are common in open, grassy country, where their plumage conceals them perfectly among grasses. They typically inhabit the weedy margins of sloughs, dugouts, wet ditches and other wetlands but tend to avoid short-grass habitats. These birds are not shy and can often be seen scurrying like feathered voles low in cover. • Savannah Sparrows are faithful to their nesting sites and often return each year to the same vicinity to breed. • Like most sparrows, the Savannah Sparrow prefers to stay out of sight. It takes flight only as a last resort and flutters only a short distance before it touches ground again. • Anyone studying songbird plumage variations might consider choosing the Savannah Sparrow as a model. At least four of the 17 North American subspecies visit Canada at various times of the year, keeping birders on their toes as they try to make positive identifications.

ID: sexes similar; small sparrow; yellow lores and eyebrow; mottled, greyish brown upperparts; dark eye line and "moustache"; pink bill; medium to reddish brown wings; finely streaked, white breast and underparts; white or buff flanks; dark "V" spot on breast; pale throat; white belly and undertail coverts. *Juvenile:* browner back; buffier head; more blurry streaking; stubby tail. *In flight:* rather long, pointed wings; squared tail; bounding, buoyant flight.

Size: *L* 13–15 cm; *W* 20–22 cm.

Habitat: wide range of open country, including agricultural fields, sand dunes, pastures, meadows, recent clear-cuts and subalpine meadows.

Nesting: on the ground, usually in a slight depression; small, well-concealed cup of coarse grasses is lined with finer grasses; pair incubates 4–5 brown-marked, pale greenish blue eggs for 10–13 days.

Feeding: forages on the ground for insects, spiders and grass seeds.

Voice: high, thin *tsit* call; male's song is a high-pitched, clear, buzzy *tzip-tzip-tzip ztreeeeeee-ip* or *tea tea tea teeeeea today*, rising on 2nd-last note and dropping at the end.

Similar Species: *Vesper Sparrow* (p. 440): chestnut shoulder patch; white outer tail feathers. *Song Sparrow* (p. 450): greyer head; lacks yellow lores and eyebrow. *Lincoln's Sparrow* (p 451): buffier overall, particularly on breast and head.

GRASSHOPPER SPARROW

Ammodramus savannarum

Odds are that, if you find yourself in prime Grasshopper Sparrow habitat, you are either a birder or a rancher. Few people stop to enjoy Canada's native grasslands, and, as a result, this region and its exceptional birdlife are generally underappreciated by the public. The Grasshopper Sparrow is named not for its diet but rather for its buzzy, insectlike song. During courtship flights, males chase females through the air, buzzing at a frequency that is usually inaudible to our ears. The males sing two completely different courtship songs: one ends in a short trill, and the other is a prolonged series of high trills that change in pitch and speed. • If a nesting female Grasshopper Sparrow is flushed from her nest, she will run quietly away instead of flying. • The Grasshopper Sparrow is an open-country bird that prefers grassy expanses free of trees and shrubs. Wide, well-drained, grassy ditches occasionally attract nesting Grasshopper Sparrows, so mowing or harvesting these grassy margins early in the nesting season can be detrimental to these birds. This species' numbers in Canada have declined at a rate of about 6 percent annually. • *Ammodramus* is Greek for "sand runner," and *savannarum* is Latin for "of the savanna."

ID: sexes similar; small sparrow; mottled brown upperparts; unstreaked, white underparts with buff wash on breast, sides and flanks; flattened head profile; dark crown with pale central stripe; buff cheek; white eye ring; beady, black eyes; pale legs; may show small, yellow patch on edge of forewing. *Immature:* less buff on underparts; faint streaking across breast and sides. *In flight:* short, rounded wings; pale underwings contrast with buff underparts; sharply pointed tail; very whirry flight close to ground, diving into cover.
Size: *L* 11–13 cm; *W* 19–20 cm.
Habitat: grasslands and grassy fields with little or no shrub or tree cover; native prairies and sandhills.
Nesting: in a shallow depression on the ground, usually concealed by grass; small cup nest woven of grasses is lined with rootlets, fine grasses and hair; female incubates 4–5 variably spotted, creamy white eggs for 11–13 days.
Feeding: gleans insects and seeds from the ground and grass; eats a variety of insects, including grasshoppers.
Voice: high, sharp *tip* call; male's song is a high, faint, buzzy trill preceded by 1–3 high, thin whistled notes: *tea-tea-tea zeeeeeeeeee.*
Similar Species: *Le Conte's Sparrow* (p. 447): buff and black head stripes divided by white central crown stripe; grey cheeks; dark streaking on sides and flanks. *Nelson's Sparrow* (p. 448): orange-buff face and breast; grey central crown stripe; grey cheeks and shoulders. *Henslow's Sparrow* (p. 446): similar to immature Grasshopper but has darker breast streaking, rusty wings and small, dark ear and "whisker" marks.

BAIRD'S SPARROW

Ammodramus bairdii

Baird's Sparrows are frustrating grassland denizens. Just when the form of the sparrow drifts into focus through your binoculars, the bird dives out of view from its perch. Baird's Sparrows sing atop grass stems and low shrubs, but when they stop singing, they are almost impossible to find. • The Baird's Sparrow is a bird of native grasslands with an unusually small nesting distribution in the southern Canadian Prairies and extreme northern Great Plains states. It favours lush areas rather than grazed, short-grass prairie habitat. During years of drought, the Baird's Sparrow will seek out new foods found growing in the bottom of dried-out sloughs. • Once considered among the most common birds on the prairies, the Baird's Sparrow is now uncommon and declining throughout its range. Agriculture has eliminated much of the prairie it is reliant on and continues to reduce remaining grassland tracts. • Baird's Sparrow was the last bird collected and described by Audubon for his classic *Birds of America*. If you think this species is difficult to find and recognize now, consider that it took 30 years for a second ornithologist to "rediscover" Baird's Sparrow after it was first described.

ID: sexes similar; small sparrow; dark brown upperparts with buffy feather edges; faint chestnut on the wing coverts; whitish underparts; finely streaked "necklace"; 2 black stripes bordering white throat; buff head and nape; pale legs and bill. *In flight:* pale edges on brown tail.
Size: *L* 13–14 cm; *W* 20–22 cm.
Habitat: native grasslands and lightly grazed pastures with clumps of tall grasses and sparse weeds; occasionally among sparse, short shrubs.
Nesting: occasionally semicolonial; on the ground; often under bent grasses or a small bush; small cup nest woven of grasses and other plant fibres is lined with fur and seed fluff; female incubates 4–5 chestnut-spotted, pale grey eggs for 11–12 days.

Feeding: gleans the ground, pecking and running through tall grass; eats mainly grass seeds; also eats other plant seeds and occasionally insects.
Voice: harsh *chip* call; male's song is a tinkling, musical trill: *zip-zip-zip-zrrr-r-r-r* or *tink tink a tl tleeeeee.*
Similar Species: *Savannah Sparrow* (p. 443): yellow lores; lacks buff head and nape. *Vesper Sparrow* (p. 440): white outer tail feathers; lacks delicate "necklace" streaking. *Grasshopper Sparrow* (p. 444): sloped forehead; unstreaked breast.

HENSLOW'S SPARROW

Ammodramus henslowii

Predicting when you'll see the next Henslow's Sparrow in southeastern Canada is difficult—this bird makes irregular visits, often appearing one year but not the next. Some males have been known to occupy a field for a few weeks before suddenly disappearing, probably owing to the lack of a potential mate. The Henslow's unpredictability has made it a difficult species to study, so there's much more to learn about its habitat requirements and the reasons for its recent, widespread decline. • Watch the male Henslow's Sparrow as he throws back his streaky, greenish head while hurling a distinctive song from atop a tall blade of grass. Without the male's lyrical advertisements, the Henslow's Sparrow would almost be impossible to observe—this bird spends most of its time foraging alone along the ground. When disturbed, it may fly a short distance before dropping into cover, but usually it prefers to run through dense, concealing vegetation. • Henslow's Sparrows are known for their unusual habit of singing at night. A breeding species in Québec till the late 1960s, the Henslow's Sparrow is now confined to southern Ontario. Currently classified as endangered in Canada, the population is estimated to be less than 10 pairs.

ID: sexes similar; small sparrow; flattened head profile; olive green face, central crown stripe and nape; white eye ring; dark crown and "whisker" stripes; rusty tinge on back, wings and tail; white underparts; darkly streaked, buff breast, sides and flanks; thick bill. *Juvenile:* buff wash on underparts; faint streaking only on sides. *In flight:* deeply notched, sharply pointed tail.

Size: *L* 12–13 cm; *W* 16–17 cm.

Habitat: large, fallow or wild grassy fields and meadows with a matted ground layer of dead vegetation and scattered shrub or herb perches; often in moist, grassy areas.

Nesting: on the ground at the base of a grass clump or herbaceous plant; open cup nest of grasses and weeds is lined with fine grasses and hair; female incubates 3–5 variably spotted, whitish to pale greenish white eggs for about 11 days.

Feeding: gleans insects and seeds from the ground.

Voice: sharp *tsik* call; weak, liquid, cricketlike *tse-lick* song, often given during periods of rain or at night.

Similar Species: *Other sparrows* (pp. 433–45, 447–62): lack olive green face, central crown stripe and nape. *Grasshopper Sparrow* (p. 444): lacks dark "whisker" stripe and prominent streaking on breast and sides. *Savannah Sparrow* (p. 443): lacks buff breast. *Le Conte's Sparrow* (p. 447): buff and black head stripes with white central stripe; grey cheeks. *Nelson's Sparrow* (p. 448): grey central crown stripe and nape; white streaks on dark back.

LE CONTE'S SPARROW

Ammodramus leconteii

With sputtering wingbeats, a flushed Le Conte's Sparrow flies weakly over its marshy home before it noses down and seemingly crashes into a mass of grasses. Le Conte's Sparrows are found among damp, undisturbed meadows or tall-grass habitats. Because of their local breeding habitats, patchy distribution and secretive behaviour, they are difficult birds to find. Even singing males typically choose low, concealing perches from which to offer their weak love ballads. The song of the Le Conte's Sparrow is similar to that of a Grasshopper Sparrow but is even weaker, briefer and more buzzy. Skilled birders following the buzzy tune to its source may catch a fleeting glimpse of the singer before it dives into tall vegetation and disappears from view. This bird is also reluctant to flush from observers, instead resorting to creeping like a mouse under the matted grasses. • This bird's namesake, John Le Conte, is best remembered as one of the preeminent American entomologists of the 19th century, but he was interested in all areas of natural history.

ID: sexes similar; small sparrow; mottled, brown-and black upperparts; buff-streaked back; buff-orange face; grey cheeks; short, black line behind eye; black-bordered, pale central crown stripe; buff-orange upper breast, sides, flanks and undertail coverts; dark streaking on sides and flanks; white throat, lower breast and belly. *In flight:* sharply pointed tail feathers.

Size: *L* 11–13 cm; *W* 18 cm.

Habitat: *Breeding:* open sedge fields and dense, grassy meadows, usually near water.

Nesting: on or near the ground, concealed by lush vegetation; open cup nest woven of grasses is tied to standing plant stems and lined with finer grasses; female incubates 3–5 grey-and-brown-spotted, greyish white eggs for 12–13 days.

Feeding: gleans the ground and low vegetation for insects, spiders and seeds.

Voice: high-pitched, thin, descending *tseeez* alarm call; male's song is a weak, short, raspy, insectlike buzz: *t-t-t-zeeee zee* or *take-it ea-zeee.*

Similar Species: *Nelson's Sparrow* (p. 448): grey central crown stripe and nape; white streaks on dark back. *Grasshopper Sparrow* (p. 444): lacks buff-orange face and dark streaking on underparts.

NELSON'S SPARROW

Ammodramus nelsoni

Formerly known as "Nelson's Sharp-tailed Sparrow," this species is now known simply as the Nelson's Sparrow. This relatively colourful sparrow lives in marshy areas and will unexpectedly pop out of a soggy hiding place to perch completely exposed at a close distance. As with most sparrows, the best way to identify a Nelson's is to listen for its song—a fading *tup tup-sheeeeeeeee*. However, this sparrow's buzzy song is probably seldom recognized, and some say it should not even be considered a song! Unlike many songsters, the male Nelson's may sing throughout the day and night. • Nelson's Sparrows have an unusual breeding strategy among songbirds—the males rove around the marsh mating with all the available females, which are also promiscuous. Not surprisingly, these sparrows do not establish pair bonds or territories. • This species is named for Edward William Nelson, who, in 1916, wrote the Migratory Bird Treaty Act, an international convention between the United States, Canada and Mexico to protect migratory birds in North America that is still in effect today.

ID: sexes similar; small, secretive songbird; flattened forehead; distinctive orange triangle on face; grey cheek, central crown stripe and nape; black-and-white-streaked back; pale throat; buff orange breast and flanks with faint streaking; white belly. *Juvenile:* bright orange buff with few streaks. *In flight:* short, rounded wings; pointed tail; pale underwings contrast with buff sides; weak, buzzy flight close to the ground.
Size: *L* 13 cm; *W* 18 cm.

Habitat: *Breeding:* wet grass and sedge meadows and open, riparian willow patches; in wet years, vegetated ditches.
Nesting: in a clump of dry grasses; compact cup nest of coarse grasses is lined with finer grasses; female incubates 2–6 darkly speckled, greenish eggs for 11–12 days.
Feeding: gleans vegetation for invertebrates and seeds.
Voice: call is a short, insectlike *tup tup-sheeeeeeeee.*
Similar Species: *Le Conte's Sparrow* (p. 447): dark brown stripes on sides; white median crown stripe; streaked, grey nape.

FOX SPARROW

Passerella iliaca

Scratching away leaf litter and duff to expose seeds and insects, the Fox Sparrow forages much like a towhee. Its loud, whistled courtship songs are easily recognized, and to an attentive listener, they can be as moving as a loon's wail or a wolf's howl. • Unlike other songbirds, which may pass through in a series of lingering waves, most Fox Sparrows appear in southern Canada for only a week or two before moving on to their nesting grounds. They are among the first sparrows to appear in spring and one of the last to leave in autumn. • This bird's overall reddish brown appearance inspired taxonomists to name it after the red fox. However, Fox Sparrows in western Canada are sooty brown, which can cause confusion for eastern birders. • One of North America's largest sparrows, the Fox Sparrow has at least 18 subspecies. Eight occur in Canada as breeders, and another four are winter visitors. The subspecies vary not only in plumage characteristics but also in habitat preferences, songs and calls and migratory routes. It is likely that ornithologists will soon split the Fox Sparrow and its subspecies into two or more distinct species.

ID: sexes similar, but plumage highly variable; small sparrow; reddish brown wings and tail; grey nape, crown stripe and eyebrow; reddish brown cheeks; mostly white jaw line; streaked, grey-and-brown back; heavily streaked breast with ragged-looking central breast spot; mostly white underparts; large, conical, grey or yellowish bill; pinkish legs. *In flight:* gray rump and brownish red tail.

Size: *L* 18 cm; *W* 27 cm.

Habitat: *Breeding:* moist riparian willow and alder thickets; also dense vegetation in mixed-wood forests. *In migration* and *winter:* dense shrub tangles, shrubby clear-cuts and urban yards.

Nesting: low in a shrub or on the ground; bulky nest of grasses, mosses, leaves and small twigs is lined with fine grasses; female incubates 2–5 russet-marked, pale green eggs for 12–14 days.

Feeding: scratches the ground to uncover seeds and invertebrates; also eats small fruits, especially blackberries, in fall and winter; visits feeders in winter and in migration.

Voice: explosive *tak* call; burry, whistled songs of male coastal birds sound patchy and uninspired; interior birds have a purer, ringing song with noticeable trills.

Similar Species: *Song Sparrow* (p. 450): streaked upperparts; breast and sides more streaked than spotted; brown-striped, greyish face; slimmer bill. *Hermit Thrush* (p. 381): lighter breast spotting; thin bill. *Swainson's Thrush* (p. 380): olive tail; light breast spots; prominent eye ring.

SONG SPARROW

Melospiza melodia

When amateur birder Margaret Morse Nice began studying plumage variations in the familiar and widespread Song Sparrow in the 1920s, she had no idea how much interest she would eventually trigger among professional ornithologists. We now know that the dowdy Song Sparrow probably has the greatest variation in plumage of any North American songbird, with 31 recognized subspecies. Given time and isolation, these subspecies could become so different that they are reproductively incompatible and thus may become distinct species. • Other sparrows might have more beautiful songs, but the complexity, rhythm and sweetness of the male Song Sparrow's springtime rhapsodies justify the name. Song Sparrows learn to sing by listening to their fathers or to rival males. By the time a young male is a few months old, he will have the basis for his own courtship tune. • In recent decades, mild winters in southern Canada and an abundance of backyard feeders have enticed an increasing number of Song Sparrows to overwinter here. • Most songbirds are lucky if they are able to produce one brood per year; some Song Sparrows in extreme southern Canada have been known to successfully raise three broods.

ID: sexes similar; large sparrow; streaked, rufous-and-brown upperparts; brown or rufous crown with pale centre stripe; greyish face; pale eye line; heavy, grey bill; white jaw line bordered by dark "whisker" and "moustache" stripes; whitish underparts with heavy, brown streaking that converges into central breast spot; long, rounded tail; pinkish legs. *Juvenile:* finely streaked. *In flight:* broad, rounded wings; pale grey underwings; often pumps tail in flight.

Size: *L* 14 cm; *W* 21 cm.

Habitat: shrubby areas, usually near water, including willow shrublands, riparian thickets, forest openings, fence lines and lakeshores; also brushy edges of gardens, fields and roads.

Nesting: usually in a grass tuft or low shrub; bulky cup nest of coarse grasses is lined with finer grasses; female incubates 3–6 russet-marked, pale blue or green eggs for 10–13 days; 2–3 broods per year.

Feeding: gleans the ground for insects and seeds; coastal birds eat intertidal crustaceans and soft-shelled molluscs; also eats berries in winter and visits urban feeders.

Voice: short *tsip* and nasal *tchep* calls; male's song is 1–4 bright introductory notes, such as *sweet, sweet, sweet,* followed by a buzzy *towee* and a short, descending trill.

Similar Species: *Fox Sparrow* (p. 449): plainer head and upperparts; lines of spots or chevrons rather than streaks. *Lincoln's Sparrow* (p. 451): daintier; grey or olive face and back; buff "moustache"; finer bill; brown-streaked, buff breast and flanks; more white on belly. *Savannah Sparrow* (p. 443): yellow eyebrow; notched tail; lacks dark "moustache."

LINCOLN'S SPARROW

Melospiza lincolnii

The subtle beauty of the Lincoln's Sparrow's plumage is greater than the sum of its feathers. The colours are not spectacular, yet they give the bird a distinctive, well-groomed look that is lacking in most other sparrows. • Males will sometimes sit openly on exposed perches and sing their bubbly, wrenlike song, but as soon as they're approached, they tend to slip under the cover of nearby shrubs. When they're not singing their courtship songs, Lincoln's Sparrows remain well hidden in tall grass and dense brush. • During the breeding season, Lincoln's Sparrows are vigilant defenders of their young, uncharacteristically exposing themselves to intruders and chirping noisily. The male guards his territory until about a week after his family leaves the nest. Because this species breeds in boreal and alpine regions, preferring to nest in boggy sites in dense shrub cover, and is elusive in nature, its nesting biology has been poorly documented. • In Canada, two subspecies of the Lincoln's Sparrow are found: the *gracilis* race occurs from Alaska to central British Columbia, and the *lincolnii* race occurs from the Yukon eastward to Newfoundland. • This sparrow bears the name of Thomas Lincoln, a young companion to John J. Audubon on his voyage to Labrador.

ID: sexes similar; small, pale sparrow; grey face; buffy jaw stripe; buffy eye ring; dark cheeks; peaked, dark brown cap with grey median stripe; grey or olive nape and eyebrow; mottled, grayish brown to reddish brown upperparts; often rufous on wings and tail; crisp streaking on breast and flanks; white belly; buff-washed breast and sides. *In flight:* rounded tail tip.

Size: *L* 13 cm; *W* 19 cm.

Habitat: *Breeding:* open, brushy swamps, bogs, meadows, clear-cuts and roadsides. *In migration* and *winter:* brushy forest edges, fields, wetlands and subalpine areas.

Nesting: usually on the ground; well-hidden nest of grasses or sedges is lined with finer grasses and hair; female incubates 4–5 russet-marked, pale green eggs for 10–13 days.

Feeding: scratches on the ground for insects, spiders, millipedes and seeds; occasionally visits feeders.

Voice: calls include a buzzy *zeee* and a *tsup*; male's song is a musical mixture of buzzes, trills and warbled notes; often sings at night.

Similar Species: *Song Sparrow* (p. 450): dark central breast spot; white upper breast; dark "whisker" and "moustache" stripes. *Swamp Sparrow* (p. 452): grey head with rufous crown; unstreaked, white throat. *Savannah Sparrow* (p. 443): yellow lores; white eyebrow and jaw line.

SWAMP SPARROW

Melospiza georgiana

nonbreeding

Sharing its marshy habitat with blackbirds, wrens and yellowthroats, the Swamp Sparrow is one of the most common wetland inhabitants in eastern and northern parts of North America. These sparrows are well adapted to life near water and skulk about the emergent vegetation of cattail marshes, foraging for a variety of invertebrates, including beetles, caterpillars, spiders, leafhoppers and flies. Like other sparrows, the Swamp Sparrow is unable to swim, but because it has slightly longer legs than its kin, it is perfectly suited to wading through shallow water and picking invertebrates from the surface. Because it eats mostly insects and fewer hard seeds than other sparrows, its bill and associated muscles are comparatively small. • The male Swamp Sparrow's song won't win any awards, but, in this bird's open habitat, a simple, dry, rattling trill is all that is required to gain the attention of females. • Swamp Sparrows are most easily seen in spring, when males sing their familiar trills from atop cattails or shoreline shrubs.

breeding

ID: small sparrow; generally brownish upperparts; rufous wings and tail; black and buff stripes on back; dark eye line; small, greyish bill with yellow base; buff-chestnut flanks; greyish breast; paler, unstreaked belly; dark-bordered, whitish throat. *Breeding male*: rufous crown with pale median stripe; mostly olive grey face and nape. *Breeding female*: darker brown crown.

Size: *L* 14 cm; *W* 18 cm.

Habitat: cattail marshes, open wetlands, wet meadows and open, riparian deciduous thickets; various habitats with other sparrows in migration.

Nesting: in a small shrub or dense clump of wetland vegetation; cup nest of coarse grasses is lined with finer grasses; female incubates 3–5 heavily marked, greenish or bluish eggs for 12–14 days.

Feeding: forages on the ground, on mud and in low marshland vegetation for insects and seeds.

Voice: emphatic *tchip* call; male's song is a slow, musical trill: *chinga chinga*, fading at the end.

Similar Species: *Chipping Sparrow* (p. 436): clean white eyebrow; complete black eye line; uniformly grey underparts; white wing bars. *American Tree Sparrow* (p. 435): dark central breast spot; white wing bars; 2-tone bill. *White-throated Sparrow* (p. 453): larger; black or darker brown crown; more clearly defined white throat; yellow lores; white or light buff eyebrow; indistinct, white wing bars. *Lincoln's Sparrow* (p. 451): more defined streaking on breast and flanks; greyer back; fine eye ring.

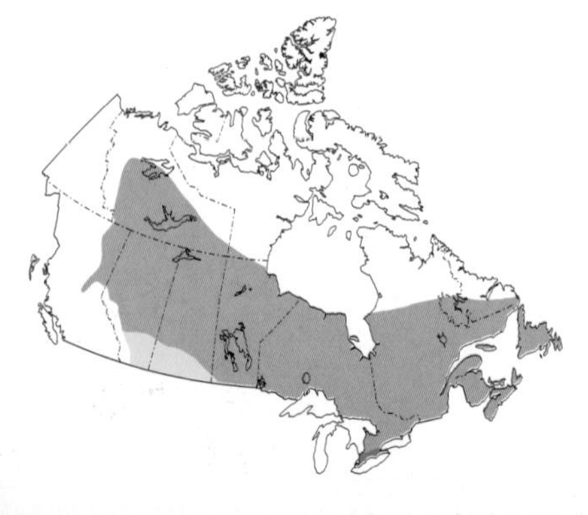

WHITE-THROATED SPARROW

Zonotrichia albicollis

"Tan-striped" adult

This handsome sparrow's simple *dear sweet Canada Canada Canada* song is freely offered throughout the spring months, and though many weekend cottagers and campers might not know the bird, most know the song. The male's purposeful phrases are occasionally delivered late into the night. • Early risers might be treated to the sight of a White-throated Sparrow scratching at patches of leaf litter in search of food. • During the nesting season, the White-throated Sparrow shuns large tracts of forest in favour of second-growth woodlands and forest clearings. In migration, however, it commonly inhabits small shrubs or thickets, often intermingling with other sparrows. • Two colour phases are common throughout Canada: one has black and white stripes on its head, whereas the other has brown and tan stripes. White-striped males are more aggressive than their tan-striped counterparts, and tan-striped females are more nurturing than are white-striped birds. • This sparrow's breeding range is limited to Canada and the northeastern United States. Most White-throated Sparrows winter in the U.S. and northern Mexico, but milder winters and an abundance of backyard feeders has enticed an increasing number to overwinter in southern Canada.

"White-striped" adult

ID: sexes similar; large sparrow; mottled brown upperparts; rufous-marked, brown wings; white throat; dark eye line; finely streaked, greyish brown breast and flanks; whitish belly; pale brown rump and tail. *"White-striped" adult:* black crown; bold, white median stripe and eyebrow; bright yellow lores; sharply defined, white throat patch. *"Tan-striped" adult:* light tan median crown stripe and eyebrow; smudgier facial markings; fainter yellow lores; brown-tinged neck and flanks; divided white throat patch. *In flight:* greyish brown tail.

Size: *L* 15–18 cm; *W* 23 cm.

Habitat: *Breeding:* shrubby, semi-open forests, regenerating clearings and shrubby forest edges. *In migration* and *winter:* shrubby edge habitats and suburban yards.

Nesting: on the ground, concealed by vegetation; open cup nest of grass, weeds, twigs and conifer needles is lined with rootlets, fine grasses and hair; female incubates 4–5 variably marked, pale blue or green eggs for 11–14 days.

Feeding: scratches the ground for insects, spiders, millipedes, snails and seeds; frequently visits feeders.

Voice: distinctive, sharp *chink* or slurred *tseet* call; male's variable song is a clear and distinct whistled *dear sweet Canada Canada Canada.*

Similar Species: *White-crowned Sparrow* (p. 455): pink bill; lacks distinct, white throat and eyebrow.

HARRIS'S SPARROW

Zonotrichia querula

An unassuming migrant with a rather small nesting and winter range, the Harris's Sparrow passes through the central provinces during spring and fall, frequently mixing with flocks of White-throated Sparrows and White-crowned Sparrows. Occasionally, a few Harris's Sparrows pick through the seed offerings at backyard feeders, and each year, a few successfully overwinter at well-stocked feeders in southern Canada. • The Harris's Sparrow breeds exclusively in the North, in the ecozone where Canada's northern coniferous treeline forest (taiga) fades into treeless tundra. Its remote breeding habitat and rather secretive nesting behaviour resulted in it being one of the last songbirds in North America to have its nest and eggs located—in northern Manitoba in 1931. Breeding in the extreme northwestern corner of Ontario was not confirmed until 1983. • The largest member of the sparrow family, the Harris's Sparrow is also the only endemic breeding passerine in Canada.

nonbreeding

breeding

ID: sexes similar; large sparrow; mottled, brown-and-black upperparts; white underparts; pink-orange bill. *Breeding:* black crown, ear patch, throat and "bib"; grey face; black-streaked sides and flanks; white wing bars. *Nonbreeding:* brown face; brownish sides and flanks; mostly black "bib" and crown. *Immature:* white throat; mostly brownish crown with some black streaking.

Size: *L* 18–19 cm; *W* 26 cm.

Habitat: *Breeding:* near treeline, where stunted spruce trees border tundra. *In migration:* brushy roadsides, shrubby vegetation, forest edges and riparian thickets.

Nesting: on or near the ground, usually under cover of a low shrub or tree; open cup nest of twigs, lichen and moss is lined with fine grasses and hair; female incubates 3–5 brown-marked, pale green eggs for 12–15 days.

Feeding: gleans the ground and vegetation for seeds, fresh buds, insects and berries; occasionally visits feeders.

Voice: *jeenk* or *zheenk* call; male's song is a series of 2–4 long, quavering whistles; each series may be offered at the same or different pitch; flocks in flight may give a rolling *chug-up chug-up*.

Similar Species: *House Sparrow* (p. 496): male is brownish overall with grey crown, broad brown band behind eyes, broad, whitish jaw stripe and dark bill. *White-throated Sparrow* (p. 453): greyish bill; yellow lores; black-and-white-striped crown. *White-crowned Sparrow* (p. 455): black-and-white-striped crown; grey collar; immature has broad, grey eyebrow bordered by brown eye line and crown.

WHITE-CROWNED SPARROW

Zonotrichia leucophrys

Although most of Canada's sparrows can quite honestly be described as LBJs (little brown jobs), the White-crowned Sparrow always gives the impression of being smartly turned out. Aside from its good looks, this large, bold sparrow is one of our finest singers. Highly visible and audible on its breeding grounds, the ubiquitous White-crowned Sparrow is also a regular member of winter sparrow flocks in extreme southern British Columbia and Ontario. Of the five subspecies described on the continent, three occur in Canada. Up close, song, bill colour and variations in plumage can separate the races. • The White-crowned Sparrow is one of North America's most studied sparrows, but northern populations are the least known. This elegant sparrow has given scientists an intriguing, and somewhat confusing, insight into avian speciation and the geographic variation in song dialects. White-crowned Sparrows are tireless singers, even bursting into song under the light of the moon. • A regular at feeders during migration, most of these sparrows winter throughout much of the U.S. and Mexico.

ID: sexes similar; large sparrow; dark-streaked, brownish upperparts; plain grey underparts; buff or brown flanks; 2 white wing bars; black-bordered, white crown stripe; broad, white eyebrow; black eye line; pinkish to orange bill; white, grey or black lores. *Juvenile:* pale bill; brown and buff head stripes. *In flight:* short, rounded wings; fairly long tail; pale grey-brown rump.

Size: *L* 18 cm; *W* 24 cm.

Habitat: *Breeding:* open, shrubby meadows, bogs, forest edges, forest clearings, riparian thickets and willow clumps on tundra. *In migration* and *winter:* brushy forest edges, transmission corridors, alpine meadows and lakeshores; also suburban yards.

Nesting: on the ground or in a low shrub or conifer; neat grass nest is lined with finer grasses; female incubates 3–5 russet-marked, pale greenish blue eggs for 11–12 days.

Feeding: forages on the ground for seeds and insects; also eats plant buds and fruit; visits feeders.

Voice: high *seet* and crisp *pink* calls; male's song is a repeated *O see me pretty pretty me*; coastal birds sing a cleaner, more rapid, 4-part song.

Similar Species: *Golden-crowned Sparrow* (p. 456): immature has yellowish forecrown and grey bill. *White-throated Sparrow* (p. 453): yellow lores; grey bill; more rufous in wings; white throat.

GOLDEN-CROWNED SPARROW

Zonotrichia atricapilla

Along with the spectacular scenery, one of the rewards of climbing western Canada's interior peaks is to walk through the world of the Golden-crowned Sparrow. Breeding in open, scrubby areas near treeline, the Golden-crowned Sparrow offers its sad, flat song to the rugged landscape. • Birders are not the only ones to attribute words to bird songs—Alaskan prospectors thought this bird's descending, three-note song, *oh dear me*, sounded as tired as they felt. Ironically, the only gold most of the prospectors were to encounter was on the crown of this mountain songbird. • Unlike many wintering birds, male Golden-crowns also regularly sing throughout the colder months, when they migrate to the more temperate climes of the southern coast of British Columbia. • While foraging for seeds and invertebrates, these sociable sparrows team up with other ground-feeding birds, especially Fox Sparrows and Dark-eyed Juncos, as they scour shrubby open ground, dense brush and the ground below backyard feeders. • The species name *atricapilla* is Latin for "black hair," in reference to the broad outline of the golden crown.

nonbreeding

breeding

ID: sexes similar; large sparrow; bicoloured bill; plain grey face and chest; buff and black streaks on brown back; rufous-tinged, brownish wings; 2 white wing bars; light brown, unstreaked rump; unstreaked, grey underparts. *Breeding:* broad, black cap with yellow median stripe. *Nonbreeding:* less black in crown; duller yellow central crown stripe; faint eye line.
Size: *L* 18 cm; *W* 24 cm.
Habitat: *Breeding:* subalpine willow, birch and alder patches. *In migration* and *winter:* brushy thickets and woodland edges; also suburban yards.
Nesting: on the ground or in a low shrub; bulky cup nest of coarse grasses is lined with finer grasses, feathers and hair; female incubates 3–4 heavily marked, pale blue eggs for 11–13 days.
Feeding: scratches the ground for insects and seeds; also eats berries and other small fruits, moss capsules, buds, blossoms and fresh leaves; visits feeders, often with Dark-eyed Juncos.
Voice: sharp, loud *seek* call; male's usual song is a 3-note whistle: *oh dear me*; birds in Canadian Rockies add a slow trill.
Similar Species: *White-crowned Sparrow* (p. 455): immature has pinkish bill and brown-and-tan-striped head. *House Sparrow* (p. 496): shorter tail; female has white upperwing bar, buffy eyebrow and no yellow on head. *Harris's Sparrow* (p. 454): black "bib"; orange-pink bill.

DARK-EYED JUNCO

Junco hyemalis

"Oregon"

♂

The Dark-eyed Junco is a common, widespread sparrow, with 15 easily recognizable subspecies, six of which are found in Canada. Western Canada boasts a great diversity of junco subspecies, some of which occasionally visit eastern Canada. • Dark-eyed Junco subspecies in Canada are grouped into either the brownish-sided, black-headed "Oregon" Junco or the mostly grey "Slate-colored" Junco. These subspecies frequently interbreed, producing many confusing variations. • Juncos usually congregate in backyards with feeders and sheltering conifers—with such amenities at their disposal, more and more juncos overwinter in our country. • Juncos spend most of their time on the ground and are readily flushed from roadsides, wooded trails and backyard feeders. Their distinctive, white outer tail feathers will flash in alarm as they rush for the cover of a nearby tree or shrub. • When Dark-eyed Juncos aren't nesting, they are gregarious, flocking in shrubby openings that offer open ground on which to search for seeds. • Juncos rarely perch at feeders, preferring to snatch up seeds that are knocked to the ground by other visitors.

♂

"Slate-colored"

ID: sexes similar; medium-sized sparrow; grey to black "hood"; white belly and outer tail feathers; pink bill; unstreaked body. *"Slate-colored":* grey head, back and breast (brown on female). *"Oregon":* black "hood" and breast (grey on female); reddish brown back; paler sides and flanks; grey-and-brown wings. *In flight:* broad, rounded wings; prominent white outer tail feathers; dark flanks contrast with pale belly and underwings.

Size: *L* 15 cm; *W* 24 cm.

Habitat: *Breeding:* mixed coniferous and hardwood forests with openings and shrubby understorey; also subalpine forests and suburban yards. *In migration* and *winter:* brushy edges of forests, fields and lakeshores; rural and suburban yards.

Nesting: in a depression on the ground or low in a shrub or tree; deep cup nest of coarse grasses, mosses, rootlets and twigs is lined with fine grasses and hair; female incubates 3–5 chestnut-marked, pale blue eggs for 12–13 days.

Feeding: gleans the ground and low vegetation for seeds, insects and spiders; also eats berries and regularly visits feeders.

Voice: call is a high, often-repeated, clicking *stip*; male's song is a brief trill; wintering flocks keep up a constant, ticking chorus.

Similar Species: none; well-defined plumage with white outer tail feathers is distinctive.

MCCOWN'S LONGSPUR

Calcarius mccownii

The next time you visit southern Alberta or Saskatchewan, do yourself a favour and pull off onto one of the gravel roads that snake across the grasslands. If you've entered a good piece of native prairie, you might see a McCown's Longspur retreating from a fence post. This longspur, which has extremely small nesting and winter ranges, favours the sparse vegetation provided by native short-grass prairie or overgrazed pastures. As with so many other grassland species, its range has been shrinking—a condition directly linked to the cultivation of native grasslands. The loss of habitat has contributed to an annual 9-percent decline of this longspur's population in Canada. • The male McCown's Longspur delivers his courtship song on the wing, tinkling as he sails to the ground with his tail fanned and his wings held high. The male's soft voice does not dominate the morning chorus of prairie birds; rather, it is a complementary song that adds dimension to the musical feast. • Female McCown's Longspurs are persistent incubators, leaving their nests only when they are almost stepped upon.

nonbreeding

♀ ♂

breeding

ID: *Breeding male:* black cap, "whisker" stripe and "bib"; rufous shoulder; light grey and white face; grey underparts; black bill; white outer tail feathers. *Breeding female:* similar patterning to male, but not as bold; mostly tan-coloured and white plumage; white belly. *In flight:* white tail has dark tip and central stripe that form an inverted "T."
Size: *L* 14–16 cm; *W* 27–28 cm.
Habitat: short-grass prairie, native grasslands, overgrazed pastures and agricultural areas.
Nesting: on the ground, at the base of a clump of vegetation; cup nest is woven of coarse grasses and lined with finer materials; female incubates 2–5 darkly marked, pale buff eggs for 12–13 days.
Feeding: gleans the ground for seeds and invertebrates, especially grasshoppers, beetles and moths; occasionally drinks at shallow ponds.
Voice: calls include a *poik* and a soft rattle; male's song is a fast, twittering warble delivered on the wing.
Similar Species: *Chestnut-collared Longspur* (p. 461): breeding male has black underparts; female has drab, brown-streaked underparts. *Lapland Longspur* (p. 459): narrow, white outer tail feathers; male has black face and rufous nape. *Vesper Sparrow* (p. 440): chestnut wing patch; more white in face. *Savannah Sparrow* (p. 443): lacks white outer tail feathers.

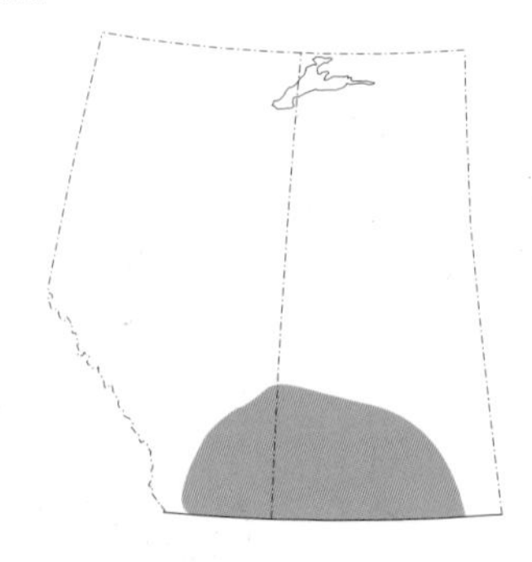

LAPLAND LONGSPUR

Calcarius lapponicus

breeding

Longspurs are medium-sized, sparrowlike birds of open, treeless country, including grasslands, fields and tundra. Nesting in the High Arctic, the Lapland Longspur is a common migrant and occasional winter resident in open areas of southern Canada. Migrating and wintering Lapland Longspurs are often spotted among roaming flocks of Horned Larks, Snow Buntings and American Pipits. From day to day, their movements are largely unpredictable, but they typically appear wherever open fields offer an abundance of seeds or waste grain. Flocks of longspurs can be surprisingly inconspicuous until they suddenly erupt into the sky, giving flight calls and flashing their white outer tail feathers. • A Lapland Longspur's extremely long claws are not suited to gripping branches, so this bird prefers to perch on the ground or on flat-topped boulders, fallen logs and posts. • The male Lapland Longspur's courtship display includes a conspicuous, tinkling flight song, issued as the male rises into the air and then floats downward with outstretched wings and a spread tail. • Males arriving in fall have long since moulted out of their spectacular breeding plumage, but by the time they migrate through again in spring, generally in April or early May, most males will already be sporting their unmistakable black-and-rufous breeding colours.

nonbreeding

ID: medium-sized songbird; white outer tail feathers; rufous wing patch bordered by 2 thin, white bars; streaked sides; back and wings mottled with black, white and brown; streaked, grey rump; white belly; long claws. *Male:* black face, throat and mid-breast; yellowish eye line; yellow bill; dark-streaked flanks; chestnut nape. *Female* and *nonbreeding male:* similar to breeding male, but rufous and black areas appear washed out; mottled, brown-and-black upperparts; finely streaked breast and sides on otherwise pale underparts; pinkish bill. *In flight:* mostly grey underwings; white outer tail feathers.

Size: *L* 15 cm; *W* 29 cm.

Habitat: *Breeding:* wet meadows and scrub on coastal Arctic tundra. *In migration* and *winter:* ploughed grain fields, pastures, grassy lakeshores, estuaries and upper marine beaches.

Nesting: on open tundra; at the base of a clump of vegetation; cup nest of grasses and moss is lined with finer materials; female incubates 4–7 brown-mottled, pale greenish or buff eggs for 10–14 days.

Feeding: forages in flocks on the ground for invertebrates and seeds; eats mostly seeds and waste grain in winter.

Voice: mellow, whistled contact calls; rattled *tri-di-dit* and descending *teew* flight calls; male's flight song is a rapid, slurred warble.

Similar Species: *Smith's Longspur* (p. 460): 2 white outer tail feathers; female has buffy chest and belly. *Horned Lark* (p. 341): black face patch; black tail with white outer feathers. *Chestnut-collared Longspur* (p. 461): smaller; breeding male has black belly and pale buff throat; other plumages show little colour. *Snow Bunting* (p. 462): nonbreeding bird has unstreaked, tan head and mostly white wings with black tips.

SMITH'S LONGSPUR

Calcarius pictus

The Smith's Longspur is an uncommon species normally seen in small numbers during spring and fall migration through the Canadian Prairies and occasionally east into Ontario. These uncommon and secretive birds might be seen briefly as they retreat south for the winter, but you'll have to look closely through larger flocks of Lapland Longspurs to spot them. In fall, the two species look so similar that the exercise can be a daunting one. Occasionally, if you are lucky enough to be at the right place at the right time, which in the Prairies is generally early to mid May in spring and mid-September to mid-October in fall, flocks of migrant Smiths may be found feeding in stubble fields or other open areas. • Smith's Longspurs spend their summers on a narrow band of tundra that runs from the Yukon and the northwest NWT to the Hudson Bay coast. • This species has an unusual breeding system, and it may be the only songbird that is polygynandrous—each female pairs and copulates with two or three males for a single clutch of eggs. Males defend females rather than territories, following the females around and guarding them, and usually pair and copulate with two or more females.

breeding

ID: brown upperparts with black, white and buff streaking; small, white shoulder patch (often concealed); yellowish orange bill is dusky on top. *Breeding male:* black crown; black-and-white face; buff-orange underparts and collar; faint streaking on sides and flanks. *Female, nonbreeding male* and *immature:* streaked crown and nape; buff underparts; faint brownish breast streaking. *In flight:* medium grey underwings; dark-centred tail with white outer feathers.

Size: *L* 15 cm; *W* 27 cm.

Habitat: *Breeding:* dry sedge tundra, usually close to raised sandy-gravel ridges. *In migration:* grassy shorelines, pastures, stubble fields and airports.

Nesting: on the ground; on a dry tundra hummock, often concealed by a low shrub; open cup nest of grass and sedges is lined with softer material; female incubates 2–5 darkly marked, pale green to tan eggs for 11–14 days.

Feeding: gleans insects and from the ground and low vegetation.

Voice: calls include a buzzy *goeet* and a dry, clicking rattle; alarm call is a slow *tick tick tick*, like a watch; male's song is a warbling *switoo-whideedeedew, whee-tew.*

Similar Species: *Lapland Longspur* (p. 459): less white in tail; whitish belly and undertail coverts; often shows black on neck or breast; breeding male has black throat and "bib" and chestnut nape. *Vesper Sparrow* (p. 440): chestnut shoulder patch; white underparts. *American Pipit* (p. 390): thinner bill; streaked breast; lacks white shoulder patch.

CHESTNUT-COLLARED LONGSPUR

Calcarius ornatus

Come spring, in areas where the stale prairie dust hangs in the air, cock your ear for the tinkling song of the Chestnut-collared Longspur. The colourful males can occasionally be seen in flight or atop boulders, shrubs or fence posts that rise out of the dancing waves of grass. • The male Chestnut-collared Longspur is easily the most colourful of the grassland sparrows. However, the female shares only the male's white-edged, black tail and can be readily confused with many of the plains sparrows. • Once one of the most abundant birds on the Prairies before agriculture altered the landscape, this larkspur historically chose breeding locations where herds of bison had recently grazed. Now the species is found only in areas that have escaped cultivation or where nature has restored once-ploughed fields. • Since 1970, this species has been declining at almost 5 percent annually. In recent decades, the Chestnut-collared Longspur has disappeared from many of its former nesting sites on the eastern edge of its range. • Longspurs are so named because they have an extremely long hind claw. The elongated appendage is an advantage for a bird that spends so much of its life on the ground.

ID: *Breeding male:* mottled brown, buff and white upperparts; chestnut nape; black underparts; yellow throat; black-and-white-striped face with black cap and white eyebrow; white outer tail feathers; black central and terminal tail feathers; white undertail coverts. *Female* and *nonbreeding male:* mottled, greyish brown overall; sometimes has pale chestnut nape; light, brownish breast streaks; pale, often pinkish bill. *In flight:* pale grey underwings; white tail shows black "Y" from above.
Size: *L* 14–16 cm; *W* 25 cm.
Habitat: short-grass prairie or similar-structured grasslands; tall or dense grass not usually tolerated.

Nesting: well concealed by grass in a depression or scrape; small cup nest woven of grass is lined with feathers and seed fluff; female incubates 3–5 darkly marked, off-white eggs for 10–13 days.
Feeding: gleans the ground for plant seeds and invertebrates.
Voice: calls include a hiccupped *deedle* and a quick, downslurred *phew*; males's song is a musical warble.
Similar Species: breeding male is distinctive. *McCown's Longspur* (p. 458): female has hint of chestnut in wing and unstreaked, white underparts. *Vesper Sparrow* (p. 440): chestnut wing patch; more white in face. *Savannah Sparrow* (p. 443): lacks white outer tail feathers.

SNOW BUNTING

Plectrophenax nivalis

Early winter brings flocks of Snow Buntings to fields and roadsides in open areas of southern Canada, their startling black-and-white wings flashing in contrast to the snow-covered landscape. Snow Buntings endure the cold season in tight, small, wandering flocks, often with a few Lapland Longspurs mixed in, scratching away the snow in search of seeds and waste grains. On occasion, they will ingest small amounts of roadside sand or gravel as a source of minerals and to aid digestion. • It might seem strange that Snow Buntings are whiter in summer than in winter, but the darker winter plumage is thought to help these birds absorb heat on clear, cold days. • As spring approaches, the warm rufous tones of the winter plumage fade, leaving the male with a decidedly formal, black-and-white breeding plumage. • Snow Buntings nest in the rock-strewn tundra of the far North. One bunting took the trek to its Arctic breeding grounds a little too far and was reported close to the North Pole, thus setting a record apparently unbeaten and unmatched by any other songbird.

♂ ♀ *breeding*

♂ ♀ *nonbreeding*

ID: small songbird; large, white wing patches; dark wing tips; white underparts. *Breeding male:* black back; brilliant white head, nape, breast and belly; black bill. *Breeding female:* streaked, grey head; streaky, brown-and-whitish back and crown; dark bill; more extensive white mottling on wing. *Nonbreeding male:* warm golden rufous crown, cheek and partial breast band; yellow bill; dark-streaked, rufous back and rump. *Nonbreeding female:* similar to male; larger patches and deeper hues of golden brown; blackish forecrown. *In flight:* white wings are black or grey on outer third and part of upper leading edge; dark tail with white outer feathers.

Size: *L* 15–18 cm; *W* 30 cm.

Habitat: *Breeding:* tundra with rocky outcroppings. *In migration* and *winter:* open country, especially farmlands, grasslands, roadsides and beaches.

Nesting: in a crevice among boulders; cup nest of grasses, moss and lichens is heavily lined with finer grasses and feathers; female incubates 4–6 variably marked, pale blue or greenish blue eggs for 10–14 days.

Feeding: forages on the ground for insects and spiders in summer and seeds at other times.

Voice: call is a whistled *tew*; flocks keep up a constant twittering on the ground and especially in flight; male's song is a repetitive warble.

Similar Species: *Lapland Longspur* (p. 459): brown, mottled back; streaked head pattern in winter plumage; lacks black-and-white wings in flight.

SUMMER TANAGER

Piranga rubra

The northern limits of the Summer Tanager's breeding range lie south of Canada, but each year, small numbers make their way here, dazzling birders. More than 75 percent of all Summer Tanager sightings in Ontario have been recorded in the extreme southwestern corner of the province, and it is even possible that the species may occasionally nest in this region. Individuals (especially males) are routinely observed during spring migration in most provinces, but particularly in eastern Canada. It is a regular vagrant into the Maritime provinces. • Summer Tanagers thrive on a wide variety of insects, but they are best known for their courageous attacks on wasps. In the U.S., these birds snatch flying wasps from menacing swarms, and they are also known to raid wasp nests in search of larvae. When not foraging, Summer Tanagers tend to be hard to spot as they move slowly through leafy treetop concealment. • As with other tanagers, male Summer Tanagers are brilliantly coloured, whereas females and immatures are a drabber greenish yellow. Male Summer Tanagers do not attain full adult plumage until their second fall; first-spring males have a patchy green-and-red coloration.

ID: robust, pale yellowish or orangey bill. *Male:* rose red overall; sometimes shows crest; 2nd-year male has patchy red and greenish plumage. *Female:* greyish to greenish yellow upperparts; dusky yellow underparts; may have an orange or reddish wash overall. *In flight:* wing lining matches underparts.

Size: *L* 18–19 cm; *W* 30–31 cm.

Habitat: pine and open mixed woodlands, especially those with oak or hickory; also riparian woodlands with cottonwoods.

Nesting: not confirmed to nest in Canada.

Feeding: eats mainly insects, especially bees and wasps; gleans insects from the tree canopy; may hover-glean or hawk insects in midair; known to raid wasp nests; also eats berries and small fruits.

Voice: *pit* or *pit-a-tuck* call; song is a series of 3–5 sweet, clear, whistled phrases, like a faster version of American Robin's song.

Similar Species: *Scarlet Tanager* (p. 464): smaller bill; male has black tail and wings; female has darker wings, brighter underparts and uniformly olive upperparts. *Northern Cardinal* (p. 466): red bill; prominent head crest; male has black mask and "bib." *Western Tanager* (p. 465): wing bars. *Orchard Oriole* (p. 481) and *Baltimore Oriole* (p. 483): female has wing bars and sharper bill.

SCARLET TANAGER

Piranga olivacea

Few birds can match the tropical splendour of the tanagers. The Scarlet Tanager and the closely related Western Tanager are the only two tanagers that routinely nest in Canada. However, in Central and South America, more than 200 tanager species represent every colour of the rainbow. • Each spring, birders eagerly await the return of the brilliant red male to wooded ravines and traditional migratory sites. Like most tanagers, the male Scarlet Tanager does not acquire full breeding plumage until its second year; a male in first-summer plumage is dull red to orange overall with brownish black wings. • During spring migration, you might observe a Scarlet Tanager at eye level as it forages in the forest understorey during cold and rainy weather. At other times, however, this bird can be surprisingly difficult to spot as it darts through the forest canopy in pursuit of insect prey. • Scarlet Tanagers need extensive mature forests for nesting; fragmentation of these forests can lead to the absence of this area-sensitive species. Breeding bird survey data has shown this species to be declining in Canada at an annual rate of 3 percent since 1968.

♂ ♀

breeding

ID: *Breeding male:* bright red overall; pure black wings and tail; pale bill. *Female:* uniformly olive upperparts; yellowish green underparts; greyish brown wings. *Nonbreeding male:* yellowish underparts; olive upperparts; black wings and tail. *In flight:* white wing linings.

Size: *L* 17–19 cm; *W* 29 cm.

Habitat: *Breeding:* mature upland deciduous and mixed forests and large woodlands. *In migration:* shade trees and coastal shrubbery.

Nesting: usually high in a deciduous tree; flimsy, shallow cup nest of grass, weeds and twigs is lined with rootlets and fine grasses; female incubates 2–5 brown-spotted, pale blue-green eggs for 12–14 days.

Feeding: gleans insects from the tree canopy; also hover-gleans or hawks insects in midair; may forage at lower levels during cold weather; occasionally eats berries.

Voice: *chip-burrr* or *chip-churrr* call; male's song is a series of 4–5 sweet, clear, whistled phrases, like a slurred version of the American Robin's.

Similar Species: *Summer Tanager* (p. 463): larger bill; male has red tail and red wings; female has paler wings and is duskier overall, often with orange or reddish tinge. *Northern Cardinal* (p. 466): red bill, wings and tail; prominent head crest; male has black mask and "bib." *Orchard Oriole* (p. 481) and *Baltimore Oriole* (p. 483): females have wing bars and sharper bills.

WESTERN TANAGER

Piranga ludoviciana

The colourful Western Tanager is the most northerly breeding tanager species. The male, with his golden body accentuated by black wings and tail, is responsible for converting many people to birding. Western Tanagers are tropical for most of the year—they fly to Canada for only a few short months to raise a new generation on the seasonal explosion of food in our forests. The difficult-to-learn song of the male Western Tanager resembles the phrases of an American Robin's song, but the notes are somewhat hoarser, as if the bird has a sore throat, and it ends with a distinctive, hiccuplike *pit-a-tik.* • Similar to their eastern cousins, the Western Tanager can be difficult to spot, preferring to remain in the upper canopy of large shady trees and often moving in a rather sluggish fashion. Generally, female tanagers are more cryptically coloured than males, and the female Western Tanager is no exception. • Although its nesting range extends as far east as Saskatchewan, it has been recorded as a vagrant eastward to Newfoundland. • The name "tanager" is derived from *tangara*, the Tupi name for this group of birds in the Amazon Basin.

ID: *Breeding male:* red head; yellow nape, underparts and rump; black back, wings and tail; 1 yellow and 1 white wing bar; pale bill. *Breeding female:* olive green overall, yellower below; dark wings; 2 yellowish white wing bars. *Nonbreeding male:* yellowish head with orange wash near bill; duller yellow underparts; duller back; typically only forehead and chin areas are reddish. *In flight:* pale olive yellow wing linings.

Size: *L* 18 cm; *W* 29 cm.

Habitat: *Breeding:* open, mature, coniferous or mixed coniferous-hardwood forests. *In migration:* almost any forested habitat; visits bird baths and feeders in gardens.

Nesting: usually in a conifer; in a fork of a horizontal branch; loose cup nest of twigs, rootlets and mosses is lined with hair and fine rootlets; female incubates 3–5 brown-marked, blue eggs for 13 days.

Feeding: gleans vegetation and catches flying insects on the wing in summer; eats fruits and berries in migration.

Voice: soft, whistled calls; fast, rattling notes when agitated; male's robinlike song is a series of somewhat hoarse phrases and slurred whistles, often delivered very rapidly.

Similar Species: none; male is distinctive. *Orioles* (pp. 481–83): females have thinner, sharper bills. *Summer Tanager* (p. 463) and *Scarlet Tanager* (p. 464): females lack white wing bars.

NORTHERN CARDINAL

Cardinalis cardinalis

A bird as beautiful as the Northern Cardinal rarely fails to capture our attention and admiration, and most Canadians can easily recognize this delightful year-round resident without the help of a field guide. • Although these birds prefer the tangled, shrubby edges of woodlands, they are easily attracted to backyards with feeders and sheltering trees and shrubs. • Cardinals form one of the bird world's most faithful pair bonds, the male and female remaining in close contact year-round, singing to one another through the seasons with soft, bubbly whistles. • Few realize that the Northern Cardinal is a relative newcomer to Canada. Its range has expanded northward over the years into southern regions of Manitoba east to Nova Scotia, probably because of the wealth of backyard feeders, the warm microclimate of our urban centres and forest fragmentation, which has created ideal habitat for this bird. • The Northern Cardinal owes its name to the vivid red plumage of the male, which resembles the robes of Roman Catholic cardinals.

♀

♂

ID: *Male*: red overall; pointed crest; black mask and throat; red, conical bill. *Female:* shaped like male; brown-buff to buff-olive overall; blackish mask; red bill, crest, wings and tail. *Juvenile male:* similar to female, but has dark bill and crest.
Size: *L* 19–23 cm; *W* 30–31 cm.
Habitat: brushy thickets and shrubby tangles along forest and woodland edges; also backyards and suburban and urban parks.

Nesting: in a dense shrub or thicket, or low in a conifer; open cup nest of twigs, bark shreds, grasses, leaves and rootlets is lined with hair and fine grasses; female incubates 2–5 variably marked, whitish to bluish or greenish white eggs for 12–13 days.
Feeding: gleans the ground and shrubs for seeds, insects and berries; frequents feeders.
Voice: metallic *chip* call; male's song is a variable series of clear, bubbly whistled notes: *What cheer! What cheer! birdie-birdie-birdie What cheer!*
Similar Species: *Summer Tanager* (p. 463) and *Scarlet Tanager* (p. 464): lack head crest, black mask and throat and red, conical bill; Scarlet Tanager has black wings and tail.

ROSE-BREASTED GROSBEAK

Pheucticus ludovicianus

The male Rose-breasted Grosbeak is an involved parent, helping not only with choosing the location of the nest and its construction but also with incubating the eggs. Once the young are hatched, the male helps feed them insect larvae and seeds and continues to assist them in finding food after they have left the nest. Grosbeaks often raise two broods per year, and family groups usually stay together until migration. • Many songbirds quietly incubate their eggs, but male Rose-breasted Grosbeaks often sing while they are on the nest. • This bird's songs are quite similar to those of the American Robin, but the grosbeak runs its phrases together without pausing to take a breath. Although the female lacks the male's magnificent colours, she shares his vocal flair. • Rose-breasted Grosbeaks only rarely leave the green treetop canopy, descending cautiously to drink at water puddles. • This species has experienced significant declines throughout its range, including Canada. The loss of tropical rainforests may be contributing to these declines, though the loss of mature deciduous forests in eastern North America has also depleted forest bird numbers.

♀ ♂

breeding

ID: medium-sized songbird; large, pale, conical bill. *Breeding male:* black head and back; red triangle on breast; white underparts and rump; black wings with bold, white markings; black tail with white outer feathers; red underwings and white "wrist" patches in flight. *Female* and *nonbreeding male:* sparrowlike plumage; brown-streaked upperparts; darkly streaked, white underparts; boldly striped head with wide, white eyebrow; buffy underwings and dark tail in flight.

Size: *L* 19 cm; *W* 30 cm.

Habitat: second-growth to mature deciduous forests.

Nesting: in a fork or crotch of a tree or tall shrub; flimsy cup nest of twigs, bark strips, weeds, grasses and leaves is lined with rootlets and hair; pair incubates 3–5 brown-blotched, bluish green eggs for 11–14 days.

Feeding: gleans tree foliage for insects, seeds, buds, blossoms, berries and some fruit; may visit feeders.

Voice: distinctive, metallic *eek!* call; male's song is a rich warble, like a fast and more varied version of an American Robin's song.

Similar Species: *Black-headed Grosbeak* (p. 468): western species; female has yellowish head and wing linings, less streaked underparts and lacks white in wings. *Purple Finch* (p. 486): smaller; female has dark bill and "moustache" stripe and lacks white in wings.

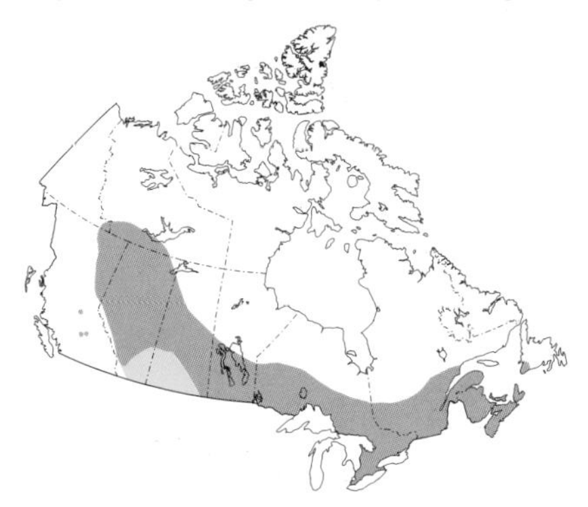

BLACK-HEADED GROSBEAK

Pheucticus melanocephalus

Almost any summer visit to brush, riparian areas and transmission corridors in extreme southern British Columbia and Alberta or southeastern Saskatchewan will reveal Black-headed Grosbeaks. They arrive in May, the first sign of their presence the robinlike songs of the males advertising their territories with extended bouts of complex, accented carolling. Meanwhile, the females forage and conduct the household chores within dense foliage cover, frequently betraying their presence with sharp, woodpecker-like calls to reassure the males. • Males and females look so dissimilar that they can be mistaken for different species. The males do not attain full breeding plumage until their second breeding season. • Black-headed Grosbeaks eat many insects, even dining on monarch butterflies, which are distasteful and even toxic to most birds. • Within Canada, Black-headed Grosbeak populations have been increasing 5 percent per year since the late 1960s. Most of this increase has been in British Columbia, where the species is now found as far north as the Skeena River. • The word *gros* means "big" in French, and true to their name, grosbeaks are songbirds with large, conical bills.

ID: medium-sized songbird; large, bicoloured, conical bill; grey legs. *Male:* black head; orange-chestnut collar, underparts and rump; white-streaked, dark brown back and wings; 2 white wing patches; yellowish belly. *Female:* brown upperparts and wings; dark face; white crown stripe, eyebrow and "moustache" stripe; finely streaked, buff to whitish neck and underparts; 2 white wing bars. *In flight:* yellow wing linings; male has white "wrist" patches and black-centred, white-cornered tail; female has all-dark tail.
Size: *L* 18–20 cm; *W* 30–33 cm.
Habitat: brushy, wet or dry areas near wetlands, lakeshores, mixed deciduous woodlands and transmission corridors.

Nesting: in a tall shrub or small tree, often near water; loosely woven cup nest of twigs is lined with fine grasses; pair incubates 3–5 chestnut-spotted, bluish green eggs for 12–14 days.
Feeding: gleans vegetation for invertebrates, seeds, buds and fruit; visits feeders.
Voice: high, woodpecker-like *pik* contact call; male's leisurely, whistled, warbling song is a long series of robinlike phrases without breaks.
Similar Species: male is distinctive. *Rose-breasted Grosbeak* (p. 467): female has pale bill and streaked breast. *Purple Finch* (p. 486): smaller; female has dark bill and "moustache" stripe and lacks white in wings. *Bullock's Oriole* (p. 482): male has orange face with dark eye line, slimmer, longer bill, larger white patch on upperwing and brighter underparts.

LAZULI BUNTING

Passerina amoena

Small flocks of Lazuli Buntings work their way northward from Mexico and Arizona into southwestern Canada each spring, bringing a splash of welcome colour to open, dry habitats with scattered bushlands and small trees. • From mid-April to late July, the brightly coloured male's crisp and varied songs punctuate the hot "siesta hours" during which only a handful of other species regularly vocalize. Singing intensity diminishes as broods near fledging. • By late August, most Lazuli Buntings have deserted their summer homes. They undergo a partial moult before they leave, completing their change of plumage on their wintering grounds. • Where human activities have made habitat more favourable for Brown-head Cowbirds, Lazuli Bunting nests suffer from increased nest parasitism. Overall in Canada, populations of this charming species are on the increase. • This bird is named for the blue gemstone lapis lazuli. The generally accepted pronunciation of the name is *LAZZ-you-lie*, but personal variations are plentiful. The scientific name *amoena* is from the Latin for "charming," "delightful" or "dressy," all of which characterize this bird.

ID: small, beautifully coloured songbird; short, pale to greyish, conical bill; 2 white wing bars; whitish belly and undertail coverts; bluish tail; dark legs. *Breeding male:* bright blue head, back and rump; grey-and-blue wings; orangey chestnut breast; buffy sides. *Female*: buff eye ring; dull greyish brown upperparts; off-white underparts; buff breast. *Nonbreeding male*: tan-mottled, blue upperparts; duller breast. *In flight:* whitish (male) or drab buff (female) wing linings; hints of blue on dark rump and tail.

Size: *L* 14 cm; *W* 23 cm.

Habitat: shrubby, weedy areas on dry hillsides and brushy draws; also riparian habitats, wooded valleys and open scrub; patches of brambles and broom on the Pacific Coast.

Nesting: in a crotch in a shrubby tangle; coarsely woven nest of dried grasses and weed stems is lined with finer grasses and hair; female incubates 3–4 pale bluish white eggs for 11–14 days.

Feeding: gleans the ground and low shrubs for invertebrates and seeds.

Voice: buzzy, trilling flight calls and woodpecker-like *pik* notes; male's song is a brief, varied series of warbling notes.

Similar Species: *Eastern Bluebird* (p. 373) and *Western Bluebird* (p. 374): larger; lack wing bars. *Indigo Bunting* (p. 470): no wing bars; male lacks orangey chestnut breast.

INDIGO BUNTING

Passerina cyanea

Shadowed in a towering tree, a male Indigo Bunting can look almost black. If this happens on your first encounter with this bird, reposition yourself quickly so the sun strikes and enlivens this bunting's incomparable indigo colour—the rich shade of blue is rivalled only by the sky. • Males sing constantly through the heat of the day, as long as their young are in the nest. Indigo Buntings can have a second clutch, the young still in the nest in September. • Raspberry thickets are a favoured nesting location for this species in southeastern Canada. The dense, thorny stems provide nestlings with protection from predators, and the berries are a convenient source of food. • The Indigo Bunting employs a clever and comical foraging strategy to reach the grass and weed seeds on which it feeds: the bird lands midway on a stem then shuffles slowly toward the seed head, which eventually bends under the bird's weight, giving the bunting easier access to the seeds. • With this neon blue species currently increasing within its range, more Canadians may soon be able to enjoy this gift from the avian world.

♀ ♂ *breeding*

ID: stout, grey, conical bill; beady, black eyes; black legs; no wing bars. *Breeding male:* blue overall; black lores; wings and tail may show some black. *Female:* soft brown overall; faintly brown-streaked breast; whitish throat. *Nonbreeding male:* patchy blue and grey overall; whitish belly and undertail coverts.

Size: *L* 14 cm; *W* 20 cm.

Habitat: deciduous forest and woodland edges, regenerating forest clearings, shrubby fields, orchards, abandoned pastures and hedgerows; occasionally along mixed woodland edges.

Nesting: usually in an upright fork of a small tree or shrub; cup nest of grasses, leaves and bark strips is lined with rootlets, hair and feathers; female incubates 3–4 white to bluish white eggs for 12–13 days.

Feeding: gleans low vegetation and the ground for insects, especially grasshoppers, beetles, weevils, flies and larvae; also eats the seeds of thistles, dandelions, goldenrods and other native plants.

Voice: quick *spik* call; male's song consists of paired, warbled whistles: *fire-fire, where-where, here-here, see-it see-it.*

Similar Species: *Blue Grosbeak* (p. 506): larger overall; larger, more robust bill; 2 rusty wing bars; male has black around base of bill; female lacks breast streaking. *Mountain Bluebird* (p. 375): larger; slimmer bill; male has pure blue wings and tail. *Lazuli Bunting* (p. 469): 2 pale wing bars; male has orangey chestnut breast.

DICKCISSEL

Spiza americana

Arriving in suitable nesting habitat before the smaller females, breeding male Dickcissels bravely announce their presence with stuttering, trilled renditions of their own name. Territorial males perch atop tall blades of grass, fence posts or rocks to scour their turf for signs of potential mates or rival males. Dickcissels are polygynous, and males may mate with as many as eight females in a single breeding season. This breeding strategy means that the male gives no assistance to the females in nesting or brooding. • Dickcissels are sometimes seen foraging in small flocks during migration but are most regularly seen among flocks of House Sparrows at backyard feeders over winter. • Dickcissels are sporadic breeders within their Canadian range. This species is not only renowned for movements within its breeding range from year to year, but also for moving outside its core range to nest in surrounding areas where suitable grassland habitat exists. In some years, it may occur in reasonable numbers into southern Saskatchewan, southern Manitoba and through much of southern Ontario. • The Dickcissel has experienced major habitat changes on both its breeding and nonbreeding ranges, as natural tall-grass prairie and savannas have been largely replaced by agriculture, but it appears to have adapted well and even thrives in some modified landscapes.

♂ ♀

breeding

ID: brown upperparts; rufous shoulder patch; grey head and nape; yellow eyebrow; pale greyish brown underparts; yellow-washed breast; whitish undertail coverts; dark, conical bill. *Male:* white "chin"; black "bib"; duller colours in nonbreeding plumage. *Female*: duller version of male; white throat; buffy eyebrow; lacks black "bib." *In flight:* greyish underwing with brown-edged flight feathers.

Size: *L* 15–18 cm; *W* 24 cm.

Habitat: abandoned fields dominated by forbs, weedy meadows, croplands, taller grasslands and grassy roadsides.

Nesting: on or near the ground; well concealed amid tall, dense vegetation; bulky, open cup nest of forbs, grasses and leaves is lined with rootlets, fine grasses or hair; female incubates 3–6 pale blue eggs for 12–13 days.

Feeding: gleans insects and seeds from the ground and low vegetation; small flocks may visit feeders, especially in winter.

Voice: flight call is a buzzerlike *bzrrrrt*; male's song consists of 2–3 introductory notes followed by a trill, often paraphrased as *dick dick dick-cissel.*

Similar Species: *Eastern Meadowlark* (p. 474) and *Western Meadowlark* (p. 475): much larger; long, pointed bills; yellow "chin" and throat with black "necklace." *American Goldfinch* (p. 494): white or yellow-buff bars on dark wings; may show black forecrown; lacks black "bib."

BOBOLINK

Dolichonyx oryzivorus

breeding

The plumage of the male Bobolink is unlike that of any other North American bird—Roger Tory Peterson described it as a "backward tuxedo." The male's courting tactics are straightforward, however, and he flashes his yellowish buff nape and white wing patches at any female within view. If this doesn't work, the male launches into a bubbly, zestful song in a trembling song-flight or from exposed grassy perches. • Although male Bobolinks commonly mate with more than one female, they do assist in feeding and raising the young. Once his early summer duties have been performed, the male Bobolink moults into the same cryptic, sparrow-like colouring as the female and prepares for fall migration. • Bobolinks once benefited from forest clearing and agricultural activity, but modern practices, such as harvesting hay early in the season, have had negative effects on this species' reproductive success. Breeding bird survey data shows this species to be declining in Canada at a rate of 5 percent annually. • The male's unique song and looks have earned it various nicknames including "Bubbling Bob, the Bobolink" and "Skunk-bird"!

ID: medium-sized songbird; short, pointed tail; conical bill. *Breeding male:* mostly black plumage; yellowish buff nape; white shoulder patches and rump; black bill. *Female:* resembles large sparrow; buff brown overall; dark-streaked back, sides, flanks and rump; pale eyebrow; dark eye line; dark-bordered, pale central crown stripe; pink bill. *Nonbreeding male:* similar to breeding female; richer golden buff on upperpart feather edges, underparts and face. *In flight:* wing linings match sides.

Size: *L* 15–20 cm; *W* 28–30 cm.

Habitat: *Breeding:* hayfields and tall-grass agricultural fields with a variety of forbs; occasionally croplands (especially rye fields).

Nesting: in a small, loose colony; on the ground in dense plant cover; coarse grass nest is lined with finer grasses; female incubates 2–6 purple-marked, greyish eggs for 11–13 days.

Feeding: gleans the ground for adult and larval invertebrates, including spiders; also eats seeds.

Voice: *pink* flight call; male's song, often issued in flight, is a series of bubbly notes and banjolike twangs: *bobolink bobolink spink spank spink.*

Similar Species: *Red-winged Blackbird* (p. 473): female is darker, with heavily streaked breast and longer, narrower bill. *Grasshopper Sparrow* (p. 444): smaller; shorter tail. *Lark Bunting* (p. 442): breeding male has black nape and rump; female and nonbreeding male are browner, with dark breast spot and unevenly streaked underparts.

RED-WINGED BLACKBIRD

Agelaius phoeniceus

A birder's winter blahs might be remedied by the sound of the season's first Red-winged Blackbird. Arriving in Canada before the females, male Red-winged Blackbirds get an early spring start on staking out territories. Few wetlands and cattail marshes are free from the classic calls of these bossy, aggressive birds. Swaying atop cattail stalks, males display their bright red shoulder patches while singing their loud *konk-a-ree* or *ogle-reeeee* songs. A male's shoulder patches, or "epaulettes," and song are his most important tools in the often-intricate strategy required to defend his territory from rivals. A richly voiced male with a large and productive territory can attract several mates to his cattail kingdom. In field experiments, males whose red shoulders were painted black soon lost their territories to previously defeated rivals. • Once his breeding-season duties are completed, the male Red-winged Blackbird becomes just another member of the enormous flocks of foraging blackbirds that roam agricultural fields, grasslands and marshes. • Red-winged Blackbirds are among the most abundant birds in North America.

ID: medium-sized songbird; longish tail; slender, black bill. *Male:* all-black, non-glossy plumage; orangey red shoulder patch with partly hidden, yellowish lower border. *Female:* variable brown overall; heavily streaked, pale underparts; reddish-tinged throat; pale eyebrow stripe; rufous-edged back and upperwing feathers. *In flight:* black (male) or brown (female) underwing.

Size: *L* 18–23 cm; *W* 30–33 cm.

Habitat: *Breeding:* wetlands with standing aquatic vegetation (especially cattails), including willows; also ditches and flooded fields. *In migration* and *winter:* agricultural fields and marshes.

Nesting: colonial and polygynous; usually in cattails; also bulrushes, sedges and shrubs; woven nest of sedges, dried cattail leaves and grasses is lined with fine grasses; female incubates 3–6 darkly marked, pale bluish green eggs for 11–13 days.

Feeding: gleans the ground and wetland vegetation for emerging insects in summer; eats seeds, including waste grain, and invertebrates in winter; occasionally visits feeders.

Voice: calls include a harsh *check* and a high *tseert*; male's song is a loud, raspy *konk-a-ree* or *ogle-reeeee*; female may give a *che-che-che chee chee chee*.

Similar Species: male is distinctive. *Rusty Blackbird* (p. 477) and *Brewer's Blackbird* (p. 478): females lack streaked underparts. *Brown-headed Cowbird* (p. 480): juvenile is smaller with stubbier, conical bill.

EASTERN MEADOWLARK

Sturnella magna

The Eastern Meadowlark's trademark *See-you at school-today* tune rings throughout spring and early summer from fence posts and power lines, wherever grassy meadows and pastures are found. • The male's bright yellow underparts, black, V-shaped "necklace" and white outer tail feathers help attract mates. Females share these colourful attributes for a slightly different purpose—when a predator approaches too close to a nest, the incubating female explodes from the grass in a burst of flashing colour. Most predators cannot resist chasing the moving target, and once the female has led the predator some distance from the nest, she simply folds away her white tail flags, exposes her camouflaged back and disappears into the grass without a trace. • Meadowlarks are not actually larks but members of the blackbird family, even though they don't seem to fit in because of their bright plumage. When theses birds are seen in silhouette, however, the similarities become apparent. • Urbanization, intensive agricultural practices and natural succession of habitat have all contributed to the declines of grassland species throughout Canada. Eastern Meadowlarks have been declining at an annual rate of approximately 5 percent for the last 40 years.

ID: sexes similar; large, stocky songbird; long, sharp bill; pinkish legs; short, banded tail with white outer feathers. *Breeding male:* dark-mottled, pale sandy brown upperparts; yellow throat and underparts; black, V-shaped breast band; boldly black-streaked, whitish flanks and undertail coverts; dark-bordered, pale central crown stripe; pale eyebrow; blackish eye line; yellow lores. *Female* and *nonbreeding male:* less boldly marked; paler overall. *In flight:* drab, white-lined underwing.

Size: *L* 23–24 cm; *W* 34 cm.

Habitat: grassy meadows and pastures; also croplands, weedy fields, grassy roadsides and old orchards.

Nesting: in a depression or scrape on the ground, concealed by dense grass; domed grass nest with side entrance is woven into surrounding vegetation; female incubates 3–7 white eggs, heavily spotted with brown and purple, for 13–15 days.

Feeding: gleans grasshoppers, crickets, beetles and spiders from the ground and vegetation; extracts grubs and worms by probing bill into soil; also eats seeds.

Voice: rattling flight call and a high, buzzy *dzeart*; male's song is a rich series of 2–8 melodic, clear, slurred whistles: *see-you at school-today* or *this is the year.*

Similar Species: *Western Meadowlark* (p. 475): paler upperparts, especially crown stripes and eye line; yellow on throat extends onto lower cheek; different song and call. *Dickcissel* (p. 471): much smaller; all-dark crown; white throat; lacks dark streaking on sides and flanks.

WESTERN MEADOWLARK

Sturnella neglecta

The Western Meadowlark displays a combination of cryptic coloration on its upperparts and bright courtship elegance on its underparts. To show off his plumage and vocal abilities to potential mates, the male finds a prominent perch or indulges in a brash display flight. If he notices human observers, the male will often turn away, hiding his brightly coloured underparts. • Many westerners think of the Western Meadowlark as the true harbinger of spring, because it returns to the frozen prairies early in the season, usually in late March. This bird's rich and varied song—males have up to 12 song types—is recognized as one of the most beautiful voices of the prairies. • Western Meadowlarks benefit from the conversion of sagebrush to grassland, and their numbers have declined wherever expanses of grassland habitat have been degraded or replaced. Unlike the significant declines seen in Eastern Meadowlark populations throughout Canada, the Western's numbers appear to be stable, except in Ontario, where declines have made it a rare breeding species. • This bird was overlooked by the Lewis and Clark expedition because it was confused with the Eastern Meadowlark. This oversight is acknowledged in its species name, *neglecta.*

ID: sexes similar; large, stocky songbird; long, sharp bill; pinkish legs; short, banded tail with white outer feathers. *Breeding:* darkly mottled, pale sandy brown upperparts; yellow throat, lower cheek and underparts; black, V-shaped breast band; boldly black-streaked, whitish sides and flanks; pale eyebrow; brown eye line; yellow lores. *Female* and *nonbreeding male:* less boldly marked; paler overall. *In flight:* drab, white-lined underwing.

Size: *L* 20–25 cm; *W* 35–38 cm.

Habitat: *Breeding:* open country, including grasslands, pastures, rangeland, roadside ditches and edges of grassy agricultural field. *In migration* and *winter:* cultivated fields, beaches, airports and estuaries.

Nesting: on the ground; domed canopy, with a side entrance, of grasses, weed stalks, leaves and plant stems is sometimes interwoven with surrounding vegetation; female incubates 3–7 white eggs, marked with purple and brown, for 13–15 days.

Feeding: gleans ground for grasshoppers, crickets, sowbugs, snails and spiders; probes soil for grubs, worms and other insects; also eats seeds.

Voice: calls include a low, loud *chuck* or *chup*, rattling flight call or a few clear, whistled notes: *who-who are you?*; male's song is a rich series of flutelike warbles.

Similar Species: *Eastern Meadowlark* (p. 474): darker upperparts, especially crown stripes and eye line; yellow on throat does not extend onto lower cheek; different song and calls. *Dickcissel* (p. 471): much smaller; all-dark crown; white throat; lacks dark streaking on sides and flanks.

YELLOW-HEADED BLACKBIRD

Xanthocephalus xanthocephalus

One of the first signs of spring in many freshwater marshes of western Canada is the arrival of the male Yellow-headed Blackbird. In a perfect world, the Yellow-headed Blackbird's song would match its splendid plumage, but unfortunately, when the male arches his golden head back to sing, all he produces is a pathetic, metallic grinding noise. • The later-arriving male Yellow-headed Blackbirds bully their way to the choicest bulrush areas, forcing the Red-wings to peripheral stands in shallow water that are more vulnerable to predation. • Yellow-headed Blackbirds routinely leave their marsh habitat to forage for seeds and insects in upland fields and pastures a couple of kilometres or more from their nesting marsh. Unlike other blackbirds, however, their nests are only found in areas of deep water, away from the shoreline. • The bright yellow heads of the males make large colonies of these birds resemble fields of flowering mustard plants. After the colony completes its breeding season, flocks gather to feed with other blackbirds in nearby agricultural fields. Although most Yellow-heads winter in the southern U.S. and Mexico, a few stragglers may remain behind with wintering flocks of other blackbird species.

ID: medium-sized, gregarious songbird; black, pointed bill; bright yellow feathers around cloaca. *Male:* black body; yellow head, neck and breast; black lores; white wing patches. *Female:* dark brown overall; dull yellow eyebrow, throat and breast; variable yellow on head; all-dark wings. *In flight:* dark underwing and tail.

Size: *L* 20–28 cm; *W* 33–38 cm.

Habitat: *Breeding:* larger wetlands with emergent vegetation, especially cattails, bulrushes and reed-grasses; also sewage ponds. *In migration* and *winter:* agricultural fields, grassy uplands and lakeshores.

Nesting: loosely colonial and polygynous; nest is attached mainly to bulrushes or cattails over water; bulky nest is firmly woven of leaves and stems of marsh plants and lined with finer grasses; female incubates 3–5 variably marked, greenish or greyish white eggs for 11–13 days.

Feeding: eats mainly insects and seeds in summer and seeds and waste grain in migration and winter.

Voice: calls are harsher than those of Red-winged Blackbird, except for a whistled trill and soft *kruk* or *kruttuk* flight call; male's song is a strained, metallic grating, similar to an old chainsaw.

Similar Species: male is distinctive. *Rusty Blackbird* (p. 477) and *Brewer's Blackbird* (p. 478): females lack yellow throat and face.

RUSTY BLACKBIRD

Euphagus carolinus

The Rusty Blackbird owes its name to the colour of its rusty fall plumage, but the name could just as well reflect this bird's grating, squeaky song, which sounds very much like a rusty hinge. • Overshadowed by all other blackbirds in Canada in both abundance and aggressiveness, the Rusty Blackbird is the most subdued member of the group. Unlike many blackbirds, the Rusty Blackbird nests in isolated pairs or very small, loose colonies. • During fall migration, Rusty Blackbirds often intermingle with flocks of other blackbirds, sometimes blackening the skies of rural southern Canada. Their days are spent foraging along the wooded edges of fields and wetlands, and they occasionally pick through the manure-laden ground of cattle feedlots. The Rusty Blackbird may also be found in isolated woodlands, which it favours more than other blackbird species do. Unlike other blackbirds, it seldom depredates crops or is a nuisance at communal roosts. • Rusty Blackbirds are generally less abundant and less aggressive than their relatives, and they generally avoid human-altered environments. Since the mid-1960s, this blackbird is believed to have declined by 85 percent and is now classified as a species of special concern in Canada.

breeding

ID: medium-sized songbird; yellow eyes; dark legs; long, sharp, dark bill. *Breeding male:* black overall; subtle green gloss on body; subtle bluish or greenish gloss on head and breast. *Breeding female:* dark brown overall, without gloss. *Nonbreeding male:* rusty upperparts and head; rust-mottled underparts. *Nonbreeding female:* paler than male; buffy underparts; rusty cheek. *In flight:* dark underwing.

Size: *L* 23 cm; *W* 35–36 cm.

Habitat: *Breeding:* bogs, fens, beaver ponds, wet meadows and the shrubby shorelines of lakes, rivers and swamps. *Nonbreeding:* marshes, open fields, feedlots and woodland edges near water.

Nesting: low in a shrub or small conifer, often above or very near water; bulky nest of twigs, grass and lichens is lined with mud and decaying vegetation; female incubates 4–5 pale blue-green eggs, spotted with grey and brown, for about 14 days.

Feeding: walks along shorelines gleaning waterbugs, beetles, dragonflies, snails, grasshoppers and occasionally small fish; also eats waste grain and seeds.

Voice: harsh *chack* call; male's song is a squeaky, creaking *kushleeeh ksh-lay.*

Similar Species: *Brewer's Blackbird* (p. 478): male has glossier, more iridescent plumage, greener body and more purple on head; female has dark eyes; nonbreeding bird lacks conspicuous rusty highlights. *Common Grackle* (p. 479): longer, keeled tail; larger body and bill; more iridescent plumage. *European Starling* (p. 389): speckled appearance; dark eyes; bill is yellow in summer.

BREWER'S BLACKBIRD

Euphagus cyanocephalus

The dry-land counterpart to its marsh-loving Rusty relative, the Brewer's Blackbird looks like a Red-winged Blackbird without the red shoulder patches, and the Brewer's feathers show an iridescent quality as reflected rainbows of sunlight move along the feather shafts. • This blackbird is commonly mistaken for a Common Grackle because both birds jerk their heads back and forth while walking. • Rather than gathering in large colonies, Brewer's Blackbirds nest in small, loose groups or as widely dispersed solitary pairs. Following nesting, adults and juveniles form small flocks, which then team up with roving mobs of other blackbirds and European Starlings, to form large flocks that spread out over lowland fields and farmlands during autumn. • Our network of highways provides a bounty of vehicle-struck insects for Brewer's Blackbirds, which exploit this roadkill resource better than any other songbird. Brewer's have benefited from human habitat alteration to invade open brushy areas. • John J. Audubon named this bird after Thomas Mayo Brewer, a friend and prominent 19th-century oologist (a person who studies eggs).

breeding

ID: slender, medium-sized songbird; short, sharp bill; long tail. *Male:* black overall with iridescent, bluish green body and purplish head; yellow eyes; nonbreeding male may show some faintly rusty feather edgings. *Female:* dull greyish brown overall; brown eyes; dark eye line.

Size: *L* 20–23 cm; *W* 40 cm.

Habitat: agricultural areas and fields with brushy edges, roadsides, wet meadows, grasslands, landfills and stockyards; also beaches, recent clear-cuts, urban parks and suburbs.

Nesting: loosely colonial; in a tree or shrub or on the ground; sturdy basket of small twigs and grasses is lined with finer grasses, rootlets and hair; female incubates 4–6 brown-marked, greyish eggs for 11–13 days.

Feeding: picks up a wide variety of invertebrates, seeds and fruits; often wades in shallow water for marine and freshwater invertebrates.

Voice: unusually quiet for a blackbird; plain *chek* call; male's song is a creaking, 2-note *k-sheee.*

Similar Species: *Common Grackle* (p. 479): larger body and bill; long, wedge-shaped tail. *Brown-headed Cowbird* (p. 480): smaller; stouter bill; dark eyes; male has brown head; sandy brown female has some streaking on underparts. *Rusty Blackbird* (p. 477): more slender bill; rusty wash overall in nonbreeding plumage; breeding male has more subtle green and blue gloss in plumage; breeding female is dark overall with yellow eyes.

COMMON GRACKLE

Quiscalus quiscula

A large, mostly terrestrial blackbird, the male Common Grackle is a poor but spirited singer. Perched in a shrub, the male grackle will slowly take a deep breath, inflating his breast and causing his feathers to rise, then close his eyes and issue a loud, strained *tssh-schleek*. Despite our perception of his musical shortcomings, the male proudly poses with his bill held high after his "song." • The Common Grackle is easily distinguished from the Rusty Blackbird and Brewer's Blackbird by its long, heavy bill and its long, wedge-shaped tail. In flight, the grackle's long tail trails behind it like a hatchet blade. • Although nesting grackles consume a wide variety of harmful insects and other pests, many urban homeowners feel nothing but disdain for the Common Grackle because of its raucous call, its habit of robbing other birds' nests and its overall cheekiness. • Common Grackles nest in isolated pairs in riparian habitat and forested regions of Canada, but they are especially common in cities, towns and farmsteads, particularly in open, mature conifer plantings. Over the past seven decades, this bird has expanded its breeding range westward into BC and NWT; however, breeding bird surveys reveal overall declines in its numbers since the late 1960s.

purple adult

ID: large songbird; yellow eyes; long, sturdy, dark bill; long, keeled tail. *Male:* dark, iridescent plumage; bronze back and sides; purple wings and tail; purplish blue breast and head. *Female:* similar to male but smaller, with duller, less iridescent body.
Size: *L* 28–33 cm; *W* 43 cm.
Habitat: *Breeding:* ponds, marshes, swamps and slow-moving creeks, especially where cattails and willows abound. *In migration* and *winter:* partly open areas with scattered trees, coniferous forest edges, hedgerows, riparian woodlands, shrubby parks and gardens and around human habitations.
Nesting: singly or in a small, loose colony; among cattail stems or branches of shrubs or in a building or tree cavity, usually near water; bulky cup nest of cattail leaves, twigs and plant fibres is lined with finer grasses; female incubates 4–6 brown-blotched, pale blue eggs for 11–15 days.
Feeding: opportunistic; gleans the ground for insects, earthworms, seeds, grain, aquatic invertebrates, small vertebrates and fruit; also eats bird eggs and nestlings.
Voice: loud *swaaaack* or *chack* call; male's song is a split rasping *tssh-schleek* or *gri-de-leeek*.
Similar Species: *Rusty Blackbird* (p. 477) and *Brewer's Blackbird* (p. 478): much smaller; lack heavy bill and long, keeled tail.

BROWN-HEADED COWBIRD

Molothrus ater

The male Brown-headed Cowbird's song, a bubbling, liquid *glug-ahl-whee*, might well be interpreted by other bird species as "here comes trouble!" The Brown-headed Cowbird once followed the great bison herds across the Great Plains and prairies of central North America—they now regularly associate with cattle, hence the name "cowbird." Because its nomadic lifestyle made building and tending nests impractical, cowbirds adopted a successful alternative: the deceptive art of nest parasitism. These birds lay eggs in the nests of other songbirds, and, in spring, female Brown-headed Cowbirds are frequently seen seeking out potential host nests. Some songbirds recognize the odd eggs, but many species simply incubate cowbird eggs along with their own. • Cowbird chicks typically hatch first and develop much more quickly than their nestmates, which are pushed out of the nest or outcompeted for food. • The expansion of ranching and the fragmentation of forests has significantly increased the Brown-headed Cowbird's range, and it now parasitizes more than 140 bird species in North America, including species that probably had no contact with it prior to widespread human settlement. Cowbird numbers have declined on breeding bird surveys, however, especially in Canada where it has declined by 2 percent per year over the last 40 years.

ID: sparrowlike songbird; dark eyes; thick, conical bill; short, squared tail is held upraised when foraging on the ground. *Male:* iridescent, greenish blue body plumage usually looks glossy black; dark brown head and neck. *Female:* greyish brown overall; pale throat; finely streaked underparts. *Juvenile:* clearly streaked underparts.

Size: *L* 15–20 cm; *W* 30 cm.

Habitat: *Breeding:* agricultural areas, mixed forests with shrubby openings, farmland, brushy transmission corridors, landfills, parks and rangeland with cattle, horses and bison.

Nesting: female lays up to 40 eggs in other birds' nests (usually 1 egg per nest) in a single breeding season; brown-marked, greyish eggs hatch in 10–13 days.

Feeding: gleans the ground for seeds and invertebrates, the latter often flushed by livestock.

Voice: squeaky, high-pitched *seep*, *psee* and *wee-tse-tse* calls, often given in flight; also gives a fast chipping note series; male's song is a high, liquid, gurgling *glug-ahl-whee* or *bubbloozeee.*

Similar Species: *Rusty Blackbird* (p. 477) and *Brewer's Blackbird* (p. 478): larger; lack contrasting brown head and darker body; slimmer, longer bills; longer tails; all have yellow eyes except female Brewer's Blackbird. *European Starling* (p. 389): immature has longer bill and shorter tail.

ORCHARD ORIOLE

Icterus spurius

The Orchard Oriole is the smallest oriole in North America and the male is the only one with brick red plumage. Orchards may have once been a favoured haunt of this bird, but since modern orchards are heavily sprayed and manicured, they tend to be avoided. Orchard Orioles also avoid heavily forested regions, preferring scrubby woods, hedgerows with isolated tall trees, gardens and roadside shade trees, especially in areas with nearby water. • This oriole is one of the last birds to return from its neotropical wintering grounds in the spring, showing up no earlier than mid to late May in southern Canada. The wooded edges of Ontario's Carolinian forest zone are by far the most productive areas to find oriole males engaged in their romantic serenades. • Orchard Orioles have one of the shortest stays of any songbird in Canada, nesting promptly and departing on their long migration by mid-August. Although they breed in the southern fringes of Ontario, Manitoba and Saskatchewan, they have expanded their nesting range north and west substantially during the last couple decades. As the Orchard Oriole moves into new habitats and increases densities in old habitats, its abundance is increasing at a rate of 5 percent annually.

ID: medium-sized songbird. *Male:* black "hood" and tail; chestnut underparts, shoulder and rump; dark wing has 1 white wing bar and white feather edgings. *Female* and *immature:* dull olive upperparts; yellow to olive yellow underparts; 2 faint, white bars on dusky grey wings. *In flight:* reddish chestnut (male) or mostly white (female) wing linings.
Size: *L* 15–18 cm; *W* 24 cm.
Habitat: open woodlands, suburban parklands, forest edges, hedgerows and groves of shade trees.
Nesting: in the fork of a deciduous tree or shrub; hanging, pouch nest is woven from grass and other fine plant fibres; female incubates 4–5 variably marked, pale bluish white eggs for about 12–15 days.
Feeding: gleans insects and berries from trees and shrubs; probes flowers for nectar.
Voice: quick *chuck* call; male's song is a rich, lively warble ending with a distinctive *pli titi*.
Similar Species: *Bullock's Oriole* (p. 482) and *Baltimore Oriole* (p. 483): males have bright orange plumage and orange-and-black tails; female Baltimore has orange overtones; female Bullock's has greyish belly. *Summer Tanager* (p. 463) and *Scarlet Tanager* (p. 464): females have thicker, pale bills and lack wing bars.

BULLOCK'S ORIOLE

Icterus bullockii

Although bright and flashy, the plumage of the male Bullock's Oriole blends remarkably well with the sunlit and shadowed upper-canopy summer hardwood foliage in which it spends much of its time. When clearly seen in bright sunlight, however, the male never fails to elicit a "Wow!" from observers, particularly new birders. Drab plumage camouflages the female, and an elaborate hanging nest provides a safe haven for the offspring. • The Bullock's Oriole is the common oriole of western North America. It was once thought to interbreed freely with the Baltimore Oriole of eastern North America, and in 1973, the two birds were lumped together by the American Ornithologist's Union as a single species, the "Northern Oriole." Just as the birding community was adjusting to the change, new studies concluded that interbreeding was sufficiently restricted, and the American Ornithologists' Union decided in 1995 that these birds should once again be considered two distinct species. • In Canada, the Bullock's nesting range includes southern extremes of Saskatchewan, Alberta and British Columbia, but it has been recorded as a vagrant eastward to the Maritimes.

♂

♀

ID: medium-sized songbird; long tail; pointed bill. *Male:* black crown, upperparts and throat patch; bright orange face, underparts and rump; black eye line; large, white wing patch; black-tipped, orange tail. *Female:* olive grey upperparts; dusky yellow upper breast, throat and face; grey back and wings; 2 white wing bars; greyish underparts; yellowish orange rump and tail.

Size: *L* 18–23 cm; *W* 28–30 cm.

Habitat: *Breeding:* edges of deciduous woodlands near open areas; riparian cottonwoods, poplars and willows; also tall windbreaks in farmland, suburban yards and along roadsides; sometimes in orchards.

Nesting: attached to a high, drooping hardwood branch; woven hanging nest of plant fibres is lined with finer materials; female incubates 4–5 darkly marked, greyish or bluish white eggs for 12–14 days.

Feeding: gleans upper-canopy foliage and shrubs for caterpillars and other invertebrates; also eats small fruits and plant nectar; sometimes visits feeders.

Voice: 2-note whistles; dry chatters; male's song is a rich, sharply punctuated series of 6–8 whistled and guttural notes.

Similar Species: *Baltimore Oriole* (p. 483): male lacks orange cheek and large, white wing patch; female's orange-yellow is brightest on breast rather than face. *Black-headed Grosbeak* (p. 468): heavy, conical bill; black cheek; darker underparts. *Western Tanager* (p. 465): yellow body plumage; lacks black cap and throat.

BALTIMORE ORIOLE

Icterus galbula

Although Baltimore Orioles are fairly common in central and southeastern Canada, they are often difficult to find because they inhabit the forest heights. The striking males are visible when they sing from perches, but the females typically remain near the nest, which is hidden in the upper canopy of a large shade tree. • Baltimore Orioles prefer open stands of deciduous trees over extensive forest. In urban areas, such requirements are met by shade trees; in the grasslands, by riparian woodlands and wooded coulees; and in the parklands, by aspen groves. They are some of the last birds to return each spring, arriving in May. Like many songbirds that nest in Canada, the Baltimore Oriole is really only a visitor; it spends at least half of each year in the tropical forests of Central and South America. • Baltimore Orioles and Bullock's Orioles hybridize where their ranges overlap. • The Baltimore Oriole was named for the black-and-orange coat of arms of George Calvert, who established the colony of Baltimore, Maryland. • Baltimores Orioles have undergone slight but significant declines throughout their range since the late 1960s, with an annual decline in Canada of 3.4 percent since 1980.

ID: medium-sized songbird; long tail; pointed bill. *Male:* black "hood," back and wings; bright orange shoulder bar and underparts; black tail with orange outer feathers. *Female:* olive brown upperparts (darkest on head); 2 white wing bars; orange-yellow breast; greyish belly; brownish tail.
Size: *L* 18–20 cm; *W* 28 cm.
Habitat: hardwood and mixed forests, particularly riparian woodlands, natural openings, shorelines, roadsides, orchards, gardens and parklands.
Nesting: suspended in a deciduous tree near the end of a branch; hanging pouch nest of grasses, bark shreds, rootlets and plant stems is lined with fine grasses, rootlets and hair; female incubates 4–6 greyish white eggs for 11–14 days.
Feeding: gleans canopy vegetation and shrubs for caterpillars, beetles, wasps and other invertebrates; eats some fruit and nectar; may visit hummingbird feeders.

Voice: calls include 2-note *tea-too*, rapid chatter and a low, whistled *hu-lee* note; male's 2- or 3-note phrases are strung together into a robust song: *peter peter here here peter.*
Similar Species: *Bullock's Oriole* (p. 482): male has large, white wing patch, black throat patch and orange face with black eye line; female has greyer back and white belly. *Orchard Oriole* (p. 481): smaller; male has darker chestnut plumage; female and immature have olive yellow underparts and lack orange overtones. *Scarlet Tanager* (p. 464): female has thicker, paler bill and lacks wing bars and orange underparts.

GRAY-CROWNED ROSY-FINCH

Leucosticte tephrocotis

The remarkable Gray-crowned Rosy-Finch spends summer high in the Rockies, where the air is thin and the wind is frigid. Here, these alpine breeders nest near the summits and higher slopes of mountains with permanent snowfields and glaciers. At the end of summer, family groups assemble into larger flocks of sometimes hundreds of birds, which remain in the alpine zone until driven into the lowlands by the first major winter storms. Most Gray-crowned Rosy-Finches overwinter in the American Northwest, but a few winter from British Columbia eastward to southwestern Saskatchewan, with occasional vagrants showing up at feeders eastward to the Maritimes. • There are two fairly distinct races of Gray-crowned Rosy-Finch that appear in Canada: the grey-cheeked "Coastal" or "Hepburn's" Rosy-Finch (*L. t. littoralis*) and the brown-cheeked "Interior" race (*L. t. tephrocotis*), for which the species is named. • The Gray-crowned Rosy-Finch is the most widely distributed of the three rosy-finch species. • During summer, rosy-finches develop rodent-style cheek pouches that allow them to carry larger numbers of insects to their rapidly developing young.

♂ *"Interior"*

♂ *"Coastal" or "Hepburn's"*

ID: medium-sized finch. *Male:* grey cap or "hood"; dark forehead and throat; conical, black bill; plain, dark brown upperparts; brown back, wings and underparts; rosy wing coverts, flanks and rump; dark tail. *Female* and *nonbreeding male:* generally paler with less pink in plumage; conical, yellow bill. *Juvenile:* uniformly grey-brown body; lacks grey crown, black forehead and pink on underparts.

Size: *L* 15 cm; *W* 33 cm.

Habitat: *Breeding:* rock talus slopes and piles below snowfields, often near seepage areas with some vegetation. *In migration* and *winter:* grasslands, roadsides and agricultural land.

Nesting: among rocks and boulders; bulky nest consists of mosses, grasses and lichens and is lined with feathers; female incubates 4–5 white eggs for 12–14 days.

Feeding: walks and hops on the ground or snow gleaning seeds and insects.

Voice: flight call is a soft *jeewf* or a buzzy *jeerf*; male's song is a slow, descending series of whistles: *jeew jeew jeww.*

Similar Species: none.

PINE GROSBEAK

Pinicola enucleator

Pine Grosbeaks are colourful nomads of boreal and mountain forests. Here, they feed and breed at leisure, occasionally introducing themselves with a distinctive loud, whistling call uttered from the highest point they can find. Early summer hikers might be surprised by what seems to be the song of a robin high atop a mountain, only to find that the songster is a particularly beautiful finch. Sometimes, you can walk right past one as it sits still on a perch in the forest. • Pine Grosbeaks do not regularly migrate from their nesting range, and their winter visits to southern Canada are erratic—it is a great moment when Pine Grosbeaks emerge from the wilds to settle on your backyard feeder. They might hang around for the entire winter or, in typically irruptive fashion, disappear for months. These invasions are not completely understood, but it is thought that cone crop failures or changes to forest ecology caused by logging, forest fires or climatic factors may force these hungry birds southward in search of food. • Pine grosbeaks appear to be in significant decline in the southern areas of their breeding range.

♂

♀

ID: large, plump-looking finch; long, black, notched tail; dark wings with 2 white bars; dark, conical bill. *Male:* mostly pinkish red overall; pinkish grey underparts; streaked back; white undertail coverts. *Female:* generally grey; yellowish olive or russet crown, face and rump.

Size: *L* 20–23 cm; *W* 35–38 cm.

Habitat: *Breeding:* open subalpine and northern boreal coniferous forests. *Winter:* forested and shrubby habitats, including urban and residential areas, especially with berry-laden mountain-ash trees; also visits feeders.

Nesting: in a conifer or tall shrub; loose, bulky nest is made of mosses, twigs, grasses and lichens; female incubates 4 heavily marked, bluish green eggs for 13–14 days.

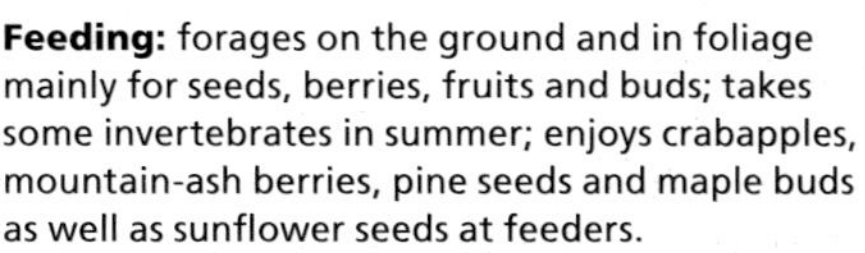

Feeding: forages on the ground and in foliage mainly for seeds, berries, fruits and buds; takes some invertebrates in summer; enjoys crabapples, mountain-ash berries, pine seeds and maple buds as well as sunflower seeds at feeders.

Voice: call is a 3-note *tew* whistle with a higher middle note; short, muffled trill in flight; male's song is a short, sweet, musical warble: *fillip illy dilly didalidoo*.

Similar Species: *White-winged Crossbill* (p. 490): much smaller; lacks stubby bill and prominent grey coloration. *Red Crossbill* (p. 489): lacks stubby bill and white wing bars. *Evening Grosbeak* (p. 495): female has pale bill, dark "whisker" stripe, tan underparts and broad, white wing patches.

PURPLE FINCH

Carpodacus purpureus

The courtship of Purple Finches is a gentle and appealing ritual. The liquid, warbling song of the male bubbles through conifer boughs announcing his presence to potential mates. Upon the arrival of an interested female, the colourful male dances lightly around her, beating his wings until he softly lifts into the air. • The Purple Finch prefers the cool, northern coniferous and mixed forests of the western mountains and the Canadian Shield. Flat, raised, table-style feeding stations with nearby tree cover are sure to attract Purple Finches, and erecting one may keep a small flock in your area over winter. • Purple is a poor description of this bird's reddish coloration. Roger Tory Peterson said it best when he described the Purple Finch as "a sparrow dipped in raspberry juice." • Although males do not attain breeding plumage until their second year, first-year birds still practise their songs. Females occasionally also sing softly during nest building. • House Finches outcompete native Purple Finches in many areas of Canada. At feeding stations, Purples usually eat on the ground at the periphery of the feeders because of the aggressive behaviour of House Finches. • Breeding bird surveys since the 1960s have revealed significant declines in Purple Finch numbers both in the U.S. and in Canada.

ID: medium-sized finch; thick, pointed bill; black eyes. *Male:* raspberry red (rarely yellow to salmon pink) head, throat, breast and nape; brownish red cheeks; streaked, brown-and-red back; reddish-tinged flanks; pale, unstreaked belly and undertail coverts. *Female:* dark brown cheek and jaw line; white eyebrows and lower cheek stripe; heavily streaked underparts; unstreaked undertail coverts. *In flight:* reddish (male) or olive (female) rump; short, notched tail.

Size: *L* 14 cm; *W* 25 cm.

Habitat: *Breeding:* moist coniferous and mixed forests. *In migration* and *winter:* mixed woods, edges of coniferous forests and trembling aspen forests; also suburban feeders.

Nesting: on a conifer branch, away from the trunk; cup nest of twigs, grasses and rootlets is lined with mosses and hair; female incubates 4–5 darkly marked, pale greenish blue eggs for 12–13 days.

Feeding: gleans the ground and vegetation for seeds (often from mountain-ash trees), buds, berries and some insects; readily visits table-style feeders.

Voice: single, metallic *cheep* or *weet* call; male's song is a bubbly, continuous warble ending with descending trill *cheerrr.*

Similar Species: *House Finch* (p. 488): female has indistinct facial pattern; male has brighter red on crown, throat, breast and rump and streaked underparts. *Cassin's Finch* (p. 487): BC only; streaked undertail coverts; male has brighter red forecrown; female's streaks are darker and clearer. *Red Crossbill* (p. 489): bill has crossed tips; male is richer red overall and has darker wings.

CASSIN'S FINCH

Carpodacus cassinii

The Cassin's Finch occupies conifers throughout North America's western interior mountains. In Canada, it is a regular breeder in the Rockies of southeastern British Columbia and southwestern Alberta. For a few short days in spring, the Cassin's Finch is the Waterton Lakes National Park townsite's most common bird. Some days it seems that scarcely a tree, not to mention a feeder, is without one of these birds. Once the Cassin's Finch moves to the high country to breed, it is more difficult to find, but its bubbling courtship song serves as a pleasant reminder of its time in the lowlands. • The Cassin's Finch is not a terribly energetic bird—individuals might perch in place for minutes at a time. British Columbia is the only province in Canada to have Cassin's, Purple and House finches. Cassin's Finches claim the dry, open coniferous forests in the south-central interior, Purple Finches the moist coniferous forests on the south coast and trembling aspen forests in the Peace River area and House Finches the open-country and suburban areas of southern BC. • As with Purple Finches, immature male Cassin's Finches require more than a year to change from their drab, streaky plumage to their reddish adult plumage.

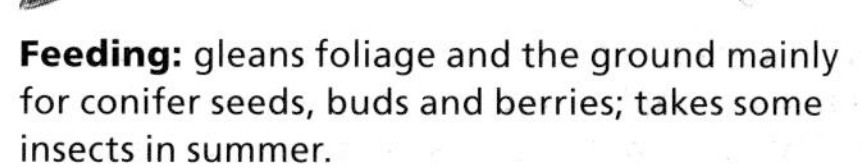

ID: medium-sized finch; finely streaked undertail coverts and rear flanks; notched tail; pale "spectacles"; heavily dark-streaked, greyish brown upperparts; pale, conical bill. *Male:* pinkish wash on wings and back; whitish sides, flanks and belly; pinkish red rump, breast and head; bright red crown. *Female:* short, well-defined dark streaks on white underparts.
Size: *L* 16 cm; *W* 30 cm.
Habitat: *Breeding:* open coniferous forests, especially ponderosa pine and Douglas-fir. *In migration* and *winter:* similar habitat but moves to lower foothills and valleys.
Nesting: near the end of a large conifer limb, rarely in a shrub; nest is made of grasses, mosses, bark shreds, rootlets and lichens; female incubates 3–5 darkly spotted, bluish green eggs for 12–14 days.

Feeding: gleans foliage and the ground mainly for conifer seeds, buds and berries; takes some insects in summer.
Voice: flight call is a warbling *kiddileep;* male's rich, warbling song is higher and livelier than Purple Finch.
Similar Species: *Purple Finch* (p. 486): male has no pale eye ring and pink is more widespread; female has prominent white eyebrow, less clear, browner streaks and unstreaked undertail coverts. *House Finch* (p. 488): squared tail; male has less contrasting orange-red on crown and brown, striped flanks; female has plainer face and shows more brown on underparts.

HOUSE FINCH

Carpodacus mexicanus

Formerly restricted to the arid American Southwest and Mexico, the House Finch has, with the help of humans and introductions in California, spread across all the lower 48 states and southern Canada. The House Finch was originally brought to eastern parts of the continent as an illegally captured cage bird known as the "Hollywood Finch." In the early 1940s, New York pet shop owners released their birds to avoid prosecution and fines, and it is the descendants of those birds that are now found in the West. • The male House Finch's colour depends on how well it processes the carotenoids in its diet; plumage can range from the usual reddish to a yellowish orange. • The male's cheerful song can be heard year-round, and no berry bush, grassy patch or weedy shrub is beyond the interest of family parties in late summer and fall. • The resourceful House Finch usually gains the upper hand on the more even-tempered Purple Finch, and it is the only bird aggressive and stubborn enough to successfully outcompete the House Sparrow. Like the House Sparrow, this finch has prospered in urban environments. Both birds often build their messy nests among eaves, rafters, chimneys and other human-fashioned habitats, and both thrive on seeds.

♂ ♀

ID: medium-sized finch; streaked, brown upperparts; white flanks and undertail coverts heavily streaked with greyish brown; 2 pale wing bars; grey, conical bill. *Male:* reddish orange (sometimes yellow) eyebrow, throat, breast, cheeks and rump; brown crown; weakly streaked back. *Female:* plain head; drab brown overall. *In flight:* red (male) or greyish brown (female) rump; square tail.
Size: *L* 12–15 cm; *W* 22–25 cm.
Habitat: human-altered areas, including agricultural lands, orchards, suburbs, weedy roadsides and fields, parks and airports; visits feeders in winter.

Nesting: in a cavity, building, dense foliage or abandoned nest; nest is made of twigs, grasses, leaves, rootlets and hair; female incubates 4–5 lavender- and black-dotted, pale blue eggs for 12–14 days.
Feeding: mainly eats seeds, buds and berries; sometimes visits hummingbird feeders with sugar water.
Voice: sweet *cheer* flight call, given singly or in series; male's song is a bright, disjointed warble lasting about 3 seconds, often ending with a harsh *jeeer* or *wheer.*
Similar Species: *Purple Finch* (p. 486): male has more widespread, pinker coloration; female has prominent white eyebrow, less clear, browner streaks and unstreaked undertail coverts. *Cassin's Finch* (p. 487): BC only; pale "spectacles"; male has bright red crown, and streaks on underparts are limited to flanks and undertail coverts; female's streaks are darker and clearer.

RED CROSSBILL

Loxia curvirostra

With crossed bill tips and nimble tongues, crossbills are uniquely adapted for extracting seeds from conifer cones. Owing to the patchy distribution of appropriate cone crops, these birds are nomadic, becoming numerous or even abundant in one area, while remaining scarce or absent in another. These nomads scour mountain forests for pine cones, and if they discover a bumper crop, they might breed regardless of the season—it is not unusual to hear them singing in midwinter or to find newly fledged juveniles begging for food in November or February. • Red Crossbills can often be located by listening for their loud calls wherever the boughs of conifers are laden with ripe cones. • Ongoing research, based on different vocalizations, bill sizes and molecular analysis, suggests that the Red Crossbill may actually be composed of several different species. • Although irruptive species are notoriously difficult to assess trends for, breeding bird surveys since 1980 have revealed declines of over 3 percent per year in Red Crossbill populations in Canada.

ID: curved bill is crossed at tip; dark lores; plain, dark brown wings. *Male:* dull orangey red to brick red plumage. *Female:* olive grey to dusky yellow plumage. *Juvenile:* streaked, brown body; double wing bars. *In flight:* short, notched, dark tail.
Size: *L* 12–15 cm; *W* 25–28 cm.
Habitat: mature coniferous forests; favours red and white pines; also uses other pine and spruce-fir forests and plantations.
Nesting: loosely colonial; saddled on a high conifer branch; loose, bulky nest of twigs, grasses, mosses, rootlets and bark strips is lined with finer materials; female incubates 2–4 darkly spotted, pale blue or green eggs for about 14 days.

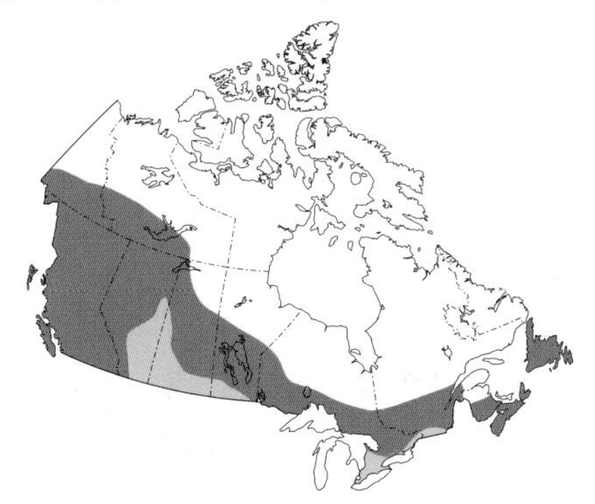

Feeding: eats mainly conifer seeds; also tree buds, berries and some insects; gathers at mineral licks and recently salted highways.
Voice: distinctive *jip-jip* call note, often in flight; male's typical finch song is a varied series of warbles, trills and *chip* notes.
Similar Species: *White-winged Crossbill* (p. 490): 2 prominent, white wing bars; male is usually pinker; female has blurry underpart streaking. *Pine Grosbeak* (p. 485): larger; shorter, uncrossed bill; 2 prominent, white wing bars; more grey in plumage. *Purple Finch* (p. 486) and *House Finch* (p. 488): conical bills; less red overall; lighter brownish wings; lack red on lower belly.

WHITE-WINGED CROSSBILL

Loxia leucoptera

People are often amazed by the colourful and bizarre shapes of bird bills. Although this bird's bill lacks the colourful flair of the tropical toucan's and the massive proportions of the hornbill's, its cross-mandible design is shared by only two bird species in North America. • White-winged Crossbills primarily eat conifer seeds, favouring spruce and tamarack, and their oddly shaped bills are adapted to prying open cones. They then use their nimble tongues to extract the soft, energy-rich seeds nestled within. • Crossbills are the nomads of the bird world, wandering far and wide in search of ripe conifer cones. There is no telling when or where they will find the next bumper crop, and they breed regardless of the season. The presence of a foraging group of these birds high in a spruce tree creates an unforgettable shower of conifer cones and crackling chatter. Like many finches, White-winged Crossbills can be abundant one year, then absent the next. • When not foraging in spruce spires, White-winged Crossbills often drop to ground level, where they drink water from shallow forest pools or lick salt from winter roads, which often results in crossbill fatalities. • Crossbills can be either right-billed or left-billed.

ID: curved bill is crossed at tip; dark lores; dark wings; 2 bold, white wing bars. *Male:* pinkish red overall. *Female:* greenish yellow body; streaked, brown upperparts; dusky yellow underparts; slightly brown-streaked flanks. *Immature:* heavily brown-streaked overall. *In flight:* short, notched, all-dark tail.

Size: *L* 38–43 cm; *W* 68 cm.

Habitat: coniferous forests, primarily spruce, Douglas-fir, tamarack and less frequently fir; occasionally visits townsites, urban parks and mixedwood forests.

Nesting: saddled on a conifer branch; open cup nest of twigs, grasses, bark shreds and forbs is lined with mosses, lichens, rootlets, hair and soft plant down; female incubates 2–4 variably marked, whitish to pale blue-green eggs for 12–14 days.

Feeding: prefers conifer seeds, especially tamarack and spruce; also eats deciduous tree seeds and occasionally insects; often licks salt and minerals from roads when available.

Voice: call is a series of harsh, questioning *cheat* notes, often given in flight; male's song is a high-pitched series of warbles, trills and chips.

Similar Species: *Red Crossbill* (p. 489): lacks white wing bars; male is deeper red (less pinkish). *Pine Grosbeak* (p. 485): much larger; long tail; stubby, conical bill; thinner wing bars; male has grey sides; female is very grey.

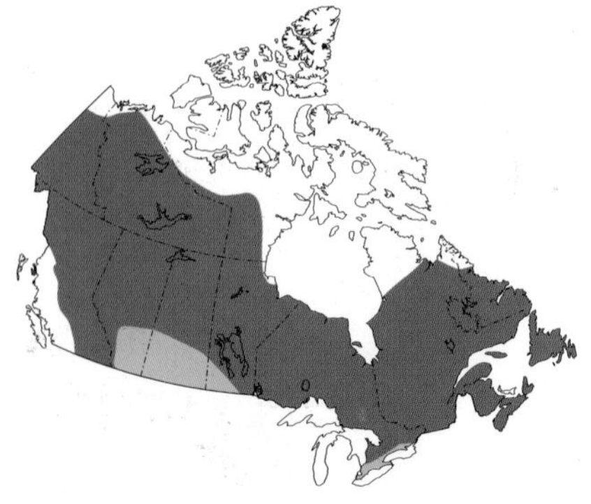

COMMON REDPOLL

Acanthis flammea

A predictably unpredictable winter visitor, the Common Redpoll is seen in varying numbers, depending on the year. It might appear in flocks of hundreds or in groups of a dozen or fewer. • Renowned for their effective winter adaptations, redpolls can endure colder temperatures than any other songbird. Highly insulative feathers enable these birds to withstand bitter cold, especially when the feathers are fluffed out to trap layers of warm air. The Common Redpoll's light, fluffy body allows it to "float" on the softest snowbanks without sinking beyond its belly. Because they do not have much body fat, to avoid dying from hypothermia, they maintain a high metabolic rate and eat almost constantly, gleaning seeds from willows, birches and alders. They also take advantage of winter feeders. Their focus on food helps make wintering redpolls remarkably fearless of humans. • The Common Redpoll breeds at the northern edge of the boreal forest and in the Arctic taiga across Canada and throughout the Northern Hemisphere. Large movements south of its breeding range occur in years when there is low seed production among spruce, birch, alder and other northern-latitude trees and shrubs.

ID: small finch; black "chin"; red forecrown; tiny, yellowish bill; streaked upperparts, including rump; lightly streaked sides, flanks and undertail coverts; notched tail; dark wings; 2 white wing bars. *Male:* pinkish red breast (brightest in breeding plumage); less streaking. *Female:* lacks pinkish breast; heavier body streaking.

Size: *L* 13 cm; *W* 23 cm.

Habitat: *Breeding:* near wetlands with birch and alder thickets. *In migration* and *winter:* semi-open areas, including shrubby edges of woodlands, brushy hedgerows, roadsides, fields and transmission corridors; also residential yards.

Nesting: low in a shrub or dwarf spruce; open cup nest of fine twigs, grasses, plant stems, lichens and mosses is lined with feathers, hair or plant down; female incubates 4–5 darkly speckled, pale blue eggs for 10–13 days.

Feeding: gleans the ground, snow surface and deciduous trees and shrubs for seeds; often visits feeders; takes some insects in summer.

Voice: calls are a soft *chit-chit-chit-chit* and a faint *swe-eet*; song is a twittering series of trills; calls and songs indistinguishable from Hoary Redpoll.

Similar Species: *Hoary Redpoll* (p. 492): generally paler and more plump overall; unstreaked or partly streaked rump; usually has faint or no streaking on sides and flanks; bill may look stubbier; lacks streaking on undertail coverts. *Pine Siskin* (p. 493): slender, pointed bill; heavily streaked overall; yellow highlights in wings and tail; lacks black "chin."

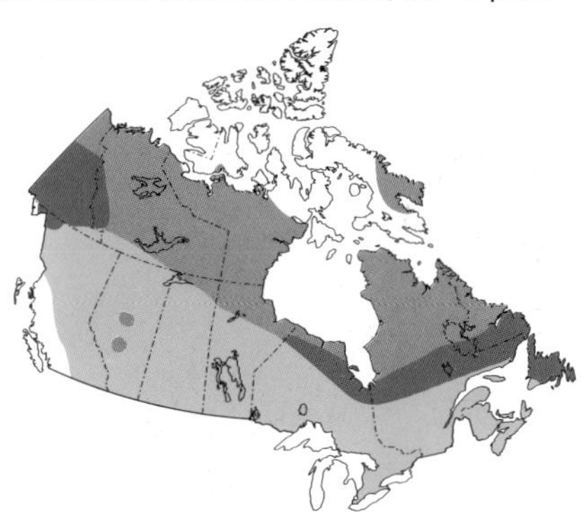

HOARY REDPOLL

Acanthis hornemanni

Common Redpolls and Hoary Redpolls can be confusing to differentiate because they look alike, share similar ranges and have similar calls and songs. The Hoary Redpoll is almost always seen with flocks of Common Redpolls and is generally paler and less streaked, with a stubbier bill. Some Hoary Redpolls are nearly impossible to differentiate from Commons, however, because of individual variation within both species and some overlap in identifying features. Both redpoll species are remote northern nesters and "predictably unpredictable" winter visitors to southern Canada. Like many finches, they can be abundant one year and almost absent the next. In the south, they tend to be more abundant every second winter, reflecting cyclic lows and highs in seed availability in their normal (northern) range. Things were much simpler when the two redpolls were considered a single species. • Redpolls are well adapted to life in the cold. They have a high level of food intake, in part because of a special storage pouch in the esophagus (the esophageal diverticulum), which allows them to carry large quantities of energy-rich seeds.

ID: small finch; frosty white overall; red forecrown; black "chin"; stubby, yellowish bill; lightly streaked upperparts, except for unstreaked rump; unstreaked underparts (flanks may have faint streaking); notched tail. *Male:* pinkish-tinged breast. *Female:* white to light grey breast; heavier body streaking.

Size: *L* 13–14 cm; *W* 23 cm.

Habitat: *Breeding:* Arctic taiga and coastal tundra. *Winter:* open fields, meadows, roadsides, utility cutlines, railways, forest edges and backyards with feeders.

Nesting: on or near the ground; open cup nest of grasses is lined with plant down, feathers and hair; female incubates 4–5 darkly speckled, pale blue eggs for about 14 days.

Feeding: gleans the ground, snow and vegetation for seeds and buds; occasionally visits feeders in winter; also takes some insects in summer.

Voice: calls are a soft *chit-chit-chit-chit* and a faint *swe-eet*; male's song is a twittering series of trills; calls and songs indistinguishable from Common Redpoll.

Similar Species: *Common Redpoll* (p. 491): generally darker and slimmer overall; slightly larger bill; streaked rump, sides, flanks and undertail coverts. *Pine Siskin* (p. 493): heavily streaked overall; yellow highlights in wings and tail; slender, pointed bill; lacks black "chin."

PINE SISKIN

Spinus pinus

You can spend many hours chasing flitty Pine Siskins trying to get a decent look, or you can just set up a finch feeder filled with niger seed and wait for them to appear. If the feeder is in the right location, you can expect your backyard to be visited by Pine Siskins at just about any time of year but particularly in winter. If natural food sources are abundant, however, siskins may shun seed feeders altogether. Siskins consume seeds in great quantities, and even a small lingering flock can necessitate daily feeder restocking. Keep the seed dry and disinfect the feeder regularly to prevent salmonellosis outbreaks among the feeder users. • Tight flocks of these gregarious birds are frequently heard before they're seen. Once you recognize their characteristic rising *zzzreeeee* calls and boisterous chatter, you can confirm the presence of these finches by simply looking in the treetops for a flurry of activity that shows occasional flashes of yellow. • Siskins are typically associated with pines and other conifers. Like most finches, they are highly nomadic and rarely occupy the same nesting territories in successive years. • Breeding bird surveys show declines of greater than 5 percent per year for the last 30 years in Canada.

ID: sexes similar (males may be yellower); small, heavily streaked, yellowish finch; darkly streaked, brown upperparts; 2 yellow wing stripes; yellow patches on sides at base of tail; small, sharp, dusky bill; whitish belly; short, notched tail.
Size: *L* 10–13 cm; *W* 20–23 cm.
Habitat: *Breeding:* mostly coniferous forests up to subalpine; also ornamental evergreens in parks and cemeteries. *In migration* and *winter:* lower-elevation conifers and mixed woods, weedy roadsides and fields (especially with dandelions); also residential areas with feeders.
Nesting: loosely colonial; saddled on an outer conifer branch; nest of twigs, rootlets and grasses is lined with finer rootlets, mosses, hair and feathers; female incubates 1–6 darkly spotted, pale greenish blue eggs for 14–15 days.

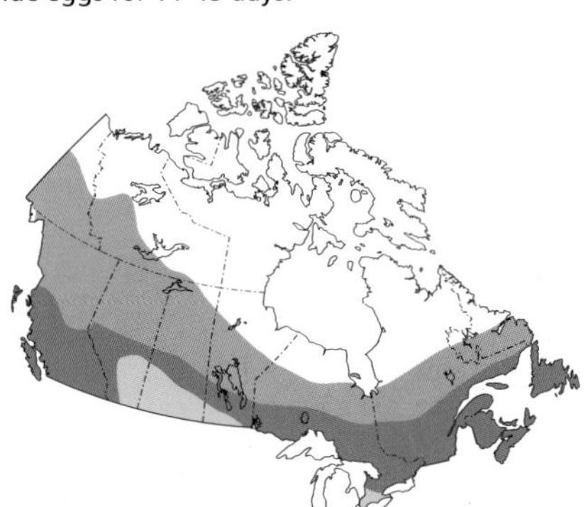

Feeding: gleans the ground, shrubby vegetation and tree foliage for seeds, buds and insects; attracted to road salt; regularly visits feeders for niger and sunflower seeds.
Voice: distinctive, buzzy, ascending *zzzreeeee* call; male's song is a variable, bubbly mix of squeaky, metallic and raspy notes, sometimes resembling a jerky laugh.
Similar Species: *Common Redpoll* (p. 491) and *Hoary Redpoll* (p. 492): red forecrown; lack yellow in wings and tail. *Purple Finch* (p. 486) and *House Finch* (p. 488): females have thicker bills and no yellow in wings or tails. *Sparrows* (pp. 435–56): all lack yellow in wings and tails.

AMERICAN GOLDFINCH

Spinus tristis

American Goldfinches, often referred to by non-birders as "wild canaries," are bright, cheery songbirds commonly seen in weedy fields, roadsides and backyards during summer. They often feed on thistle seeds and flutter over fields in a distinctive, undulating flight style, filling the air with their jubilant *po-ta-to-chip* or *per-chick-or-ree* calls. • Although not typically a regular wintering species in southern Canada, numbers of wintering goldfinches have been increasing because of the widespread use of feeders. • The American Goldfinch delays nesting until June, to ensure a dependable source of thistles and dandelion seeds to feed to its young. It is enjoyable to observe a flock of goldfinches raining down to ground level to poke and prod the heads of dandelions. These birds do their best to play up the comedy, as they attempt to step down on the flower stems to reach the crowning seeds. • Most finches have one moult per year, in the fall, but the American Goldfinch moults all of its feathers again in the spring. Goldfinches of both sexes are frequently misidentified after they moult into their muted winter plumages.

♀ ♂ *nonbreeding*

♀ ♂ *breeding*

ID: small, stocky finch; small, conical, chestnut pink (breeding) or dusky (nonbreeding) bill; whitish undertail coverts. *Breeding male:* bright yellow overall; black wings, tail and crown (extends onto forehead); white rump; mostly black wings; narrow, white wing bar; white-edged, brown tail feathers. *Breeding female:* yellowish green head and upperparts; yellow throat and breast; dark wings; 1 broad, yellowish wing bar and 1 narrower white one. *Nonbreeding:* olive brown back and head; greyish brown underparts; brownish bill; yellow-tinged face and throat. *In flight:* white "wing pits"; short, notched tail.
Size: *L* 14 cm; *W* 24 cm.
Habitat: *Breeding:* shrublands, bushy wetlands, open woodlands, agricultural lands, grasslands and orchards. *Winter:* open country with weedy fields and other low but open vegetation; also suburban feeders.
Nesting: in the fork of a shrub or small tree; tightly woven nest of mainly plant fibres, coarse grasses and plant down is lined with finer grasses and hair; female incubates 4–6 pale blue eggs for 12–14 days.
Feeding: gleans vegetation for thistle and dandelion seeds, some insects, flower buds and berries; visits feeders for niger seed.
Voice: calls include *po-ta-to-chip* or *per-chick-or-ee* (often in flight) and a whistled *dear-me, see-me*; male's song is a long, varied series of trills, twitters and warbles.
Similar Species: *Evening Grosbeak* (p. 495): much larger; massive bill; lacks black forehead. *Wilson's Warbler* (p. 430): olive upperparts; olive wings without wing bars; thin, dark bill; black cap does not extend onto forehead.

EVENING GROSBEAK

Coccothraustes vespertinus

Anyone with a winter bird feeder knows the appetite of Evening Grosbeaks—a small flock can devour a tray of sunflower seeds in no time. Scattering husks and unopened seeds, they often create an unplanned sunflower garden that is discovered the following spring. If they don't like what's offered, these picky eaters are quick to look elsewhere. • Dispersing in pairs to nest in coniferous forests in summer, Evening Grosbeaks appear scarce until after their young fledge. By late summer, flocks are assembling, ready to devour tender hardwood buds before launching an aerial assault on neighbourhood feeders. • It's hard not to notice the massive bill of this seed-eater. As any seasoned bird-bander will tell you, the Evening Grosbeak's bill can exert an incredible force per unit area—it might be the most powerful of any North American bird. • During outbreaks of spruce budworms, the larvae and pupae form as much as 80 percent of the Evening Grosbeak's diet. • Large irruptions of Evening Grosbeaks occur every two to three years, but in many parts of the country, winter populations are diminishing. Breeding bird surveys also show declines in Canada of over 5 percent per year since 1980.

ID: large, conical, yellowish bill (greenish in breeding plumage); conspicuous white wing patch; dark wings; short tail. *Male:* dark olive brown head and throat; yellow forehead and eyebrow; brownish back and breast; yellow shoulder, belly and undertail coverts; all-black tail. *Female:* greyish brown upper back and head; yellow-tinged shoulder and underparts; white undertail coverts; black lores and vertical "chin" line; black wings with grey area; olive nape and flanks; black tail with white tip and spots.

Size: *L* 18–20 cm; *W* 34 cm.

Habitat: *Breeding:* mostly mixed coniferous forests; also deciduous woods. *In migration* and *winter:* usually open hardwood forests, especially with buds and fruits; frequents feeders.

Nesting: saddled well out on a conifer limb; nest of twigs, sticks, rootlets, plant fibres and grasses is lined with finer plants; female incubates 3–4 darkly marked, blue or bluish green eggs for 12–14 days.

Feeding: gleans the ground and foliage for large seeds, insects and berries; takes sunflower seeds at feeders.

Voice: high, clear, whistled *teew* call; flocks utter a low, dry, rattling *buzz*; male's song consists of repeated call notes.

Similar Species: *American Goldfinch* (p. 494): much smaller; small bill; smaller wing bars; male has black cap. *Pine Grosbeak* (p. 485): female is grey overall with black bill and smaller wing bars. *Black-headed Grosbeak* (p. 468): female is browner overall with broad, buff eyebrow.

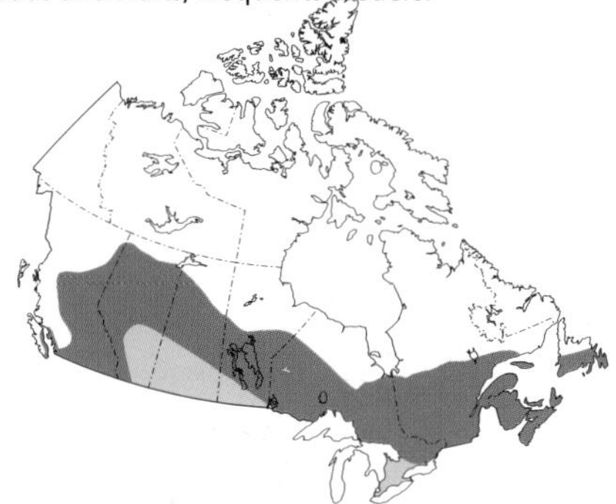

HOUSE SPARROW

Passer domesticus

Probably the most familiar songbird in the world, and one of the least prized, the House Sparrow is not a true sparrow but a weaver-finch that is native to Eurasia and northern Africa. • Introduced to Brooklyn, New York, in the 1850s, and later to California and Utah to control insect pests, the House Sparrow immediately began to exploit human-modified habitats. Nonmigratory by nature, this bird nevertheless has a knack for colonizing far-flung settled areas and competing with native species for food and nesting sites along the way. The House Sparrow will nest in any building crevice or nest box within practical flight distance of foraging sites. With its year-round breeding habits, it occupies nest boxes before most native cavity-nesting birds, such as bluebirds and swallows, arrive in spring. • This is a common bird around fast-food restaurants throughout its range. • In the early 1900s, the House Sparrow was probably the most common bird in North America. Breeding bird surveys reveal significant declines in this species' numbers in Canada over the past 40 years. • Although it was introduced to control crop pests, the House Sparrow is largely a seed eater.

♂ ♀

breeding

ID: large head; unstreaked chest; brown, streaked back; 2 indistinct, white wing bars; short tail; conical bill is yellowish, pinkish or black (breeding male). *Breeding male:* buff-streaked, chestnut brown upperparts; black throat and chest; white wing bar; chestnut nape and wing markings; pale grey underparts; grey crown; light grey cheeks. *Female:* buff-striped, brown upperparts; unstreaked, drab greyish brown underparts; buffy eyebrow; yellowish bill. *Nonbreeding male:* smaller, black "bib"; pale bill. *In flight:* rounded wings and tail.
Size: *L* 15 cm; *W* 24 cm.
Habitat: urban and rural human settlements, especially where grain is stored.

Nesting: in a nest box, building crevice, natural cavity or occasionally in a tree; nest is a domed collection of grasses, straw, feathers, string, plastics and other materials, often lined with feathers; female incubates 4–7 pale eggs, marked with grey and brown, for 10–14 days; 2–3 broods per year.
Feeding: gleans the ground, low foliage, livestock manure, grain stores and human environments for seeds, blossoms, insects, fruit and food scraps; visits feeders.
Voice: call is a hoarse *chirrup*; flocks chatter and squeak constantly; male's song is a series of similar chirps.
Similar Species: *Harris's Sparrow* (p. 454): grey face; black cap; pink-orange bill. *White-crowned Sparrow* (p. 455): immature is similar to female House Sparrow but has pink bill and stripe through crown. *Dickcissel* (p. 471): brighter, more heavily streaked plumage; yellowish or buff underparts; chestnut wing coverts; white "chin."

APPENDIX
RARE AND ACCIDENTAL SPECIES

GARGANEY
Anas querquedula

The Garganey, a Eurasian equivalent of the Blue-winged Teal, is a regular vagrant to Canada's freshwater marshes. Along the coast of BC, it is considered a casual straggler from Asia. Garganeys usually appear from late April to early June, most often with our native teals, and may remain into the breeding season. Most Garganeys identified are males. Some Garganey records are thought to be possible escapees from waterfowl collections. In nonbreeding plumage, this teal can be told apart from Blue-wings by its greyish legs as opposed to the yellow-coloured legs of our teal.

ID: fairly large, blackish bill, legs and feet. *Male:* maroon-brown head; blue-flecked cheek crossed by broad white stripe; mottled brown breast, upper back and hindquarters; black-and-white-edged, blue scapulars and tertials; grey-barred sides; white belly. *Female:* mottled and spotted, pale brown overall; striped head pattern.
Size: *L* 33–41 cm; *W* 58–64 cm.

TUFTED DUCK
Aythya fuligula

The Tufted Duck resembles other native *Aythya* ducks, but the male has a loose tuft of feathers on his head. This Eurasian species occasionally ventures into North American waters and was first seen in Alaska in 1911. Since then, there have been several sightings in British Columbia, but Atlantic Canada is now the best place to see this recent immigrant. Look for the Tufted Duck in winter and spring, in coastal bays, estuaries and freshwater habitats, especially in Nova Scotia and Newfoundland.

ID: *Male:* long, loose head tuft; dark, iridescent, purple head and neck; black back, wings, rump and breast; pure white flanks and belly; black bill tip. *Female:* short, brown head tuft; dark brown upperparts; reddish brown breast; pale grey-brown flanks; white belly.
Size: *L* 43 cm; *W* 66 cm.

GREATER PRAIRIE-CHICKEN
Tympanuchus cupido

Good numbers of Greater Prairie-Chickens inhabited the Prairie provinces and the grasslands of southwestern Ontario from the late 1800s to the 1930s, but populations declined with the coming of large-scale agriculture. The species was extirpated from Canada in 1987 but still breeds across the border in North Dakota and Minnesota. Today, populations are fragmented and local, scattered around the central and northern U.S. The "Attwater's" race of Texas is endangered, and the "Heath Hen" race of the Atlantic Coast is extinct.

ID: large, chickenlike body; black-and-tan-striped plumage; feathered legs; black terminal tail band; back of pinnae black. *Breeding male:* large, yellow air sacs on sides of neck; thick, orange eyecombs.
Size: *L* 43 cm; *W* 71 cm.

MAGNIFICENT FRIGATEBIRD
Fregata magnificens

The only Magnificent Frigatebird breeding colony in North America is found in Dry Tortugas NP, Florida. Other breeding sites are found in the West Indies and Central and South America. Vagrants are rare farther north, usually following tropical storms. The species has been recorded in Ontario, Québec and Newfoundland.

ID: long, narrow wings; long, deeply forked tail; long, hooked bill. *Male:* glossy, black plumage; red, inflatable throat patch. *Female*: blackish brown plumage; white breast.
Size: *L* 0.9–1.0 m; *W* 2.0–2.4 m.

LAYSAN ALBATROSS

Phoebastria immutabilis

The pelagic Laysan Albatross is a rare vagrant along the outer Pacific Coast between February and October. From mid-October to January, adults withdraw to breeding grounds on the Hawaiian Islands. These wide-ranging birds follow ships less often than other pelagic birds, so they are not commonly encountered. Pelagic trips off the continental shelf of BC are usually the only way one is able to see this species in Canada.

ID: sexes similar; slim body; very long, slender wings; white head and body; dark eye mark; dark back, wings and tail; pale bill.
Size: *L* 80 cm; *W* 2.0 m.

CORY'S SHEARWATER

Calonectris diomedea

The pelagic Cory's Shearwater is a regular offshore visitor on the East Coast from June to October, when birds are feeding over the continental shelf. Most birds seen in Atlantic Canada are immatures that are probably migrating up the western Atlantic from the waters off South Africa and Argentina where the species winters. It breeds in colonies in the eastern Atlantic in areas such as the Azores and the Canary Islands. Sightings are usually from pelagic boat trips in the Bay of Fundy and off the southwest coast of Nova Scotia.

ID: sexes similar; yellowish bill; grey-brown upperparts; white underparts; dark wing and tail tips; pale brown head; white neck and cheek; small, white rump patch. *In flight:* holds wings slightly bent while gliding.
Size: *L* 45–53 cm; *W* 1.0 m.

FLESH-FOOTED SHEARWATER

Puffinus carneipes

The Flesh-footed Shearwater is a rare to uncommon visitor to offshore waters of the British Columbia coast from May to mid-October, with most sightings occurring in midsummer. Unlike the Sooty Shearwater, this species follows commercial fishing vessels. Colonies breed in Australia and New Zealand and on islands in the Indian Ocean. This bird visits the northern Pacific Ocean in summer.

ID: sexes similar; dark overall; broad wings; slender, black-tipped, pale pinkish bill; pale silvery flight feathers.
Size: *L* 43 cm; *W* 1.0 m.

BROWN PELICAN

Pelecanus occidentalis

Brown Pelicans visit British Columbia during late summer and autumn, with the majority of sightings occurring most years between August and November in nearshore areas off the outer south coast. No nesting has been reported in BC yet. Vagrants occasionally visit the Maritimes, primarily in May and June.

ID: sexes similar; very large, dusky bill; greyish brown upperparts; dark brown underparts; short tail. *Breeding:* yellow head; red pouch; dark rufous brown nape and lower neck patch; white foreneck. *Nonbreeding:* yellow wash on head; pale yellowish throat pouch; white neck.
Size: *L* 1.0–1.2 m; *W* 2.0–2.1 m.

breeding

TRICOLORED HERON

Egretta tricolor

breeding

Tricolored Herons are named for their blue, white and brownish plumage. The juvenile plumage is different but it, too, has three colours. Tricolored Herons breed along the Atlantic Coast as far north as Connecticut and to southern California on the Pacific Coast. Eastern populations have expanded their range northward to the Maritimes, southern Québec and southern Ontario (Lake Erie and Toronto-area marshes). They are most partial to coastal habitats such as estuaries but may be found in many types of wetlands, including those inland.

ID: sexes similar; long, slender neck. *Breeding:* purplish to greyish blue plumage; white underparts and foreneck; pale rump; long plumes on head and back; long, slender bill. *Juvenile:* chestnut nape and wing coverts.
Size: *L* 66 cm; *W* 1.0 m.

BLACK VULTURE

Coragyps atratus

Black Vultures breed up into mid-Pennsylvania and western Connecticut, and these northern breeders may be the source of birds occasionally seen soaring above southern Ontario, Québec or the Maritime provinces from spring to fall, but most often in July and October. Look for light wing tips to distinguish it from the more common Turkey Vulture. It is becoming a more common vagrant from southern Ontario to the Maritimes. A single bird regularly visits the Kispiox Valley in northern BC.

ID: sexes similar; wholly black body; white primary feathers; short, broad tail; unfeathered, greyish head.
Size: *L* 63 cm; *W* 1.3 m.

SWALLOW-TAILED KITE

Elanoides forficatus

The Swallow-tailed Kite breeds in the southern U.S. and Mexico south through central South America, and winters in South America. During May, late August and early September, vagrants are occasionally seen farther north in the Maritimes or southern Ontario. This breathtaking species glides and swoops gracefully above the treetops, its striking black-and-white plumage accentuating its long, narrow, backswept wings and long, forked tail.

ID: sexes similar; long, slender wings and tail; white body; black back and wings; long, deeply forked, black tail. *In flight:* extremely graceful; white underwing linings; black flight feathers.
Size: *L* 58 cm; *W* 1.2 m.

MISSISSIPPI KITE

Ictinia mississippiensis

Mississippi Kites were traditionally restricted to the southern U.S., but their breeding range seems to be expanding northward, with rare spring or summer sightings now occurring in New England, the Maritimes and Point Pelee National Park, Ontario. This circum-Gulf migrant follows the western Gulf Coast while migrating between its breeding grounds in the southeastern U.S. and its wintering grounds in southern South America.

ID: sexes similar; long wings and tail; dark grey back and wings; pale head; dark grey underparts; black tail; chestnut at base of primaries (often inconspicuous).
Size: *L* 36 cm; *W* 89 cm.

PURPLE GALLINULE

Porphyrio martinica

The Purple Gallinule is closely related to the American Coot. It is largely resident (with some northern withdrawal) from North Carolina through Central America to Chile and Argentina, as well as in the West Indies. Purple Gallinules have been turning up with increasing frequency in the Maritimes, especially in southern New Brunswick and the Nova Scotia headlands. Juveniles have been found at almost any time or place, including one found in central Ontario in the dead of winter hiding under a discarded Christmas tree.

ID: sexes similar; yellow legs; very long toes; greenish upperparts; dark blue head and underparts; yellow-tipped, red bill with pale blue frontal shield; white undertail coverts.
Size: *L* 30–36 cm; *W* 56 cm.

EUROPEAN GOLDEN-PLOVER

Pluvialis apricaria

Also known as "Greater Golden-Plover" and "Eurasian Golden-Plover," this mostly European plover is irregularly seen in Newfoundland, and it may occasionally stray into other parts of Atlantic Canada. The European Golden-Plover is most likely to be seen in spring, when it moults out of its striking breeding plumage.

ID: sexes similar; short, black bill. *Breeding:* mottled, golden upperparts; black face and underparts; white band separates mottled upperparts from black underparts; white underwing coverts. *Nonbreeding:* mottled, golden upperparts and breast; creamy white eyebrow and throat; white belly and underwing coverts.
Size: *L* 28 cm; *W* 66 cm.

PACIFIC GOLDEN-PLOVER

Pluvialis fulva

Although both the Pacific Golden-Plover and the American Golden-Plover pass through BC in migration, with smaller numbers in spring, the Pacific Golden-Plover is more likely to be found in winter. The Pacific is more coastal and is rarely seen in inland locations. This species was formerly classified as a subspecies of the American Golden Plover.

ID: sexes similar; dark cap; straight, blackish bill; gold, white and brown flecking on upperparts; dark barring on rump and tail; long, bluish black legs. *Breeding:* black face and underparts; white "S" stripe extends from forehead to shoulder. *Nonbreeding:* conspicuous, dark ear spot; brown-spotted, pale buff-grey underparts; finely brown-streaked, buff neck and breast. *In flight:* fairly long, pointed wings; indistinct, pale upperwing stripe.
Size: *L* 25–28 cm; *W* 61 cm.

SNOWY PLOVER

Charadrius alexandrinus

The endangered Snowy Plover nests on sandy beaches that are increasingly invaded by humans and their dogs. The species is named for its very pale, nearly white plumage, and Snowy Plovers look like snowballs running along sandy flats and shallow dunes. This plover nested in Saskatchewan in mid-1980s and is currently an irregular vagrant there. Away from Saskatchewan, sightings of this plover are very rare.

ID: sexes similar; very pale upperparts; short, thin, black bill; white underparts; dark legs and feet. *Breeding:* black forecrown, ear patch and incomplete breast band. *Nonbreeding* and *juvenile:* pale brown replaces black on head and breast.
Size: *L* 15–17 cm; *W* 43 cm.

WILSON'S PLOVER

Charadrius wilsonia

breeding

Larger than the other "ringed" plovers and with a thick, black bill, the Wilson's Plover has a distinctly "top-heavy" look and a vertical, alert posture, quite unlike that of other plovers. In other respects, it combines the appearance of a large Semipalmated Plover and a miniature Killdeer. It is primarily associated with ocean beaches. The nearest breeding population is along the Virginia and North Carolina coast, but birds have been seen in the Maritimes, mostly from mid-May to mid-June. Occasional vagrants have been seen in southern Ontario.

ID: sexes similar; brown upperparts; long, thick, black bill; white underparts; dull pink legs and feet. *Breeding:* black forecrown and breast band. *Nonbreeding* and *juvenile:* brown replaces black on forecrown and breast band; breast band often incomplete.
Size: *L* 19 cm; *W* 48 cm.

COMMON RINGED PLOVER

Charadrius hiaticula

Common Ringed Plovers typically breed in northern Europe and have colonized the eastern Arctic islands of Canada. These plovers occasionally travel south into Atlantic Canada in autumn with parties of Semipalmated Plovers, and the two species can be next to impossible to separate. They are almost identical in plumage during fall but have different calls.

ID: *Male:* short, black-tipped, bright orange bill; black mask; large, white forehead patch and crown stripe; white throat, collar and underparts; broad, black breast band; medium brown upperparts. *Female:* narrower breast band.
Size: *L* 18–20 cm; *W* 40–48 cm.

SHARP-TAILED SANDPIPER

Calidris acuminata

This sandpiper nests in Siberia and is a rare to uncommon transient along the British Columbia coast in autumn. It is rarely found in the Interior. Watch for small flocks at staging areas such as Iona Island and the George C. Reifel Migratory Bird Sanctuary from late August to mid-October. Similar in plumage to the much more common Pectoral Sandpiper, its buffy breast helps separate the two species.

juvenile

ID: sexes similar.; brown breast streaks contrast with pale belly and undertail coverts; black bill; long, yellow legs; mottled upperparts; reddish brown crown; wing tips extend beyond sharply pointed tail. *Juvenile:* buffy breast with little streaking; rufous cap; white eyebrow.
Size: *L* 22 cm; *W* 45 cm.

CURLEW SANDPIPER

Calidris ferruginea

moulting fall adult

Hardly a migration period goes by without at least one report of a curve-billed Curlew Sandpiper somewhere in Canada. From late May to July, breeding adults have a rich dark rufous breeding plumage. Sightings have been recorded in both spring and fall migration, with August to late October sightings being most reliable. At this time, the adults are in pale grey, nonbreeding plumage.

ID: sexes similar; downcurved bill; white underparts; distinct white rump. *Breeding:* rich cinnamon overall except white lower belly and undertail. *Nonbreeding:* grey upperparts.
Size: *L* 21 cm; *W* 46 cm.

SLATY-BACKED GULL

Larus schistisagus

nonbreeding

As birders look at gulls in more detail, more rarities that were previously overlooked are now being reported. The Slaty-backed Gull is one of these species. It has been recorded across Canada, with the most sightings recorded in late fall and winter. The underside edges of this bird's wings feature a "string of pearls" pattern that can be seen in flight.

ID: dark slate grey back and wings; white underparts and tail; wide, white trailing edge to wing; black wing tips; yellow bill with red spot; pink legs. *Breeding:* white head. *Nonbreeding:* brown-streaked head.
Size: *L* 64–69 cm; *W* 1.4 m.

LEAST TERN

Sterna antillarum

breeding

True to its name, the Least Tern is the smallest tern in North America. It winters along the coasts of South America and breeds coastally from California and Maine south to Mexico, inland along major rivers and in the West Indies. During summer months, vagrants may turn up along the Great Lakes and the St. Lawrence Seaway or in the Maritimes.

ID: sexes similar; shallowly forked, white tail; pale grey upperparts; white underparts; yellow legs and feet. *Breeding:* black cap and nape; white forehead; black-tipped, yellow bill. *Nonbreeding:* white forecrown; black bill and shoulder bar. *In flight:* rapid, dashing wingbeats.
Size: *L* 23 cm; *W* 51 cm.

GREAT SKUA

Catharacta skua

The Great Skua is a fast, powerful flyer that breeds mainly on the barrens of Iceland and on islands off Scotland and north of Great Britain. Many birds arrive in the western Atlantic from August to November, unlike South Polar Skuas, which typically show up along the Atlantic Coast from June to early September.

ID: sexes similar; pale streaking on dark brown upperparts; pale brown underparts; dark, hooked bill. *In flight:* white upperwing patch.
Size: *L* 56 cm; *W* 1.4 m.

SOUTH POLAR SKUA

Stercorarius maccormicki

dark adult

The small South Polar Skua breeds in the Antarctic and visits Atlantic Canada from late May to mid-September. Most are seen from pelagic boat trips or ferries. The name "skua" is derived from an Old Norse word meaning "predatory gull."

ID: sexes similar; white wing stripe; dark, hooked bill. *Dark adult:* dark grey-brown overall; yellowish nape. *Light adult:* pale body; dark wings with pale streaking.
Size: *L* 53 cm; *W* 1.3 m.

HORNED PUFFIN

Fratercula corniculata

Horned Puffins are uncommon and local summer visitors along the outer BC coast. They breed locally from Alaska south through Haida Gwaii (Queen Charlotte Islands) to northwestern Vancouver Island. During winter, this species remains on the open ocean off Alaska and BC, but sometimes weather and food supplies bring the birds as far south as California.

ID: sexes similar; large head and bill; orange legs; black back; white front. *Breeding:* red-tipped, bright yellow bill; dark "horn" above eye; white face patch. *Nonbreeding:* darker bill with red tip; grey-smudged face.
Size: *L* 36–38 cm; *W* 58 cm.

SCISSOR-TAILED FLYCATCHER

Tyrannus forficatus

The spectacular Scissor-tailed Flycatcher breeds in the south-central U.S. and extreme northeastern Mexico, and winters in Florida and Mexico south through Central America. In recent decades, the species has expanded its breeding range as far north as Nebraska. Almost every spring or fall, a stray bird visits southern Canada. In southern Ontario, many seem to skirt Lake Huron and Lake Superior, often appearing on the Bruce Peninsula and in the Thunder Bay area.

ID: slender flycatcher; long, black tail with white outer feathers; pale grey upperparts; mostly white head and underparts. *Male:* bright pink underwing and paler undertail coverts. *Female:* pale pink underparts; shorter tail than male.
Size: *L* 25 cm (38 cm including tail); *W* 38 cm.

FORK-TAILED FLYCATCHER

Tyrannus savana

This spectacular species is the only native South American bird that often strays to North America. Fork-tailed Flycatchers show up in the United States or Canada as a result of "mirror migration," whereby a bird that intends to fly south, flies north instead. In southern Canada, it is an annual vagrant in the fall.

ID: sexes similar; dark grey upperparts; white underparts; extremely long, white-edged, black tail; black cap; nearly complete white collar.
Size: *L* 25 cm (41 cm including tail); *W* 36 cm.

FISH CROW

Corvus ossifragus

The Fish Crow has greatly expanded its range in the last few decades and is increasing in numbers in Ontario. Vagrants were first recorded in southern Ontario in the late 1970s, and the species has continued to appear in the Point Pelee area. The closest breeding populations are in Illinois, New York and Pennsylvania.

ID: sexes similar; all-black plumage; virtually identical to American Crow but smaller with longer tail; best identified by higher-pitched, hoarser voice.
Size: *L* 38 cm; *W* 90 cm.

CAVE SWALLOW

Petrochelidon fulva

Texas race

The Cave Swallow originally nested solely in caves, and some populations continue to do so, but many colonies now nest under bridges or on buildings. This bird is largely resident from southern Arizona to Texas and Mexico, and from southern Florida to the West Indies. It seems to be expanding its range and is regularly seen as far north as southern Ontario in the fall into November most years.

ID: sexes similar; orange rump; blackish back, wings, and tail. *West Indian race:* dusky nape; orange head; blackish crown and eye line join behind eye; buffy or pale orange underparts. *Mexican race:* pale head; buffy throat; pale underparts.
Size: *L* 13 cm; *W* 34 cm.

BLUETHROAT

Luscinia svecica

Sighting the unusual male Bluethroat with his beautiful throat markings makes for an exceptional birding experience. This Eurasian bird breeds on tundra and in shrubby meadows in the extreme northern Yukon and western Alaska.

ID: grey-brown upperparts; grey-white underparts; conspicuous rufous patch on outer tail feathers; whitish eyebrow. *Male:* bright blue "chin" and throat with rufous patch in centre and black, white and rufous lines below.
Size: *L* 14 cm; *W* 20–24 cm.

FIELDFARE

Turdus pilaris

When this Eurasian thrush first appeared in residential St. John's, Newfoundland, in the early 1980s, it caused quite a stir. Some birds get rerouted to North America by storms, where they join southbound flocks of American Robins and attract interest from birders and the local media. A few take up winter residence at feeders.

ID: sexes similar; bluish grey "hood" and rump; chestnut brown back and wings; pale orange-cinnamon breast with dusky chevrons; stout, straight, yellow bill with blackish tip; blackish tail. *In flight:* fairly rounded wings; pure white underwing coverts; fast, direct flight.
Size: *L* 25 cm; *W* 40 cm.

EASTERN YELLOW WAGTAIL

Motacilla flava

This energetic bird's slim body is half tail, with the other half a bright yellow chest, announcing its arrival on the Arctic breeding grounds. The Eastern Yellow Wagtail is found in western Alaska and is a rare occurrence in Canada near the Mackenzie River delta. It is a fall vagrant to British Columbia.

ID: sexes similar; persistent tail bobber; olive grey upperparts; yellowish underparts; white eye stripe, wing bars and outer tail feathers.
Size: *L* 16 cm (including long tail); *W* 22–25 cm.

HERMIT WARBLER
Dendroica occidentalis

breeding

The Hermit Warbler is a casual vagrant in extreme southwestern BC, just north of its Puget Sound, Washington, breeding area. It has been recorded in eastern Canada from Ontario to Newfoundland. Hermit Warblers hybridize with Townsend Warblers regularly along the northern edge of their range. The Hermit Warbler breeds in mature coniferous forests along mountainsides.

ID: bright yellow forehead and face; 2 bold, white wing bars; black-streaked, grey back; white underparts. *Breeding male:* black "chin," throat, crown and nape. *Female:* yellow "chin"; black throat patch; grey crown and nape.
Size: *L* 11–12 cm; *W* 19 cm.

YELLOW-THROATED WARBLER
Dendroica dominica

The strikingly plumaged Yellow-throated Warbler nests in the southeastern U.S. and overwinters in the West Indies and Central America. A few overshoot this range and appear in Atlantic Canada from early May to early June. Another invasion occurs from August to November, when birds get diverted northward by strong winds. Some may linger into January at feeders, foraging on seeds or suet. This arboreal songbird creeps up and around tree trunks and branches.

white-lored form

ID: sexes similar; plain grey upperparts, including nape and hindcrown; black forehead and forecrown; bold, white eyebrow; black cheek patch extends down side of neck; bright yellow throat and breast; dark wings; 2 white wing bars; white underparts; black flank streaking.
Size: *L* 13 cm in; *W* 20 cm.

GREEN-TAILED TOWHEE
Pipilo chlorurus

The smallest towhee, this bird typically lives in the arid American Southwest and is found from open mountain meadows down to dry, shrubby hillsides. It has been recorded as a vagrant from BC to the Maritimes. The Green-tailed Towhee tends to show up in western Canada as a spring vagrant and as a late fall and winter vagrant in the East.

ID: sexes similar; rufous orange crown; yellow-green upperparts, most intense on tail; white throat bordered by 1 dark and 1 white stripe; sooty grey face and breast; grey legs; conical, grey bill.
Size: *L* 16 cm; *W* 25 cm.

BLACK-THROATED SPARROW
Amphispiza bilineata

The Black-throated Sparrow breeds in the arid, sagebrush-dominated habitat of the western United States, from east-central Washington to Idaho and south to Mexico. Since 1959, vagrants have been recorded regularly in BC, usually between May and mid-June. Black-throated Sparrows are very rare in southwestern BC, including Vancouver Island, and casual in the southern Interior.

ID: sexes similar; black tail with white outer edges; grey cheek and cap; prominent, white eyebrow; black "chin," throat and "bib"; broad, white jaw line; unstreaked, pale underparts; dark bill.
Size: *L* 11–13 cm; *W* 20 cm.

SAGE SPARROW

Amphispiza belli

This species breeds in the western United States and is considered a rare vagrant during breeding season into southern British Columbia. It favours the drier interior sagebrush habitats of the Okanagan, which are similar to its breeding grounds in eastern Washington. Occasional vagrants have been recorded in the Lower Mainland area of southwestern BC and on Vancouver Island.

ID: sexes similar; brownish grey head; white lores and eye ring; black "whisker" stripe; unstreaked, white throat, breast and belly; dark central breast spot; dark streaking on sides; dark tail has white outer edges.
Size: *L* 11–13 cm; *W* 20 cm.

BLUE GROSBEAK

Passerina caerulea

The Blue Grosbeak breeds from the southern two-thirds of the continental U.S. to southern Central America, and winters from northern Mexico to Panama. During the 20th century, this species began expanding its eastern breeding range northward. Today, vagrant Blue Grosbeaks appear regularly in Ontario in May along the wooded sandspits and peninsulas that jut into Lake Erie, but no nesting has been recorded. On the West Coast, the Blue Grosbeak is a very rare vagrant anywhere north of California, with only 2 records in BC (2001).

ID: large, conical bill; orange wing bars; flicks tail. *Male:* blue body; black face; dark wings; 2 wide, rusty wing bars; brown-mottled plumage after breeding. *Female* and *juvenile:* rich brown overall; paler underparts; plain face; blue-washed rump and shoulders.
Size: *L* 15–18 cm; *W* 28 cm.

PAINTED BUNTING

Passerina ciris

The colourful Painted Bunting has two separate breeding ranges: along the Atlantic Coast from North Carolina to Florida, and mostly inland from Alabama to Mexico. They winter in Florida and the West Indies south to Panama. In recent years, Painted Bunting numbers have declined along the Atlantic Coast, attributed mostly to increased brood parasitism from Brown-headed Cowbirds. The species is an occasional vagrant into eastern Canada.

ID: *Male:* greenish yellow back and wings; brilliant blue head; red eye ring; wholly red underparts, including throat. *Female:* brilliant yellow-green upperparts; pale yellow underparts; yellow eye ring.
Size: *L* 13–14 cm; *W* 21–22 cm.

GREAT-TAILED GRACKLE

Quiscalus mexicanus

The Great-tailed Grackle inhabits the southwestern U.S. from southern interior California and southern Utah south to South America. Since the 1960s, the species has expanded its range, following the increase in agriculture and urbanization. It is considered accidental in southern British Columbia.

ID: very long, wide tail; yellow eyes. *Male:* all-black body with purple-blue iridescence. *Female:* grey-brown body is paler below; shorter tail; pale eyebrow.
Size: *Male: L* 45 cm; *W* 58 cm. *Female: L* 38 cm; *W* 48 cm.

GLOSSARY

accipiter: a forest hawk of the genus *Accipiter*, characterized by short, rounded wings and a long tail.

alcid: a seabird of the family Alcidae, which includes murres, auklets and puffins, characterized by wing well suited to diving and swimming and that beat rapidly in flight.

amphipod: a small, krill-like crustacean with a laterally compressed body; includes shrimp, sand fleas and sea lice.

avifauna: the community of birds found in a specific region or environment.

breeding bird survey (BBS): a survey to collect data in order to assess the status and population trends of bird species; the North American Breeding Bird Survey is a joint project between the Canadian Wildlife Service and the United States Geological Survey.

brood: *n.* a family of young from one hatching; *v.* to incubate the eggs.

brood parasite: a birds that lays its eggs in the nests of other birds.

buteo: a high-soaring hawk of the genus *Buteo*, characterized by broad wings and a short, wide tail.

Carolinian forest: the forest zone extending from southwestern Ontario to the Carolinas, dominated by deciduous or broad-leaf trees such as ash, chestnut, hickory and walnut.

cavity nester: a bird that builds its nest in a tree hollow or nest box.

cere: the fleshy area above the base of the bill on some birds.

cloaca: the posterior cavity in fish, reptiles, birds and some mammals that is used for both excretion and reproduction; also known as the "vent."

clutch: the number of eggs laid by a female bird at one time.

corvid: a member of the crow family (Corvidae); includes crows, jays, magpies and ravens.

covey: a flock of partridges, quail or grouse.

crop: an enlargement of the esophagus; serves as a storage structure and also (in pigeons) has glands that produce secretions to feed the young.

cryptic plumage: a coloration pattern that helps to conceal the bird.

dabbling: a foraging technique used by ducks in which the head and neck are submerged but the body and tail remain on the water's surface; dabbling ducks can usually walk easily on land, can take off without running and have brightly coloured speculums.

disjunct population: two groups of the same species found in widely separated regions.

diurnal: most active during the day.

drake: a male duck.

eclipse plumage: a cryptic plumage, similar to that of females, worn by some male ducks in autumn when they moult their flight feathers and are consequently unable to fly.

egg dumping: the laying of eggs in another bird's nest, leaving the "foster parent" to raise the young.

endangered: facing imminent extirpation or extinction.

endemic: native and usually limited to a particular area.

extinct: no longer existing anywhere.

extirpated: no longer existing in the wild in a particular region but occurring elsewhere.

fledge: to grow the first full set of feathers.

flush: a behaviour in which frightened birds explode into flight in response to a disturbance.

flycatching: attempting to capture insects through aerial pursuit; also known as "hawking."

forb: a herbaceous, non-woody, flowering plant that is not a grass, sedge or rush.

glean: to collect insects and other invertebrates from tree branches or leaf surfaces.

hawking: attempting to capture insects through aerial pursuit; also known as "flycatching."

Holarctic: the biogeographic region encompassing the northern continents as a whole, from the North Pole to approximately 35°N latitude, including North America, Eurasia south to the Himalayas and North Africa as far as the Sahara.

irruption: the sporadic mass migration of birds into an unusual range, often in search of food.

kettle: a large group of birds wheeling and circling in the air, often in preparation for migration.

lek: a place where males gather to display for females in spring.

lores: the area on a bird between its eyes and upper bill.

mantle: the area that includes the back and uppersides of the wings.

morph: one of several alternate colour phases displayed by a species.

neotropical: the biogeographic region that includes Mexico, Central and South America and the West Indies.

obligate: a habitat, feeding style or other factor that is essential to the survival of a species.

peep: a general name for small sandpipers of the *Calidris* genus.

pelagic: open-ocean habitat far from land.

pishing: imitating a bird call to attract birds, usually passerines.

polyandry: a mating strategy in which one female breeds with several males.

polygyny: a mating strategy in which one male breeds with several females.

polynya: open seawater surrounded by ice.

prairie potholes: shallow seasonal or permanent wetlands in the Prairies and northern Midwest that were formed by glacial activity at the end of the last ice age and provide important habitat for many migratory and breeding birds.

precocial: young that are relatively well developed at hatching, usually with open eyes, extensive down and good mobility.

raft: a gathering of birds on water.

remiges: the long, stiff flight feathers on a bird's wing.

retrices: the long flight feathers on a bird's tail.

riparian: along rivers and streams.

sexual dimorphism: a difference in plumage, size or other characteristics between males and females of the same species.

snag: a standing dead or dying tree; snags provide important shelter, nesting cavities, perches and prey items for many wildlife species.

species of special concern: a species that is particularly sensitive to human activities or natural events but is not yet considered endangered or threatened.

speculum: a brightly coloured patch on the wings of many dabbling ducks.

stage: to gather in one place during migration, usually when birds are flightless or partly flightless during moulting.

stoop: a steep dive through the air, usually performed by birds of prey while foraging or during courtship displays.

successional woodlands: the sequence of vegetation that grows after a major disturbance such as a fire.

syrinx: the vocal organ of a bird.

threatened: likely to become endangered in the near future in all or part of its range.

torpor: a temporary reduction in body temperature and metabolism in order to conserve energy, usually in response to low temperatures or a lack of food.

tubenose: a member of the Order Procellariiformes, so named for the distinctive tubular structure of the nostrils; includes albatrosses, shearwaters, fulmars, petrels and storm-petrels.

tuckamore: coastal spruce, fir or other coniferous trees that are contorted and stunted by salt spray and wind.

understorey: the shrub or thicket layer beneath a canopy of trees.

vagrant: a transient bird found outside its normal range.

vent: the posterior cavity in fish, reptiles, birds and some mammals that is used for both excretion and reproduction; also known as the "cloaca."

zygodactyl feet: feet that have two toes pointing forward and two pointing backward; found in ospreys, owls and woodpeckers.

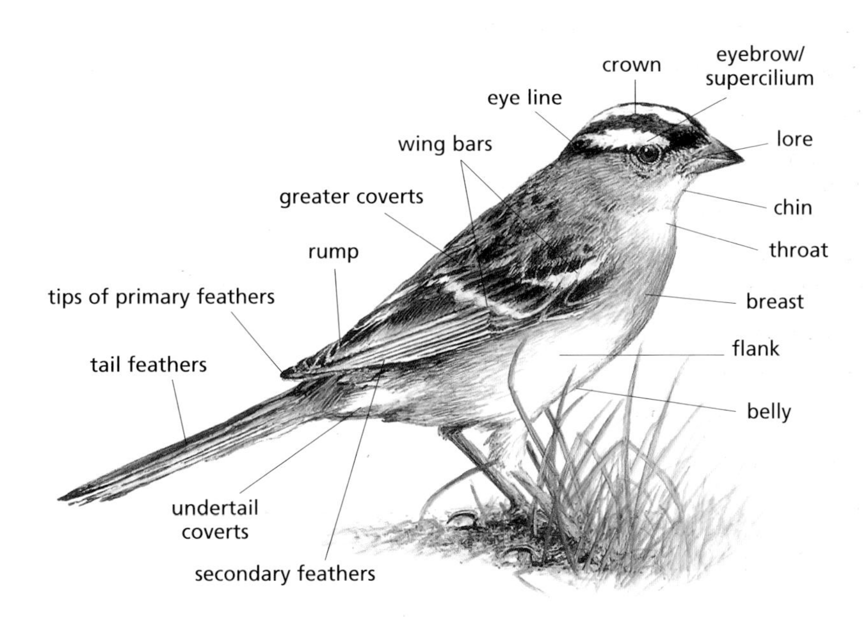

SELECTED REFERENCES

All About Birds. Cornell Lab of Ornithology. www.allaboutbirds.org/NetCommunity/Page.aspx?pid=1189.

American Birding Association. 2006. *ABA Checklist: Birds of the Continental United States and Canada.* 7th ed. American Birding Association, Colorado Springs. (Available online at www.aba.org/checklist.)

American Ornithologists' Union. 1998. *Check-list of North American Birds.* 7th ed. American Ornithologists' Union, Washington, DC. (Updates and a complete checklist are available online at www.aou.org/checklist/index.php3.)

Avibase: The World Bird Database. avibase.bsc-eoc.org/avibase.jsp.

Bird Studies Canada. www.bsc-eoc.org.

Bird, David M. 2004. *The Bird Almanac: The Ultimate Guide to Essential Facts and Figures of the World's Birds.* Rev. ed. Key Porter, Toronto.

Breeding Bird Survey. National Wildlife Research Centre. www.cws-scf.ec.gc.ca/nwrc-cnrf/Default.asp?lang=En&n=416B57CA.

Campbell, R. Wayne, Neil K. Dawe, Ian McTaggart-Cowan, John M. Cooper, Gary W. Kaiser and Michael C.E. McNall. 1990. *Birds of British Columbia, Volume 1: Nonpasserines (Introduction, Loons through Waterfowl).* University of British Columbia Press, Vancouver.

———. 1990. *Birds of British Columbia, Volume 2: Nonpasserines (Diurnal Birds of Prey through Woodpeckers).* University of British Columbia Press, Vancouver.

Campbell, R. Wayne, Neil K. Dawe, Ian McTaggart-Cowan, John M. Cooper, Gary W. Kaiser, Michael C.E. McNall and G.E. John Smith. 1997. *Birds of British Columbia, Volume 3: Passerines (Flycatchers through Vireos).* University of British Columbia Press, Vancouver.

Campbell, R. Wayne, Neil K. Dawe, Ian McTaggart-Cowan, John M. Cooper, Gary W. Kaiser and Michael C.E. McNall. 2001. *Birds of British Columbia, Volume 4: Passerines (Wood-warblers through Old World Sparrows).* University of British Columbia Press, Vancouver.

Choate, E.A. 1985. *The Dictionary of American Bird Names.* Rev. ed. Harvard Common Press, Cambridge, MA.

Clements, James F. 2007. *The Clements Checklist of Birds of the World.* 6th ed. Cornell University Press, Ithaca, NY. (Available online at http://www.birds.cornell.edu/clementschecklist/Clements%20Checklist%206.3.2%20December%202008.xls/view.)

Ehrlich, P.R., D.S. Dobkin and D. Wheye. 1988. *The Birder's Handbook: A Field Guide to the Natural History of North American Birds.* Simon & Schuster Inc., New York.

Erickson, Laura. 2006. *Ways to Help Birds.* Stackpole, Mechanicsburg, PA.

Godfrey, Earl. 1990. *The Birds of Canada.* Rev. ed. National Museums of Canada, Ottawa.

Kaufman, Kenn. 1999. *A Field Guide to Advanced Birding: Birding Challenges and How to Approach Them.* Houghton Mifflin, Boston.

Leahy, C.W. 2004. *The Birdwatcher's Companion to North American Birdlife.* Princeton University Press, Princeton.

National Geographic Society. 2006. *Field Guide to the Birds of North America.* 5th ed. National Geographic Society, Washington, DC.

———. 2005. *Complete Birds of North America.* National Geographic Society, Washington, DC.

Roth, Sally. 1998. *Attracting Birds to Your Backyard.* Rodale, Emmaus, PA.

Sibley, D.A. 2000. *National Audubon Society: The Sibley Guide to Birds.* Alfred A. Knopf, New York.

———. 2001. *National Audubon Society: The Sibley Guide to Bird Life and Behavior.* Alfred A. Knopf, New York.

Stokes, Donald, and Lillian Stokes. 2003. *Stokes Backyard Bird Book.* Rodale, Emmaus, PA.

CHECKLIST

The following checklist contains 658 species of birds that have been officially recorded in Canada. Species are grouped by family and listed in taxonomic order in accordance with the A.O.U. *Checklist of North American Birds* (7th ed.) and its supplements.

Accidental and casual species (those that are not seen on a yearly basis) are listed in *italics*. Introduced species are identified with (I). In addition, the following COSEWIC risk categories are also noted: extinct or extirpated (ex), endangered (en), threatened (th) and special concern (sc).

We wish to thank Denis Lepage of Avibase, the World Bird Database, for providing the information for this checklist. Full citation and URL can be found on page 518.

Waterfowl (Anatidae)

❑ *Black-bellied Whistling-Duck*
❑ *Fulvous Whistling-Duck*
❑ *Taiga Bean-Goose*
❑ *Tundra Bean-Goose*
❑ *Pink-footed Goose*
❑ Greater White-fronted Goose
❑ *Graylag Goose*
❑ *Emperor Goose*
❑ Snow Goose
❑ Ross's Goose
❑ Brant
❑ *Barnacle Goose*
❑ Cackling Goose
❑ Canada Goose
❑ Mute Swan (I)
❑ Trumpeter Swan
❑ Tundra Swan
❑ *Whooper Swan*
❑ Wood Duck
❑ Gadwall
❑ *Falcated Duck*
❑ Eurasian Wigeon
❑ American Wigeon
❑ American Black Duck
❑ Mallard
❑ *Mottled Duck*
❑ Blue-winged Teal
❑ Cinnamon Teal
❑ Northern Shoveler
❑ Northern Pintail
❑ *Garganey*
❑ *Baikal Teal*
❑ Green-winged Teal
❑ Canvasback
❑ Redhead
❑ Common Pochard
❑ Ring-necked Duck
❑ *Tufted Duck*
❑ Greater Scaup
❑ Lesser Scaup
❑ *Steller's Eider*
❑ *Spectacled Eider*
❑ King Eider
❑ Common Eider
❑ Harlequin Duck (sc, eastern population)
❑ Labrador Duck (ex)
❑ Surf Scoter
❑ White-winged Scoter
❑ Black Scoter
❑ Long-tailed Duck
❑ Bufflehead
❑ Common Goldeneye
❑ Barrow's Goldeneye (sc, eastern population)
❑ *Smew*
❑ Hooded Merganser
❑ Common Merganser
❑ Red-breasted Merganser
❑ Ruddy Duck

New World Quails (Odontophoridae)

❑ Mountain Quail (I)
❑ California Quail (I)
❑ Northern Bobwhite (en)

Grouse & Allies (Phasianidae)

❑ Chukar (I)
❑ Gray Partridge (I)
❑ Ring-necked Pheasant (I)
❑ Ruffed Grouse
❑ Greater Sage-Grouse (en)
❑ Spruce Grouse
❑ Willow Ptarmigan
❑ Rock Ptarmigan

- ❑ White-tailed Ptarmigan
- ❑ Dusky Grouse
- ❑ Sooty Grouse
- ❑ Sharp-tailed Grouse
- ❑ Greater Prairie-Chicken (ex)
- ❑ Wild Turkey

Loons (Gaviidae)

- ❑ Red-throated Loon
- ☒ Pacific Loon
- ☒ Common Loon
- ❑ Yellow-billed Loon

Grebes (Podicipedidae)

- ☒ Pied-billed Grebe
- ☒ Horned Grebe (sc, western population)
- ❑ Red-necked Grebe
- ❑ Eared Grebe
- ❑ Western Grebe
- ❑ Clark's Grebe

Flamingos (Phoenicopteridae)

- ❑ *American Flamingo*

Albatrosses (Diomedeidae)

- ❑ *Yellow-nosed Albatross*
- ❑ *Black-browed Albatross*
- ❑ *Laysan Albatross*
- ☒ Black-footed Albatross (sc)
- ❑ *Short-tailed Albatross* (th)

Fulmars, Petrels & Shearwaters (Procellariidae)

- ❑ Northern Fulmar
- ❑ *Mottled Petrel*
- ❑ *Black-capped Petrel*
- ❑ Cory's Shearwater
- ❑ Pink-footed Shearwater (th)
- ❑ *Flesh-footed Shearwater*
- ❑ Greater Shearwater
- ❑ Buller's Shearwater
- ☒ Sooty Shearwater
- ❑ Short-tailed Shearwater
- ❑ Manx Shearwater
- ❑ *Black-vented Shearwater*
- ❑ *Audubon's Shearwater*
- ❑ *Little Shearwater*

Storm-Petrels (Hydrobatidae)

- ❑ Wilson's Storm-Petrel
- ❑ *White-faced Storm-Petrel*
- ❑ *European Storm-Petrel*
- ❑ Fork-tailed Storm-Petrel
- ❑ Leach's Storm-Petrel
- ❑ *Band-rumped Storm-Petrel*

Tropicbirds (Phaethontidae)

- ❑ *White-tailed Tropicbird*
- ❑ Red-tailed Tropicbird

Boobies & Gannets (Sulidae)

- ❑ *Masked Booby*
- ❑ *Brown Booby*
- ❑ Northern Gannet

Pelicans (Pelecanidae)

- ☒ American White Pelican
- ❑ *Brown Pelican*

Cormorants (Phalacrocoracidae)

- ❑ Brandt's Cormorant
- ❑ *Neotropic Cormorant*
- ☒ Double-crested Cormorant
- ❑ Great Cormorant
- ❑ *Red-faced Cormorant*
- ❑ Pelagic Cormorant

Anhingas (Anhingidae)

- ❑ *Anhinga*

Frigatebirds (Fregatidae)

- ❑ *Magnificent Frigatebird*

Herons (Ardeidae)

- ❑ American Bittern
- ❑ Least Bittern (th)
- ☒ Great Blue Heron (sc, *fannini* ssp.)
- ☒ *Gray Heron*
- ☒ Great Egret
- ❑ *Little Egret*
- ❑ *Western Reef-Heron*
- ❑ Snowy Egret
- ❑ Little Blue Heron
- ❑ Tricolored Heron
- ❑ *Reddish Egret*
- ❑ Cattle Egret
- ☒ Green Heron
- ☒ Black-crowned Night-Heron
- ❑ Yellow-crowned Night-Heron

Ibises (Threskiornithidae)

- ❑ *White Ibis*
- ❑ Glossy Ibis
- ❑ White-faced Ibis

Storks (Ciconiidae)

- ❑ *Wood Stork*

Vultures (Cathartidae)

- ❑ *Black Vulture*
- ☒ Turkey Vulture
- ❑ California Condor (ex)

Hawks & Eagles (Accipitridae)

- ☒ Osprey
- ❑ *Swallow-tailed Kite*
- ❑ *White-tailed Kite*
- ❑ *Mississippi Kite*
- ❑ *Black Kite*
- ☒ Bald Eagle
- ☒ Northern Harrier
- ☒ Sharp-shinned Hawk
- ☒ Cooper's Hawk
- ☒ Northern Goshawk (th, *laingi* ssp.)
- ☒ Red-shouldered Hawk
- ☒ Broad-winged Hawk
- ☒ Swainson's Hawk
- ❑ *Zone-tailed Hawk*
- ☒ Red-tailed Hawk
- ☒ Ferruginous Hawk (th)
- ☒ Rough-legged Hawk
- ☒ Golden Eagle

Falcons (Falconidae)

- ❑ *Crested Caracara*
- ❑ *Eurasian Kestrel*
- ☒ American Kestrel
- ☒ Merlin
- ☒ Gyrfalcon
- ☒ Peregrine Falcon (sc)
- ❑ Prairie Falcon

Rails & Coots (Rallidae)

- ❑ Yellow Rail (sc)
- ❑ *Black Rail*
- ❑ *Corn Crake*
- ❑ *Clapper Rail*
- ❑ King Rail (en)
- ❑ Virginia Rail
- ☒ Sora
- ❑ *Purple Gallinule*
- ❑ Common Moorhen
- ❑ *Eurasian Coot*
- ☒ American Coot

Limpkins (Aramidae)

- ❑ *Limpkin*

Cranes (Gruidae)

- ☒ Sandhill Crane
- ❑ *Common Crane*
- ❑ Whooping Crane (en)

Plovers (Charadriidae)

- ❑ *Northern Lapwing*
- ❑ Black-bellied Plover
- ❑ *European Golden-Plover*
- ❑ American Golden-Plover
- ❑ *Pacific Golden-Plover*
- ❑ *Lesser Sand-Plover*
- ❑ *Snowy Plover*
- ❑ *Wilson's Plover*
- ❑ Common Ringed Plover
- ☒ Semipalmated Plover
- ❑ Piping Plover (en)
- ☒ Killdeer
- ❑ Mountain Plover (en)
- ❑ *Eurasian Dotterel*

Oystercatchers (Haematopodidae)

- ❑ *Eurasian Oystercatcher*
- ☒ American Oystercatcher
- ❑ Black Oystercatcher

Stilts & Avocets (Recurvirostridae)

- ❑ Black-necked Stilt
- ☒ American Avocet

Sandpipers & Allies (Scolopacidae)

- ❑ *Terek Sandpiper*
- ❑ Spotted Sandpiper
- ❑ Solitary Sandpiper
- ❑ Wandering Tattler
- ❑ *Spotted Redshank*
- ❑ Greater Yellowlegs
- ❑ *Common Greenshank*
- ❑ Willet
- ❑ Lesser Yellowlegs
- ❑ *Wood Sandpiper*
- ❑ *Common Redshank*
- ❑ Upland Sandpiper
- ❑ Eskimo Curlew (en)
- ❑ Whimbrel
- ❑ *Bristle-thighed Curlew*
- ❑ *Far Eastern Curlew*
- ❑ *Slender-billed Curlew*
- ❑ *Eurasian Curlew*
- ❑ Long-billed Curlew (sc)
- ❑ *Black-tailed Godwit*
- ❑ Hudsonian Godwit
- ❑ *Bar-tailed Godwit*

- ❑ Marbled Godwit
- ❑ Ruddy Turnstone
- ❑ Black Turnstone
- ❑ Surfbird
- ❑ *Great Knot*
- ❑ Red Knot (en, *rufa* ssp.; sc, *islandica* ssp.)
- ❑ Sanderling
- ❑ Semipalmated Sandpiper
- ❑ Western Sandpiper
- ❑ *Red-necked Stint*
- ❑ *Little Stint*
- ❑ *Temminck's Stint*
- ❑ Least Sandpiper
- ❑ White-rumped Sandpiper
- ❑ Baird's Sandpiper
- ❑ Pectoral Sandpiper
- ❑ Sharp-tailed Sandpiper
- ❑ Purple Sandpiper
- ❑ Rock Sandpiper
- ❑ Dunlin
- ❑ *Curlew Sandpiper*
- ❑ Stilt Sandpiper
- ❑ *Spoon-billed Sandpiper*
- ❑ *Broad-billed Sandpiper*
- ❑ Buff-breasted Sandpiper
- ❑ Ruff
- ❑ Short-billed Dowitcher
- ❑ Long-billed Dowitcher
- ❑ *Jack Snipe*
- ❑ Wilson's Snipe
- ❑ *Common Snipe*
- ❑ *Eurasian Woodcock*
- ❑ American Woodcock
- ❑ Wilson's Phalarope
- ❑ Red-necked Phalarope
- ❑ Red Phalarope

Gulls & Terns (Laridae)

- ❑ Black-legged Kittiwake
- ❑ Red-legged Kittiwake
- ❑ Ivory Gull (en)
- ❑ Sabine's Gull
- ☒ Bonaparte's Gull
- ☒ Black-headed Gull
- ❑ Little Gull
- ❑ Ross's Gull (th)
- ❑ Laughing Gull
- ☒ Franklin's Gull
- ❑ *Black-tailed Gull*
- ❑ Heermann's Gull
- ❑ Mew Gull
- ☒ Ring-billed Gull
- ❑ Western Gull
- ☒ California Gull
- ☒ Herring Gull
- ❑ *Yellow-legged Gull*
- ❑ Thayer's Gull
- ❑ Iceland Gull
- ❑ Lesser Black-backed Gull
- ❑ *Slaty-backed Gull*
- ☒ Glaucous-winged Gull
- ☒ Glaucous Gull
- ❑ Great Black-backed Gull
- ❑ *Sooty Tern*
- ❑ *Bridled Tern*
- ❑ *Aleutian Tern*
- ❑ *Least Tern*
- ❑ *Gull-billed Tern*
- ❑ Caspian Tern
- ❑ Black Tern
- ❑ *White-winged Tern*
- ❑ Roseate Tern (en)
- ☒ Common Tern
- ❑ Arctic Tern
- ❑ Forster's Tern
- ❑ *Royal Tern*
- ❑ *Sandwich Tern*
- ❑ *Elegant Tern*
- ❑ *Black Skimmer*

Skuas & Jaegers (Stercorariidae)

- ❑ Great Skua
- ❑ South Polar Skua
- ❑ Pomarine Jaeger
- ☒ Parasitic Jaeger
- ❑ Long-tailed Jaeger

Alcids (Alcidae)

- ❑ Dovekie
- ❑ Common Murre
- ❑ Thick-billed Murre
- ❑ Razorbill
- ❑ Great Auk (ex)
- ❑ Black Guillemot
- ❑ Pigeon Guillemot
- ❑ *Long-billed Murrelet*
- ❑ Marbled Murrelet (th)
- ❑ *Kittlitz's Murrelet*
- ❑ Xantus's Murrelet
- ❑ Ancient Murrelet (sc)
- ❑ Cassin's Auklet
- ❑ *Parakeet Auklet*
- ❑ *Least Auklet*
- ❑ *Crested Auklet*

- ❑ Rhinoceros Auklet
- ❑ Atlantic Puffin
- ❑ Horned Puffin
- ☒ Tufted Puffin

Pigeons & Doves (Columbidae)

- ☒ Rock Pigeon (I)
- ❑ Band-tailed Pigeon (sc)
- ❑ *Oriental Turtle-Dove*
- ❑ Eurasian Collared-Dove (I)
- ❑ *White-winged Dove*
- ❑ Zenaida Dove
- ☒ Mourning Dove
- ❑ Passenger Pigeon (ex)
- ❑ *Inca Dove*
- ❑ *Common Ground-Dove*

Cuckoos (Cuculidae)

- ❑ Yellow-billed Cuckoo
- ☒ Black-billed Cuckoo
- ❑ *Groove-billed Ani*

Barn Owls (Tytonidae)

- ❑ Barn Owl (en, eastern population; sc, western population)

Owls (Strigidae)

- ❑ Flammulated Owl (sc)
- ❑ Western Screech-Owl (en, *macfarlanei* ssp.; sc, *kennicottii* ssp.)
- ❑ Eastern Screech-Owl
- ❑ Great Horned Owl
- ❑ Snowy Owl
- ❑ Northern Hawk Owl
- ❑ Northern Pygmy-Owl
- ☒ Burrowing Owl (en)
- ❑ Spotted Owl (en, *caurina* ssp.)
- ❑ Barred Owl
- ❑ Great Gray Owl
- ❑ Long-eared Owl
- ❑ Short-eared Owl (sc)
- ❑ Boreal Owl
- ❑ Northern Saw-whet Owl (th, *brooksi* ssp.)

Nightjars (Caprimulgidae)

- ❑ *Lesser Nighthawk*
- ☒ Common Nighthawk (th)
- ❑ Common Poorwill
- ❑ Chuck-will's-widow
- ❑ Whip-poor-will (th)

Swifts (Apodidae)

- ❑ Black Swift
- ❑ *White-collared Swift*
- ☒ Chimney Swift (th)
- ❑ Vaux's Swift
- ❑ White-throated Swift

Hummingbirds (Trochilidae)

- ❑ *Green Violet-ear*
- ❑ *Broad-billed Hummingbird*
- ☒ Ruby-throated Hummingbird
- ❑ Black-chinned Hummingbird
- ❑ Anna's Hummingbird
- ❑ *Costa's Hummingbird*
- ☒ Calliope Hummingbird
- ❑ *Broad-tailed Hummingbird*
- ☒ Rufous Hummingbird

Kingfishers (Alcedinidae)

- ☒ Belted Kingfisher

Woodpeckers (Picidae)

- ❑ Lewis's Woodpecker (sc)
- ☒ Red-headed Woodpecker (th)
- ❑ *Acorn Woodpecker*
- ❑ Red-bellied Woodpecker
- ❑ Williamson's Sapsucker (en)
- ☒ Yellow-bellied Sapsucker
- ☒ Red-naped Sapsucker
- ❑ Red-breasted Sapsucker
- ☒ Downy Woodpecker
- ☒ Hairy Woodpecker
- ❑ White-headed Woodpecker (en)
- ❑ American Three-toed Woodpecker
- ☒ Black-backed Woodpecker
- ☒ Northern Flicker
- ☒ Pileated Woodpecker

Flycatchers (Tyrannidae)

- ☒ Olive-sided Flycatcher (th)
- ☒ Western Wood-Pewee
- ❑ Eastern Wood-Pewee
- ❑ Yellow-bellied Flycatcher
- ❑ Acadian Flycatcher (en)
- ❑ Alder Flycatcher
- ❑ Willow Flycatcher
- ☒ Least Flycatcher
- ❑ Hammond's Flycatcher
- ❑ Gray Flycatcher
- ❑ Dusky Flycatcher
- ❑ Pacific-slope Flycatcher
- ❑ Cordilleran Flycatcher
- ❑ *Black Phoebe*
- ☒ Eastern Phoebe
- ❑ Say's Phoebe

- [ ] *Vermilion Flycatcher*
- [ ] *Ash-throated Flycatcher*
- [ ] Great Crested Flycatcher
- [ ] *Sulphur-bellied Flycatcher*
- [ ] *Variegated Flycatcher*
- [ ] *Tropical Kingbird*
- [ ] *Couch's Kingbird*
- [ ] *Cassin's Kingbird*
- [ ] *Thick-billed Kingbird*
- [x] Western Kingbird
- [x] Eastern Kingbird
- [ ] *Gray Kingbird*
- [ ] *Scissor-tailed Flycatcher*
- [ ] *Fork-tailed Flycatcher*

Shrikes (Laniidae)
- [ ] *Brown Shrike*
- [ ] Loggerhead Shrike (en, *migrans* ssp.; th, *excubitorides* ssp.)
- [x] Northern Shrike

Vireos (Vireonidae)
- [ ] White-eyed Vireo
- [ ] *Bell's Vireo*
- [ ] *Black-capped Vireo*
- [ ] Yellow-throated Vireo
- [ ] *Plumbeous Vireo*
- [ ] Cassin's Vireo
- [ ] Blue-headed Vireo
- [ ] Hutton's Vireo
- [ ] Warbling Vireo
- [ ] Philadelphia Vireo
- [x] Red-eyed Vireo
- [ ] *Yellow-green Vireo*

Jays & Crows (Corvidae)
- [x] Gray Jay
- [x] Steller's Jay
- [x] Blue Jay
- [ ] *Western Scrub-Jay*
- [ ] *Pinyon Jay*
- [x] Clark's Nutcracker
- [x] Black-billed Magpie
- [x] Eurasian Jackdaw
- [x] American Crow
- [x] Northwestern Crow
- [x] Fish Crow
- [x] Common Raven

Larks (Alaudidae)
- [ ] Sky Lark (I)
- [x] Horned Lark (en, *strigata* ssp.)

Swallows (Hirundinidae)
- [ ] Purple Martin
- [x] Tree Swallow
- [ ] Violet-green Swallow
- [x] Northern Rough-winged Swallow
- [x] Bank Swallow
- [ ] Cliff Swallow
- [ ] *Cave Swallow*
- [x] Barn Swallow

Chickadees & Titmice (Paridae)
- [ ] *Carolina Chickadee*
- [x] Black-capped Chickadee
- [x] Mountain Chickadee
- [ ] Chestnut-backed Chickadee
- [x] Boreal Chickadee
- [ ] Gray-headed Chickadee
- [ ] Tufted Titmouse

Bushtits (Aegithalidae)
- [x] Bushtit

Nuthatches (Sittidae)
- [x] Red-breasted Nuthatch
- [x] White-breasted Nuthatch
- [ ] Pygmy Nuthatch

Creepers (Certhiidae)
- [x] Brown Creeper

Wrens (Troglodytidae)
- [ ] Rock Wren
- [ ] Canyon Wren
- [ ] Carolina Wren
- [ ] Bewick's Wren
- [x] House Wren
- [x] Winter Wren
- [ ] Sedge Wren
- [ ] Marsh Wren

Dippers (Cinclidae)
- [x] American Dipper

Kinglets (Regulidae)
- [x] Golden-crowned Kinglet
- [x] Ruby-crowned Kinglet

Old World Warblers & Gnatcatchers (Sylviidae)
- [ ] *Arctic Warbler*
- [ ] Blue-gray Gnatcatcher

Thrushes (Turdidae)
- [ ] *Siberian Rubythroat*

- ❑ *Bluethroat*
- ❑ *Siberian Blue Robin*
- ❑ Northern Wheatear
- ❑ *Stonechat*
- ❑ Eastern Bluebird
- ❑ Western Bluebird
- ❑ Mountain Bluebird
- ❑ Townsend's Solitaire
- ❑ Veery
- ❑ Gray-cheeked Thrush
- ❑ Bicknell's Thrush (th)
- ❑ Swainson's Thrush
- ❑ Hermit Thrush
- ❑ Wood Thrush
- ❑ *Eurasian Blackbird*
- ❑ *Dusky Thrush*
- ❑ *Fieldfare*
- ❑ *Redwing*
- ❑ *Song Thrush*
- ❑ American Robin
- ❑ Varied Thrush

Mockingbirds & Thrashers (Mimidae)

- ❑ Gray Catbird
- ❑ Northern Mockingbird
- ❑ Sage Thrasher (en)
- ❑ Brown Thrasher
- ❑ *Bendire's Thrasher*
- ❑ *Curve-billed Thrasher*

Starlings (Sturnidae)

- ❑ European Starling (I)

Accentors (Prunellidae)

- ❑ *Siberian Accentor*

Wagtails & Pipits (Motacillidae)

- ❑ Eastern Yellow Wagtail
- ❑ *Gray Wagtail*
- ❑ White Wagtail
- ❑ *Red-throated Pipit*
- ❑ American Pipit
- ❑ Sprague's Pipit (th)

Waxwings (Bombycillidae)

- ❑ Bohemian Waxwing
- ❑ Cedar Waxwing

Silky-Flycatchers (Ptilogonatidae)

- ❑ *Phainopepla*

Wood-Warblers (Parulidae)

- ❑ Blue-winged Warbler
- ❑ Golden-winged Warbler (th)
- ❑ Tennessee Warbler
- ❑ Orange-crowned Warbler
- ❑ Nashville Warbler
- ❑ *Virginia's Warbler*
- ❑ Northern Parula
- ❑ Yellow Warbler
- ❑ Chestnut-sided Warbler
- ❑ Magnolia Warbler
- ❑ Cape May Warbler
- ❑ Black-throated Blue Warbler
- ❑ Yellow-rumped Warbler
- ❑ Black-throated Gray Warbler
- ❑ Black-throated Green Warbler
- ❑ Townsend's Warbler
- ❑ *Hermit Warbler*
- ❑ Blackburnian Warbler
- ❑ *Yellow-throated Warbler*
- ❑ Pine Warbler
- ❑ Kirtland's Warbler (en)
- ❑ Prairie Warbler
- ❑ Palm Warbler
- ❑ Bay-breasted Warbler
- ❑ Blackpoll Warbler
- ❑ Cerulean Warbler (sc)
- ❑ Black-and-white Warbler
- ❑ American Redstart
- ❑ Prothonotary Warbler (en)
- ❑ Worm-eating Warbler
- ❑ *Swainson's Warbler*
- ❑ Ovenbird
- ❑ Northern Waterthrush
- ❑ Louisiana Waterthrush (sc)
- ❑ Kentucky Warbler
- ❑ Connecticut Warbler
- ❑ Mourning Warbler
- ❑ MacGillivray's Warbler
- ❑ Common Yellowthroat
- ❑ Hooded Warbler (th)
- ❑ Wilson's Warbler
- ❑ Canada Warbler (th)
- ❑ *Painted Redstart*
- ❑ Yellow-breasted Chat (en, *auricollis* ssp.; sc, *virens* ssp.)

Sparrows & Allies (Emberizidae)

- ❑ *Green-tailed Towhee*
- ❑ Spotted Towhee
- ❑ Eastern Towhee
- ❑ *Cassin's Sparrow*
- ❑ *Bachman's Sparrow*
- ❑ American Tree Sparrow
- ❑ Chipping Sparrow

❑ Clay-colored Sparrow
❑ Brewer's Sparrow
❑ Field Sparrow
❑ Vesper Sparrow (en, *affinis* ssp.)
❑ Lark Sparrow
❑ *Black-throated Sparrow*
❑ *Sage Sparrow*
❑ Lark Bunting
❑ Savannah Sparrow (sc, *princeps* ssp.)
❑ Grasshopper Sparrow
❑ Baird's Sparrow
❑ Henslow's Sparrow (en)
❑ Le Conte's Sparrow
❑ Nelson's Sparrow
❑ *Saltmarsh Sparrow*
❑ *Seaside Sparrow*
❑ Fox Sparrow
❑ Song Sparrow
❑ Lincoln's Sparrow
❑ Swamp Sparrow
❑ White-throated Sparrow
❑ Harris's Sparrow
❑ White-crowned Sparrow
❑ Golden-crowned Sparrow
❑ Dark-eyed Junco
❑ McCown's Longspur (sc)
❑ Lapland Longspur
❑ Smith's Longspur
❑ Chestnut-collared Longspur (th)
❑ *Rustic Bunting*
❑ Snow Bunting
❑ *McKay's Bunting*

Tanagers, Grosbeaks & Buntings (Cardinalidae)

❑ *Hepatic Tanager*
❑ Summer Tanager
❑ Scarlet Tanager
❑ Western Tanager
❑ Northern Cardinal
❑ Rose-breasted Grosbeak
❑ Black-headed Grosbeak
❑ *Blue Grosbeak*
❑ Lazuli Bunting
❑ Indigo Bunting
❑ *Varied Bunting*
❑ *Painted Bunting*
❑ Dickcissel

Blackbirds & Allies (Icteridae)

❑ Bobolink
❑ Red-winged Blackbird
❑ Eastern Meadowlark
❑ Western Meadowlark
❑ Yellow-headed Blackbird
❑ Rusty Blackbird (sc)
❑ Brewer's Blackbird
❑ Common Grackle
❑ *Great-tailed Grackle*
❑ *Shiny Cowbird*
❑ *Bronzed Cowbird*
❑ Brown-headed Cowbird
❑ Orchard Oriole
❑ *Hooded Oriole*
❑ Bullock's Oriole
❑ Baltimore Oriole
❑ *Scott's Oriole*

Finches (Fringillidae)

❑ *Common Chaffinch*
❑ *Brambling*
❑ Gray-crowned Rosy-Finch
❑ Pine Grosbeak
❑ Purple Finch
❑ Cassin's Finch
❑ House Finch (I)
❑ Red Crossbill (en, *percna* ssp.)
❑ White-winged Crossbill
❑ Common Redpoll
❑ Hoary Redpoll
❑ Eurasian Siskin
❑ Pine Siskin
❑ *Lesser Goldfinch*
❑ American Goldfinch
❑ *European Goldfinch*
❑ Evening Grosbeak

Old World Sparrows (Passeridae)

❑ House Sparrow (I)
❑ *Eurasian Tree Sparrow*

Lepage, D. 2009. *Avibase, the World Bird Database*. Checklist of birds of Canada, avibase.bsc-eoc.org/checklist.jsp?region=ca&list=aou&lang=EN (accessed December 2009).

INDEX

Entries in **boldface** type refer to the primary, illustrated species accounts.

PHOTO CREDITS

All photographs are copyright of the photographers as listed and are used by permission:

Christian Artuso 60, 84, 108, 113, 148, 165, 177, 178, 182, 192, 197, 203, 215, 217, 237, 261, 265, 278, 306, 326, 380, 395, 406, 408, 411, 414, 425, 452, 463, 464, 485; Michael "Mike" L. Baird / bairdphotos.com / Creative Commons - attribution 115, 116, 246; Norman Bateman / Dreamstime.com 277, 361; birdfreak.com 366, 416; Cascoly / Dreamstime.com 252; Patrick Coin 373; David Cree 461; Desert Digital Images 1, 23, 26a, 26b, 27b, 32b (bottom), 35a, 36, 37, 38a, 38b, 39, 41, 42, 46, 52, 57, 61, 62, 63, 66, 67, 72, 78, 79, 80, 81, 83, 85, 87, 88, 90, 96, 126, 131, 139, 140, 142, 149, 151, 153, 163, 164, 173, 174, 175, 186, 188, 190, 206, 221, 222, 223, 253, 266, 271, 272, 275, 284, 286, 287, 288, 289, 294, 295, 297, 305, 318, 320, 321, 323, 329, 333, 335, 336, 337, 338, 339, 341, 344, 349, 350, 351, 357, 360, 364, 367, 368, 374, 375, 384, 385, 392, 393, 400, 405, 430, 432, 433, 440, 441, 443, 450, 455, 456, 457, 460, 462, 465, 468, 469, 472, 473, 475, 476, 478, 482, 487, 492, 495, 496; Nancy B. Desilets 301, 484, 488; Sarah E. Desilets 255, 493; Anne Dorsey 458; Lydia Dotto / ImageInnovation Photography 121, 150, 179, 219, 225, 244, 251, 273, 274, 311, 313, 369, 370, 402, 479; Dreamstime 489; Laura Erickson / Creative Commons - attribution 447; Rick Fox 159; James Lloyd / Creative Commons - attribution 110; Jim Gilbert 75; Atie Grimsby 227; Anthony Hathaway 241; iStockphoto / John A. Anderson 418; iStockphoto / Robert Blanchard 429, 474; iStockphoto / Charles Brutlag 354; iStockphoto / Steve Byland 444; iStock-photo / Robert Devlin 451; iStockphoto / Steffen Foerster 111; iStockphoto / Andrew Howe 340, 372; iStockphoto /Tatiana Ivkovich 94; iStockphoto / Mr_Jamsey 439; iStockphoto / Lars Johansson 95; iStockphoto / Frank Leung 137, 342, 363, 365; iStockphoto / Jack Macilwinen 346; iStockphoto / Craig Marble 471; iStockphoto / Laure Neish 53; iStockphoto / Elizabeth Netterstrom 65; iStockphoto / John Pavel 125; iStockphoto / William Sherman 412, 424; iStockphoto / Michael D. Stubblefield, M.D. 30, 415; iStockphoto / Paul Tessier 394, 426; iStockphoto / Michael Westhoff 322; ; JupiterImages 22, 28, 31a, 32a (top), 40, 51, 86, 89, 91, 92, 98, 99, 102, 104, 122, 123, 127, 133, 141, 152, 156, 176, 211, 242, 259, 262, 264, 267, 268, 270, 293, 319, 332; Kaido Karner / Dreamstime.com 353; kuribo / Creative Commons - attribution 56; Paul Kusmin 298; Colin Marcano 27a, 29, 35b, 50, 54, 59, 76, 77, 124, 129, 130, 135, 283, 296, 302, 334, 343, 483, 494; Mdf / Creative Commons - attribution 325; Mesquite53 / Dreamstime.com 371; Tom Munson 260; Mark Nyhof 285, 486; Terry Parker 69, 93, 144, 290, 291, 407, 409, 459, 467, 481, 491; George Peck 171, 245, 254, 258, 276, 280, 324, 359, 396; M. K. Peck 33, 55, 68, 70, 74, 82, 100, 101, 103, 105, 106, 107, 128, 132, 134, 136, 143, 145, 146, 154, 160, 162, 166, 167, 168, 169, 170, 172, 180, 183, 185, 187, 191, 193, 194, 195, 196, 198, 200, 201, 202, 204, 207, 208, 209, 210, 214, 218, 220, 228, 231, 233, 235, 236, 239, 240, 243, 256, 263, 269, 299, 300, 304, 310, 331, 347, 348, 352, 356, 362, 377, 378, 381, 383, 388, 390, 397, 398, 399, 401, 403, 410, 419, 428, 431, 434, 435, 436, 446, 449, 453, 454, 466, 480, 490; Gerald Romanchuk 31b, 49, 64, 109, 147, 155, 157, 161, 189, 205, 213, 224, 226, 229, 232, 250, 282, 292, 309, 314, 317, 355, 358, 376, 386, 391, 413, 427, 438, 442, 445, 448, 477; C. Schlawe / US Fish and Wildlife Service 120; Ed Schneider 308; David Seibel 327; Dominic Sherony / Creative Commons - attribution 158; Stubblefieldphoto / Dreamstime.com 48; Linda Tanner / Creative Commons - attribution 315; Glen Tepke 45, 47, 58, 71, 73, 97, 112, 114, 118, 119, 181, 184, 216, 230, 234, 238, 247, 248, 249, 257, 279, 281, 316, 330, 345, 382, 404, 417, 420, 421, 422, 423, 437, 470; Michael Thompson / Dreamstime.com 387; Iouri Timofeev / Dreamstime.com 389; Mike Trewet / Dreamstime.com 138; Henry Trombley 379; Ian Wetton 117; Andy Wilson 212; Michael Woodruff 199, 307, 328; Michael Woodruff / Dreamstime.com 312

ABOUT THE AUTHORS

Tyler L. Hoar

At a young age, Tyler Hoar's love of nature was sparked by watching his grandfather rehabilitate an injured Common Moorhen. Since then, Tyler has been a nature enthusiast, whose passion led him to complete a biology degree from the University of Guelph and travel from Nunavut to the Caribbean while studying shorebirds, raptors and parrots. He has participated in research studies in Arizona, Québec, Ontario and the Arctic—all of which have furthered our understanding of breeding birds and their behaviours. Working with Canadian Wildlife Services, Tyler has contributed to the study of Little Gull migration at Oshawa's Second Marsh and is currently studying botulism in shorebirds along Lake Ontario. When not working, Tyler has travelled extensively through North America, Central America, the South Pacific and Australia while enjoying his passion for birding and nature appreciation.

Ken De Smet

Ken De Smet is an avian species at risk specialist with the provincial Wildlife and Ecosystem Protection Branch (Manitoba Conservation) and has conducted research and conservation efforts for a variety of declining grassland bird species in southwestern Manitoba since the mid-1980s. He participates on numerous National Recovery Teams and Provincial Recovery Action Groups for species at risk in Manitoba. Ken has also coordinated breeding bird survey efforts in the province for over 20 years and participates in a number of other volunteer wildlife-monitoring initiatives, including several recent record-breaking Big Day bird counts in southern Manitoba. He has written numerous articles related to birds and co-authored the very popular Lone Pine book, *Manitoba Birds.* Much of his spare time is spent organizing and participating in birdwatching, ecotourism and wildlife survey efforts.

R. Wayne Campbell

Biologist Wayne Campbell is the author of 54 books and over 500 scientific and popular articles and reports on mammals, birds, reptiles and amphibians. For 20 years, from 1973 to 1992, he was curator of Ornithology at the Royal British Columbia Museum. From 1993 to May 2000, when he "officially" retired, Wayne was a senior research scientist in the Wildlife Branch of the British Columbia Ministry of Environment, Lands and Parks. He has received numerous honours and awards for his lecturing, writing and conservation activities, and he was appointed to the Order of British Columbia in 1992. Wayne is a co-founding director of the Biodiversity Centre for Wildlife Studies, a registered non-profit society that houses the largest regional databases for wildlife information in North America.

Gregory Kennedy

Gregory Kennedy has been an active naturalist since he was very young. He has also been involved in numerous research projects around the world, ranging from studies in the upper canopy of tropical and temperate rainforests to deepwater marine investigations. He is the author of many books on natural history and has also produced film and television shows on environmental issues and indigenous concerns in Southeast Asia, New Guinea, South and Central America, the High Arctic and elsewhere.